CELL IMPAIRMENT IN AGING AND DEVELOPMENT

ADVANCES IN EXPERIMENTAL MEDICINE AND BIOLOGY

Recent Volumes in this Series

Volume 43
ARTERIAL MESENCHYME AND ARTERIOSCLEROSIS
Edited by William D. Wagner and Thomas B. Clarkson • 1974

Volume 44
CONTROL OF GENE EXPRESSION
Edited by Alexander Kohn and Adam Shatkay • 1974

Volume 45
THE IMMUNOGLOBULIN A SYSTEM
Edited by Jiri Mestecky and Alexander R. Lawton • 1974

Volume 46
PARENTERAL NUTRITION IN INFANCY AND CHILDHOOD
Edited by Hans Henning Bode and Joseph B. Warshaw • 1974

Volume 47
CONTROLLED RELEASE OF BIOLOGICALLY ACTIVE AGENTS
Edited by A. C. Tanquary and R. E. Lacey • 1974

Volume 48
PROTEIN–METAL INTERACTIONS
Edited by Mendel Friedman • 1974

Volume 49
NUTRITION AND MALNUTRITION: Identification and Measurement
Edited by Alexander F. Roche and Frank Falkner • 1974

Volume 50
ION-SELECTIVE MICROELECTRODES
Edited by Herbert J. Berman and Normand C. Hebert • 1974

Volume 51
THE CELL SURFACE: Immunological and Chemical Approaches
Edited by Barry D. Kahan and Ralph A. Reisfeld • 1974

Volume 52
HEPARIN: Structure, Function, and Clinical Implications
Edited by Ralph A. Bradshaw and Stanford Wessler • 1975

Volume 53
CELL IMPAIRMENT IN AGING AND DEVELOPMENT
Edited by Vincent J. Cristofalo and Emma Holečková • 1975

Volume 54
BIOLOGICAL RHYTHMS AND ENDOCRINE FUNCTION
Edited by Laurence W. Hedlund, John M. Franz, and
Alexander D. Kenny • 1975

CELL IMPAIRMENT IN AGING AND DEVELOPMENT

Edited by

Vincent J. Cristofalo
Wistar Institute of Anatomy and Biology
Philadelphia, Pennsylvania

and

Emma Holečková
Institute of Physiology
Czechoslovak Academy of Sciences
Prague, Czechoslovakia

SPRINGER SCIENCE+BUSINESS MEDIA, LLC

Library of Congress Cataloging in Publication Data

Symposium on Impairment of Cellular Functions During Aging and Development
   In Vivo and In Vitro, Castle of Hrubá Skála, 1974.
   Cell impairment in aging and development.

   (Advances in experimental medicine and biology ; v. 53)
   "Held under the auspices of the European Cell Biology Organization Study
Group on Aging and the Czechoslovak Medical Society."
   Includes bibliographical references and index.
   1. Aging—Congresses. 2. Developmental cytology—Congresses. I. Cristofalo,
Vincent J., 1933-   ed. II. Holečková, Ema, ed. III. European Cell Biology
Organization. Study Group on Aging. IV. Československá lékařska společnost
J. E. Purkyně. V. Title. VI. Series. [DNLM: 1. Aging—Congresses. 2. Cytology—
Congresses. WI AD559 v. 53 / QH608 C393 1974]
QH608.S9   1974   612.6'7   75-1310
ISBN 978-1-4757-0733-5   ISBN 978-1-4757-0731-1 (eBook)
DOI 10.1007/978-1-4757-0731-1

Proceedings of a symposium on Impairment of Cellular Functions During
Aging and Development *In Vivo* and *In Vitro* held at the Castle of Hrubá Skála
in Czechoslovakia, May 5-9, 1974. The symposium was held under the auspices
of the European Cell Biology Organization Study Group on Aging and the
Czechoslovak Medical Society J. E. Purkyně, Gerontological Society.

© 1975 Springer Science+Business Media New York
Originally published by Plenum Press, New York in 1975

ACKNOWLEDGMENTS

We would like to extend our sincere thanks to the European Cell Biology Organization Study Group on Aging and the Czechoslovak Medical Society under whose auspices the meeting was held.

Publication of the lively discussions following each session at the meeting was made possible through the devoted and untiring efforts of Dr. Holečková's staff at Hrubá Skalá.

We would also like to express our appreciation to Maryellen Farrell, Alice Niewoehner and Ginger Vaders for their expert help in the preparation of the final manuscripts.

We also acknowledge the support of USPHS research grant HD06323 from the National Institute of Child Health and Human Development.

FOREWORD

In 1969, eight papers dealing with aging of cultured cells
were presented at a small symposium that comprised part of a meeting
of the European Tissue Culture Society. These papers, subsequently
published by Plenum Press under the title Aging in Cell and Tissue
Culture, reflected the interests of a relatively small group of
researchers in Europe and the United States involved in the study
of aging at the cellular level.

Attention to this subject has now grown enormously. The
social and medical sciences are being asked to meet the demands
of communities whose members live longer and wish to spend their
later years as physically and mentally fit as possible. To this
end, an understanding of exactly what happens during the aging
process is essential, and basic research is fundamental to such an
understanding. This need is now widely realized, and the forty-
six papers presented at the present symposium of the study group
for Aging of the European Cell Biology Organization represent only
a part of the diverse research being done in dozens of laboratories
all over the world.

In a rapidly developing area of research such as experimental
gerontology, new models, findings, ideas and directions emerge in
great numbers; and, although it becomes more difficult to find a
common language among workers in different fields, it is also more
rewarding when joint efforts are successful. The present symposium
brought together people interested in various aspects of cellular
and molecular aging in vivo and in vitro, to confront their work
and exchange ideas and experiences, to find "meeting points" and
define gaps in knowledge.

In 1969, the most commonly used model was that of Hayflick's
diploid cell system. These cells, with their finite lifespan
in vitro, were a new star on the firmament of gerontological
research, a field clouded by almost too many theories, hypotheses
and speculations. Over the intervening years, attention to this
model system has grown rapidly, even as the general study of cellu-

vii

lar aging, to which this model contributes, has grown. Apart from
reports on work in this almost "classical" diploid cell system,
the symposium presents studies using different biological systems
with results that have been rewarding as information is obtained
on patterns of change that are common to more than one experimental
system.

Indeed, in recent years much more has been learned about the
fate of all different types of intermitotic and postmitotic cells
in situ. The symposium has also presented contributions dealing,
not directly with aging but with early ontogeny; such information
on early developmental changes should certainly shed light on some
of the mechanisms involved in aging. We are cognizant of the fact
that environmental influences resulting from the complexities of
modern civilization may have results that only occur much later, and
profoundly affect the lifespan of the organism.

There remain, of course, many unanswered questions. Whether
there is "physiological" as opposed to "pathological" aging; whether
"old" cultures living in unchanged, although not exhausted, medium,
are degenerating, not aging; what is involved when "old" fragment
cultures regenerate after excision by filling the wound with
"young" cells; why some tumor cells in vivo as well as in vitro die
while others live; all are questions deserving of our attention.

I often remember the tremendous bulk of work done in cyto-
genetics after the discovery of modern karyology, and the somewhat wistful exclamation of one of the leading cytogeneticists of
the day that "the honeymoon is over". The honeymoon in research
on aging at the cellular level seems to have just begun, and we look
forward to the fruitful world of sound scientific accuracy that
surely must follow.

E. Holečková

Institute of Physiology
Czechoslovak Medical Society
Prague, Cxechoslovakia

OPENING ADDRESS

Zdeněk Deyl

Physiol. Inst. Czech. Academy of Sciences

Prague, Czechoslovakia

Mister chairman, ladies and gentlemen, dear friends.

Along the path of biological research there are several crucial points which keep being attacked by investigators from various sides. Phylogenetic alterations, or origin of life may serve as good examples. Among these problems there is a single one, into which we get involved no matter whether we want to or not. This is maturation and ageing. To start with, the cellular level is perhaps the best one, since it allows us to go both upwards to cellular systems and ageing of tissues and organisms and to go downwards to what we may call chemical ageing. At this very moment I would like to warn you: our present knowledge is not complete enough to offer a synthesizing view and our research remains in the stage of data collecting. To those in this auditorium who are not experienced research workers in this field I would like to remind them that this collecting of data is not an easy job. To other branches of biological research we also have nothing to offer since the highest rewards among various heavens to which biologists and chemists may aspire, that with dancing girls, has been already completely booked up for muscle chemists. More detailed information about that can, perhaps, be obtained from Dr. Gallop in this auditorium who published this information in Neurath's book on proteins.

A brief look through our program can show you how different the approach to biological ageing may be. It is completely our responsibility to find a common language between different specialized branches of science to be able to communicate and transfer information, as this in my opinion is the starting point of synthesizing view.

As I have to vacate my place as soon as possible, allow me without further introduction to welcome all of you and our distinguished guests specifically to Hrubá Skalá. I wish you once more a pleasant stay and a fruitful meeting. Allow me to open this meeting in the name of the Czech Gerontological Society and in the name of its president, prof. Pacovský. And before I vanish, allow me also to express the thanks of the organizing committee to all those who helped this meeting to materialize, besides others to Dr. Cristofalo and Dr. Franks.

CONTENTS

Cell Ageing: Chairman's Introduction 1
    L. M. Franks

The effect of hydrocortisone on DNA synthesis and cell
        division during aging in vitro 7
    V. J. Cristofalo

Proliferative and functional impairment of pancreatic
        epithelial cells maintained in vitro 23
    Robert J. Hay

The use of arrested populations of human diploid fibroblasts
        for the study of senescence in vitro 41
    Robert T. Dell'Orco

Relationship between cell kinetic changes and metabolic
        events during cell senescence in vitro 51
    A. Macieira-Coelho, E. Loria, and L. Berumen

Do hyperplastoid cell lines "differentiate themselves to
        death"? 67
    G. M. Martin, C. A. Sprague, T. H. Norwood,
    W. R. Pendergrass, P. Bornstein, H. Hoehn and
    W. P. Arend

Time-lapse cinemicrophotographic studies of cell division
        patterns of human diploid fibroblasts (WI-38)
        during their in vitro lifespan 91
    P. Marlene Absher, Richard G. Absher and
    William D. Barnes

Characteristics of proliferative cells from young, old,
        and transformed WI-38 cultures 107
    Phillip D. Bowman and Charles W. Daniel

Alterations in chromatin functions during aging in vitro 123
    Jon M. Ryan

Growth-promoting alpha-globulin and ageing 137
    Jiri Michl, Mirosalv Tolar, Věra Spurná and
    Dagmar Řezáčová

Cell surface alterations and "in vitro" aging of animal
        cells 147
    R. Azencott, C. Hughes, and Y. Courtois

Changes in lysosomes during ageing of parenchymal and
        non-parenchymal liver cells 155
    D. L. Knook, E. C. Sleyster and M. J. van Noord

The effect of age on mitochondrial ultrastructure and
        enzymes 171
    P. D. Wilson and L. M. Franks

Creatinephosphokinase in human diploid cell lines 185
    M. Macek, H. Tomášová, J. Hurych, and D. Řezáčová

Cytochemical and cytogenetic findings in five human
        leukocyte long-term cultures (LaHL) of
        different origin 193
    V. Pössnerová, M. Macek, F. Heřmanský, J. Fortýnová,
    Š. Jeník, J. Holý, and M. Křeček

Enzymatic differences between short-lived and long-lived
        drosophila strains 207
    Jaroslava Skřivanová, Františka A. Gadirová, and
    Emma Holečková

Cell ageing in the intestinal tract 215
    C. Rowlatt

Studies on the proliferative capacity of mouse spleen
        cells in serial transplantation 219
    J. W. I. M. Simons and C. van den Broek

On the ageing of intermitotic cells -- Investigations on
        enterocytes and hepatocytes 235
    G. Leutert, W. Rotzsch, and W. Beier

Fat cells in ontogenesis 247
    Ludmila Kazdová and Pavel Fábry

Ageing and the loss of auditory neuroepithelium in
        the guinea pig 257
    Libuše Úlehlová

The phagolysosomal system of the retinal pigment
        epithelium in ageing rats 265
    P. M. Leuenberger

The role of retinol in, and the action of anti-
        inflammatory drugs on, hereditary retinal
        degeneration 281
    A. J. Dewar and H. W. Reading

Effect of age on kidney hyperplasia in the rat after
        unilateral nephrectomy 297
    Milena Soukupová, Přemysl Hněvkovský and Jiří Najbrt

Effect of age on kidney hyperplasia in the rat during
        cold acclimation 307
    Emma Holečková and Marie Baudysova

Age-related changes of cells involved in immune responses 315
    G. M. Butenko and A. F. Andrianova

Observations of age and environmental influences on
        the thymus kept in tissue culture 323
    O. Török, S. U. Nagy and G. Csaba

Fidelity in the collagen synthesized and modified by
        aging fibroblasts in culture 329
    Paul M. Gallop and Mercedes A. Paz

Biosynthesis of collagen during the life cycle of
        human diploid cell lines 339
    J. Hurych, M. Macek, K. Smetana, F. Beniac,
    and D. Řezáčová

Ageing processes in collagens from different tissues
        of rats 351
    M. Juřicová and Z. Deyl

The effect of nutritional regimes upon collagen
        concentration and survival of rats 359
    Z. Deyl, M. Juřicová and E. Stuchlíková

Bone aging 371
    U. J. Schmidt, Irmgard Kalbe and F. Sielaff

Relations between development of the capillary wall
        and myoarchitecture of the rat heart 375
    B. Oŝtadal, T. H. Schiebler and Z. Rychter

Functional capacity of neonatal mammalian myocardial
        cells during aging in tissue culture 389
    Frederick H. Kasten

Ultrastructural changes in senile muscle 421
    V. Hanzlíková and E. Gutmann

Denervation, reinnervation and regeneration of senile
        muscle 431
    E. Gutmann and V. Hanzlíková

Morphometrical and mathematical analysis of the ageing
        changes of the muscle-connective-tissue-
        relation in smooth muscles 441
    Paul Rother and Gerald Leutert

Some histochemical age changes of the smooth muscle
        cells in the veinous wall 451
    Martha Christova

Effect of lysolecithin on the transport of plasma
        cholesterol to tissues: Developmental aspects 459
    Milada Dobiáŝová and Eva Faltová

Viral infection and interferon in cell cultures aged
        in vitro 469
    Helena Libíková

Ultrastructure of in vitro aged chick embryo cell
        cultures in relation to viral infection 481
    Fedor Ciampor and Helena Libíková

Heritable cell cycle disturbances and late recovery
        in X-irradiated murine lymphoma L5178Y-S
        cell populations in vitro 497
    Janusz Z. Beer and Irena Szumiel

Studies on the sensitivity of anti-cancer agents of
        normal human cells in culture 511
    L. Morasca, G. Balconi, E. Erba, and
    E. Cvitkovic

Ribonucleases in mouse ascitic tumor cells during
        ageing in vivo and in vitro 521
    Katarína Horáková

Proliferation and morphology of ascitic cells as a
        function of age in cell culture 529
    A. Balázs, L. Holczinger, I. Fazekas and G. Turi

Subject Index 545

CELL AGEING: CHAIRMAN'S INTRODUCTION

L. M. Franks

Imperial Cancer Research Fund

44 Lincoln's Inn Fields, London WC2A 3PX, England

It is particularly appropriate that this meeting of the ECBO
Study Group on cell ageing should be held in Czechoslovakia, since
the initial discussions which led to the formation of the group took
place here at an earlier meeting organised by Dr. Holečková in 1969.
This was followed by a meeting in Norwich (U.K.) in 1972. A comparison of the programme of today's meeting with those of the two earlier
meetings is instructive in that it gives a guide to those areas in
which progress has been made, and underlines those areas in which
further effort may be profitable. The first meeting was concerned
mainly with ageing in cell and tissue culture. The second attempted
a broader survey of the whole field of cellular ageing. Today's
meeting attempts a synthesis since it considers--as you can see
from the programme--cell behaviour in ageing, cell structure and
function, effects of ageing on different cell types and the effects
of ageing on disease processes. Perhaps the most obvious omission
is that of a session primarily concerned with the molecular basis
of ageing. It seems to me that the time has come for us to change
the title of our group, to the ECBO Study Group on cell and molecular mechanisms of ageing. This is a topic which we must discuss
when we consider the organisation of our next meeting in 1976.

     I should like to begin this meeting by sketching in the back-
ground, as I see it, to the whole problem (1). I know that some
of you may disagree with some of my conclusions but I hope that
they may stimulate some discussion. At the level of the whole
individual there seems to be little doubt that the closely defined
lifespan for each species must be genetically determined although
we have very little information on the exact mechanisms involved.
At the cell and molecular level there is more confusion, due in

part at least, to the fact that observations are made on highly
selected systems: These systems have produced valuable information,
and perhaps even more importantly, valuable concepts (e.g. 2) but
we must consider the findings in a wider context. Although the
results provided by these systems are internally consistent, using
other systems, or other cell types, there is a wide range of
variability and many apparently contradictory results. This underlines the need for more specific model systems. This is another
topic which we might usefully discuss later.

There are two mechanisms which have been used to explain
ageing; random physico-chemical damage, and programming. The first
is an attractive theory to explain the wide and apparently uncon-
nected series of changes which accompany ageing, but there is such
a closely defined lifespan for each species that it is difficult
to avoid the conclusion that the first mechanism, which must occur,
operates within a time scale provided by the second.

Programmed cell death is a well-established feature in embryogenesis, and it has been shown by transplantation and tissue culture
experiments that in many cases the lifespan is cell-determined.
The process is highly specific and a part of the general process
of differentiation. A programmed lifespan associated with differentiation is a standard biological characteristic in many differentiated cells. Differences in lifespan between different types of
mammalian cells even in one species are enormous. Some white blood
cells have a lifespan measured in hours while other cells such as
the neurones or muscle may have a lifespan of years. Since all
these cells presumably have the same original genetic information,
differences in lifespan must be associated with the control of
gene expression during differentiation. Another feature which
appears to be associated with differentiation is the need for a
cell to go through one or more division cycles before a new differentiated character can be expressed (3).

If one accepts these postulates as being proved the ideal
differentiated cell lives out its programmed lifespan by passing
through a number of division cycles appropriate to its class,
although the final division cycle need not be followed by immediate
cell death. In real life, this ideal system may be interfered with
by agents which affect either (a) the programme or (b) the factors
controlling the initiation or cessation of cell division. The
programme is almost certainly, and the control of cell division
probably located in the DNA. The agents could include random damage
from whatever cause, "wear and tear", toxic chemicals, carcinogens,
physical and chemical mutagens and so on. Looked at in this light,
many observations can be "explained".

The longevity _in vivo_ of fixed post-mitotic cells such as the neurones or muscle cells may be due to the fact that the division cycle in these cells is switched off, so that a new character (cell death) cannot be expressed. Death can only occur in most of these cells as a consequence of random damage.

The programmed lifespan which Hayflick (4) and others have demonstrated in human diploid fibroblasts would represent the final completion of a programme of one particular cell type--a type which I have identified as relatively undifferentiated vasoformative mesenchyme (5). We have no information on the total division potential of most other cell types since they cannot yet be maintained in vitro (6). In the same way, in transplantation experiments, Williamson and Askonas (7) using immunocompetent cells and Daniel (8) using mouse mammary gland showed that other cell types may be carried through a large number of division cycles in vivo before entering a fixed post-mitotic stage. These cells too, like the Hayflick type cells, can be regarded as cells which have been allowed to complete their total programmed lifespan under highly artificial conditions.

The switching-off process is probably the same basic process involved in switching-off in any cell system, whether it be in development, in the transfer of cells from the dividing crypt cells of the small intestine into the non-proliferating villus cells, in neurones or in tissue culture cells at the end of their lifespan. Interference with the process may lead to premature cell death on the one side, or abnormal cell proliferation on the other. Merz and Ross (9) suggested that the decline phase (phase III) in cultured diploid cells was a consequence of a progressive loss of cells in the population capable of going into cell division. A similar process may occur in vivo. However, Macieira-Coelho (10) has now thrown some doubt on this concept. We shall no doubt hear more from him about this later.

The wide, and apparently unrelated series of changes found in ageing cells, of which the alteration in total amount or in physical structure of various enzymes are good examples, can be explained as a consequence of random damage to cells, either in DNA itself, in the site of transcription or translation, or in other cell sites. None of these changes may be age-associated in a causal sense. Death of the individual is almost certainly not due to cell loss.

The total possible lifespan of the individual may be con-trolled by small groups of essential cells in specific organs and tissues. The lifespan of these cells would be programmed during development as part of the normal process of differentiation. Selective loss of these cells would lead to death of the individual. Burnet (11) and others have suggested that cells in the immune system may play this role, but this seems unlikely in view of the

great proliferative potential and functional competence of these
cells (7). More likely sites are the cardiovascular and nervous
systems (12, 13). A search for "pacemaker" cells as suggested by
Finch (12) and Bullough (13) in these organs may be rewarding. It
would be aesthetically pleasing if these postulated groups of
"pacemaker" cells were associated with the equally hypothetical
cells which are thought to control other time-related biological
systems such as the diurnal rhythm.

I hope that we can consider the information we are to be
given against this theoretical background.

REFERENCES

1. Franks, L.M. 1974. Ageing in differentiated cells. Gerontologia,
   in press.

2. Orgel, L.E. 1973. Ageing of clones of mammalian cells. Nature,
   Lond. 243: 441.

3. Holtzer, H., Weintraub, H., Mayne, R. and Mochan, B. 1972.
   The cell cycle, cell lineages and cell differentiation. Curr.
   Top. develop. Biol. 7: 229.

4. Hayflick, L. 1965. The limited in vitro lifetime of human
   diploid cell strains. Exp. Cell Res. 37: 614.

5. Franks, L.M. and Cooper, T.W. 1972. The origin of human embryo
   lung cells in culture: a comment on cell differentiation, in
   vitro growth and neoplasia. Int. J. Cancer 9: 19.

6. Franks, L.M. 1970. Cellular aspects of ageing. Exp. Geront.
   5: 281.

7. Williamson, A.R. and Askonas, B.A. 1972. Senescence: an
   example of old age found in a clone of antibody-forming cells
   in mice. Nature, Lond. 238: 337.

8. Daniel, C.W. 1972. Ageing of cells during serial propagation
   in vivo. In: Strehler Adv. geront. Res., vol. 4 (Academic
   Press, New York).

9. Merz, G.S. and Ross, J.D. 1969. Viability of human diploid
   cells as a function of in vitro age. J. Cell. Phys. 74: 219.

10. Macieira-Coelho, A. 1974. Are non-dividing cells present in
     ageing cell cultures? Nature 248: 421.

11. Burnet, F.M. 1973. A genetic interpretation of ageing.
     Lancet ii: 480.

12. Finch, C.E. 1972. Cellular pacemakers of ageing in mammals.
     In: Harris, Allin and Viza, Cell Differentiation (Monksgaard,
     Copenhagen).

13. Bullough, W.S. 1971. Ageing of mammals. Nature, Lond. 229:
     608.

THE EFFECT OF HYDROCORTISONE ON DNA SYNTHESIS AND CELL DIVISION

DURING AGING IN VITRO

V. J. Cristofalo

Wistar Institute of Anatomy and Biology

Philadelphia, Pennsylvania 19104

Several years ago, during the course of our studies on lysosomes in aging human diploid cells, we observed a small but consistent increase in the lifespan of cultures grown in the presence of 14μM hydrocortisone (HC).

Further characterization of this phenomenon (1-4) has shown that: 1. Cell lifespan, in terms of actual population doublings, was extended 30-40% by the continuous inclusion of 14 μM HC in the medium; 2. This effect was maximal with 14 μM HC. No increase was observed at dosage concentrations between .014 and 1.4 μM and slight inhibition was evident at 140 and 280 μM; 3. There was no rescue from phase-out with the hormone; once a culture had reached the stage where it could no longer achieve confluency, HC could not reverse the effect; 4. When HC was added at different periods in the lifespan, the magnitude of the lifespan extension varied directly with the amount of time the culture was grown in the presence of the hormone; 5. The saturation density of the culture was increased in the presence of the hormone; 6. The overall effect of HC was not due to increased plating efficiency or increased adhesion to the glass or plastic surface.

The HC effect on cell lifespan, therefore, represents the action of a chemically defined modulator of cell division and population lifespan. The potential usefulness of this substance as a probe for understanding the regulation of cell division, has led to further explorations of the HC effect. This report presents the results of these studies.

MATERIALS AND METHODS

All studies were done with human diploid cell lines WI-38 and
WI-26 (5,6). These were obtained either from frozen stock maintained here at the Wistar Institute or from Dr. Leonard Hayflick
of Stanford University.

The cells were grown as previously described (7) in autoclav-
able Eagle's MEM (Auto-Pow, Flow Laboratories, Rockville, Md.)
modified by the addition of Eagle's BME vitamins. Immediately
before use, the medium was supplemented with L glutamine (2 mM),
$NaHCO_3$ (20 mM) and fetal calf serum (10% v/v). In the first
experiments, aureomycin (50 µg/ml) was included in the formulation.
In subsequent experiments, no antibiotics were used. Cultures were
grown at 37°C in an atmosphere of 5% CO_2-95% air, and were monitored for mycoplasma by the method of Levine (8).

Routine subcultivations were carried out when monolayers were
confluent. The cells were released from the glass by treatment
with trypsin (0.25%) in Ca^{++}- and Mg^{++}-free phosphate buffered
saline solution. After suspension in medium containing 10% fetal
calf serum, the cells were counted and inoculated into appropriate
vessels at a density of 1 X 10^4 cells/cm^2.

Cell population doublings were calculated by comparison of
the cell counts/vessel at seeding and when the cultures reached
confluency. All cell counts were performed electronically using a
Coulter Counter.

Autoradiography was carried out by our standard procedures
(7). In brief, cells were seeded at a density of 1-1.3 X 10^4
cells/cm^2 either in 60 mm Petri dishes containing 11 X 22 mm
coverslips, or in 2-chamber Lab-Tek dishes (Labtech, Inc.,
Westmont, Ill.) and incubated as described above. Twenty-four hr
after seeding, ^{3}H-thymidine (^{3}HdT) was added to the cultures to
a final concentration of 0.1 µCi/ml (spec. act. 2 Ci/mMole). At
appropriate time intervals (24-30 hr), the cell monolayers were
washed rapidly in a cold buffered balanced salt solution, fixed
in Carnoy's solution, hydrated through an alcohol series and air
dried overnight. The slides were then dipped in emulsion (Kodak
NTB-2). Following the exposure period (4 days), the slides were
developed (Kodak D-19 developer, 5 min), fixed (Kodak acid fixer,
5 min), stained lightly with Harris hemotoxylin, rinsed in cold
running water, and air dried.

The autoradiographs were analyzed microscopically by scoring
the percentage of cells with labelled nuclei (5 silver grains or
more) in random fields throughout the coverslips. At least 400
cells were counted on each coverslip and all coverslips were

prepared in duplicate.

For liquid scintillation counting, coverslips were prepared
as for autoradiography, removed at appropriate intervals, dipped

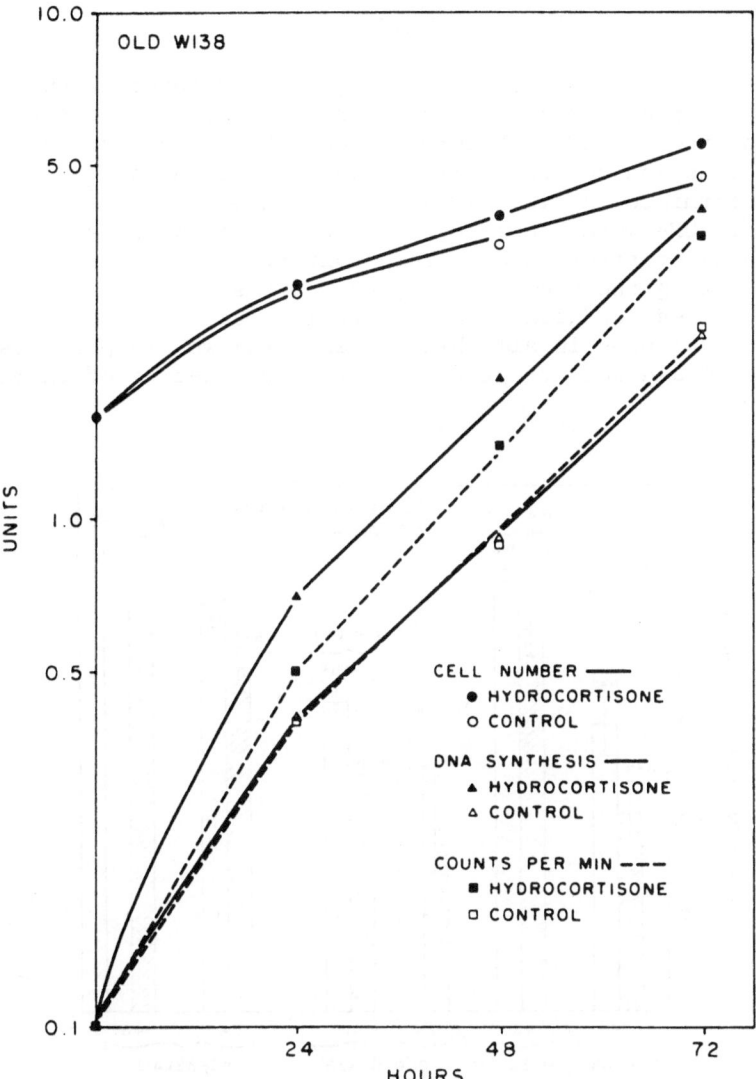

Fig. 1. The effect of hydrocortisone on cell proliferation, DNA
synthesis and ^{3}HdT incorporated into DNA in logarithmically growing
WI-38 cultures (passage 42).

in cold 10% trichloroacetic acid and placed directly into scintillation vials and counted in an Intertechnique Liquid Scintillation spectrometer.

DNA was determined by the method of Volkin and Cohn (9).

RESULTS

Initially, our studies were designed to determine how the
effect of HC in extending lifespan was expressed during a single
population growth cycle. Figure 1 shows the typical growth curves
for HC-treated and control cells in which cell number, DNA synthesis, determined chemically, and radioactivity from ^{3}HdT incorporated into DNA were monitored. The zero time point on the figure
represents 24 hr after seeding both cultures with identical cell
numbers. During the growth cycle, the curves diverged and by 72 hr,
there was a clear and consistent difference between the treated
and control cultures in all three parameters; we had previously
shown that HC did not affect the short-term kinetics of thymidine

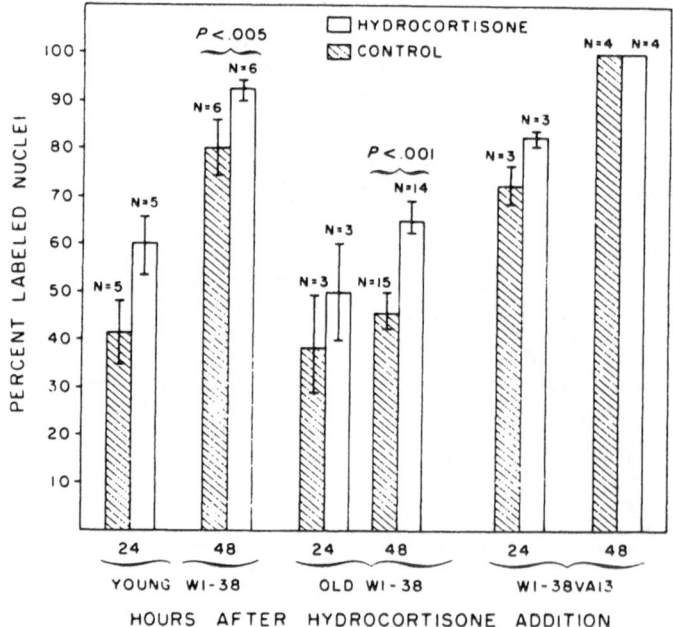

Fig. 2. The effect of hydrocortisone on nuclear labelling in young,
old and SV40-transformed human cells. The time periods along the
abcissa refer to the time on hydrocortisone.

uptake or the activity of thymidine kinase. Thus, these experiments
show that incorporation of [3]HdT can be used as a valid probe to
follow DNA synthesis in these cultures.

Figure 2 shows the results of a similar experiment in which
the effect of HC was monitored autoradiographically by determining
the percentage of labelled nuclei in the presence of a continuous [3]HdT pulse. For both young and old cells, after 48 hr
exposure to HC, there was a significant increase in the fraction
of labelled nuclei. Thus, the HC-mediated increase in DNA synthesis
seems to be due, in part at least, to an increase in the fraction
of cells in the proliferating pool, i.e. HC appears to amplify the
stimulus for proliferation. (For comparison, we included labelling
data for the permanently proliferating SV40-transformed WI-38 cell
line, WI-38VA13, in which, by 48 hr, both treated and control
cultures have essentially 100% labelled nuclei.) Note that the
differences between the HC treated and control groups are bigger
in the old cells. However, the responsiveness of the older cells
in terms of the fraction completing division was lower.

Since the loss of proliferative capacity in these diploid
populations has been shown to be due to an exponential increase
in the number of non-cycling or very slowly cycling cells in the
population (7), it was of interest to determine if cultures grown
continuously in the presence of HC showed the same heterogeneity.
Figure 3 shows the percent labelled nuclei as a function of cumulative number of population doublings (determined by direct cell count
at each subcultivation) over the lifespan of both female-derived
(WI-38) and male-derived (WI-26) cell cultures. Initially, nearly
100% of the cells were labelled in both hormone-treated and control
cultures. As the lifespan progressed, however, the accelerated
rate of proliferation of the hormone-treated population was
evidenced by the more rapid traverse of the bars representing the
hormone treated cultures across the abcissa; i.e., the hormone-
treated cultures were doubling more rapidly.

The decline in the fraction of labelled cells was more rapid
in the control cultures and they phased out well before the HC-
treated cultures; however, the pattern of decline in labelling
was the same in both. HC seemed simply to retain the cells in the
actively proliferating pool for longer periods.

Finally, it is important to note that, just as with the short
term autoradiographic experiment (Fig. 2), the differences between
HC-treated and control cultures increased with age. However, the
percentage of cells responding to the stimulus for division
declined with age. The rate of decline was slower in the HC-
treated culture. Thus, it appears from these experiments as well,
that the HC amplified the primary signal for division.

V.J. CRISTOFALO

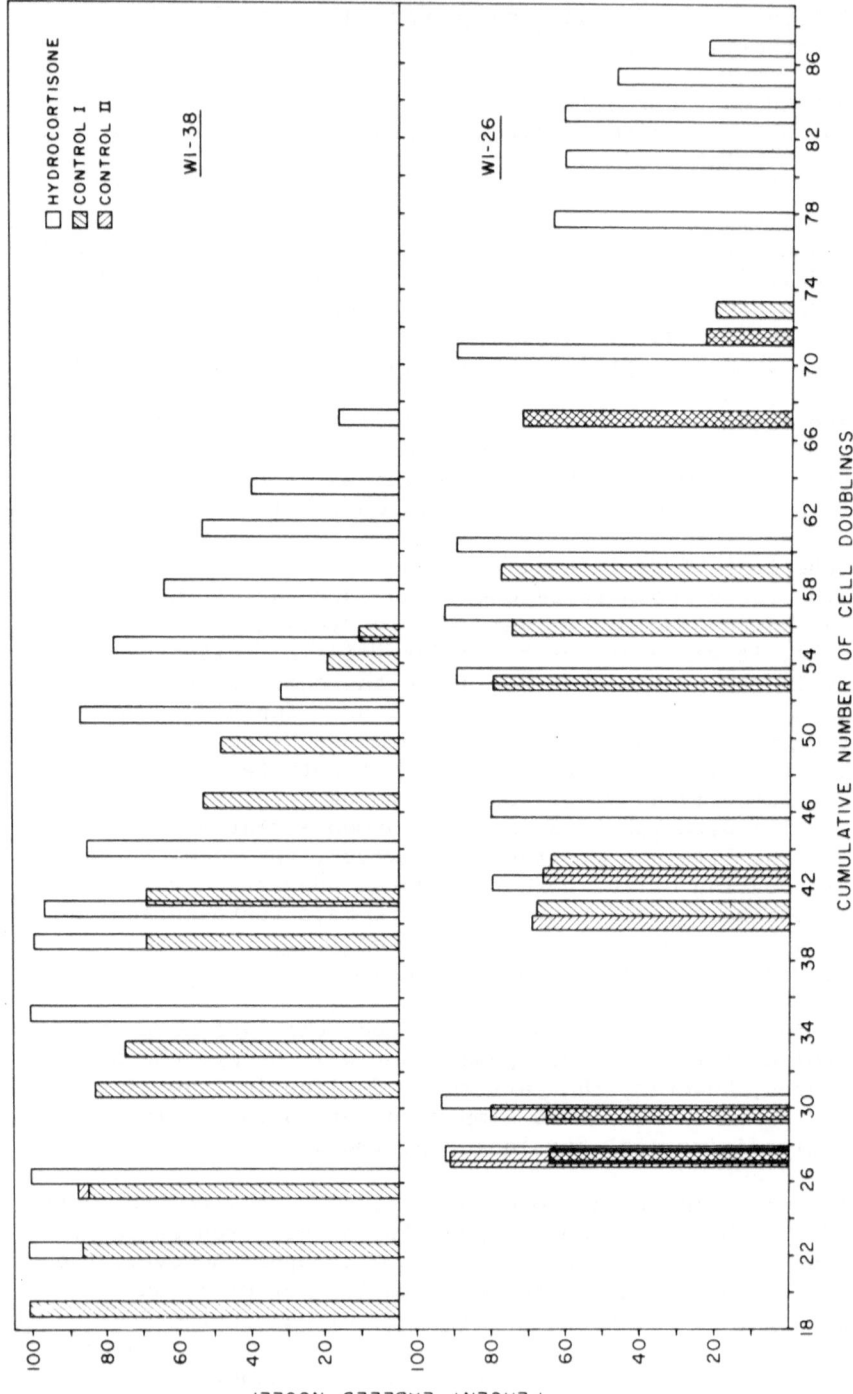

Fig. 3. The effect of hydrocortisone on the fraction of labelled nuclei in the culture
after 30 hr exposure to ³HdT at intervals throughout the lifespan of both WI-38 and WI-26
cells.

Further evidence for this interpretation is provided in Figure 4. Here, in MEM without fetal calf serum, there was essentially no division and, with graded increases in serum, there was a graded response in the fraction of cells incorporating labelled precursor; the primary signal for division was clearly serum. Note that in the presence of HC, 0.3% serum gave a response higher than 10% serum without HC, and 10% serum plus HC was equivalent to or greater than 50% serum.

When serum is added to confluent, contact-inhibited monolayers of WI-38 cells, a wave of division occurs, and the response is directly related to the amount of serum used between 5 and 30% (10, 11). In the experiments shown in Figure 5, we compared the effects of refeeding confluent monolayers with 2 different serum concentrations in the presence and absence of HC. For the youngest group (95% labelled nuclei), in the absence of refeeding, there was essentially no division. The effect of refeeding with MEM containing either 10% or 30% serum was evident from the response of 45% and

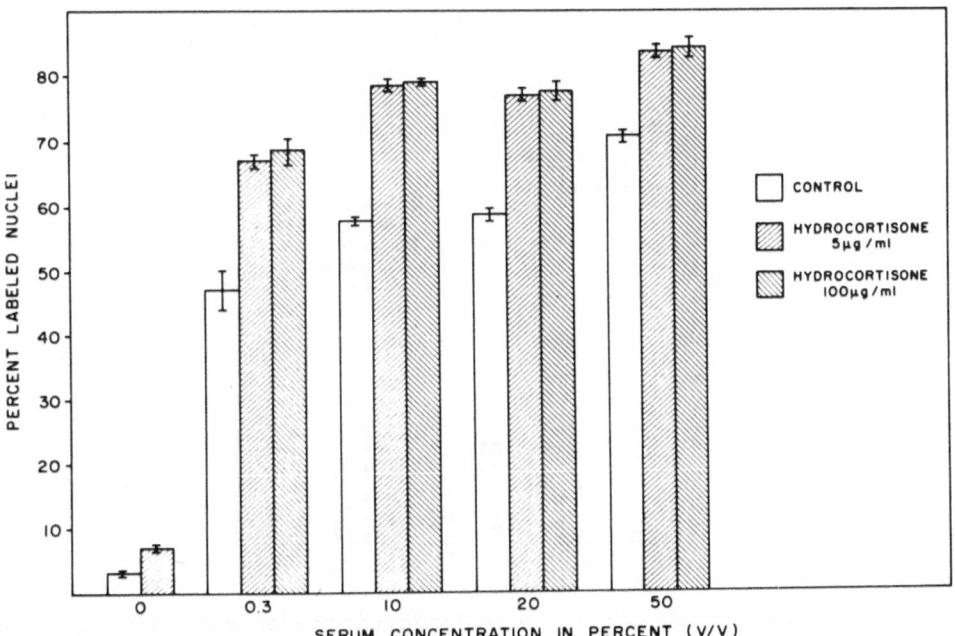

Fig. 4. The effect of serum concentration and hydrocortisone on nuclear labelling in WI-38 cultures.

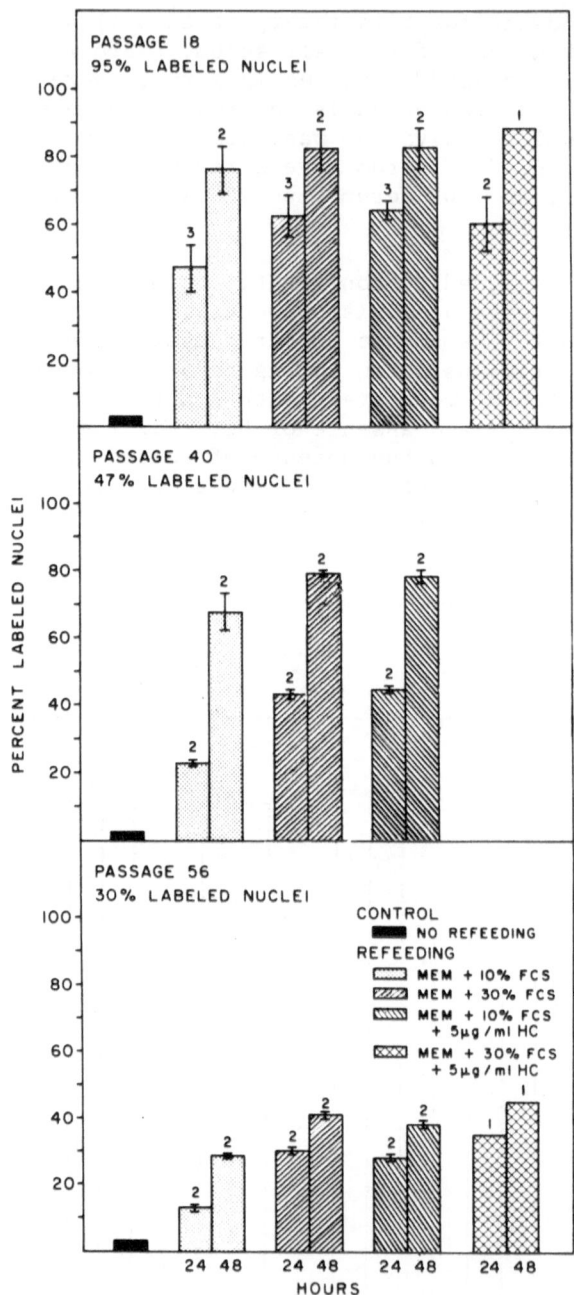

Fig. 5.   The effect of serum and hydrocortisone on cell division in
confluent cultures of WI-38 cells at different passage levels.   (The
95% labelled nuclei refers to the labelling index for parallel
logarithmically growing cultures).

60% labelled nuclei respectively in 24 hr. The response with 10%
fetal calf serum plus 5 µg/ml HC was equivalent to that of 30%
fetal calf serum. Here again, HC appeared to amplify the serum
signal. No further response was elicited when HC was added to 30%
fetal calf serum. The same pattern of division stimulation was
evident for the two older passage groups shown in Figure 5. Note
that the fraction of cells responding to refeeding continually
declined, reflecting the declining fraction of the cells still in
the proliferating pool, i.e., able to respond to the primary signal
for division. Irrespective of age, however, refeeding with 10%
serum + 5 µg/ml HC is equivalent to refeeding with 30% serum.

To approach the mechanism of action of this hormone in stimulating cell division, it was of interest to determine the time
course of induction of the effect. For these experiments, cells
were seeded at low density in duplicate so that an equal number
of vessels were available for refeeding purposes. Twenty-four hr
after seeding, the medium in half the flasks was replaced with
medium containing HC and the cells were incubated for various
periods of time. At the times indicated on Table I, the HC medium
was removed and replaced with similarly conditioned medium without
HC from the duplicate flasks and then the cells were fixed after
a total incubation of 48 hr. Controls were handled in exactly
the same way except the feeding medium did not contain exogenous
HC. The results (Table I) showed that for both experiments, the
effect was not a rapid one and required at least 12 hr for the
effect to begin. However, when the hormone was present for 12 hr,
the cells behaved, qualitatively at least, as if the hormone had
been present continually for the entire 24 hr period. Experiments
are now in progress to determine if the action of HC during this
12 hr period is sensitive to inhibition by actinomycin D and/or
cycloheximide.

TABLE I

HC Exposure Time (Hrs.)	DNA Synthesis (CPM/Cover Slip) Ratio HC/Control	
	Exp. I	Exp. II
6	.95	.90
12	1.10	1.06
18	1.24	1.26
24	1.28	1.37

DISCUSSION

The studies described above show that: one, HC increases the lifespan of diploid cultures in part, at least, by increasing the fraction of cells in the proliferating pool and two, although not a primary stimulus for cell division, HC appears to amplify the serum stimulus for division.

We have previously shown (3) that the stimulatory effect of HC on DNA synthesis is highly specific for certain molecular configurations in the substituted steroid nucleus. The recognition sites require unsaturation at the 4-5 position in ring A, the keto group at position 3, the 11β-hydroxy group and an additional keto group at position 20. Although some differences exist, this pattern of molecular structure-function relationship is similar to that reported for the stimulation of casein synthesis in mammary organ culture (12), and for glutamine synthetase induction in embryonic chick retina culture (13,14). Because of the central role of glutamine in cell culture metabolism and proliferation in general, we have investigated glutamine synthetase induction in WI-38 cells and although the enzyme could be induced by glutamine deprivation, HC had no effect on this induction (15).

In addition, since HC is generally considered to be an inhibitor of cell division in tissue culture (16), we have tested cultures of a variety of cell types. Our results (3) have shown that the stimulation of cell division is highly specific for human diploid fetal cells and is not a general phenomenon.

HC, cortisone and other corticosteroids have been reported to prolong the in vitro survival time of several cell types (17-19). These studies have been concerned with post-mitotic maintenance of the cultures. Macieira-Coelho (20), however, was the first to report the increase in proliferative lifespan of diploid, human fibroblast-like cells in culture.

Division stimulating effects with HC have also been reported by Castor and Prince (21) for cartilage cells and by Smith et al. (22) for human fetal lung cells and Donaldson and Moorhead (23) for WI-38 cells. Recently, Thrash and Cunningham (24) have shown the stimulation of division by HC in density-inhibited 3T3 cells and Armelin (25) and Gospodarowicz (26) have shown that HC amplifies the activity of the pituitary and brain-derived polypeptides which stimulate cell division.

In considering possible explanations for these results, we must consider that although changes in transcription represent the best defined action of the glucocorticoids, there are a wide range of other, less well understood, effects of steroids, many involving various membrane effects, that must be considered. As

shown by Dr. Absher and Martin et al. (these proceedings), there
is now ample evidence to indicate that aging of the population is
reflected at the cellular level by a transition from a rapidly
cycling state to one, or a series of, more slowly cycling states,
and finally, to a sterile state, i.e., a state in which the cells
are arrested or are cycling so slowly as to be unable to repopu-
late the culture vessel. These three transitions are shown in
the center (shaded) portion of Figure 6. One simple interpretation of the data is that the steroid delays these transitions.

Another possible interpretation is shown in the upper portion
of the figure and is based by analogy with the work of Bresciani
(27) in which he showed that as mammary cells differentiate, they
lose responsiveness to one hormone but acquire or maintain responsiveness to other hormonal stimuli. It is possible that young cells
are responsive to 10% serum initially, but eventually undergo a
transition to a second state where 10% serum plus HC is required
to elicit a division response. Alternatively, there may be a
second population in the culture which succeeds the first and which
will only proliferate in the presence of serum plus HC.

Finally, a third possibility is shown in the lower portion of
the figure and presupposes that slowly cycling or arrested cells
inhibit the growth of cells that are capable of division either
simply by contact or by the space they occupy or by eliciting an
inhibitor or chalone into the environment. In these rapidly
dividing cells, HC has no effect. In the slowly cycling cells,
however, HC could be working either by somehow inactivating the
hypothetical inhibitor or by killing the cells responsible for its
secretion. Thus, the young cells in the population would not be
inhibited to the same extent. Our current work is designed to
clarify these possibilities.

SUMMARY

Hydrocortisone (14 µM) added to cultures of human diploid
fibroblast-like cells extends the lifespan of the population.
This effect is expressed during a single growth cycle by an
increased rate of proliferation and a higher rate of incorporation
of ^{3}HdT into DNA. The hormone appears to exert its effect by
increasing the fraction of cells in the proliferating pool, and
this increase is expressed at all levels of serial subcultivation
(population ages). The effect requires a minimum of between 12
and 18 hr exposure of the cells to hydrocortisone. Studies with
varying amounts of serum suggest that the hormone affects the
cells by amplifying the serum signal which is the primary signal
for cell division.

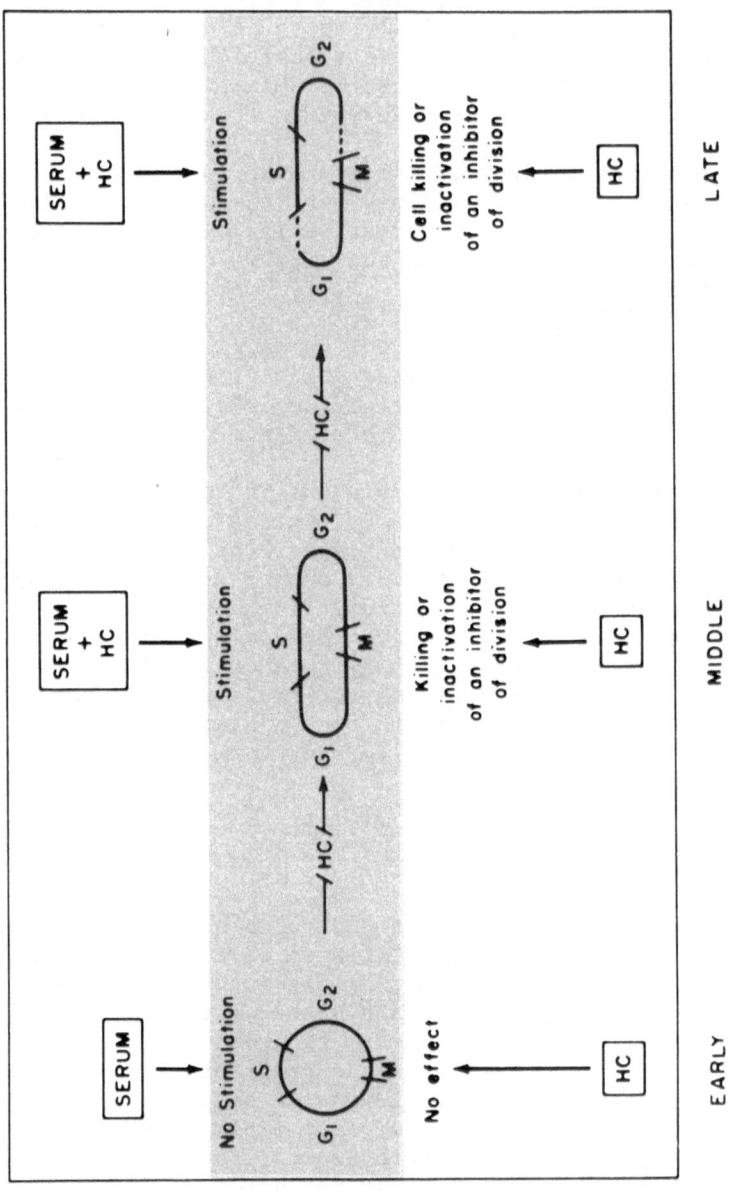

Fig. 6. Possible effects of hydrocortisone on human diploid fibroblast-like cultures.

ACKNOWLEDGEMENTS

The author is grateful to Mrs. Barbara Sharf for expert technical assistance in various phases of this work. The work reported herein was supported by U.S. Public Health Service Research Grants HD02721 and HD06323 from the National Institute of Child Health and Human Development.

REFERENCES

1. Cristofalo, V.J. 1970. Metabolic aspects of aging in diploid human cells. In: Aging in Cell and Tissue Culture. E. Holečková and V. Cristofalo, eds., Plenum Press, N.Y., p. 83.

2. Cristofalo, V.J. 1972. Animal cell cultures as a model system for the study of aging. Adv. Gerontol. Res. 4: 45.

3. Cristofalo, V.J. 1973. Cellular Senescence: Factors modulating cell proliferation in vitro. Molec. Cell. Mech. Aging 27: 65.

4. Cristofalo, V.J. and Kabakjian, J. 1974. Lysosomal enzymes and aging in vitro. Mech. Aging and Develop. (in press).

5. Hayflick, L. and Moorhead, P. 1961. The serial cultivation of human diploid cell strains. Exp. Cell Res. 25: 585.

6. Hayflick, L. 1965. The limited in vitro lifetime of human diploid cell strains. Exp. Cell Res. 37: 614.

7. Cristofalo, V.J. and Sharf, B.B. 1973. Cellular senescence and DNA synthesis. Thymidine incorporation as a measure of population age in human diploid cells. Exp. Cell Res. 76: 419.

8. Levine, E.M. 1972. Mycoplasma contamination of animal cell cultures: A simple, rapid detection method. Exp. Cell Res. 74: 99.

9. Volkin, E. and Cohn, W.E. 1954. Estimation of nucleic acids. Meth. Biochem. Anal. 1: 287.

10. Farber, J., Rovera, G. and Baserga, R. 1971. Template activity of chromatin during stimulation of cellular proliferation in human diploid fibroblasts. Biochem. J. 122: 189.

11. Todaro, G.J., Lazar, G.K. and Green, H. 1965. The initiation of cell division in a contact-inhibited mammalian cell line. Comp. Physiol. 66: 325.

12. Turkington, R.W. and Topper, Y.J. 1967. Androgen inhibition
     of mammary gland differentiation in vitro. Endocrinol. 80:
     329.

13. Moscona, A.A. and Piddington, R. 1966. Stimulation by
     hydrocortisone of premature changes in the developmental
     pattern of glutamine synthetase in embryonic retina. Biochim.
     Biophys. Acta 121: 409.

14. Reif-Lehrer, L. and Chader, G. 1972. Hormonal effects on the
     neural retina: corticoid uptake, specific binding and
     structural requirements for the induction of glutamine
     synthetase. Biochim. Biophys. Acta 264: 186.

15. Viceps, D. and Cristofalo, V.J. 1974. Glutamine synthetase
     activity in WI-38. Expt. Cell Res. (submitted for
     publication).

16. Ruhmann, A.G. and Berliner, D.L. 1965. Effect of steroids
     on growth of mouse fibroblasts in vitro. Endocrinol. 76:
     916.

17. Arpels, C., Babcock, V.J. and Southam, C.M. 1964. Effect
     of steroids on human cell cultures; sustaining effect of
     hydrocortisone. Proc. Soc. Exp. Biol. Med. 115: 102.

18. Yuan, G.C. and Chang, R.S. 1969. Effect of hydrocortisone
     on age-dependent changes in lipid metabolism of primary
     human amnion cells in vitro. Proc. Soc. Exp. Biol. Med.
     130: 934.

19. Yuan, G.C., Chang, R.S., Little, J.B. and Cornil, G. 1967.
     Prolongation of postmitotic lifespan of primary human amnion
     cells in vitro by hydrocortisone. J. Gerontol. 22: 174.

20. Macieira-Coelho, A. 1966. Action of cortisone on human
     fibroblasts in vitro. Experientia 22: 390.

21. Castor, C.W. and Prince, R.K. 1964. Modulation of the
     intrinsic viscosity of hyaluronic acid formed by human
     fibroblasts in vitro: The effects of hydrocortisone and
     colchicine. Biochim. Biophys. Acta 83: 165.

22. Smith, B.T., Torday, J.S. and Giroud, C.J.P. 1973. The
     growth promoting effect of cortisol on human fetal lung
     cells. Steroids 22: 515.

23. Donaldson, C. and Moorhead, P. 1974. Lifespan in vitro
     of fibroblasts supplemented with hydrocortisone: Thymidine
     incorporation and cell size. Exp. Cell Res. (submitted for
     publication).

24. Thrash, C.R. and Cunningham, D.D. 1973. Stimulation of
     division of density inhibited fibroblasts by glucocorti-
     coids. Nature 242: 399.

25. Armelin, H. 1973. Pituitary extracts and steroid hormones
     in the control of 3T3 cell growth. Proc. Nat. Acad. Sci.
     70: 2702.

26. Gospodarowicz, D. 1974. Localisation of a fibroblast growth
     factor and its effect alone and with hydrocortisone on 3T3
     cell growth. Nature 249: 123.

27. Bresciani, F. 1968. Topography of DNA synthesis in the
     mammary gland of the C3H mouse and its control by ovarian
     hormones; an autoradiographic study. Cell Tissue Kinet. 1:
     51.

QUESTIONS TO DR. CRISTOFALO

Dr. Libíková: Is the inhibitor of cellular division – which you
have mentioned – a hypothetical substance or a known one? If known,
does it belong to the class of chalones?

Dr. Cristofalo: At present we must consider this a hypothetical
substance since we have no evidence for the specific, hydrocortisone
modulated inhibitor. However, it certainly is reasonable to suppose
that such a substance would belong to the general class of chalones.

Dr. Balász: You supposed that arrest in the old cells may be
caused by chalone-like inhibitors. These were experimentally
demonstrated in contact inhibited fibroblast cultures by Yeh and
Fisher, by Salas and Green and Houck et al. In this case do you
regard the hydrocortisone effect as a "derepression", causing an
increase in the growth fractions?

Dr. Cristofalo: I agree, this is a possibility although other
possibilities also exist such as direct interaction of the hypothetical chalone with the hormone or stimulation of various serum
factors which modulate growth.

Dr. Franks: We have found an effect of hydrocortisone with low
dose levels (0.1 µg/ml). Do you find a difference in responsiveness
in different cell strains?

Dr. Cristofalo: We have not studied this in a systematic way for
cells other than WI-38. However, your value of 0.1 µg/ml is much
closer to the level in vivo than the 5 µg/ml we have been using.
Perhaps the difference may be a function of the fact that you use
calf serum and we use foetal calf serum. Since the hormones

travel in combination with serum proteins then the difference in serum protein may be important.

PROLIFERATIVE AND FUNCTIONAL IMPAIRMENT OF PANCREATIC EPITHELIAL CELLS MAINTAINED IN VITRO

Robert J. Hay

Department of Biological Sciences, Wright State

University, Dayton, Ohio   45431

The use of animal cells in monolayer culture for model studies of the ageing process has increased extensively during recent years. In this regard, comparatively more attention has been directed towards the limited division potential of human fibroblasts than to limits in specific functional capability of this or other cell types with time in culture (1). This may be understandable when one considers the dearth of information concerning factors, environmental or otherwise, which affect expression of differentiated function by cells in culture. Nonetheless, for many experimental gerontologists, the age-associated loss of ability to synthesize tissue-specific products is thought to represent a major consequence of tissue ageing: one which, in most cases, cannot be accounted for by concomitant loss of individual cells (2,3).

Interestingly, evidence is available indicating that differentiated function in some cell systems is lost during maintenance or serial passage in vitro. Clonally derived cell strains from chick embryonic cartilage exhibit markedly reduced product synthesis after several subcloning steps, yet they continue to proliferate for some time after this apparent "dedifferentiation" (4). The production of tissue specific proteins by human liver cells in clonal culture also decreases as sensitive immunological and radioimmunological techniques are required for their detection (5). Collagen synthesis by human fibroblasts reportedly declines with approach of the degenerative phase (6,7).

The work to be described below involves study of colonial aggregates and clonally-derived epithelial cells from human and guinea pig pancreatic tissue. A rapidly decreasing functional capability and a comparatively brief proliferative growth span are characteristic

of these cells under the conditions imposed. Thus a qualitative
relationship exists between this and other systems exhibiting degenerative change with time in vitro. The culture technique is proposed
as a useful and interesting model system for the study of factors
controlling expression of differentiated function in exocrine cells
from this organ.

MATERIALS AND METHODS

Pancreatic tissue was obtained from aborted human fetuses,
usually of 10 to 20 weeks gestation, from guinea pig embryos near
term or from postnatal guinea pigs 250 grams or less. Each organ
was rinsed in Hank's saline and thoroughly minced with lens scissors.
The fragments were dissociated in a mixture of crude trypsin (0.125%)
and collagenase (0.5%) by rapid agitation (120 rpm) for 15 min at
37°C. The collagenase stock solution was prepared from crude
material (Sigma Type I) as described by Cahn et al (8). Preliminary
studies indicated that this procedure gave optimum cell yields per
unit time when compared to several other dissociation mixtures. It
also compared favorably when percent viability (by erythrosin B exclusion) and overall plating efficiencies were determined. The
growth medium used throughout was Coon's modification of Ham's F12
(8) supplemented with 15% selected fetal calf serum (for clonal
culture or aggregation) or 20% calf serum for colonial aggregate
isolation. Any deviation from this is stated in the text. Standard
techniques for clonal culture and for physical isolation of clones
were utilized (8).

Fractionation and Aggregation Procedure for Acinar Cells

A procedure for increasing the ratio of acinar to non-acinar
cells was adapted from that of Amsterdam and Jamieson (9). Dissociated cells from guinea pig tissue were collected in F12 with 15%
fetal calf serum and filtered through nylon mesh of 20 μ pore size.
The resultant cell suspension was dispersed further by mixing through
a syringe fitted with an 18 guage needle. The suspension was then
layered over a solution containing 4% bovine serum albumin in Hank's
saline. In most cases the latter was set up in 30 ml amounts in 40
ml centrifuge tubes and 3 to 4 ml of cell suspension (10^7 viable
cells/ml) were added at the top. The tubes were then spun at 100
x g for 5 min.

The above manipulation was performed two more times in succes-
sion and the resulting cell pellet, consisting of up to 95% acinar
cells, was resuspended in F12 containing 15% fetal calf serum and
soybean trypsin inhibitor (0.1 mg/ml). The concentration of cells
was adjusted to 2×10^6 per ml and the suspension was dispensed to
25 ml conical flasks in 5 ml aliquots for subsequent aggregation.

The flasks were then placed in a gyrating water bath set at 37°C
and the contents were permitted to aggregate for 16 to 18 hr at
80 rpm. The aggregates so formed were then used experimentally or
were transferred to F12 supplemented with 20% calf serum for seeding
at 10^3 to 10^4 aggregates per cm of culture surface area.

Isotope Labeling Studies

Aggregates of acinar cells were collected by centrifugation
and were resuspended in F12 supplemented with 15% fetal calf serum
and containing 1 C of tritiated leucine (S.A 55/cMole) per ml. A
pulse labeling period of 2 hr at 37°C was followed by 5 washes in
medium without isotope and a chase period of one hr in medium with
ten times the usual concentration of non-labeled leucine. This
was followed again by washing, and the aggregates were separated
equally to two conical flasks. At zero time pancreozymin (0.2 U/ml)
was added to the test flask and an equal volume of vehicle to the
control. Samples were taken at intervals thereafter and aggregates
were separated from the medium by centrifugation. Radioactivity
in 20 l samples of medium was determined by addition first to the
solubilizer Soluene 100 (Packard). Scintillator consisting of
Permablend I (Packard) in toluene was added and radioactivity was
determined using a Nuclear Chicago Scintillation counter.

Biochemical and Histochemical Studies

Release of amylase by aggregates either in suspension or in
colonial form was examined biochemically and histochemically. The
micromethod of Jamieson et al. (10) was used unchanged in the former
case while the histochemical assay of Shear and Pearse (11) was
adapted for the latter. In this case, starch at 0.25% was dissolved
in borate buffer (0.01 M) at pH 6.9. This incubation medium was
also made to contain 1% Noble agar. Plates to be examined were
washed several times with Hank's saline and the complete histochemical medium at 45°C was added. The agar was allowed to solidify and
the plates were incubated at 37°C in a moist chamber for 3 to 18 hr.

Protein estimations were performed by the standard method of
Lowry et al. (12).

RESULTS

Clonal Analyses

Epithelial cell colonies appeared at much lower frequency than
did fibroblast clones when human embryonic pancreas served as source

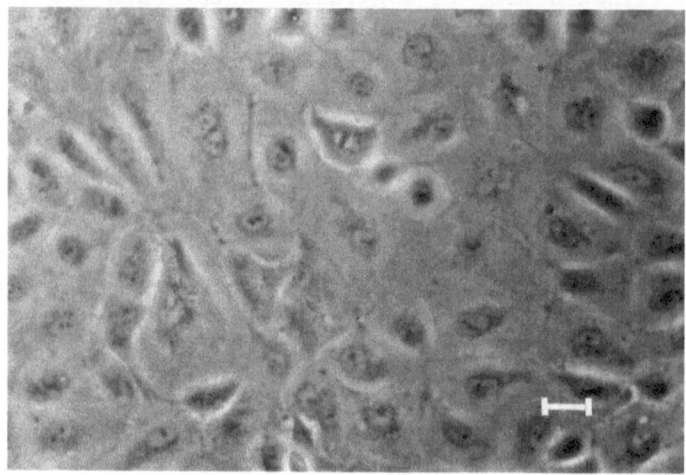

Fig. 1. An epithelial cell clone from human embryonic pancreas
photographed 14 days after initial plating. Note the presence of
actively dividing cells. Bar is 30μ.

material. At this writing over 20 fetuses have been received and
processed. Typical results are presented in Table I but occasional
low overall plating efficiencies (0.2%) have been obtained. In no
case did the number of epithelial clones exceed 10% of the total
clonable population.

     Attempts were made initially to isolate cell strains from
epithelial clones but subcultivation has not yet been possible.
This is true despite obvious presence of active cell division during
the first two weeks (Fig. 1). Isolated clones have never developed
sufficiently to form a monolayer within their confining penicylinders.
Although clones were fed every 3 to 4 days mitoses appeared to decrease by the third to fourth week and by the ninth or tenth week
initial signs of degeneration were recognized. Individual cells
became detached from one another to gradually assume a rounded,
refractile appearance (Fig. 2). Total degeneration was apparent by
twenty weeks in all cases studied. Transformation of epithelial
cells during such maintenance, although anticipated, has not occurred.

     Clonal analyses with cells isolated from guinea pig pancreas
yielded similar results. A preponderance of fibroblast clones was
observed with epithelial clones representing no more than 10% of
the clonable population (Figs. 3 and 4). It has not been possible
to subcultivate such clones with retention of epithelial morphology.
In contrast fibroblasts and clones which give rise to strains of
endothelial morphology have been isolated and propagated for over
50 generations (13).

Table I

Clonal Isolation of Cells from the Human Pancreas

Plate Treatment	Clone Type by Gross Morphology		Average Total Plating Efficiency (%)
	Fibroblast	Epithelial	
Uncoated	104;94 108;115	3;3 3;2	10.9
Collagen Coated	91;112	0;2	10.2

Cells were dissociated with the trypsin-collagenase mixture. They were seeded to 9 cm. plates at 10^3 viable cells per dish in 5 ml. modified F12 with 10% fetal calf serum. Cultures were fed every 4 days and were fixed and stained on the 15th day.

Aggregates and Aggregate Colonies

The successful isolation of epithelial cell groupings by the plating of cell aggregates from various chick embryonic tissues (14) prompted similar trials with cells from the human and guinea pig pancreas. Aggregates from human embryonic source material were formed by gyration and permitted to attach to Falcon plastic plates in F12 with 15% fetal calf serum. That this technique can be used for effective isolation is evidenced in Figures 5, 6 and 7. Human pancreatic cell aggregates adhered comparatively slowly and even by 24 hr after plating the majority had not firmly attached. This tendency can be used to advantage in selecting against fibroblasts. Aggregates which have not adhered after 18 to 24 hr (Fig. 5) can be replated leaving fibroblastic elements on the original culture surface. By 48 hr after plating human aggregates had attached and were beginning to spread over the substrate in radial fashion. Figure 6 depicts the edge of such an aggregate colony after 6 days. Cell groupings similar to these were isolated physically with standard penicylinders essentially as was the case with the clonal work described earlier. Figure 7 is a photomicrograph of such an aggregate colony which was maintained for 20 days. Note that individual epithelial cells were less contiguous and many had become rounded and refractile. As was observed with epithelial clones, the cells of human colonial aggregates never proliferated sufficiently to form complete monolayers within the confines of their respective penincylinders. In most cases total degeneration occurred during the second month in culture. No instance of spontaneous transformation has yet been observed.

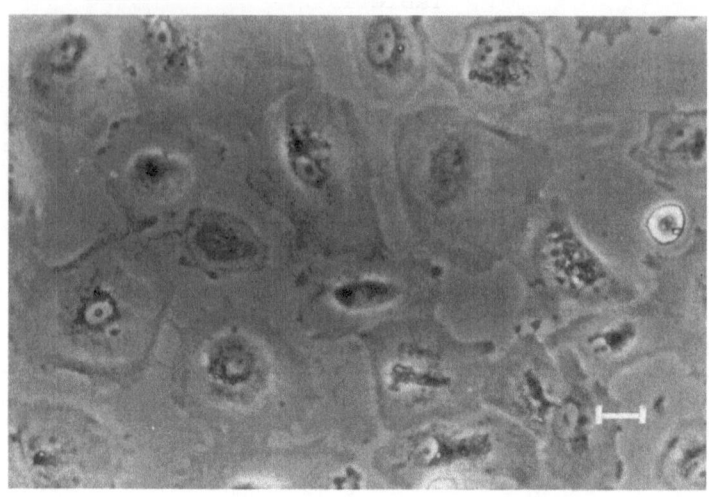

Fig. 2. A sister colony to that of Fig. 1 photographed 65 days
after initial plating. Bar is 30µ.

Consideration of functional changes which occur with time in
culture has been directed mainly towards aggregates and colonial
aggregates formed from guinea pig pancreatic cells. Preliminary
work established that freshly dissociated cells would readily
aggregate, and that aggregates could be selected for by plating in
F12 supplemented with calf serum rather than fetal calf serum.
Table II summarizes data supporting this contention.

More recent experiments have been performed using suspensions
highly enriched in acinar cells by the fractionation procedure outlined in the Materials and Methods section above. The resulting
aggregates were plated for determination of characteristics in
culture or were utilized directly for functional tests. The appearance of a typical guinea pig pancreatic cell aggregate, with prominent zymogen droplets, is shown in Figure 8. After a few days cultivation a colonial aggregate similar to that depicted in Figure 9
developed. Such aggregates ultimately degenerated in much the same
manner and over a similar time span as indicated above for the human
cell groupings.

Tests for functional integrity of the system have been of two
types. Colonial cell groupings have been examined histochemically
to detect the presence of amylase. This enzyme was found to be
associated with colonial aggregates at 72 hr after dissociation but
was almost totally absent by 120 hr (Fig. 10).

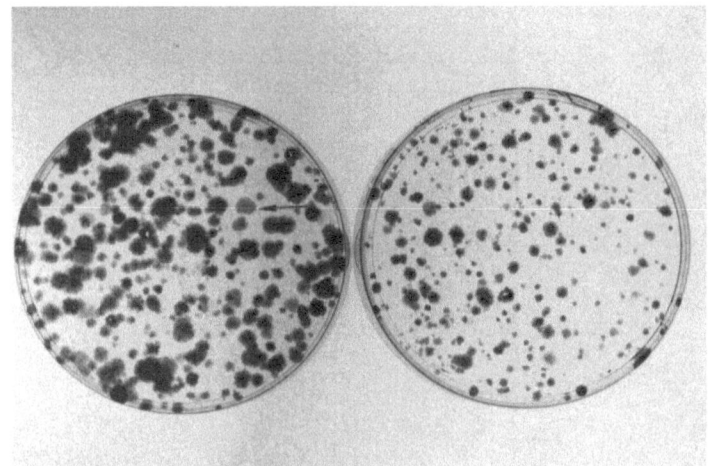

Fig. 3. A photograph of two clonal plates containing epithelial
(arrow) and fibroblastic colonies from the embryonic guinea pig
pancreas. The plate with larger colonies was set up and fed with
F12 containing 15% fetal calf serum. Medium supplemented with 5%
fetal calf serum was used in the other plate. The cultures were
both fixed on the 14th day and were stained with 1% aqueous
toluidene blue for photography.

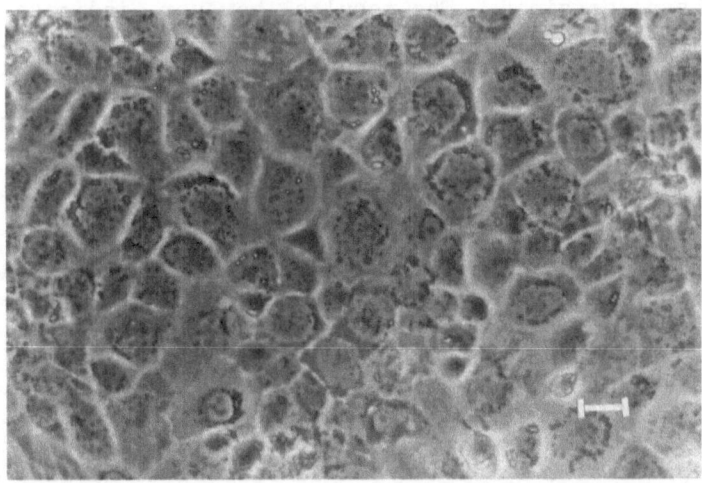

Fig. 4. An epithelial cell clone from guinea pig embryonic
pancreas photographed 12 days after initial plating. Bar is 30μ.

Table II

Selective Effect of Calf Serum For Isolation of Pancreatic
Aggregate Colonies

Serum Type & Lot No.	Inoculation Density (viable cells/plate)	Average No. Colonies/9 cm Plate Gross Morphology	
		Fibroblast	Aggregate
Fetal E0203K	10^3	132	12
Fetal A1226K	10^3	136	8
Fetal J76006	10^3	92	7
Calf A9243C	10^4	5	27
Calf 9442K	10^4	4	46
Calf C7293G	10^4	9	67

Plates were fed every 4 days with 10 ml of the appropriate medium and were fixed and stained on the 14th day.

One possible explanation for the above findings would be that the pancreatic acinar cells were irrepairably damaged during dissociation. This premise was examined by exposing aggregates, formed through incubation and gyration for 16 hr, to the secretogogue pancreozymin. Our rationale was that altered cells comprising aggregates would loose responsiveness to this agent if extensive surface damage had been sustained. The results, however, indicate that pancreatic aggregates were still responsive to the hormone (Figs. 11 and 12). Further experimentation involving isotope labeling and the measurement of subsequent release of radioactivity by stimulated versus control aggregates, substantiated this conclusion (Fig. 13).

DISCUSSION

The proliferative potential of cloned pancreatic epithelial cells and acinar cells of colonial aggregates is clearly very limited under the conditions imposed. One central question can be considered. Does this represent a restriction involving lack of appropriate factors in the medium or, alternatively, an intrinsic and irreversible change generated by the dissociation and cultivation of cells in a two dimensional system? Ample evidence exists indicating that pancreatic epithelia can be induced to divide, develop and function in organ culture (15-17). All attempts to duplicate conditions reported to be effective in this latter case with application of these to the pancreatic cell culture systems have met with failure. To site only a few trial studies, the altering of methionine concentrations (15); the addition of tissue extracts (16);

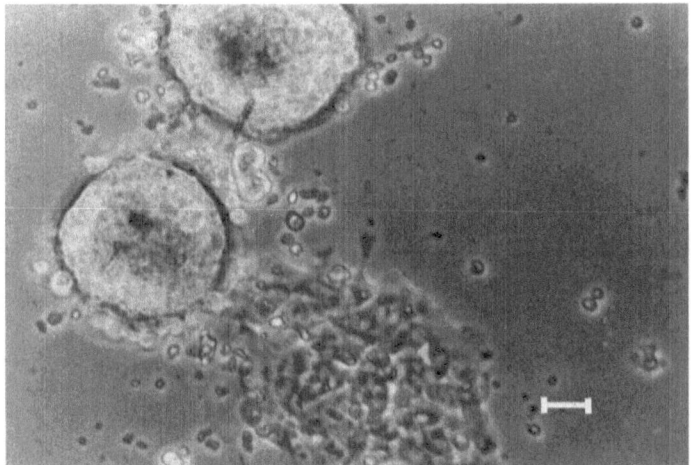

Fig. 5. Aggregates and an early colonial aggregate of epithelial cells from human embryonic pancreas photographed 24 hr after initial seeding. Bar is 30µ.

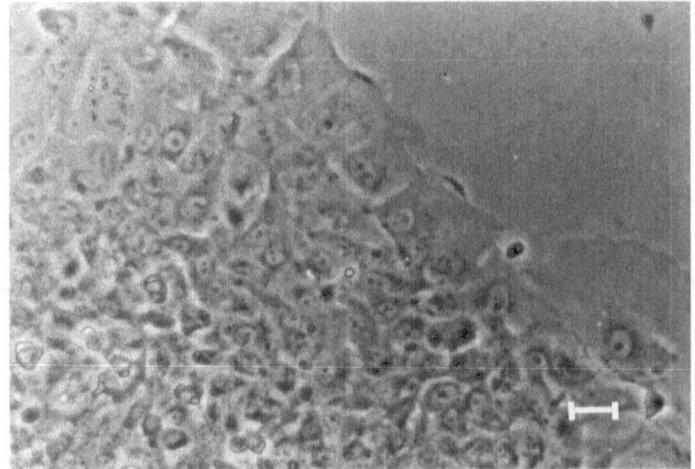

Fig. 6. The edge of a colonial aggregate of epithelial cells from human embryonic pancreas photographed 6 days after seeding. Bar is 30µ.

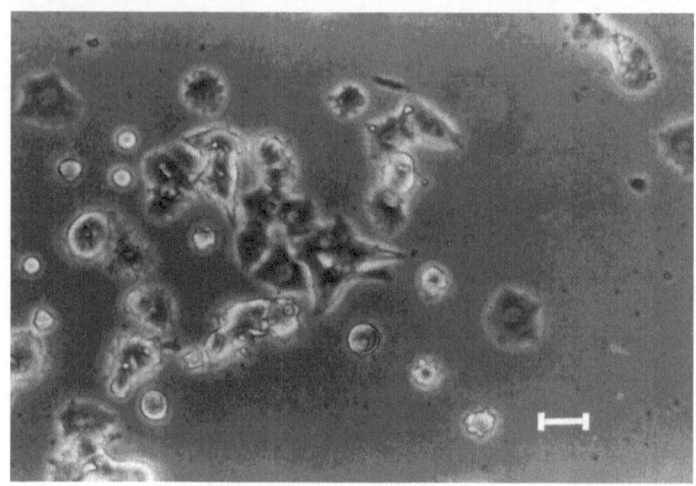

Fig. 7. A colonial aggregate similar to those shown in Figs. 5
and 6 but photographed 20 days after seeding. Bar is 30µ.

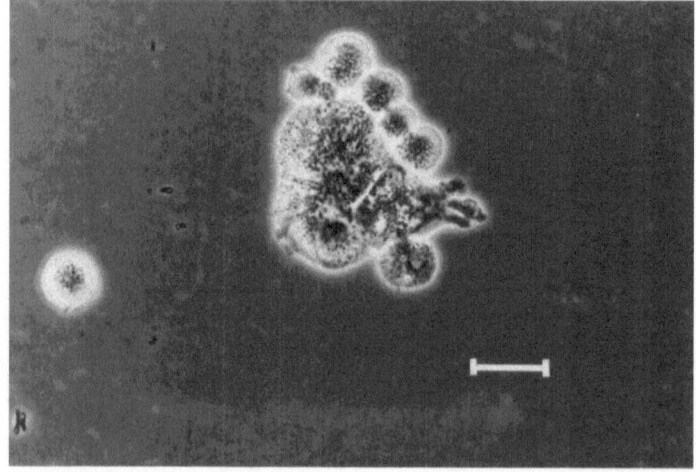

Fig. 8. An aggregate formed using suspensions enriched with acinar
cells from postnatal guinea pig pancreas. The photograph was taken
24 hr after dissociation at the time of primary plating. Bar is
20µ.

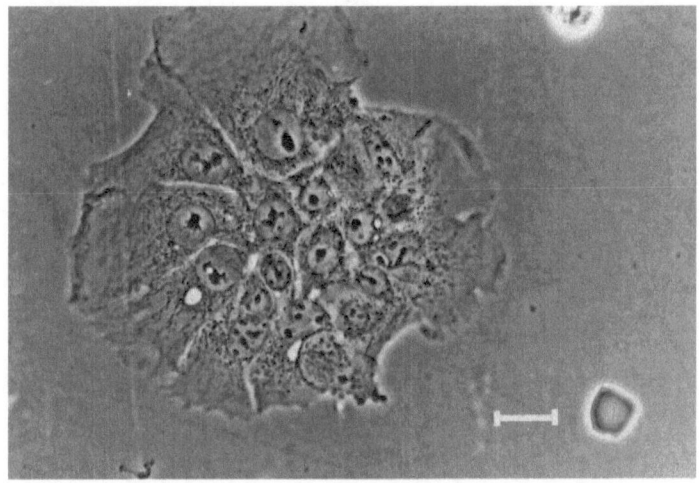

Fig. 9. A colonial aggregate which developed from an aggregate
similar to that shown in Fig. 8. The change in morphology was
followed by marking the plastic plate using a substage scribe.
The photograph was taken 3 days after primary plating. Bar is 20μ.

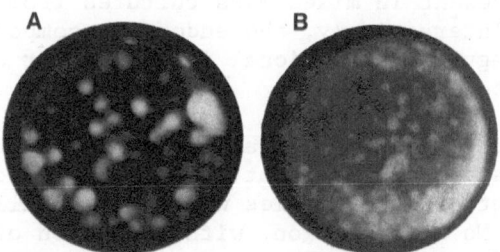

Fig. 10. Histochemical assay to detect the presence of amylase in
colonial aggregates of guinea pig acinar cells. The presence of
amylase is indicated by clear zones in the starch-agar over each
colony. Note the drop in amylase when 72 hr (A) versus 120 hr
cultures (b) are compared.

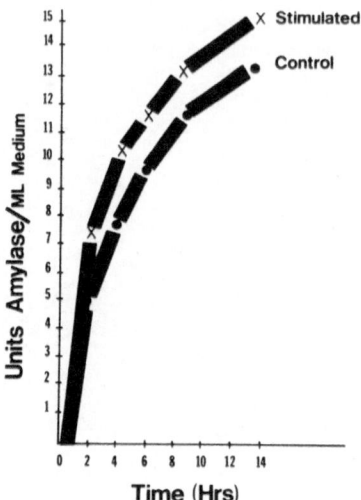

Fig. 11. Release of amylase into the suspending medium by aggregates of guinea pig acinar cells. Test aggregates were stimulated by addition of pancreozymin to give 0.2 Units/ml. Zero time is actually 16 hr after dissociation since this interval was allowed for aggregation and cell recovery.

and the cocultivation of pancreatic epithelial cells with mesenchyme (17) all have met with little success. Criteria chosen to indicate positive response included morphology, histochemistry and rates of overall protein synthesis determined quantitatively with radioactive leucine (unpublished data). Recent reports from other laboratories have indicated similar findings of "dedifferentiation" in the exocrine component present in mixed mass cultures from pancreatic tissue (18,19). Interestingly, the endocrine component in some cases retains a degree of functional integrity for considerable time intervals after isolation.

The lack of response in this acinar cell culture system may be due to inadequate conditions but is also reminiscent of experience in this and other laboratories with the so-called senescent cell in culture. No manipulation, with exception of induced or spontaneous transformation, has yet permitted recovery of cell strains from the degenerative phase. One might speculate that common mechanisms are inherent to both loss of proliferative ability and loss of function.

It is also interesting to comment on the marked difference in proliferative properties of epithelial cells in culture versus those

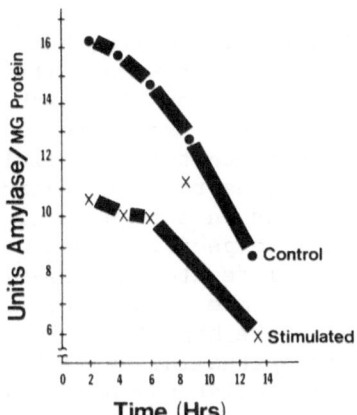

Fig. 12. Retention of amylase by aggregates of guinea pig acinar cells. The procedure used was identical to that of Fig. 11 but the amylase content of stimulated versus control aggregates was determined.

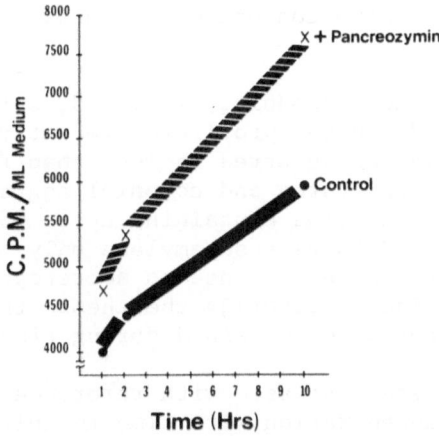

Fig. 13. Release of radioactivity by aggregates of guinea pig acinar cells. Aggregates collected 16 hr after dissociation were subjected to pulse-chase treatment as described in the Methods section. The release of labelled protein in the presence and absence of pancreozymin (0.2 U/ml) was then determined.

of cells having a fibroblastic morphology. The human amnion cell
is perhaps the only euploid epithelial cell from our species which
has been systematically studied over an extended time interval in
vitro. Like pancreatic epithelial cells, human amnion cells exhibit
a progressive decline in proliferative ability during the first few
weeks in culture. Marked alterations in metabolism also occur (20).
One explanation for the fact that there are no other euploid human
epithelial cell strains may be that this cellular component of human
tissue generally behaves as does the pancreatic acinar cell under
culture maintenance conditions in current use. This is an interesting and perplexing problem, especially when we are confronted by
the fact that such cells proliferate extensively in vivo. This
enigma is one which will continue to attract attention of experimen-
talists in the years ahead. We hope that the colonial aggregate
technique for isolation of epithelial cells in general and of pancreatic acinar cells in particular may play some role in its solution.

SUMMARY

Studies on the proliferative and functional properties of
epithelial cells from human and guinea pig pancreatic tissue were
presented and discussed. A novel technique for the isolation of
epithelial cell groupings as colonial aggregates in two dimensional
culture was utilized. More conventional clonal analyses were also
performed.

Irrespective of the methodology employed, epithelial cells from
both species exhibited reduced proliferative activity with time in
vitro. Total degeneration occurred in less than five months in
every case studied. Aggregates and colonial aggregates, formed
using fractionated suspensions containing up to 95% guinea pig
acinar cells, were found to release amylase only during the first
few days in vitro. Positive response to secretogogue stimulation
was interpreted as evidence favoring the thesis that permanent
cellular damage had not been sustained during tissue dissociation.

These findings were discussed with reference to the phenomenon
of human fibroblast degeneration with time in culture. Differences
between fibroblast and epithelial cell behavior in vitro and in vivo
were emphasized.

ACKNOWLEDGEMENTS

The author wishes to express his appreciation to Mr. Jim Miller
and Mrs. Paula Rookstool for their expert technical assistance with
various aspects of this research. This work was supported by con-
tracts E-72-3281 and NOI-CP-43231 from the US National Cancer Insti-

tute and in part by a grant from the American Cancer Society, Greene County Unit, Ohio.

REFERENCES

1. Cristofalo, V.J. 1972. Animal cell cultures as a model system for the study of aging. Adv. Gerontol. Res. 4: 45.

2. Adelman, R.C. 1971. Age-dependent effects in enzyme induction-- a biochemical expression of aging. Exp. Gerontol. 6: 75.

3. Curtis, H.J. and Tilley, J. 1971. The life-span of dividing mammalian cells in vivo. J. Gerontol. 26: 1.

4. Coon, H.G. 1966. Clonal stability and phenotypic expression of chick cartilage cells in vitro. Proc. Nat. Acad. Sci. USA 55: 66.

5. Kaighn, M.E. and Prince, A.M. 1971. Production of albumin and other serum proteins by clonal cultures of normal human liver. Proc. Nat. Acad. Sci. USA 68: 2396.

6. Houck, J.C., Sharma, V.V. and Hayflick, L. 1971. Functional failures of cultured human diploid fibroblasts after continued population doublings. Proc. Soc. Exp. Biol. Med. 137: 331.

7. Macek, M., Hurych, J. and Chvapil, M. 1967. The collagen protein formation in tissue cultures of human diploid strains. Cytologia 32: 426.

8. Cahn, R.D., Coon, H.G. and Cahn, M.B. 1966. Cell culture and cloning techniques. In: Methods in Developmental Biology. F.H. Wilt and N.K. Wessells, eds., Crowell, p. 493.

9. Amsterdam, A. and Jamieson, J.D. 1972. Structural and functional characterization of isolated pancreatic exocrine cells. Proc. Nat. Acad. Sci. USA 69: 3028.

10. Jamieson, A.D., Pruitt, K.M. and Caldwell, R.C. 1969. An improved amylase assay. J. Dental Res. 48: 483.

11. Shear, M. and Pearse, A.G.E. 1963. A starch substrate film method for the histochemical localization of amylase. Exp. Cell Res. 32: 174.

12. Lowry, O.H., Rosebrough, N.J., Farr, A.L. and Randall, R.J. 1951. Protein measurement with the Folin phenol reagent. J. Biol. Chem. 193: 265.

13. Buckley, B., Rookstool, P. and Hay, R.J.   Unpublished observa-
     tions.

14. Hay, R.J., Yoshikawa-Fukada, M., Yuyama, S. and Ebert, J.D.
     1968. Cell differentiation and viral susceptibility. Carnegie
     Inst. Year Book 67: 429.

15. Parsa, I., Marsh, W.H. and Fitzgerald, P.J.   1972.   Pancreas
     acinar cell differentiation. V. Significance of methyl groups
     in morphologic and enzymatic development. Exp. Cell Res. 73:
     49.

16. Ronzio, R.A. and Rutter, W.J.   1973.   Effects of a partially
     purified factor from chick embryos on macromolecular synthesis
     of embryonic pancreatic epithelia. Dev. Biol. 30: 307.

17. Wessels, N.K. and Cohen, J.H.   1967.   Early pancreas organo-
     genesis: morphogenesis, tissue interactions, and mass
     effects. Dev. Biol. 15: 237.

18. Orci, L., Like, A.A., Amherdt, M., Blondel, B., Kanazawa, Y.,
     Marliss, E.B., Lambert, A.E., Wollheim, C.B. and Renold, A.E.
     1973. Monolayer cell culture of neonatal rat pancreas: An
     ultrastructural and biochemical study of functioning endocrine
     cells. J. Ultrastruct. Res. 43: 270.

19. Chick, W.L., Lauris, V., Flewelling, J.H., Andrews, K.A. and
     Woodruff, J.M.   1973.   Effects of glucose on beta cells in
     pancreatic monolayer cultures. Endocrinol. 92: 212.

20. Chang, R.S.   1962.   Metabolic alterations with senescence of
     human cells: Some observations in vitro. Arch. Int. Med.
     110: 563.

QUESTIONS TO DR. HAY

Dr. Franks: Have you used adult tissue in your system?

Dr. Hay: We have not tried to use adult guinea pig tissue and have
been totally unsuccessful despite many attempts with human autopsy
samples. The problem in the latter case is presumably due to tissue
autolysis.

Dr. Kasten: What evidence is there that these pancreatic cells
remain diploid throughout their life in vitro?

Dr. Hay: We have done no karyological studies on the pancreatic
epithelial cells. Division occurs in cell clones and is limited
in time only to the first few weeks.

Dr. Ryan: Loss of functional and proliferative capacity while maintaining pancreatic cells _in vitro_ may result because of lack of appropriate endogenously supplied hormonal stimulation. Have you tested the effects of any mammalian hormones on pancreatic cells?

Dr. Hay: Yes, we have looked extensively for active factors using tritiated thymidine to detect stimulation of DNA synthesis and tritiated leucine to estimate protein synthesis. These studies are presently being continued but, as yet, no hormone, tissue extract or change of low molecular weight nutrient has had a definite stimulatory effect. These experiments have been performed using the exocrine pancreatic cell culture system (colonial aggregate technique) exclusively.

THE USE OF ARRESTED POPULATIONS OF HUMAN DIPLOID FIBROBLASTS FOR THE STUDY OF SENESCENCE IN VITRO

Robert T. Dell'Orco

Biomedical Division, The Samuel Roberts Nobel

Foundation, Inc. Ardmore, Oklahoma 73401

INTRODUCTION

In recent years attempts have been made to equate in vivo senescence with the finite lifespan demonstrated by human diploid cells in vitro (1). While preliminary evidence indicates that it may be possible to use cultured cells as an in vitro model system for cellular aging, several critical questions need to be resolved before any true parallel can be drawn between the in vitro phenomenon and the in vivo process.

One of these basic questions is whether cell death in culture is actually dependent upon the number of doublings that have occurred or is merely a result of the total time that the cells have been maintained in vitro (2,3). We have attempted to gain more insight into this problem by maintaining human diploid cells in as essentially nonmitotic state for extended periods of time during their in vitro lifespan. By returning these cells to an actively growing state, we were able to compare the number of population doublings and the total in vitro culture time of arrested cells to these parameters in cells which were allowed to proliferate under normal growth conditions for their entire lifespan (4,5).

MATERIALS AND METHODS

A number of human diploid cell strains, as shown in Table I, have been successfully maintained in the arrested state. These cells were passaged routinely at a 1:4 split ratio in McCoy's Medium 7a (6) supplemented with 10% fetal bovine serum and antibiotics. For the induction of the arrested state, the growth medium

41

TABLE I.   Cell Strains Successfully Maintained With 0.5% Serum

Cell Strain	Origin	Normal In Vitro Lifespan[a]
WI-38	Embryonic lung	50 ± 10
FeSin	Embryonic skin	-
CF-1	Newborn foreskin	40 ± 10
CF-3	Newborn foreskin	60 ± 5
HFMD	Newborn foreskin	60 ± 5
WLW	Adult (35 years) foreskin	15-20

[a]The characteristic number of population doublings occurring

in vitro before entrance into Phase III.

on confluent cultures was replaced with McCoy's Medium 7a containing 0.5% serum and antibiotics. This medium was replaced twice weekly during the experimental periods. Cells have been maintained in this manner for as long as 177 days and can be induced to reenter a proliferative state by subculturing the population with medium containing 10% serum (4,5).

Cells were prepared and analyzed by flow microfluorometry as described by Tobey et al. (7). Protein was determined according to the method of Lowry et al. (8).

RESULTS AND DISCUSSION

Figure 1 shows the DNA distribution patterns obtained by flow microfluorometric analysis from cells taken at confluency (0 day) and after 28 days of exposure to medium containing 0.5% serum. The major peak represents cells that have a DNA content consistent with the G_1 phase of the division cycle. The peak at twice the mode of the G_1 DNA content represents cells in G_2 + M. Cells in S phase with varying degrees of completion of DNA replication are distributed between the G_1 and the G_2 + M peaks. These results showed that exposure to low serum medium maintained the population distribution that was present when cells attained confluency in medium containing 10% serum. Approximately 90% of this population was made up of cells with a G_1 DNA content, i.e., through 28 days only

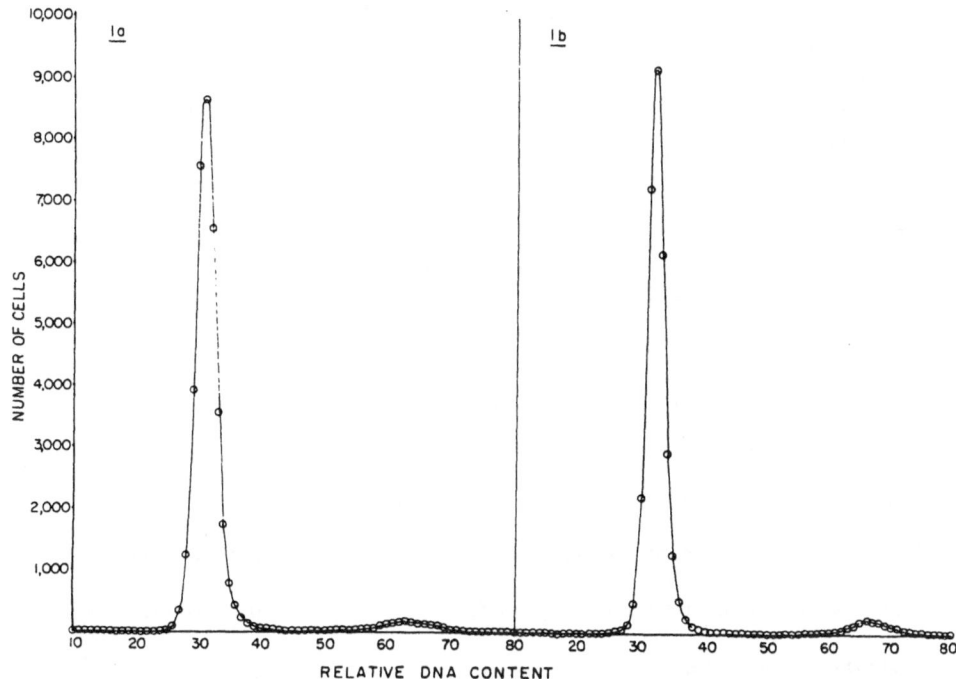

Fig. 1.   DNA distribution patterns from CF-1 cells examined by
flow microfluorometry.   1a is the distribution pattern obtained
at confluency (0 time); 1b is the distribution pattern obtained
after 28 days of exposure to medium containing 0.5% serum.

10% of the cells at any time were beyond the G_1/S boundary.   Parallel
experiments indicated that this was not a static population but
that there was a minimum amount of division occurring, less than
1% of the cells in the G_2 + M fraction were in mitosis, also the
time needed to traverse the cell cycle was prolonged (9).

Cell numbers and the protein content of CF-1 cells were followed
through a series of 10 experiments with a 21-day exposure to 0.5%
serum.   The results from these experiments are shown in Table II.
Both cell numbers and total proteins decrease during exposure to
0.5% serum.   These losses amounted to 10-20% of the cells and 40%
of the total proteins.   Thirty percent of the protein loss occurred
during the first 7 days of incubation with 0.5% serum.   Subsequent
protein losses paralleled the cell losses which occurred during the
remainder of the experimental period.   On a protein per cell basis,
a similar 10% decrease was observed during the first 7 days; however,
this value then remained constant during the subsequent 14 days.
While the absolute values for these parameters varied between cell

TABLE II.   Effect of Incubation With 0.5% Serum on Cell
Number and Protein Content of CF-1 Cells[a]

Day	Cell Counts	Total Protein	μg Protein 10^6 Cells
0	100.0	100.0	100.0
7	103.4 ± 3.1	73.1 ± 1.6	71.4 ± 2.5
14	91.0 ± 2.8	63.4 ± 1.3	70.4 ± 2.1
21	86.8 ± 3.0	59.2 ± 3.3	68.3 ± 3.3

[a]Taking 0 time as 100%, values are percent 0 time, mean ±
SEM. Data taken from 10 experiments using cells placed on low
serum at passage numbers ranging from 13 to 31. Taken from
Dell'Orco et al. (1973).

strains, the pattern of loss for both protein and cell numbers was
similar with all cell strains investigated (4,5).

The 10 to 20% loss of cells was attributed to culture manipu-
lations, such as feeding and sample preparation, but the possibility
does exist that the conditions of low serum cultivation caused cell
selection with a constant alteration in the population characteristics. In addition to this, the 30 to 40% loss of cellular protein
may reflect a nutritional deficiency brought about by reducing the
serum concentration to 0.5%. On the other hand, the protein loss
may be a normal cellular response to the inhibition of the synthetic
mechanisms necessary for cellular division.

Table III contains results from a series of experiments using
three cell strains, WI-38, CF-1, and WLW, designed to determine if
population doublings or calendar time was the primary determinant
of in vitro lifespan. Regardless of what passage number the cells
were placed on low serum medium and for incubation periods up to
177 days, arrested cells could be returned to a proliferative
state by subcultivation with growth medium containing 10% serum.
In all cases, cells which had been arrested subsequently achieved
passage levels equivalent to control cells which were continuously

TABLE III. Extention of In Vitro Lifespan of WI-38, CF-1, and WLW Cells After Incubation With Medium Containing 0.5% Serum

Cell Strain	Expt.	P^a	A-time[b] (days)	P at Phase III[c] Arrested	Control	Days in Culture[d] Arrested	Control
WI-38	1	23	21	42	41	96	75
		27	21	50		130	65
	2	25	39	50	41	116	49
		25	66	49		133	49
		27	102	51		175	45
CF-1	1	19	21	40	36	108	69
		23	21	38		90	54
		29	21	39		73	34
	2	13	21	34	32	116	77
		23	21	32		80	41
		27	21	37		66	27
		15	77	37		161	70
		13	177	36		287	77
WLW	1	8	21	16	17	64	37
		15	21	17		36	9
	2	8	21	18	19	80	50
		13	21	18		56	31
	3	11	21	20	18	79	42
		11	59	18		109	42
		11	99	16		133	42

[a]Passage number of cells when exposed to medium containing 0.5% serum.

[b]Days of incubation with medium containing 0.5% serum.

[c]Passage number at entrance into Phase III.

[d]Days in culture from indicated passage number (a) until entrance into Phase III. Taken from Dell'Orco et al. (1973 and 1974).

subcultured with medium containing 10% serum. Also, in all cases the experimental cells were maintained in culture for proportionately longer periods of time than the controls. These results indicate that in cells from three different donor ages, embryonic, newborn, and adult, the primary determinant of _in vitro_ lifespan was related to the cumulative number of population doublings (mitotic events) that have taken place and not the length of time that these cells have been maintained in culture.

Whereas no difference between cell strains was detected with respect to extension of calendar time after recovery from the arrested state, a notable difference was apparent in the ultimate passage level achieved by experimental cells as compared to growth controls. Table IV shows that a gradient of attained passage number according to donor age was present. Arrested cells from younger donors have the ability to reach a significantly higher number of population doublings than do those from older donors when compared to suitable controls.

While the reason(s) for this observed gradient is not clear, it has been shown that by placing human diploid cells in an arrested,

TABLE IV. Change in Attained Population Doublings After Exposure to Medium Containing 0.5% Serum

Cell Strain[a]	Donor Age	ΔP[b]	Significance[c]
WI-38 (5)	Embryonic	7.4	
			$P < 0.02$
CF-1 (8)	Newborn	3.1	
			$P < 0.01$
WLW (7)	Adult (35 years)	-0.4	

[a] Number in parenthesis is the number of observations.

[b] Mean change in achieved population doublings (experimental minus control).

[c] P values obtained from Student t-test taken from Dell'Orco et al. (1974).

essentially nonmitotic state the division potential of populations derived from younger donors is effected to a greater extent than it is in those derived from older donors. It should be noted, however, that in no case did the passage level attained by arrested popula- tions exceed that which has been reported as the maximum for these cell strains (Table I). Therefore, by arresting the populations we may be improving their ability to reach the limit of their in vitro division potential but not necessarily to exceed that limit.

SUMMARY AND CONCLUSIONS

It has been shown that human diploid cells from various donor ages can be arrested in an essentially nonmitotic state by reducing the serum concentration of the incubation medium from 10 to 0.5%. Cells incubated at this serum level maintained the population dis- tribution that was present when the cells reached confluency. The population, which has 90% of the cells in the G_1 phase of the divi- sion cycle, was not static and exhibited a low level of mitotic activity with prolonged interdivision times. These cells also exhibited a greatly reduced (30%) protein content which occurred within the first 7 days of cultivation with 0.5% serum. Cells arrested by incubation with low serum medium and subsequently recovered to a proliferative state underwent an equivalent or greater number of population doublings with a concomitant extension of in vitro calendar time when compared to growth controls. This indicated that the number of mitotic events and not the length of time in culture was the primary determinant of in vitro lifespan. The ultimate passage level achieved by experimental cells as com- pared to controls was different in the three cell strains studied. A gradient of attained passage number according to donor age was established with cells from younger donors reaching a significantly greater number than those from older donors.

The use of arrested cell populations for the study of cellular senescence offers a unique opportunity to have an in vitro system which may more closely approximate those in vivo tissues which nor- mally do not exhibit a rapid rate of proliferation. By the use of this system numerous biochemical parameters can be investigated at various cell ages without the interference of proliferative processes.

REFERENCES

1. Cristofalo, V.J. 1972. Animal cell cultures as a model system for the study of aging. Adv. Gerontol. 4: 45.

2. Hayflick, L. 1965. The limited in vitro lifetime of human diploid cell strains. Exp. Cell Res. 37: 614.

3. Hay, R.J. 1970. Cell strain senescence in vitro: cell culture
    anomaly or an expression of a fundamental inability of normal
    cells to survive and proliferate. In: "Aging in Cell and
    Tissue Culture", (E. Holeckova and V.J. Cristofalo, eds.), p. 7,
    Plenum Press, New York.

4. Dell'Orco, R.T., Mertens, J.G. and Kruse, P.F. Jr. 1973.
    Doubling potential, calendar time, and senescence of human
    diploid cells in culture. Exp. Cell Res. 77: 356.

5. Dell'Orco, R.T., Mertens, J.G. and Kruse, P.F. Jr. 1974.
    Doubling potential, calendar time, and donor age of human
    diploid cells in culture. Exp. Cell Res. 84: 363.

6. Kruse, P.F. Jr., Whittle, W.L. and Miedema, E. 1969. Mitotic
    and nonmitotic multiple-layered perfusion cultures. J. Cell
    Biol. 42: 113.

7. Tobey, R.A., Crissman, H.A. and Kraemer, P.M. 1972. A method
    for comparing effects of different synchronizing protocols on
    mammalian cell cycle traverse. J. Cell Biol. 54: 638.

8. Lowry, O.H., Rosebrough, N.J., Farr, A.L. and Randall, R.J.
    1951. Protein measurement with the Folin phenol reagent.
    J. Biol. Chem. 193: 265.

9. Dell'Orco, R.T., Crissman, H.A., Steinkamp, J.A. and Kraemer,
    P.M. Submitted for publication, 1974. Population analysis
    of arrested human diploid fibroblasts by flow microfluorometry.

QUESTIONS TO DR. DELL'ORCO

Dr. Ryan: Have you made any biochemical comparisons of cells
treated with low serum concentrations to confluent nondividing
cell populations maintained with presumptively exhausted medium?

Dr. Dell'Orco: We have not done any comparisons of this nature.

Dr. Martin: Are there tetraploid cells arrested in G2 in your
cell populations?

Dr. Dell'Orco: It is impossible to distinguish between tetraploid
G_1 cells and G_2 cells from our analysis of these populations by
flow microfluorometry. However, it is possible to isolate this
fraction by a "Cell Sorter" technique (J.A. Steinhamp et al., Rev.
Sci. Instr. 44: 1301, 1973) and to determine the number of cells
that are in mitosis. When this was done with the arrested populations, it was determined that less than 1% of the cells that

made up this $G_2 + M$ peak were mitotic.

Dr. Kasten: I believe you could answer Dr. Martin's question as to whether the peak is due to tetraploid G_1 cells or diploid G_2 cells by treating the cultures with colcemid. If there is no effect on the peak, then it is due to G_1 cells. If there is a decrease in the peak, then it is due to a reduced G_1 population.

Dr. Macieira-Coelho: If you label the arrested populations continuously with ^{3}H-thymidine for several days, what is the percentage of cells that becomes labelled?

Dr. Dell'Orco: We did this in association with the flow microfluorometry experiments. The cells were exposed to ^{3}H-thymidine for 72 hr before harvest and over this period of time between 2.6 and 6.0% of the cells were labelled, as determined by autoradiography.

Dr. Hay: Do fibroblasts maintained on low serum have the same metabolic activity as cells maintained on standard medium?

Dr. Dell'Orco: We have looked at collagen synthesis and glucose utilization in arrested populations. The cells continue to synthesize collagen at what appears to be amounts equivalent to confluent cultures in medium containing 10% serum. The utilization of glucose falls to approximately 50% of that which is observed in confluent growth cultures.

RELATIONSHIP BETWEEN CELL KINETIC CHANGES AND METABOLIC EVENTS DURING CELL SENESCENCE IN VITRO

A. Macieira-Coelho, E. Loria, and L. Berumen

Institut de Cancérologie et d'Immunogénétique de
l'I.N.S.E.R.M. (U50), 94800-Villejuif, France

It was previously shown that human fibroblasts originated from adult human donors go through less generations in vitro than fibroblasts originating from human embryos (1). These findings led to the suggestion that the division potential of cells declines with aging. This hypothesis was further supported by the findings that cells obtained from human adults, early during their life in vitro are kinetically similar to embryonic cells during the last stages of their in vitro lifespan (2); that the potential number of doublings in vitro is inversely proportional to the age of the donor (3); that the cells originating from individuals with premature aging have a reduced division potential (5); and that one of the manifestations of aging in cell culture, the prolongation of the G_2 period (6), is also found in tissue of old animals (7).

We have previously described the kinetics of the growth decline of human fibroblasts in vitro (2,6,8,9).

An attempt was now made to correlate the changes found in cell kinetics with functional failures observed at the molecular level. At the moment our data suggest that the failure of division during cell senescence could be due to an impairment of ribosomal RNA synthesis leading to a decline in protein synthesis.

KINETIC MODEL OF CELL SENESCENCE IN HUMAN FIBROBLASTS

The cell kinetic studies previously performed on fibroblasts throughout their lifespan in vitro have shown a decrease in the number of cells entering the division cycle between subcultivation

51

and resting stage, a lower saturation density and a prolonged
generation time (2,6,8). The fraction of cells that does not enter
the S period between subcultivation and resting phase increases
logarithmically at each passage (10). Although there is a secondary
change in the S period (6) it seems that the primary changes leading
to the growth decline are located in the periods preceding DNA synthesis and mitosis (G_1 and G_2) rather than in these periods themselves (2). It was further shown that in human (9) and chick fibroblasts (11) all cells are capable of division up to the end of their
lifespan. In chick cultures most cells enter division at each
passage (11) and in human cells, those that do not finish the division
cycle during one passage can do so during the next one (9). Hence
cell senescence in vitro is not accompanied by an accumulation of a
non-dividing population. Is is also possible that non-dividing cells
are lost during subcultivation. These data led to the kinetic
model for human fibroblasts, illustrated schematically in Figure 1.

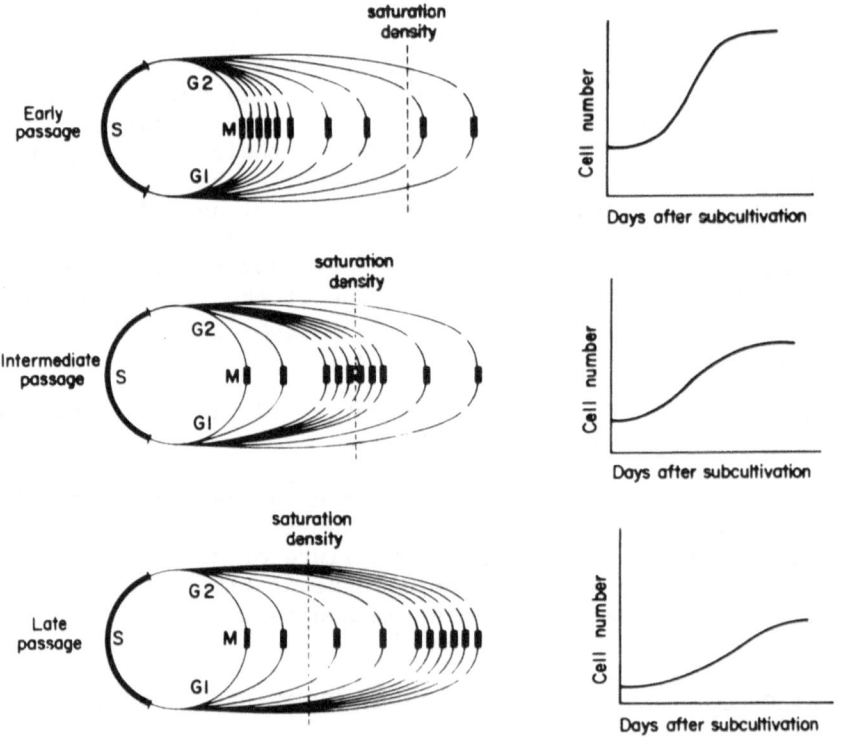

Fig. 1. Cell kinetics of human embryonic fibroblasts at 3 different
passage levels. Each circle or ellipse represents the generation
time of a cell. The vertical hatched line corresponds to the

saturation density. On the right hand side are represented the
growth curves that are obtained at the corresponding passage levels.

Three different periods of the lifespan are represented. On
the left side of the figure each circle and ellipse represent the
generation time of a cell. One can see that in early passages most
cells in the population have sort division cycles. Some cells already exist with long generation times (ellipses) which are stretched
at the expense mainly of G_1 and G_2. However, since the saturation
density is high (vertical hatched line) most cells are able to
complete their cycle before this density is reached. At this stage
the growth curve of the population (right hand side of the figure)
has a short lag period, a phase of logarithmic growth with a steep
slope and a high stauration density reached within 3-4 days.

Towards the middle of the population lifespan (intermediate
passage), there are still cells with short generation times, but
most cells have division cycles between the two extremes. Since
the saturation density is lower, an increased number of cells will
not have time to complete their cycles. Thus during a continuous
labelling with tritiated thymidine (^{3}H-dT) an increased number of
cells will appear as unlabelled. At this stage the slope of the
logarithmic portion of the growth curve will be less steep and the
cells will reach a lower density although still within 3-4 days.

Towards the end of the lifespan (late passage) most cells have
very long generation times, and since the saturation density is very
low, an increased number of cells do not have time to complete the
cycle between subcultivation and confluency. Thus an increased number
of cells will appear as unlabelled after growing continuously in
the presence of ^{3}H-dT. The growth curve will be characterized by
a long lag phase, a period of short logarithmic growth with a
shallow slope, and a low saturation density reached only after 7-10
days.

The slowdown of division as cells reach confluency is a slow
process that aggravates the delay of cells in G_1 and G_2 (12) and
thus should accentuate the indetics illustrated in Figure 1. In
the Figure the prolongation of G_1 and G_2 is identical. In reality
however, the G_1 component should be more important early during the
lifespan. Later on, as cells become delayed in G_2 during aging, the
G_2 component may become more important.

CHRONOLOGY OF EVENTS DURING SENESCENCE OF FIBROBLASTS IN
VITRO

At the present stage it seems to us that the important question

is no longer to know what happens when cells have aged, but rather
to investigate what are the events that lead to the growth decline,
i.e., to distinguish between primary and secondary events.

This was attempted with chick embryo fibroblasts which have
a shorter lifespan and hence facilitate the performance of the
experiments. Different parameters were followed throughout the
lifespan of chicken embryonic fibroblastic cultures. The first
change was observed in protein synthesis (13) measured by pulse-
labelling the cells at different times after subcultivation with
radioactive aminoacids (Fig. 2). A decline in total protein synthesis, already seen after the 5th passage, continued throughout
the whole lifespan. The pattern at each passage, however, never
changed, i.e., there was a peak at the 12th hr and then a decline
until resting phase.

The next change observed was an increase in cell volume (11)
after the 12th passage (Fig. 3).

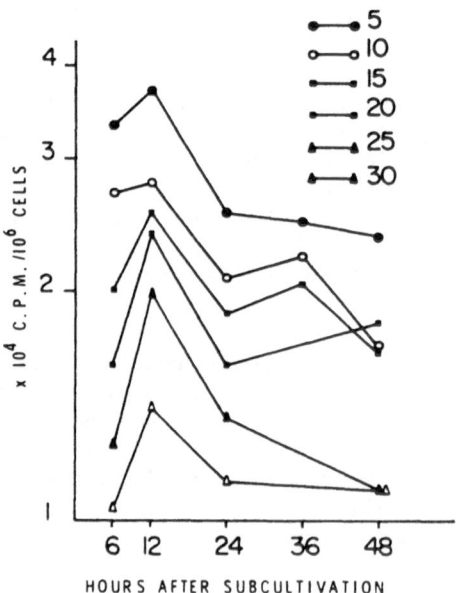

Fig. 2   Incorporation of ^{14}C-labeled aminoacids into the acid-
insoluble fraction, after pulse-labeling chicken embryonic fibroblasts at different times after subcultivation at the indicated
passage levels.

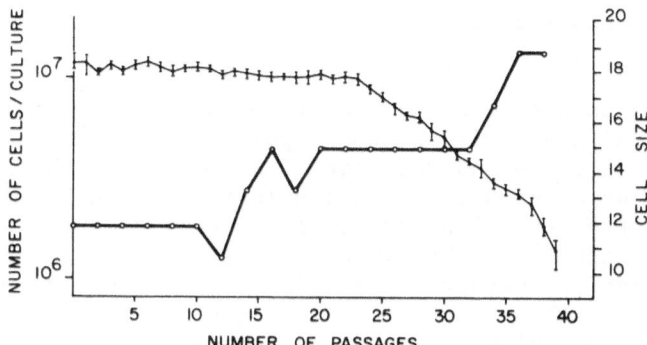

Fig. 3.   Saturation densities before each subcultivation (●—●) and modes of the cell size distributions at each second passage (○—○) during the lifespan of chicken embryonic fibroblasts. The vertical bars indicate the extreme values found within samples.

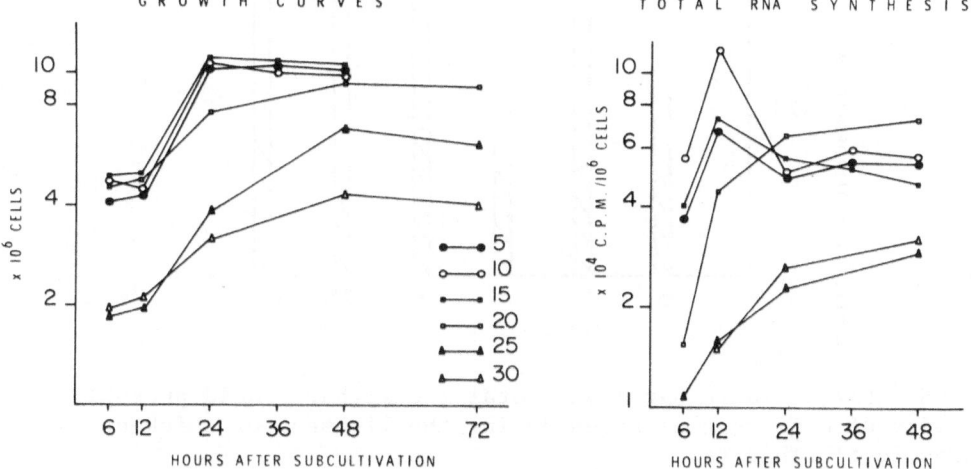

Fig. 4.   Growth curves and total RNA synthesis at different times

after subcultivation of chicken embryonic fibroblasts at the indicated passage levels.   Total RNA synthesis was obtained by pulse
labeling the cultures with ^{3}H-UR and measuring the radioactivity in
the acid-insoluble fraction (13).

The third change occurred simultaneously after the 15th passage
in the total RNA synthesis and the population doubling time (Fig. 4):
both the resting phase and the peak of total RNA synthesis were
reached after 48 hr of subcultivation instead of 24 hr (13).   This
was followed after the 20th passage by a decline in the saturation
density and in the total RNA synthesized (Fig. 4).   Finally, after
the 30th passage, a further increase in cell volume was observed
(Fig. 3), along with an augmentation of the total acid phosphatase
activity (Fig. 5), which was the last parameter to change out of
all those we measured (13).

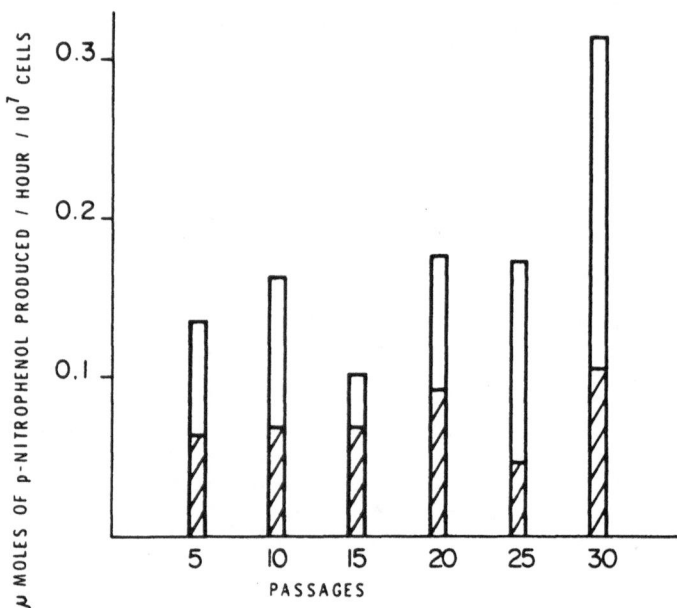

Fig. 5.   Latent (dark bars) and total (total bars) acid phosphatase
activity at different passages during the lifespan of chicken
embryonic fibroblasts.

POSSIBLE UNDERLYING MECHANISMS AT THE MOLECULAR LEVEL

One of the constant features of aging at the cellular level
is an increase in cell size (2,14,15). Although this change does
not appear to be a primary event as reported above, it does seem
seem to accentuate the growth decline (16). Since protein synthesis
declines throughout the population lifespan, the increased size
could be due to changes in cell permeability leading either to a
decreased excretion of proteins or to an increased uptake. For
this reason cell permeability was evaluated at different passage
levels by measuring the uptake of I^{125}-albumin with methods previously described (17). Results expressed in Table I show that
late passage cultures take up more albumin than early passage
cultures. Poly-ornithine, which is known to stimulate the transport of albumin (17), has a more pronounced effect on late passage

TABLE I. Uptake of I^{125}-albumin, alone or in the presence of
10 g/ml poly-ornithine (M.W. 170,000), by adult lung fibroblasts
during resting phase at different passages, expressed as the increase
in cell bound radioactivity (c.p.m./mg protein) during the first
hr after adding the labeled protein

Passage level	Without poly-ornithine	With poly-ornithine
9	5,336	22,697
22	11,887	61,676
28	35,290	1,331,878
31	35,352	1,819,680

than on early passage cultures. The cell line used died at the
33rd passage. Thus from the results, human fibroblasts show an
increased permeability to proteins during their growth decline
which could be associated with the increase in cell size. It
is also possible that it is the increase in cell size which
leads to changes in the transport of albumin.

Another constant feature of aging in vitro is the decline of
the density at which cells reach resting phase (saturation density).
It is now well known that polyribosome assembly declines in cells
reaching the saturation density (18) and accumulating evidence
has suggested that ribosomal RNA synthesis is the site of an
important control mechanism of cell division in vitro (19-23). The
lifespan and the saturation density of human fibroblasts can be
increased by addition of cortisone or hydrocortisone to the nutrient
medium (24, 25). Since one of the actions of cortisone seems to
be the induction of ribosomal RNA synthesis (26-28), we decided to
measure ribosome formation in phase II and III cells and in cultures
carried in medium supplemented with 5 g/ml hydrocortisone (25).

Figure 6 illustrates the incorporation of ^{3}H-uridine (^{3}H-UR) in
polyribosomes and ribosomes of early and late passage human embryonic
fibroblasts 48 hr after subcultivation. A peak corresponding to
polyribosomes (10th fraction) is visible only in early passage cells
and the incorporation of the radioactive precursor is significantly
increased in the ribosomes (20-25th fractions) of young cultures.

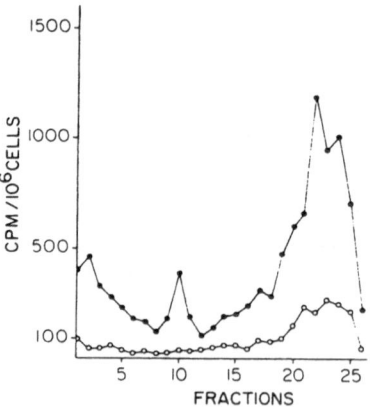

Fig. 6. Radioactivity found in the different fractions collected
from a 7-52% sucrose gradient after centrifugation of cytoplasma
extracts from young (●——●) and old (0——0) fibroblasts labeled with
^{3}H-UR. The 10th fraction corresponds to polyribosomes and the
20-25th fractions to ribosomes (29).

Uridine incorporation in the ribosomes and polyribosomes was also measured in the same type of cultures carried with and without hydrocortisone. Two series of cultures were carried separately from the 30th passage on, with and without the hormone. Six passages later the cells in each group were pooled, subcultivated and each day thereafter labelled during 3 hr with ^{3}H-UR. Cytoplasm extracts were prepared and centrifuged on a 7-25% sucrose gradient (29). The areas corresponding to the polyribosome and ribosome peaks found each day after subcultivation were plotted in Figure 7 where it can be seen that the radioactivity incorporated was always significantly higher in the cells carried with hydrocortisone supplemented medium. No difference was found in the acid soluble radioactivity. During the same experiment identical cultures were pulse-labelled each day after subcultivation with ^{14}C aminoacids and the acid insoluble and soluble radioactivity was measured (29). Results illustrated in Figure 8 show that there was a significant increase in the acid insoluble radioactivity of hydrocortisone-treated cells on the first day after subcultivation. Then, as protein synthesis declined, no difference could be detected between the two groups of cultures. The acid soluble radioactivity was always identical in both groups, suggesting that the increased protein synthesis induced by hydrocortisone was not due to an increased transport of the radioactive precursors into the cell.

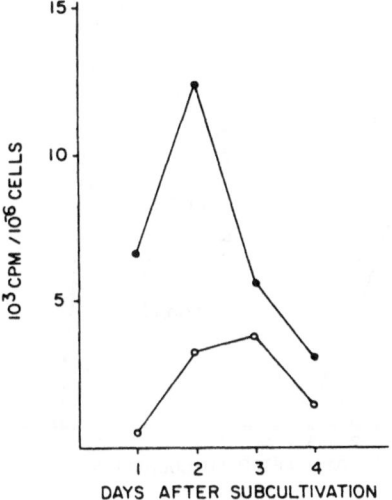

Fig. 7. Incorporation of ^{3}H-UR in ribosomes and polyribosomes of human fibroblasts carried without (0——0) and with (●——●) hydrocortisone. Each point corresponds to the area found under the peaks obtained after centrifugation of cytoplasm extracts from pulse-labeled cells on sucrose gradients (29), each day after subcultivation.

A. MACIEIRA-COELHO ET AL.

It seems reasonable to infer from the above that the decline
of the saturation density during cell senescence is due to an
impaired ribosome synthesis and that hydrocortisone delays the
decline of the saturation density by its sustaining action on
ribosome synthesis. The long-term effect of the hormone on the
lifespan of human fibroblasts could also be explained by its regulatory effect on ribosomal synthesis. In fact, it has been previously reported that RNA has a small turnover in cells with un-

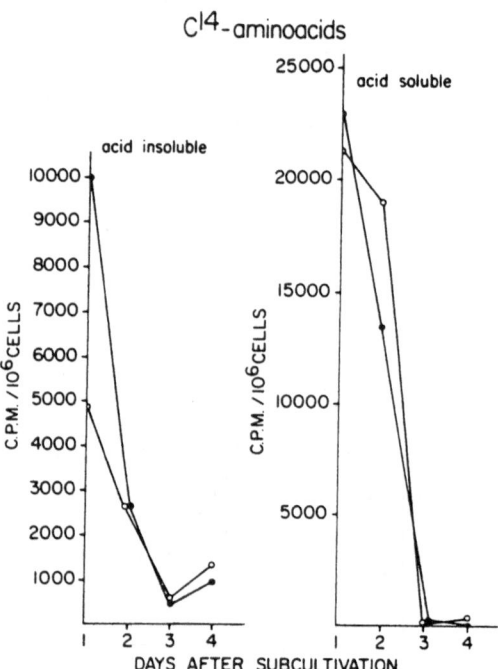

Fig. 8. Radioactivity found in the acid-insoluble and soluble
fractions of human fibroblasts maintained without (O——O) and with
(●——●) hydrocortisone, pulse-labeled each day after subcultivation
with ^{14}C-labeled aminoacids.

limited growth potential while it is very unstable in cells with a
finite lifespan (30). An association between the long-term effect
of hydrocortisone and its effect on ribosomal synthesis fits the
cell kinetic data which show that the mechanisms involved in the
loss of the division potential are located in the preparation of
DNA synthesis and mitosis. This association would also fit the
theory that aging is due to an accumulation of errors in protein
synthesis (31). Hydrocortisone could delay the deleterious effect
of faulty molecules through an overall stimulation of protein
synthesis.

SUMMARY

Cell kinetic studies performed throughout the lifespan of
fibroblasts with a limited lifespan in vitro have led to the
conclusion that although division slows down, almost all cells are
able to divide until the last subcultivation. The prolongation of
the division cycle is primarily due to the impariment of mechanisms
preceding DNA synthesis and mitosis. An attempt was made to
distinguish between primary and secondary changes and to correlate
the findings concerning cell kinetics with alterations observed at
the molecular level. A decline in protein synthesis was the
first modification detected. The two parameters that are always
present during cell senescence in vitro, i.e., and increase in
cell volume and a decrease in saturation density could be due
respectively to a change in cell permeability and a decline in
ribosome synthesis. The latter could also be the step responsible
for the limited potential of division.

REFERENCES

1. Hayflick, L. 1965. The limited in vitro lifetime of human
    diploid cell strains. Exp. Cell Res. 37: 614.

2. Macieira-Coelho, A., and Ponten, J. 1969. Analogy in growth
    between late passage human embryonic and early passage human
    adult fibroblasts. J. Cell Biol. 43: 374.

3. Martin, G., Sprague, C.A., and Epstein, J. 1970. Replicative
    life span of cultivated human cells. Effects of donor's
    age, tissue and genotype. Lab. Invest. 23: 86.

4. Goldstein, S. 1969. Lifespan of cultured cells in progeria.
    Lancet 1: 242.

5. Goldstein, S., Littlefield, J.W., and Soeldner, J.S. 1969.
    Diabetes mellitus and aging. Diminished plating efficiency
    of cultured human fibroblasts. Proc. Natl. Acad. Sci. USA
    64: 155.

6. Macieria-Coelho, A., Ponten, J., and Philipson, L. 1966. The
    division cycle and RNA synthesis in diploid human cells at
    different passage levels in vitro. Exp. Cell Res. 42: 673.

7. Pederson, T., and Gelfand, S. 1966. Mitosis in regeneration
    and aging in relation to cell division cycle. J. Cell Biol.
    81: 84A.

8. Macieria-Coelho, A., Ponten, J., and Philipson, L. 1966b.
    Inhibition of the division cycle in confluent cultures of
    human fibroblasts in vitro. Exp. Cell Res. 43: 20.

9. Macieira-Coelho, A. 1974. Are non-dividing cells present
    in ageing cell cultures? Nature 248: 421.

10. Cristofalo, V.J., and Sharf, B.B. 1971. Cellular senescence
    and DNA synthesis: Thymidine incorporation as a measure
    of population age in human diploid cells. Exp. Cell Res.
    76: 419.

11. Lima, L., and Macieira-Coelho, A. 1972. Parameters of aging
    in chicken embryo fibroblasts cultivated in vitro. Exp.
    Cell Res. 70: 279.

12. Macieira-Coelho, A., and Berumen, L. 1973. The cell cycle
    during growth inhibition of human embryonic fibroblasts
    in vitro. Proc. Soc. Exp. Biol. Med. 144: 43.

13. Macieira-Coelho, A., and Lima, L. 1973. Aging in vitro. In-
    corporation of RNA and protein precursors and acid phos-
    phatase activity during the lifespan of chick embryo fibro-
    blasts. Mech. Age. Dev. 2: 13.

14. Simons, J.W.I.M. 1967. The use of frequency distributions
    of cell diameters to characterize cell populations in tissue
    culture. Exp. Cell Res. 45: 336.

15. Stenkvist, B. 1966. Effects of Rous sarcoma virus in vitro
    on cells of human and bovine origin. Uppsalal dissertations
    in medicine. Acta. Univ. Uppsal. 30: 16.

16. Macieira-Coelho, A. 1970. The decreased growth potential in

vitro of human fibroblasts of adult origin. IN: Aging in
Cell and Tissue Culture (E. Holeckova and V.J. Cristofalo,
editors), p. 121, Plenum Press, New York.

17. Ryser, H.J.P. 1963. The measurement of I^{131}-serum albumin
uptake by tumor cells in tissue culture. Lab. Invest. 12:
1009.

18. Levine, E.M., Becker, Y., Boone, C.W., and Eagle, H. 1965.
Contact inhibition, macromolecular synthesis and polyribo-
somes in cultured human diploid fibroblasts. Proc. Natl.
Acad. Sci. USA 53: 350.

19. Ward, G.A., and Plagemann, P.G.W. 1969. Fluctuations of DNA-
dependent RNA polymerase and synthesis of macromolecules
during the growth cycle of Novikoff rat hepatoma cells in
suspension culture. J. Cell Physiol. 73: 213.

20. Stanners, C.P., and Becker, H. 1971. Control of macromolecular synthesis in proliferation and resting Syrian hamster
cells in monolayer culture. I. Ribosome function. J. Cell
Physiol. 77: 31.

21. Becker, H., Stanners, C.P., and Kudlow, J.E. 1971. Control
of macromolecular synthesis in proliferating and resting
Syrian hamster cells in monolayer culture. II. Ribosome
complement in resting and early G_1 cells. J. Cell Physiol. 77:
31.

22. Emerson, C.P. 1971. Regulation of the synthesis and the
stability of ribosomal RNA during contact inhibition of growth.
Nature New Biol. 232: 101.

23. Hodgson, J.R., and Fisher, H.W. 1971. Formation of polyribosomes during recovery from contact inhibition of replication.
J. Cell Biol. 49: 845.

24. Macieira-Coelho, A. 1966. Action of cortisone on human
fibroblasts in vitro. Experientia 22: 390.

25. Cristofalo, V.J. 1970. Metabolic aspects of aging in diploid
human cells. IN: Aging in Cell and Tissue Culture (E.
Holeckova and V.J. Cristofalo, editors), p. 83, Plenum Press,
New York.

26. Greengard, O., and Feigelson, P. 1961. A difference between
the modes of action of substrate and hormonal inducers of
rat liver tryptophan pyrrolase. Nature 190: 446.

27.   Feigelson, M., Gross, P.R., and Feigelson, P. 1962. Early
      effects of cortisone on nucleic acid and protein metabolism
      of rat liver. Biochem. Biophys. Acta 55: 495.

28.   Jerwell, Kr. F., and Osnes, J.B. 1963. Further studies on
      the effects of cortisone on the ribonucleic acid metabolism
      in rat liver. Life Sci. 12: 975.

29.   Brouty-Boye, D., Macieira-Coelho, A., Fixman, M.Y., and Gresser,
      I. 1973. Interferon and cell division. VII. Effect of
      interferon on macromolecular synthesis in L1210 cells in
      vitro. Int. J. Cancer 12: 250.

30.   Michl, J., and Svobodova, J. 1967. RNA turnover and the
      growth potential of human cells in culture. Exp. Cell Res.
      47: 616.

31.   Orgel, L.E. 1963. The maintenance of the accuracy of protein
      synthesis and its relevance to ageing. Proc. Natl. Acad.
      Sci. USA 49: 517.

QUESTIONS TO DR. MACEIRA-COELHO

Dr. Cristofalo: I would first like to express my appreciation to
Dr. Macieira-Coelho for trying to clarify some of the confusion that
has arisen concerning the use of the term "non-dividing cells".
Secondly, I would like to ask how one reconciles the apparent random
arrest of slowly dividing cells with the fact that in confluent
cultures refed with serum, the burst of division takes place within
about 24 hr.

Dr. Macieira-Coelho: I showed previously that when cells approach
resting phase, the cell cycle does not stop immediately. The arrest
of cells at confluency is rather a progressive slow down than a sudden
and complete stop.

Dr. Sova: Have you studied or do you know about the qualitative
changes in RNA synthesis during ageing?

Dr. Macieira-Coelho: So far we measured only ribosomal RNA synthesis
in young and old cells. We know, however, that hydrocortisone
stimulates more ribosomal RNA than transfer RNA. We are now check-
ing what happens with messenger RNA.

Dr. Dell'Orco: Have you looked at the uptake of specific metabolites
such as amino acids or their analogues, i.e., AIB?

Dr. Macieira-Coelho: No, we measured only albumin uptake.

Dr. Hay: Does the uptake of radioactive albumin by cells in your system reflect uptake rate or total uptake at some equilibrium state?

Dr. Maceira-Coelho: It reflects total uptake.

Dr. Macek: Do you think that the enlarged cell size in your senescent cultures could be at least partially caused by the increase of endomitosis and heteroploidy, as we observed earlier in our human diploid cell lines?

Dr. Maceira-Coelho: No, I do not think it can be explained that way because the increase in size starts well before any chromosome changes can be seen.

DO HYPERPLASTOID CELL LINES "DIFFERENTIATE THEMSELVES TO DEATH"?

G.M. Martin, C.A. Sprague, T.H. Norwood,
W.R. Pendergrass, P. Bornstein, H. Hoehn and W.P. Arend

Depts. of Pathology, Biochemistry and Medicine
University of Washington, Seattle, Washington

Hayflick and Moorhead (1) clearly differentiated between two classes of mammalian cell lines: 1) Those typified by HeLa are apparently immortal and may serve as models for the study of neoplastic cell proliferation; we refer to them as "neoplastoid." 2) Those typified by WI-38 and by human skin fibroblast cultures eventually cease replicating and may be useful as models for the study of hyperplastic cellular proliferation or wound healing; consequently, we refer to them as "hyperplastoid." Martin and Sprague (2) have recently tabulated some 21 parameters which have been claimed to differentiate between these two classes of cell lines. In mass cultures, the replicative life-span is currently among the most unambiguous differential parameters. Individual clones of either type of culture may cease proliferating, however, and it is this phenomenon which we refer to as "clonal senescence." In the case of human diploid somatic cells, it is probable that some thousands of such clones have been followed in many different laboratories and to the best of our knowledge, all of them eventually stop growing, unless they are induced to undergo malignant transformation. Curiously, much less is known about the replicative life histories of individual clones and sub-clones of neoplastoid cells, even though they are comparatively easy to clone.

There are currently two major theories which have been put forth to explain clonal senescence. In its original form, the Orgel hypothesis (3) ascribed the loss of proliferative potential to an "error catastrophe"—an exponentially increasing cascade of mistakes in protein synthesis. It is reasonable to assume that a few erroneous molecules of a structural protein like collagen, for example, would do a cell lineage no great harm, even if those defective molecules were passed on to the progeny. On the other

hand, if the faulty protein molecules were themselves utilized in
the synthesis of other proteins, they would be in a position to
greatly amplify errors in protein structure, with the potential
involvement of all types of proteins.

An alternative to the Orgel hypothesis (4) is that such cultures
undergo a sort of terminal differentiation in vitro, analogous to
the kinds of terminal differentiation one observes with many different types of stem cells in vivo, such as hematopoietic cells or
myoblasts. Once a cell ceases to replicate, however, abnormal
proteins could well accumulate as a secondary degenerative phenomenon
and thus the cell could truly undergo senescence. It is conceivable
that most or all such abnormal proteins result from post-translational
modifications. We therefore propose, in Figure 1, a two stage model
of clonal senescence which we believe is consistent with the observations being made in several different laboratories. Stage I, we
believe, proceeds via a process of clonal attenuation (to be des-

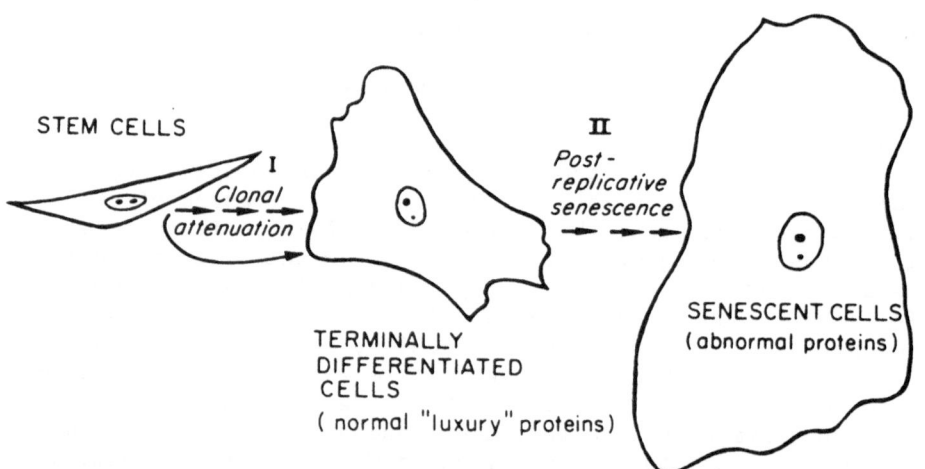

Fig. 1. A diagramatic representation of a two stage model of
clonal senescence. In Stage I a stem cell gradually loses its
"stemness" via a process of clonal attenuation (4) (Fig. 4) with
the resulting accumulation of terminally differentiated cells which
synthesize characteristic sub-sets of normal proteins, possibly
including specific "luxury" proteins. In Stage II, secondary degenerative phenomena intervene in post-replicative cells, with the
appearance of abnormal proteins; the latter is used in the generic
sense and includes abnormal post-translational modifications as well
as mistakes in transcription and translation. The diagram is over-
simplified, in that it does not show potential cell-cell interactions.

cribed below) to a terminally differentiated cell, best defined as
a cell which synthesizes a special sub-set of normal proteins,
possibly including so-called "luxury" proteins (5,6)--namely those
characteristic of the particular differentiated cell. However, in
the present state of our ignorance of the nature of the contrasting
varieties of gene action which can bring about a differentiated
cell, it would seem prudent to keep the definition as broad as
possible so as to include quantitative as well as qualitative changes
in gene expression (7). Stage II would then involve the actual
senescence of such post-replicative cells, perhaps best defined
by the accumulation of abnormal proteins and associated degenerative
morphologic alterations.

METHODS

Details of the cell culture methods, cloning techniques and
media have been previously described (8-11). Assessment of the
heat labile component of glucose-6-PO_4 dehydrogenase (G6PD) was
via a kinetic analysis comparable to that employed by Holliday
and Tarrant (12); details of the enzyme assay will be published
elsewhere (13). Methods used in preparing and evaluating heterokaryons have been described (14).

The assay for receptors for IgG (Fc portion) or for C_3 (third
component of complement) employed rosette techniques (15). Anti-
sheep RBC sera (Baltimore Biological Lab., Cockeysville, Md.) was
separated into IgG and IgM fractions by ion exchange chromotography.
Monomeric IgG and IgM anti-sheep RBC antibodies were obtained by
subsequent gel filtration on Sephadex G-200 or Sepharose 6B columns,
respectively (Pharmacia Fine Chemicals, Inc., Piscataway, N.J.).
The dilution of each antibody solution giving maximal complement
fixation by sensitized sheep RBC was determined using pig serum.
These dilutions were used to prepare IgG (erythrocyte antibody [EA]
rosettes) and IgM sensitized sheep RBC for rosette experiments. The
IgM sensitized cells were incubated with fresh mouse serum as a
source of complement (erythrocyte antibody complement [EAC] rosettes).
IgM sensitized cells without complement and unsensitized sheep RBC
were included as controls in each experiment. Bovine albumin
(2 mg/ml) was added to the freshly prepared sensitized and unsensitized sheep RBC preparations to minimize non-specific adherence in
the slide rosette experiments. Positive controls included human
and rabbit peripheral blood monocytes (prepared by Ficoll-Hypaque
sedimentation) (16) and rabbit alveolar or peritoneal macrophages
(17) both in suspension and adherent to glass slides. In some
experiments, serum was not added to the incubation medium for
variable periods of time before the cells were used on rosette
experiments.

RESULTS AND DISCUSSION

Using a line of human embryo lung fibroblast-like cells, Holliday and Tarrant (12) and Lewis and Tarrant (18) have published evidence indicative of an accumulation, in late passages, of heat labile and/or immunologically unreactive protein for three different dehydrogenases: glucose-6-phosphate dehydrogenase (G6PD), 6-phosphogluconate dehydrogenase and lactic acid dehydrogenase. Holliday and his colleagues (19) have also observed unusually large (21-23%) proportions of heat labile G6PD in skin fibroblast cultures from a patient with Werner's syndrome.

In our own studies, however, no such accumulation of heat-labile or of immunologically unreactive G6PD could be found in late passages of a line of neonatal skin fibroblasts (13,20). Table I summarizes the results of the heat lability studies; both old and young cultures gave around 5% heat labile enzyme. Two types of positive controls were carried out: 1) cells, grown in the presence of the proline analogue, azetidine carboxylic acid (21-23) showed up to 25-30% heat labile G6PD and 2) cultures from a female heterozygous for a heat labile G6PD mutation showed some 59% heat labile enzyme in early passages. Because of what we believe to be in vitro clonal selection for cells with an X chromosome bearing the normal allele

Table I

Summary of G6PD heat lability experiments with a line of human neonatal normal diploid skin fibroblasts (line 71-95).

	"Young"	"Old"
No. Expts.	26	20
C.P.D.	5-22	49-60
L.I.	75-85	5
T.E.	4.8 ± 1	5.0 ± 1

C.P.D. = cumulative population doublings. L.I. = labeling index (% cells synthesizing DNA as detected autoradiographically). T.E. = % thermolabile enzyme ($\pm$ standard error of the mean). Methodologic details are given by Pendergrass et al. (13).

for G6PD, late passages gave the paradoxical effect of a reduction in the level of heat labile enzyme to around 8%. We conclude that it is possible to observe the virtual cessation of replicative activity (less than 5% of cells incorporating thymidine under the experimentally defined conditions) without any detectable accumulation of heat labile G6PD.

An extensive study of viral replication in aging WI-38 cells by Holland and co-workers (24) also failed to provide evidence for the accumulation of defective proteins. Similar conclusions were reached by Tomkins et al. (25). There are several possible explanations for these discrepancies. One possibility, for example, is that with certain cell lines, different gene loci are switched on in terminally differentiated cells giving enzymes with altered properties. There is in fact evidence for a hexose-6-phosphate dehydrogenase in human liver which is immunologically distinct from human liver glucose-6-phosphate dehydrogenase (26). This could be the human counterpart of the autosomal "G6PD" locus demonstrated in other species (27-29). We currently favor the view, however, that a variety of post-translational modifications of proteins may occur as secondary degenerative phenomena in post-replicative cells (Stage II of Fig. 1), but that the rate at which these occur may vary considerably depending upon the genotype and cell type being cultured, the media employed and various techniques of cell culture. In the case of the dehydrogenases, for example, variations in the amounts of nicotinamide in the media may influence the stability of G6PD, since this compound is a precursor of NADP, the stabilizing co-factor for the enzyme (30). Genotype, cell type and conditions of culture may also influence the stability of lysosomal membranes, so that proteases may be liberated in situ to different degrees. Abnormal proteins may therefore prove of interest as an approach to the study of the senescence of post-replicative cells both in vitro and in vivo (31-34). There is as yet no compelling evidence, however, to indicate that fibroblast-like cells stop replicating in culture because they accumulate abnormal proteins.

Now let us turn to an examination of the evidence for our alternative view that cells stop replicating in vitro because they "differentiate themselves to death." In order to provide definitive evidence in support of the terminal differentiation theory, we must demonstrate that our alleged terminally differentiated cells in fact synthesize a differential spectrum of normal proteins, presumably some of which are associated with some specialized function; this has not yet been demonstrated. We must also remember, however, that the tissue culture environment differs substantially from that which exists in vivo, so that one might expect to observe more of a caricature of normal terminal differentiation than an exact phenocopy of the analogous in vivo process.

We might first of all inquire as to the nature of the stem
cells which grow out of skin explants. Are they really fibroblasts?
In the case of WI-38, Franks (35) has suggested that they are not,
and that many such cells may be derived from pericytes or endothelial cells. In the case of human adult skin explants, our preliminary autoradiographic experiments show comparatively little DNA
synthesis in dermal fibroblasts; much more striking is the labeling
of sweat gland units, probably including myo-epithelial cells. This
does not, of course, mean that such cells ultimately predominate in
established skin "fibroblast" cultures. Basal epidermal cells, for
example, are also comparatively heavily labeled but they make little
contribution to the cultures which are eventually passaged. In any
case, we should keep an open mind as to the origin of tissue culture
"fibroblasts." Much more work is needed, for example, in the quantitative and qualitative characterization of the collagens synthesized
by cultures from various fetal, neonatal and adult tissues (36).

Let us now consider the morphology of the principal type of
clone which does emerge from human neonatal or adult skin explant
cultures. In the case of those clones which grow vigorously for
periods of some three weeks (we call them "megaclones"), it is clear
from the macroscopic photographs (Fig. 2a,b,c) that such colonies
have a very special histology characterized by periodic ribs or
whirls of dense cellular multilayering. Furthermore, such a histology is characteristic for this "F" (fibroblast) type of megaclone.
The morphology of megaclones which develop from the normal diploid
somatic cells which predominate in the amniotic fluid of human
second trimester pregnancies ("AF" cells) is quite different (Fig.
2d,e,f) (37). A third variety of megaclone, derived from amniotic
fluid epithelioid cells, is illustrated in Figure 2g,h,i.

The finding of a characteristic histologic pattern immediately
suggests to us that we are dealing with a differentiating system.
In vivo, such systems consist of two or more interacting cell types.
What is the evidence that there are two or more interacting cell
types in our skin fibroblast cultures? First of all, there is the
observation of asymmetric mitoses, in which daughter cells of
strikingly different size, shape and growth potential are observed
(4). In such cases, the larger of the two daughter cells may undergo
no further divisions over a period of at least several days, during
which time many progeny develop from the smaller cell. In order to
explain this observation on the basis of the error theory, one
would have to postulate a mechanism whereby there was preferential
segregation of abnormal proteins and abnormal protein synthesizing
machinery to only one of two daughter cells during telophase. On
the other hand, such asymetric mitoses are consistent with a "quantum
mitosis" proposed for differentiating stem cells (38), whereby
differential gene action is initiated in one of the two progeny.
There are of course alternative explanations, such as an unequal

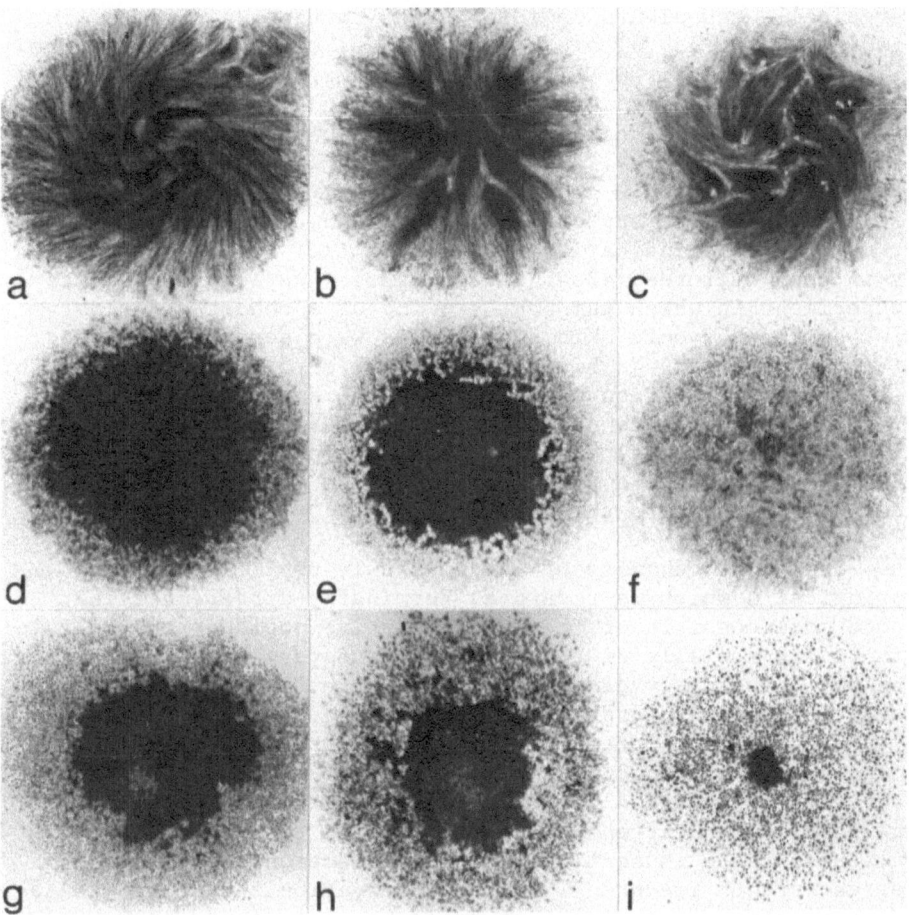

Fig. 2. Macroscopic appearances of three examples each of three
different classes of megaclones. a-c: "F" or fibroblast types,
derived by dilute plating from early passages of a line (72-17)
established from skin explants (upper mesial arm of 19 week old
female); d-f: "AF" types, the most frequent class of primary clone
which can be cultured from second trimester amniotic fluid (37);
g-i: "E" or epithelial types, derived by primary cloning from
second trimester amniotic fluid (37). 1% crystal violet in 20%
ethanol. X 3.2 (a-e,g,h), and X 3.8 (f,i).

distribution of cytoplasmic organelles (mitochondria would be a
good candidate) or, perhaps more likely, of plasma cell membrane
receptors for serum mitogens (39-41).

Large cells can always be identified, to varying degrees, in
our megaclones (Fig. 3b,d) and clearly can be cloned out as a pure
type (Fig. 3e,f). It is not possible, however, to obtain a pure
clone of uniformly small, actively replicating "young" cells; evidently, when young cells divide, they have a certain probability of
"segregating" more or less larger "old" cells which have more limited
replicative life-spans. A growth kinetic analysis of clones and
their serial sub-clones supports this view and confirms and extends
earlier observations indicative of a substantial amount of heterogeneity in such cultures (42-45). In Figure 4 we see the results
of an experiment in which the best growing of a cohort of primary
clones of human embryonic skin fibroblasts were serially recloned.
In all cases, we observe a bimodal distribution of cell growth. This
is a consistent observation made with several different human skin
fibroblast lines and is consistent with the evidence of Smith and
Hayflick (46) of a bimodal distribution of clonal life-spans of WI-
38 cells. (However, the data of Herz and Ross [47] do not suggest
bi-modality of clonal growth rates in WI-38.)

The first mode consists of large, slowly replicating or non-
replicating cells which we regard as terminally differentiating.
The second mode consists of the "proliferative pool" of the culture;
since it is log normally distributed, we discover that a comparatively
small number of actively growing clones apparently sustain our cultures and that actively growing cells are continually segregating
slowly growing cells. Moreover, there is a gradual attenuation of
the clonal growth with proportionately greater numbers of terminally
differentiated cells and a less vigorous proliferative pool. Thus,
assuming that we can extrapolate these results to the conditions of
mass cultures, there is continual clonal selection for cells with
the greatest growth potential. (This observation leads, incidentally,
to the interesting prediction that small healing wounds or hyperplastic foci may by oligoclonal or even monoclonal.)

At the cellular level, clonal attenuation appears to be a
stochastic process, in that there seems to be a certain probability,
at any given mitosis, of a more or less asymmetric mitosis, with
one (or possibly both) of the progeny destined to leave the proliferative pool. How can one reconcile this behavior with the concept
of genetically programmed, orderly, histogenetic (Fig. 2) differentiation? This is exactly the problem which Till et al. (48) were
faced with in their investigations of hematopoietic stem cells.
They suggested an analogy with radioactive nuclides, in which there
is a predictable, reproducible pattern of decay in large populations
of atoms; at the level of the individual atom, however, decay is
random and unpredictable. In the clonal attenuation model which we

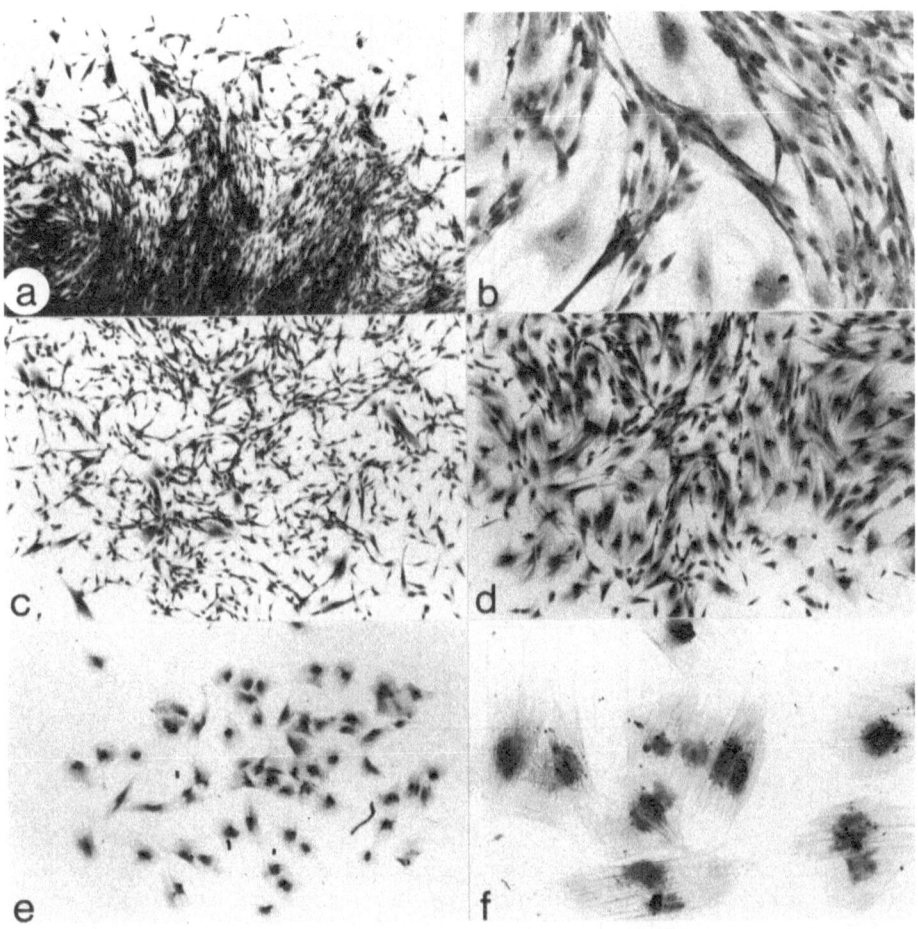

Fig. 3. Microscopic heterogeneity of "F" type megaclones (line
72-17, Fig. 2) a: margin of rapidly growing clones; b: high
power of marginal area of clone shown in a (note "segregation" of
large "senescent" cells); c and d: clones with intermediate
growth potential, with greater proportion of large cell type and,
e: very slow growing clone consisting exclusively of large
"senescent" cells; f: high power of e. 1% crystal violet in 20%
ethanol. X 22.5 (a,c-e) and X 90 (b,f).

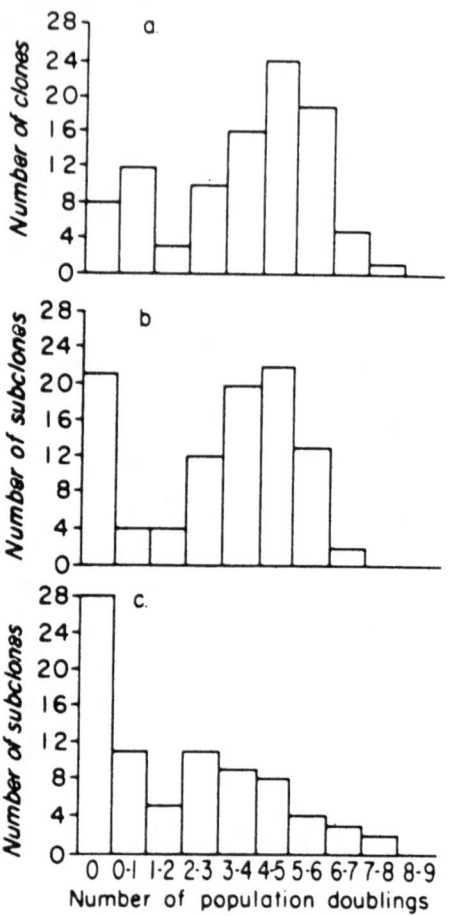

Fig. 4. Distributions of population doublings achieved by clones
from human embryonic skin fibroblasts   a: primary clones, 5 days
after isolation, unfed;   b: secondary clones from best growing
primary clone, 5 days after isolation, unfed;   c: quaternary
clones serially derived from best growing secondary and tertiary
clones, 17 days after isolation, with feeding on day 9 (4).

have described, it may also be that it is the population as a whole
which is closely regulated, rather than the individual cell.   We
suspect that cell interactions play important roles in such regulation.   Recent studies on human skeletal muscle differentiation in
vitro support these speculations (49,50).

One might further ask whether sub-clones of neoplastoid cell

lines also differ greatly with respect to their rates of growth.
In the case of L cells and Chang liver cells, this seems to be the
case. In Figure 5, we illustrate a bimodal distribution for the
diameters of a cohort of sub-clones of Chang liver cells grown
under identical conditions. A control experiment with normal human
diploid fibroblasts, using identical methods, is shown in Figure 6;
as expected, it gives a bimodal distribution. Could the mode of
non-replicating and slowly replicating cells in neoplastoid cell
lines also represent a pool of "terminally differentiating" cells?
Certainly, it is almost the rule that, in vivo, neoplasms segregate
variously differentiated daughter cells--for example, foci of keratin-
synthesizing "pearls" in otherwise undifferentiated squamous cell
carcinomas (51). A much more likely explanation, however, is that
the extensive chromosomal segregation which characterizes such neoplastoid cell lines is responsible for the poorly growing cells.
To the best of our knowledge, such processes as nondisjunction do

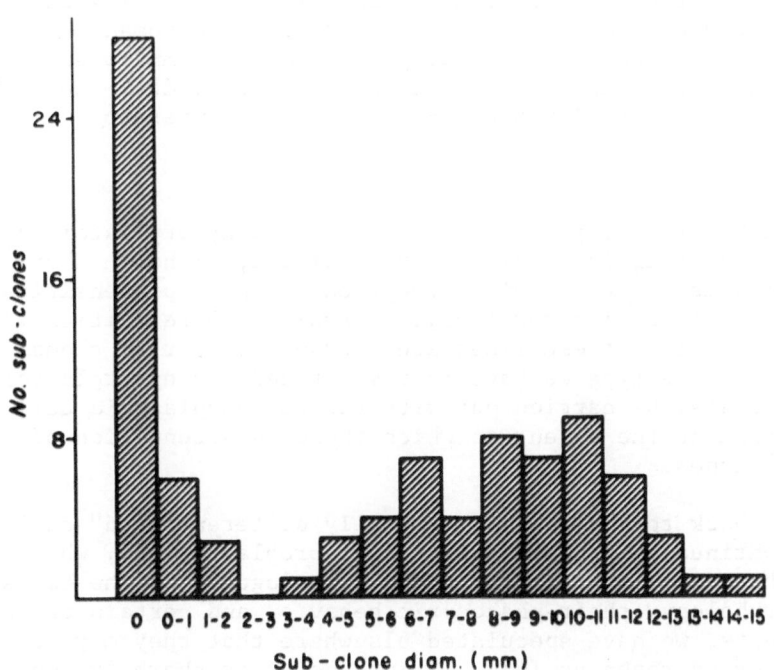

Fig. 5. Distributions of sub-clonal diameters achieved (21 days)
by secondary clones from a strain of Chang liver cells. Cells
were individually isolated on 3 mm coverslips (9) which were then
transferred to 3" X 2" glass slides. Measurements were carried
out with a Nikon Shadowgraph Comparator.

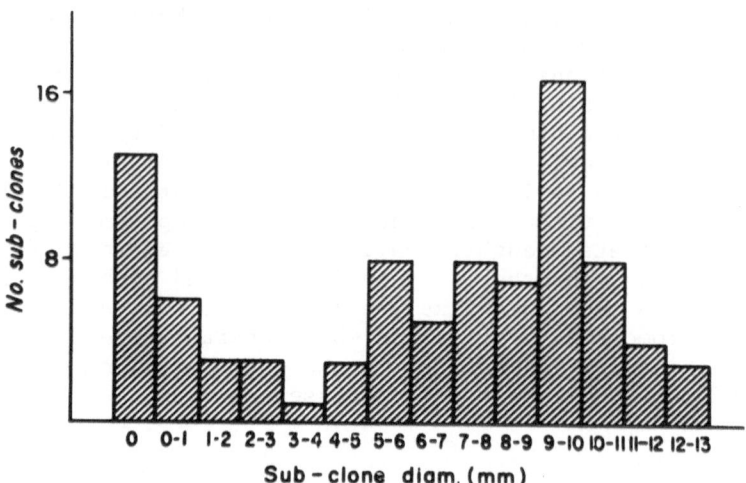

Fig. 6. Distribution of sub-clonal diameters achieved (24 days)
by secondary clones from a strain of normal diploid human skin
fibroblasts. Cells were individually isolated on 3 mm coverslips
(9) which were then transferred to 3" X 2" glass slides. Measurements were carried out with a Nikon Shadowgraph Comparator.

not occur with sufficiently high frequencies in hyperplastoid cell
lines, even when individual clones are examined, although there do
appear to be some genotypes with exceptionally high propensities
for mitotic chromosomal errors (52). Certainly, more critical cytogenetic studies along these lines are warranted. Serial clonal
experiments, of the type we have just described for hyperplastoid
lines, should also be carried out with various neoplastoid cell
lines to determine the extent to which there is attenuation of
various sub-clones.

      Getting back to the large "terminally differentiated" cells
which are continually appearing in skin fibroblast lines, what
clues are there to suggest their in vivo analogue? On the basis
of their morphology, their phagocytic behavior and certain teleo-
logic arguments, we have speculated elsewhere that they may be
histiocytes or macrophages (4). We know now that there are no
detectable cell surface receptors for at least two well-known
markers for macrophages—the third component for complement and
the Fc component of the IgG molecule. Therefore, if they are
macrophages, they are presumably not of the usual immunological
variety. Virtually nothing is known concerning the specific gene
action which might characterize the soft tissue histiocyte, so it

is difficult to know how to pursue that possibility. There is, however, one aspect of the morphology of these cells which suggests that there may be occurring the constitutive synthesis of a given group of macromolecules; I refer to the aggregates of cytoplasmic microfibrils which are so often observed in "aging" fibroblasts (4) and which may also occur in various cell types, including macrophages (53). Tumilowicz and Sarkar (54) have described massive accumulations of microfilaments with the electron microscope in aging fibroblast cultures from breast tumors and in WI-38 cells. These filaments measure 110 A in diameter and appear to be composed of sub-units around 20 A in diameter. It would be of interest to characterize these filaments biochemically and to compare them with the microfilaments and/or microtubules which are presumably induced in certain cell lines by raising the intracellular levels of cyclic AMP (55). The phenotype of our "old" or "terminally differentiated" cells shows some parallels with 3T3 cells treated by dibutyryl cyclic AMP (56). Other EM studies emphasize aggregates of lysosomes (57-59); these may also be markers for a histiocytic type of cell; alternatively, they may be interpreted as evidence of post-replicative degeneration.

Finally, we wish to briefly review some fascinating heterokaryon studies recently carried out in our laboratory. You may recall that some years ago, Professor Henry Harris and his colleagues (60,61) showed that DNA synthesis could be reinitiated in three different types of terminally differentiated cells--the rabbit peritoneal macrophage, the rat thoracic duct lymphocyte and the nucleated hen erythrocyte. We reasoned that if our "old" fibroblasts were actually terminally differentiated cells then we too should be able to "rescue" them in heterokaryons with actively replicating cells. On the other hand, if these cells were full of abnormal proteins, it would seem likely that some such molecules would have qualitatively aberrant functions and behave as dominants extinguishing or at least diminishing DNA synthesis in the replicating partner. In order to identify heterokaryons, we took advantage of the elegant technique of double layer autoradiography developed by Baserga and Nemeroff (62). In brief outline, old cells were prelabeled with ^{3}H methionine and young cells with ^{14}C thymidine; after fusion of the two with chemically inactivated Sendai virus, 3 serial "test pulses" of tritiated thymidine were given to culture aliquots in order to follow the course of DNA synthesis. Figure 7 illustrates such heterokaryons at focal levels which reveal silver grains predominantly attributable to either low energy tritium emissions (a,c) or high energy carbon 14 emissions (b,d).

In the first such experiments which were carried out (63) involving crosses of old and young homologous and isologous diploid skin fibroblasts, the results indicated that the senescent phenotype was dominant. Not only did the young fibroblast nuclei fail to rescue DNA synthesis in the old nuclei with which they were fused,

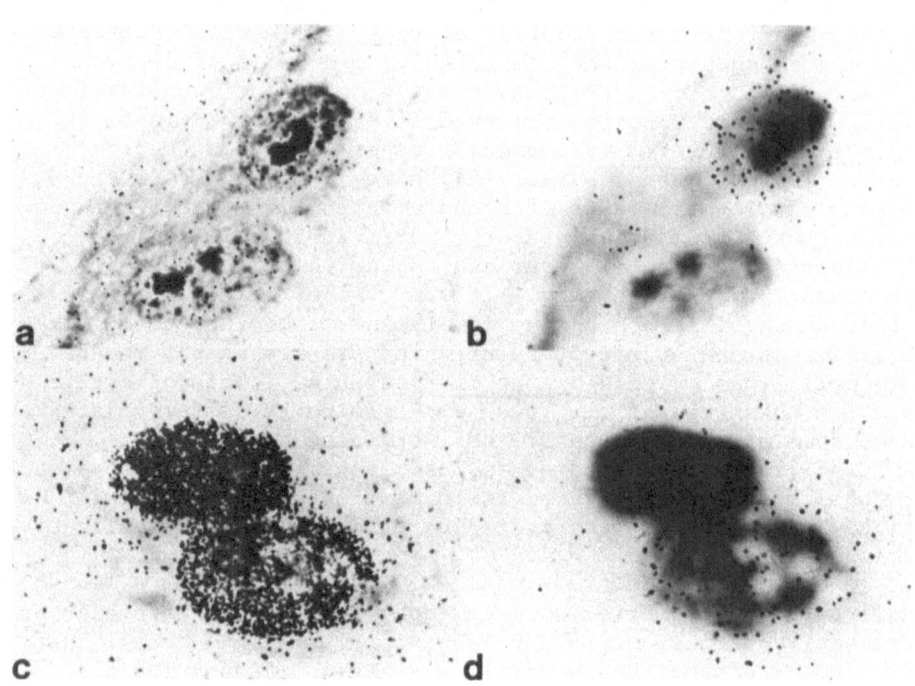

Fig. 7. Photomicrographs demonstrating the results of double layer
autoradiography of heterokaryons. a,c: focal planes of lower
layers of emulsions revealing silver grains primarily attributable
to 1) tritiated methionine (cytoplasmic and nuclear thus identifying the prelabeled old cells, and 2) tritiated thymidine (Nuclear)
thus identifying nuclei actively synthesizing DNA during the test
pulse period. b,d: focal planes of upper layer of emulsion revealing
silver grains primarily attributable to higher energy ^{14}C thymidine,
thus identifying the prelabeled actively proliferating cells.
a,b: "old" fibroblast X "young" fibroblast illustrating dominance
of the senescent phenotype in such a cross. c,d: "old" fibroblast
X HeLa cell illustrating dominance of the HeLa phenotype in that
cross.

but its own DNA synthesis was suppressed. The results of an isologous cross are shown in Figure 8. We then realized that, with one
possible exception (64), the actively replicating partner in all of
Harris' experiments were neoplastoid cells--either HeLa, A9 or
Ehrlich ascites--and we therefore predicted that, in heterokaryons
with neoplastoid cells, DNA synthesis would in fact be reinitiated
in the old nucleus. This prediction has now been confirmed for two
different neoplastoid lines--HeLa, and a SV40 transformed fibroblast
line (Table II) (63,65).

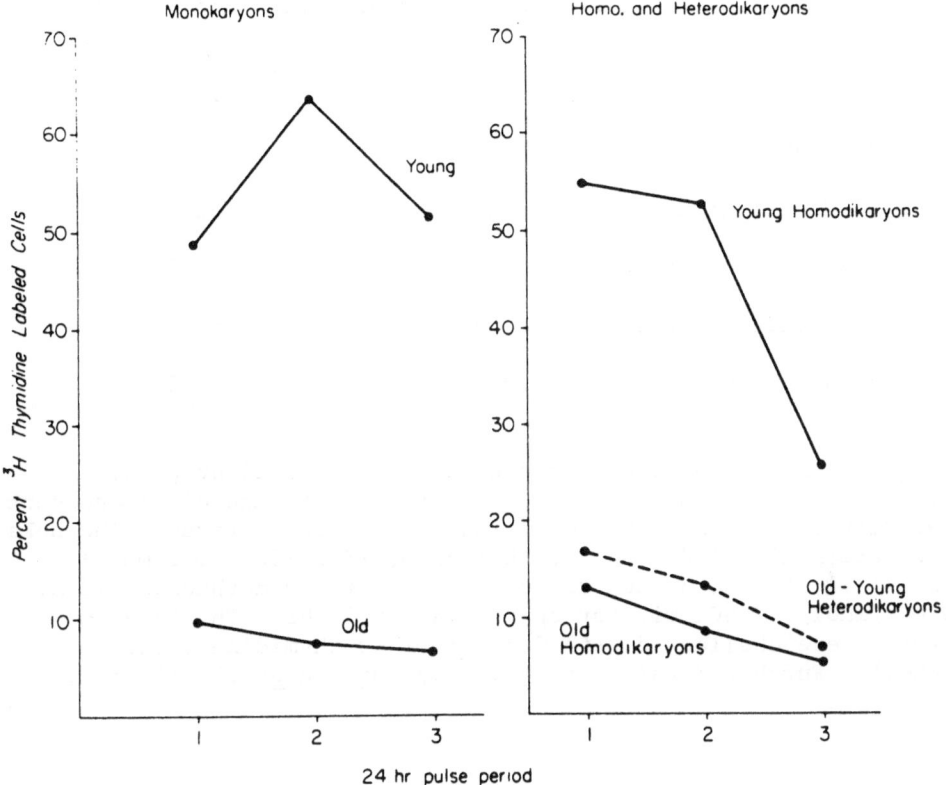

Fig. 8. Labeling indices of nuclei in isologous "heterokaryons" and controls derived by Sendai virus-mediated fusion between "old" (late passage) and "young" (early passage) fibroblasts of the same line of male neonatal normal diploid fibroblasts (line 71-95). Details of the methods are given by Norwood et al. (14).

        We believe that these experiments make an error catastrophe mechanism unlikely and strengthen the case for terminal differen-tiation. However, one might still argue that the rescue by neoplas-toid cells is attributable to highly efficient "scavenger" enzyme systems (66) in the cytoplasm of these cells and that such enzymes preferentially degrade abnormal proteins. One could also argue that an error catastrophe may not necessarily be expressed as a dominant under the conditions of our experiments since we were merely scoring for one round of DNA synthesis. The failure of normal cells to rescue could be attributable to the fact that they have only two copies of relevant genes, whereas the neoplastoid cells may have three or more copies.

TABLE II. Labeling Indices of Nuclei in HeLa-Old Heterokaryons

HeLa Monokaryons	93.2%
Old Monokaryons	1.6%
HeLa Nuclei in Heterodikaryons	82.0%
Old Nuclei in Heterodikaryons	64.2%
HeLa Nuclei in Heteropolykaryons	51.6%
Old Nuclei in Heteropolykaryons	27.3%

Percent ^{3}H thymidine labeled nuclei during first 24 hr pulse period in cultures containing co-cultivated HeLa and old (senescent) human fibroblasts which had been fused with Sendai virus. The HeLa cells were prelabeled with ^{14}C thymidine, (S.A. 52.7 mCi/mMol) 0.75 µCi/ml, for 30 hr and the old cells with ^{3}H methionine (S.A. 190 mCi/mMol), 2 µCi/ml, for approximately 15 hr. The fused cultures were challenged with 0.5 µCi/ml ^{3}H thymidine (S.A. 67 Ci/mMol). Further details are given by Norwood et al. (14,65).

In summary, we can say that the answer posed by our title: "Do Hyperplastoid Cell Lines Differentiate Themselves to Death?" is that we do not know, but strongly suspect that this is the case. There is obviously a great deal of interesting work ahead for all of us, work which we hope will not only contribute to our under-standing of cellular aging, but which may also provide us with important insights into the control mechanisms underlying growth and differentiation, including the nature of neoplastic prolifera-tion.

ACKNOWLEDGEMENTS AND DEDICATION

This research was supported by NIH Research Grants AM 04826, HD 04872 and Training Grants GM 00100 and GM 00052. Paul Bornstein was a recipient of Research Career Development Award K-4-AM42582 from the USPHS. We wish to dedicate this paper to the memory of our beloved friend, colleague and co-author, Curtis A. Sprague, who lost a long and heroic battle with Hodgkin's Disease in February, 1974.

REFERENCES

1. Hayflick, L. and Moorhead, P.S.   1961.   The serial cultivation
    of human diploid cell strains.   Exp. Cell Res. 25: 585.

2. Martin, G.M. and Sprague, C.A.   1973.   Symposium on in vitro
    studies related to atherogenesis.   Life histories of hyper-
    plastoid cell lines from aorta and skin.   Exp. Molec. Path.
    18: 125.

3. Orgel, L.E.   1963.   The maintenance of the accuracy of protein
    synthesis and its relevance to aging.   Proc. Nat. Acad. Sci.
    USA 49: 517.

4. Martin, G.M., Sprague, C.A., Norwood, T.H. and Pendergrass, W.R.
    1974.   Clonal selection, attenuation and differentiation in
    an in vitro model of hyperplasia.   Am. J. Path. 74: 137.

5. Holtzer, H. and Abbotto, J.   1968.   Oscillations of the chon-
    drogenic phenotype in vitro.   In: The Stability of the
    Differentiated State.   H. Ursprung, ed., Springer-Verlag,
    Berlin, Heidelberg, p. 1.

6. Ephrussi, B.   1972.   Hybridization of Somatic Cells.   Princeton
    University Press, New Jersey.

7. Delcour, J. and Papaconstantinou, J.   1974.   A change in the
    stoichiometry of assembly of bovine lens α-crystallin subunits
    in relation to cellular differentiation.   Biochem. Biophys.
    Res. Commun. 57: 134.

8. Martin, G.M.   1973a.   Human skin fibroblasts.   In: Tissue
    Culture-Methods and Applications.   P.F. Kruse and M.K.
    Patterson, eds., Academic Press, Inc., New York, p. 39.

9. Martin, G.M.   1973b.   Dilution plating on coverslip fragments.
    In: Tissue Culture-Methods and Applications.   P.F. Kruse
    and M.K. Patterson, eds., Academic Press, Inc., New York,
    p. 264.

10. Martin, G.M., Sprague, C.A. and Epstein, C.J.   1970.   Replica-
    tive life-span of cultivated human cells.   Effects of donor's
    age, tissue and genotype.   Lab. Invest. 23: 86.

11. Ginsburg, H. and Lagunoff, D.   1967.   The in vitro differen-
    tiation of mast cells.   Cultures of cells from immunized
    mouse lymph nodes and thoracic duct lymph on fibroblast
    monolayers.   J. Cell Biol. 35: 685.

12. Holliday, R. and Tarrant, G.M. 1972. Altered enzymes in
     ageing human fibroblasts. Nature 238: 26.

13. Pendergrass, W.R., Martin, G.M. and Bornstein, P. 1974b.
     Evidence contrary to the protein error hypothesis for in
     vitro senescence, in preparation.

14. Norwood, T.H., Pendergrass, W.R., Sprague, C.A. and Martin,
     G.M. 1974a. Dominance of the senescent phenotype in
     heterokaryons between replicative and post-replicative human
     fibroblast-like cells. Proc. Nat. Acad. Sci. USA, in press.

15. Shevach, E.M., Jaffe, E.S. and Green, I. 1973. Receptors
     for complement and immunoglobulins on human and animal
     lymphoid cells. Transplant. Rev. 16: 3.

16. Boyum, A. 1968. Separation of leucocytes from blood and
     bone marrow. Scand. J. Clin. Lab. Invest. 21: Suppl. 97.

17. Arend, W.P. and Mannik, M. 1972. In vitro adherence of
     soluble immune complexes in macrophages. J. Exp. Med. 136:
     514.

18. Lewis, C.M. and Tarrant, G.M. 1972. Error theory and
     ageing in human diploid fibroblasts. Nature 239: 316.

19. Holliday, R., Porterfield, J.S. and Gibbs, D.D. 1974.
     Premature ageing and occurrence of altered enzyme in
     Werner's syndrome fibroblasts. Nature 248: 762.

20. Pendergrass, W.R., Martin, G.M. and Bornstein, P. 1974a.
     Evidence contrary to the protein error hypothesis for in
     vitro senescence. Gerontologist, in press (abstract).

21. Takeuchi, T. and Prockop, D.J. 1969. Biosynthesis of abnormal
     collagens with amino acid analogues I. Incorporation of L-
     azetidine-2-carboxylic acid and cis-4-fluoro-L-proline into
     protocollagen and collagen. Biochim. Biophys. Acta 175: 142.

22. Lane, J.M., Dehon, P. and Prockop, D.J. 1971a. Effect of
     the proline analogue azetidine-2-carboxylic acid on collagen
     synthesis in vivo I. Arrest of collagen accumulation in
     growing chick embryos. Biochim. Biophys. Acta 236: 517.

23. Lane, J.M., Parkes, L.J. and Prockop, D.J. 1971b. Effect of
     the proline analogue azetidine-2-carboxylic acid on collagen
     synthesis in vivo II. Morphological and physical properties
     of collagen containing the analogue. Biochim. Biophys. Acta.
     236: 528.

24. Holland, J.J., Kohne, D. and Doyle, M.V. 1973. Analysis of
     virus replication in ageing human fibroblasts cultures.
     Nature 245: 316.

25. Tomkins, G.C., Stanbridge, E.J. and Hayflick, L. 1974. Viral
     probes of aging in the human diploid cell strain WI-38. Proc.
     Soc. Exp. Biol. Med. in press.

26. Srivastava, S.K., Blume, K.G., Beutler, E. and Yoshida, A.
     1972. Immunological differences between glucose-6-P dehydro-
     genase and hexose-6-P dehydrogenase from human liver. Nature
     New Biol. 238: 240.

27. Shaw, C.R. and Barto, E. 1965. Autosomally determined
     polymorphism of glucose-6-phosphate dehydrogenase in
     peromyscus. Science 148: 1099.

28. Ohno, S., Payne, H.W., Morrison, M. and Beutler, E. 1966.
     Hexose-6-phosphate dehydrogenase found in human liver.
     Science 153: 1015.

29. Ruddle, F.H., Shows, T.B. and Roderick, T.H. 1968. Autosomal
     control of an electrophoretic variant of glucose-6-phosphate
     dehydrogenase in the mouse (mus musculus). Genetics 58: 599.

30. Bonsignore, A. and Deflora, A. 1972. Regulatory properties
     of glucose-6-phosphate dehydrogenase. In: Current Topics
     in Cellular Regulation. B.L. Horecker and L. Stadtman, eds.,
     v. 6, p. 21.

31. Marks, P.A. 1964. Glucose-6-phosphate dehydrogenase: Its
     properties and role in mature erythrocytes. In: The Red
     Blood Cell. C. Bishop and D.M. Surgenor, eds., Academic
     Press, Inc., New York, p. 211.

32. Fornaine, G. 1967. Biochemical modifications during the life
     span of the erythrocyte. Ital. J. Biochem. 16: 257.

33. Park, C. 1971. A study on the properties of glucose-6-phosphate
     dehydrogenase from young and old human erythrocytes. Yon. J.
     Med. Sci. 4: 118.

34. Yip, L.C., Dancis, J., Mathieson, B. and Balis, M.E. 1974.
     Age-induced changes in adenosine monophosphate: pyrophosphate
     phosphoribosyl-transferase and iosine monophosphate: pyro-
     phosphate phosphoribosyl-transferase from normal and Lesch-
     Nyhan erythrocytes. Biochemistry 13: 2558.

35. Franks, L.M. and Cooper, T.W. 1972. The origin of human
    embryo lung cells in culture: a comment on cell differen-
    tiation, in vitro growth and neoplasm. Int. J. Cancer 9: 19.

36. Trelstad, R.L. 1973. The developmental biology of vertebrate
    collagens. J. Histochem. Cytochem. 21: 521.

37. Hoehn, H., Bryant, E.M., Karp, L.E. and Martin, G.M. 1974.
    Cultivated cells from diagnostic amniocentesis in second
    trimester pregnancies. I. Clonal morphology and growth
    potential. Ped. Res., in press.

38. Holtzer, H., Weintraub, H., Mayne, R. and Mochan, B. 1972.
    The cell cycle, cell lineages, and cell differentiation.
    In: Current Topics in Developmental Biology. A. A. Moscona
    and A. Monroy, eds., v. 7, Academic Press, Inc., New York,
    p. 229.

39. Dulak, N.C. and Temin, H.M. 1973. Multiplication stimulating
    activity for chicken embryo fibroblasts from rat liver cell
    conditioned medium: a family of small polypeptides. J. Cell
    Physiol. 81: 161.

40. Houck, J.C. and Cheng, R.F. 1973. Isolation, purification,
    and chemical characterization of the serum mitogen for diploid
    human fibroblasts. J. Cell Physiol. 81: 257.

41. Ross, R., Glomset, J., Kariya, B. and Harker, L. 1974. A
    platelet dependent serum factor that stimulates the prolifera-
    tion of arterial smooth muscle cells in vitro. Proc. Nat.
    Acad. Sci. USA 71: 1207.

42. Pious, D.A., Hamburger, R.N. and Millis, S.E. 1964. Clonal
    growth of primary human cell cultures. Exp. Cell Res. 33:
    495.

43. Macieira-Coelho, A., Ponten, J. and Philipson, L. 1966.
    Inhibition of the division cycle in confluent cultures of
    human fibroblasts in vitro. Exp. Cell Res. 43: 20.

44. Merz, G.S. and Ross, J.D. 1969. Viability of human diploid
    cells as a function of in vitro age. J. Cell Physiol. 74:
    219.

45. Cristofalo, V.J. and Sharf, B.B. 1973. Cellular senescence
    and DNA synthesis. Thymidine incorporation as a measure of
    population age in human diploid cells. Exp. Cell Res. 76:
    419.

46. Smith, J.R. and Hayflick, L. 1974. Variation in the lifespan of clones derived from human diploid cell strains, unpublished.

47. Merz, G.S. and Ross, J.D. 1973. Clone size variation in the human diploid cell strain, WI-38. J. Cell. Physiol. 82: 75.

48. Till, J.E., McCulloch, E.A. and Siminovitch, L. 1964. A stochastic model of stem cell proliferation, based on the growth of spleen colony-forming cells. Proc. Nat. Acad. Sci. USA 51: 29.

49. Hauschka, S.D. 1974a. Clonal analysis of vertebrate myogenesis II. Environmental influences upon human muscle differentiation. Dev. Biol. 37: 329.

50. Hauschka, S.D. 1974b. Clonal analysis of vertebrate myogenesis III. Developmental changes in the muscle-colony-forming cells of the human fetal limb. Dev. Biol. 37: 345.

51. Willis, R.A. 1967. The Pathology of Tumours. Butterworths, London.

52. Martin, G.M., Sprague, C. and Bryant, J.S. 1967. Mitotic nondisjunction in cultivated human cells. Nature 214: 612.

53. Allison, A.C., Davies, P. and dePetris, S. 1971. Role of contractile microfilaments in macrophage movement and endocytosis. Nature 232: 153.

54. Tumilowicz, J.J. and Sarker, N.H. 1972. Accumulating filaments and other ultrastructural aspects of declining cell cultures derived from human breast tumors. Exp. Molec. Path. 16: 210.

55. Puck, T.T., Waldren, C.A. and Hsie, A.W. 1972. Membrane dynamics in the action of dibutyryl adenosine 3':5'-cyclic monophosphate and testosterone on mammalian cells. Proc. Nat. Acad. Sci. USA 69: 1943.

56. Johnson, G.S. and Pasten, I. 1972. Role of 3', 5'-adenosine monophosphate in regulation of morphology and growth of transformed and normal fibroblasts. J. Nat. Cancer Inst. 48: 1377.

57. Robbins, E., Levine, E.M. and Eagle, H. 1970. Morphologic changes accompanying senescence of cultured human diploid cells. J. Exp. Med. 131: 1211.

58. Brandes, D., Murphy, D.G., Anton, E.B. and Barnard, S. 1972. Ultrastructural and cytochemical changes in cultured human lung cells. J. Ultrastruct. Res. 39: 465.

59.   Lipetz, J. and Cristofalo, V.J. 1972. Ultrastructural
      changes accompanying the aging of human diploid cells in
      culture. J. Ultrastruct. Res. 39: 43.

60.   Harris, H., Watkins, J.F., Ford, C.E. and Schoefl, G.I.
      1966. Artificial heterokaryons of animal cells from
      different species. J. Cell Sci. 1: 1.

61.   Harris, H. 1967. The reactivation of the red cell nucleus.
      J. Cell Sci. 2: 23.

62.   Baserga, R. and Nemeroff, K. 1962. Two-emulsion radioauto-
      graphy. J. Histochem. Cytochem. 10: 628.

63.   Norwood, T.H., Pendergrass, W.R., Sprague, C.A. and Martin,
      G.M. 1974b. A heterokaryon study on in vitro senescence.
      In Vitro 9: 351.

64.   Harris, H., Sidebottom, E., Grace, D.M. and Bramwell, M.E.
      1969. The expression of genetic information: a study with
      hybrid animal cells. J. Cell Sci. 4: 499.

65.   Norwood, T.H., Pendergrass, W.R., Sprague, C.A. and Martin,
      G.M. 1974c. Reinitiation of DNA synthesis in senescent
      human fibroblasts upon fusion with immortal cells. Submitted
      for publication.

66.   Goldberg, A.L. 1972. Degradation of abnormal proteins in
      Escherichia coli. Proc. Nat. Acad. Sci. USA 69: 422.

QUESTIONS TO DR. MARTIN

Dr. Macek: The epitheloid cells forming epitheloid colonies in
amniotic fluid cultures grow slowly. Cells are lost after several
passages. These contrasts with the longer lifespan of fibroblastoid cultures derived from amniotic fluid. Is there any prospect
of improving the degree of proliferation of the epitheloid cultures
or does it represent some sort of terminal differentiation of this
type of cells?

Dr. Martin: Quantitative parameters of cell growth and colonal
longevity for the several types of clones which can be grown from
amniotic fluid are given by Hoelm et al. in a forthcoming publication (Paediatric Research, in press); these studies support your
experience and the experience of others concerning the comparatively
short lifespans of epitheloid cells. The mechanism by which their
growth ceases would well be a sort of terminal differentiation and
is investigated from this point of view. If so, perhaps a media

could be designed to extend the probability of mitoses leading to a tendency for terminal differentiation. As a more practical approach we could suggest exploring variations of the trypsin harvest procedure, since, in our experiments these cells are comparatively resistant to trypsinization and some may be influenced irreversibly. Another approach would be to explore various substances. For example, in our experiments with the epidermal cells emerging from skin explants, they grow much more prolifically on plastic than on glass.

Dr. Kasten: What is the frequency of asymetric mitoses and with what type of cell is it associated (fibroblastic, epitheloid, etc)?

Dr. Martin: We observed varying degree of asymetric mitoses; the photograph I showed illustrated an unusually striking example in a skin fibroblast mitosis. We have no quantitative data on the frequencies of the different morphologic degrees of asymmetry in any cell type.

Dr. Courtois: Did you try to fuse old separated or anucleated old cells with HeLa cells?

Dr. Martin: The experiments along those lines are currently being carried out by Dr. Thomas Norwood in our laboratory using cytochalasin B-anucleated cells. These methods were first applied to the analyses of the senescence phenotype by Wright and Hayflick at Stanford; their experimental design, however, is quite different and in the experiments they have discussed to date (Fed. Proc. Abstracts, 1974) they didn't use neoplastoid cells such as HeLa.

Dr. Deyl: If the second part of your working hypothesis about specific "senescent proteins" is true, then one should see structurally altered proteins when analysing old tissues, at least in trace amounts. To my knowledge nothing like that has been reported yet. Or alternatively, do you expect in terminally differentiated cells proteins being synthesized in altered proportions as compared with the preceeding stages?

Dr. Martin: I hope I did not imply the necessity for "specific" types of abnormal or "senescent" proteins in my hypothetical stage II of clonal senescence; my guess would be that one would find a variety of abnormal proteins in degenerative cells, most of them on a post-translational basis. It would of course be of great interest to systematically seek out evidence for such altered proteins in predominantly post-replicative mammalian cells in vivo. Very little work has been done along these lines beyond that reported by the Gershons. As regards the second part of your question, it is of course possible that a differentiated cell type

could be characterized by quantitative differences in the proportion
of normal proteins also synthesized by precursor stem cells, but
by analogy with such in vivo models as the hematopoietic system,
one would anticipate the gradual build up of entirely unique
proteins, such as hemoglobin, not synthesized by the stem cells, or
synthesized in only trace amounts.

Dr. Gallop: Have you performed any G6PD heat lability studies
with cells from patients with Werner's syndrome?

Dr. Martin: Dr. Pendergrass has only completed two experiments
with cells from such a patient and the range of percent labile
enzyme is from about 4-12%. Further experiments must be done
before any conclusion can be drawn since a statistical analysis
is required. Dr. Robin Hollyday and his colleagues have in fact
discovered comparatively high levels of thermolabile enzyme in
skin fibroblast from a patient of theirs.

TIME-LAPSE CINEMICROPHOTOGRAPHIC STUDIES OF CELL DIVISION PATTERNS

OF HUMAN DIPLOID FIBROBLASTS (WI-38) DURING THEIR IN VITRO LIFESPAN

P. Marlene Absher, Richard G. Absher and
William D. Barnes

Department of Medical Microbiology, The University
of Vermont College of Medicine, Burlington, Vermont

Since Hayflick and Moorhead (1) and Hayflick (2) reported on the finite lifespan of cultured diploid fibroblasts, researchers have used these cell types to study aging at the cellular level. The decline in proliferative capacity of diploid fibroblasts with increasing in vitro age has been attributed to a gradual increase in the proportion of non-dividing cells (3,4,5,8).

Loss of division potential with increasing culture age has been shown by a decline in cloning ability (3,6), prolonged generation time (7) and smaller fraction of cells synthesizing DNA (7,8). The studies reported here were directed at an assessment of cell division patterns of human embryonic lung diploid fibroblasts (WI-38) at various stages of in vitro propagation. Using the technique of time-lapse cinemicrophotography we have studied cell division behavior of progeny derived from single cells at different passage levels. Certain direct measurements can be made using this technique: interdivision time through successive generations, attachment of cells, cell death, migration patterns andcell motility, cell morphology, stemline or clonal division patterns. Somewhat more subjective measurements include cell-cell interactions and fate of "non-dividing" cells in the clones. In this paper we report genealogies of clones, and other aspects of cell behavior of WI-38 diploid fibroblasts from phase II and phase III cultures.

MATERIALS AND METHODS

Human embryonic lung fibroblasts (WI-38) were obtained from

91

Dr. Leonard Hayflick. Stock cultures were maintained in Eagle's
basal medium supplemented with 10% fetal calf serum and 50 g/ml
aureomycin ($EBME_{90}FC_{10}$), and grown in an atmosphere of 5% CO_2 and
95% air at 37°C. Routine subcultivation was done at a 1:2.5 split
ratio using a trypsinization procedure. Phase II cultures were
transferred twice weekly and phase III cultures were transferred
once weekly, or when confluent.

Cultures for filming were prepared from confluent monolayers.
Cloning medium consisted of equal parts of fresh $EBME_{90}FC_{10}$ and
conditioned medium (taken from the confluent monolayer used as
a source of cells). This mixture was found to be satisfactory
for support of growth of clones being filmed. A 60 mm Falcon
Cooper type tissue culture dish with depressed lid was used for
the microscope chamber. To contain the cells within the photographic field, sterile silicone (Dow Corning high grade vacuum
grease) was used to fashion small ponds (approximately 0.25 mm^2) on
the growing surface of the dish. A seeding density of approximately
between 1.6 and 2.5 x 10^4/ml in a total volume of 8 ml cloning
medium allowed for deposition of one or two cells in the photographic filed. Filming was begun when cells had settled on the
growing surface of the dish (approximately 1 hr after seeding).

The equipment for filming consists of a Zeiss GFL 658 microscope with long working distance condensor, 10X phase contrast
objective, and built in light source. The Sage time-lapse #501
apparatus was fitted with a Bolex 16 mm camera. Kodak Plus-X
reversal film was used and filming was done at the rate of one
frame per min.

                              RESULTS

The genealogies reported here were obtained from films of
clones of WI-38 fibroblasts at passages 20, 28 and 53. The calculated population doubling level (PDL) for these cultures were 21,
32 and 65 respectively. We have derived genealogies from filmed
sequences of 63 clones of WI-38 at passage levels ranging from
19 to 53 (PDL range = 20 to 65). The clones reported here were
selected as representative of the more highly proliferating cells
at these passage levels.

Fig. 1. Genealogy of passage 28 WI-38 clone (PDL = 32). Numbers
indicate interdivision time in hr. Lines ending in arrows indicate
cells followed to end of filming without dividing. Dashed lines
indicate cells that could not be followed beyond that point.

WI-38, Passage 28

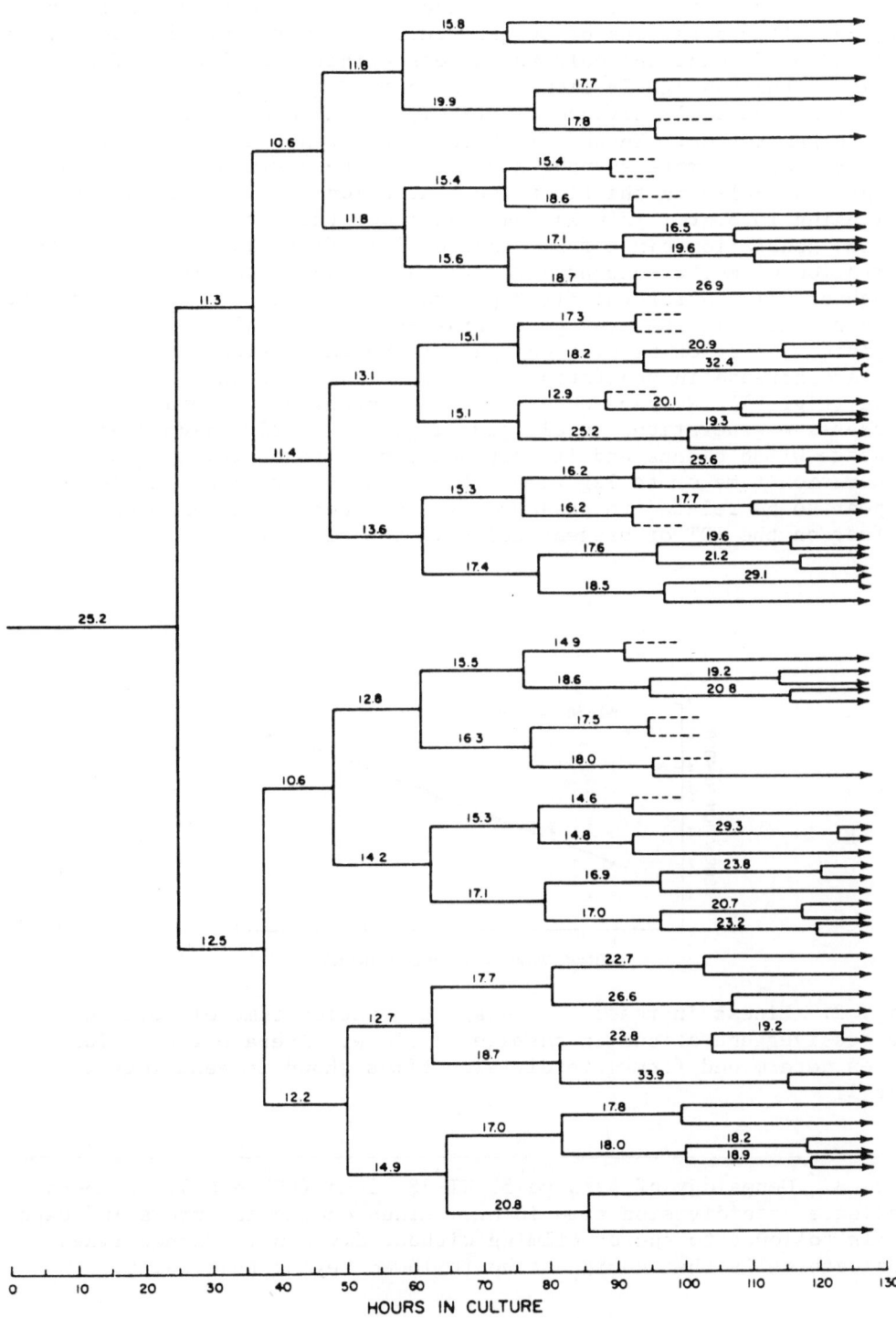

HOURS IN CULTURE

The passage 28 clone (PDL = 32) was the most highly proliferative clone yet studied. This culture began with a single cell which
yielded six generations of progeny cells in approximately 130 hr. At
the end of filming the culture was essentially confluent. The genealogy of the passage 28 clone is shown in Figure 1. It can be
seen that the cells divided relatively synchronously through the
fourth generation. In the fifth and sixth generations the division
pattern becomes more asynchronous. The interdivision time (IDT)
of progeny cells in the first few generations is low, the lowest
IDT being 10.6 hr. This is somewhat lower than previously reported generation times from human diploid fibroblasts (7,9). Observation of mother-daughter relationships shows the IDT of most
daughter cells after the first generation is longer than that of the
mother cell. If the average generation time for all cells in
a given generation is plotted against the generation number a
linear increase in generation time is noted after the first generation (Fig. 2). We have observed this lengthing of average IDT with
successive generations in all genealogies of early, middle and late
passage WI-38 clones and it does not appear to be related to initial
generation time or to lag time. Lag time is variable and does not
appear to be related to passage level nor does it seem to have an
effect on the IDT of progeny cells in the clones.

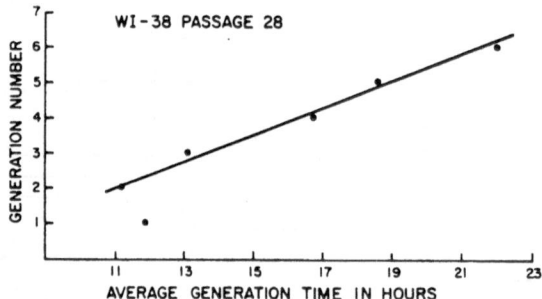

Fig. 2. Linear increase of average generation time of cells with
successiveggenerations in passage 28 clone. Average generation
times determined from interdivision times shown in genealogy in
Fig 1.

Fig. 3. Genealogy of passage 53 WI-38 clone (PDL = 65). Numbers
indicate interdivision time in hr. Lines ending in arrows indicate
cells followed to end of filming without dividing. Dashed lines
indicate cells that could not be followed beyond that point.

WI-38, Passage 53

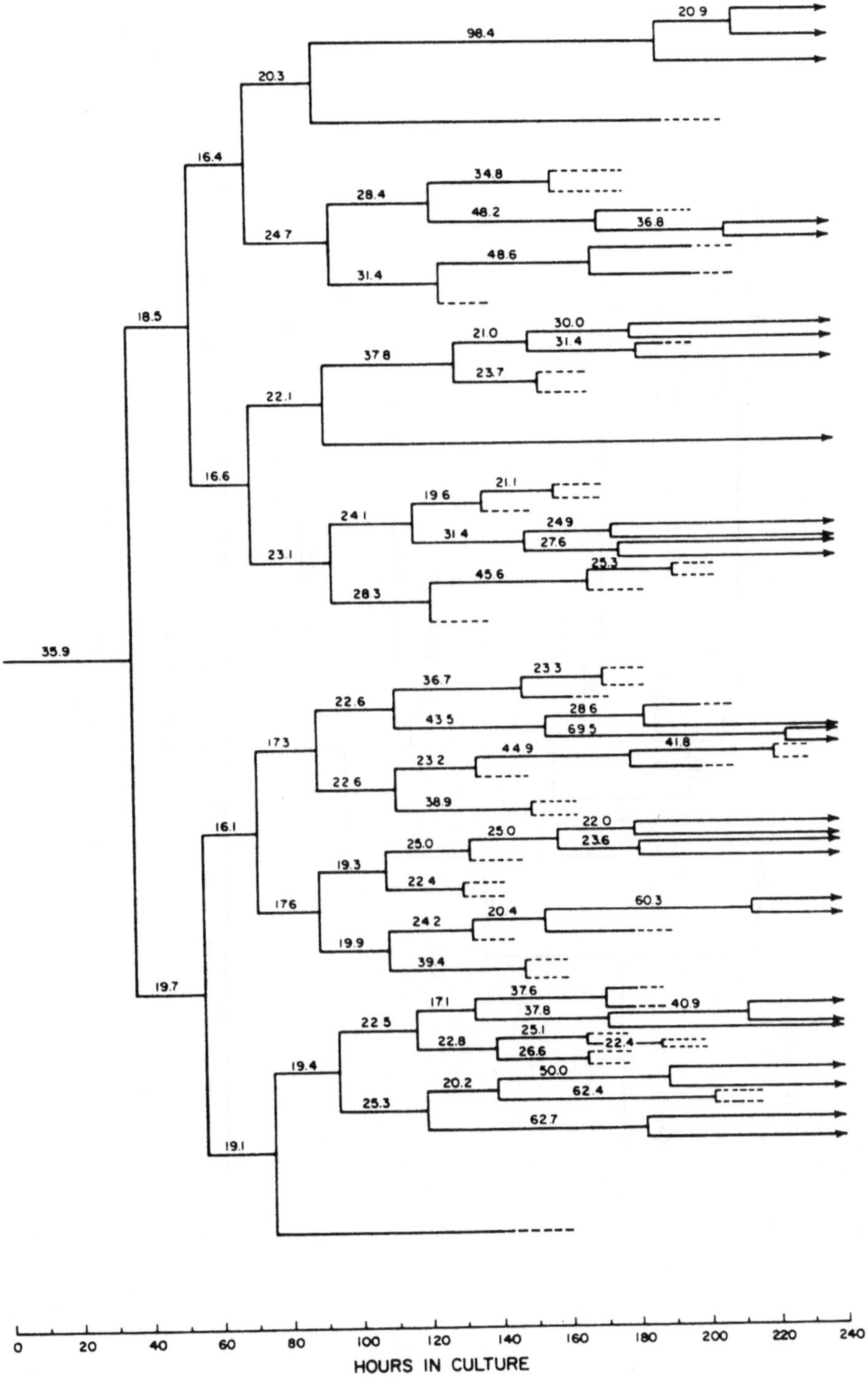

HOURS IN CULTURE

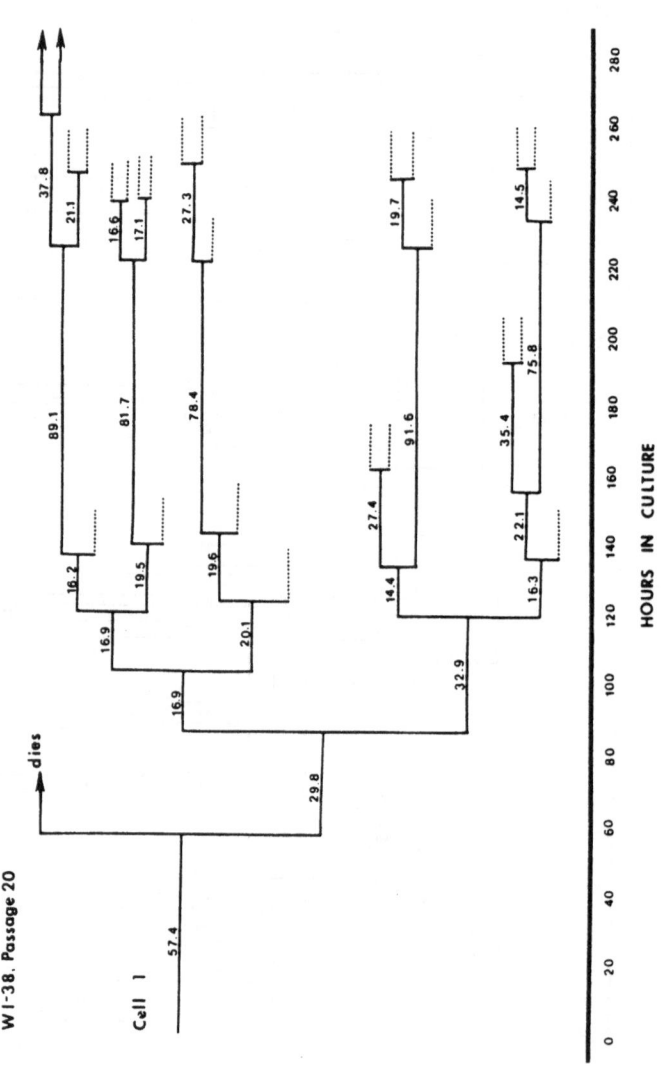

Fig. 4A. (caption on p. 98)

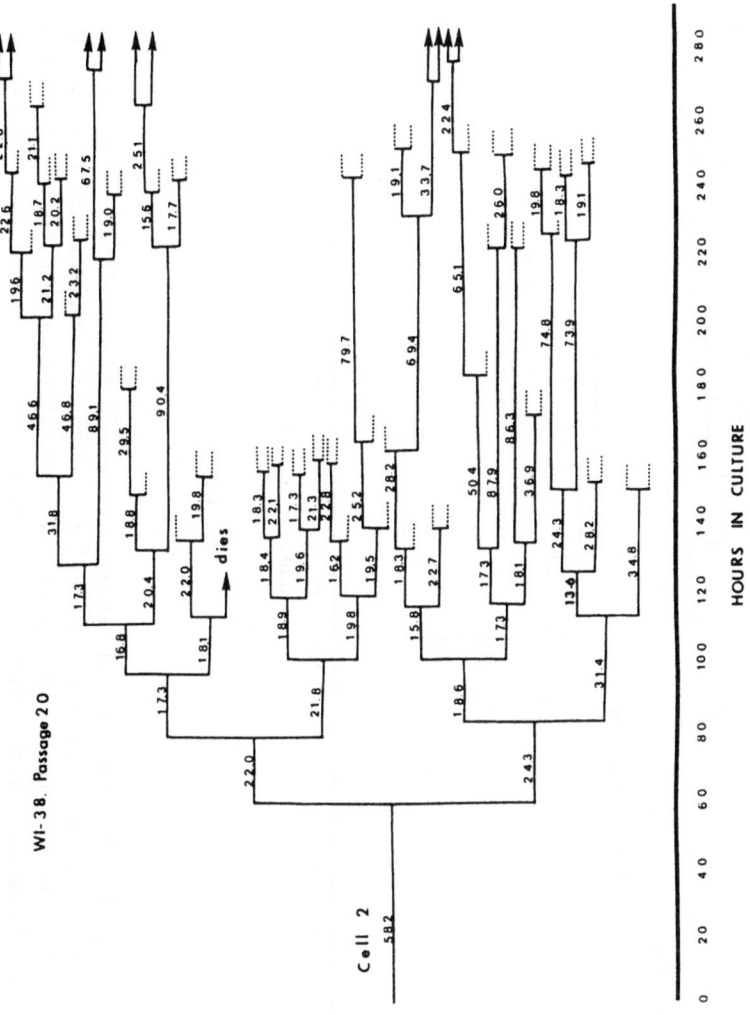

Fig. 4B. (caption on p. 98)

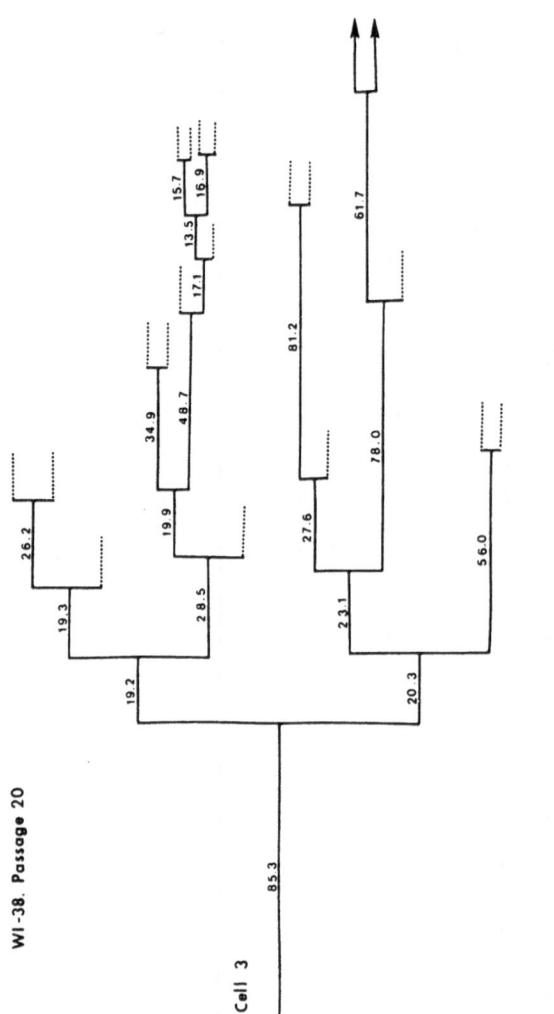

Fig. 4C.

Fig. 4. Genealogies of passage 20 WI-38 clones (PDL = 21); (A)
cell #1, (B) cell #2, (C) cell #3. Numbers indicate interdivision
time in hr. Lines ending in arrows indicate cells followed to end
of filming without dividing. Dashed lines indicate cells that
could not be followed beyond that point. Culture was fed fresh
medium at 192 hr.

Compared to the passage 28 clone, the passage 20 (PDL = 21) and 53 (PDL = 65) clones are more heterogeneous in regard to IDT and division pattersn (Figs. 3 and 4). Division patterns tend to be less synchronous in these clones than in the passage 28 clone. Mother-daughter, as well as sister-sister relationships, are more variable in the passage 20 and 53 clones than in the passage 28 clone. IDT of many cells in the passage 20 and 53 clones are considerably longer than those observed in the passage 28 clone.

Genealogies of clones of passage 20 cells are shown in Figures 4a, b, and c. This culture started with three cells which gave rise to clones. In addition, present from the beginning was a large cell appearing somewhat similar to a macrophage in morphology, which did not divide through the 12 days of filming. In each of the three clones there is a somewhat asynchronous division pattern with variable IDT between mother-daughter and sister-sister pairs. Heterogeneity is also noted between the three clones in the culture. Cells 1 and 2 (Fig. 4a and b) behaved similarly in terms of IDT and division pattern, while cell 3 (Fig. 4c) generally had longer IDT and a less synchronous division pattern. The asynchrony of the cell 3 clone appears similar to that observed in the passage 53 clone (Fig. 3).

As noted in the passage 28 clone, average IDT increases with successive generations (Fig. 2). This pattern is also evident in the passage 20, and 53 clones. In the passage 53 clone (Fig. 3), IDT of several daughter cells increased dramatically in the fourth generation. In the fifth and sixth generations of the passage 53 clone, however, approximately 25% of the daughter cells exhibit an IDT that is lower than that of the mother cell. This same phenomena occurs in the fifth and sixth generations of the passage 28 clone (Fig. 1), but the decreases in IDT are not as great as in the passage 53 clone.

The same pattern of increasing, then decreasing IDT may also be observed in the passage 20 clones (Figs. 4a, b and c). Striking increases in IDT, averaging approximately 80 hr, were noted in the fourth and fifth generations. This particular culture was fed with a one-half change of medium at 192 hr (8 days). Approximately 24 hr later there was a burst of division giving rise to 2 or 3 more generations from some of the cells with IDT comparable to those of the first 3 to 4 generations. The passage 28 and passage 53 clones were not fed during the filming time of 130 hr and 240 hr respectively. Thus, in these 2 clones, reasons for decreased IDT in the fifth and sixth generations are unclear. The passage 20 clone was seeded at a density of 2.5×10^4 cells/8 ml while both the passage 28 and 53 clones were seeded at 1.6×10^4 cells/8 ml. The prolonged IDT in the fifth and sixth generations of the passage 20 clones may have resulted from depletion of essential nutrients or concen-

tration of toxic products in the medium since the seeding density
was higher in this culture. Refeeding would supply fresh nutrients
and dilute any toxic substances present, thus resulting in a shifting
back to shorter IDT.

In addition to heterogeneity in division patterns and IDT in
the various clones studied, we have found variations in migration
patterns and migration rates. Any relationship of migration to
cell division behaviour is unclear at this time. We have observed
cultures in which the progeny cells are actively motile with
short IDT, and other cultures of actively motile cells and longer
IDT. Migration rates were meaured from tracings of cell's migration between divisions. Distance travelled between division was
measured in arbitrary units using a map meter, and migration rate
was expressed as units travelled per hr. The average migration
rate for the first three generations of each clone were: 0.84 for
the passage 28 clone; 0.45 for the passage 53 clone; 0.22, 0.24,
and 0.59 for cells 1, 2 and 3 of the passage 20 clones. We see no
unique relationships between migration behaviour and cell division
patterns. However, it is interesting that the three clones of
passage 20 behaved differently both in cell division patterns and
in migratory activity.

DISCUSSION

Assessment of cell division patterns of human diploid fibroblasts during their in vitro lifespan has been facilitated by the
technique of time-lapse cinemicrophotography. Genealogies of WI-38
embryonic lung fibroblasts at selected population doubling levels
have been obtained using this technique. The technique has the
advantage of yielding accurate quantitation of interdivision time,
cell attachment, cell death, and cell migration behaviour. Of less
certainty is assessment of those cells in the clones which appear
to be non-dividers. These cells may be arrested at some stage in
their cycle and repassaging the clones might trigger them into the
dividing cycle.

Our studies indicate heterogeneity in proliferative capacity
and interdivision time of clones of diploid fibroblasts. The
genealogies reported here are felt to be representative of the
more highly proliferating cells in populations of WI-38 fibroblasts at selected passage levels. Heterogeneity of clonal distribution of WI-38 cells has been noted by Merz and Ross (10) and
Smith and Hayflick (6). It has been suggested that populations
of WI-38 fibroblasts at any given passage level are mixtures of
cells with varying proliferative capacities. Thus, cells capable
of establishing large or small clone sizes would be present
throughout the lifespan of the cultured cells. Bimodal distri-

butions of clones of human skin fibroblasts have been recently reported by Martin et al.(11), thus illustrating the heterogeneity
in proliferative capacities of these cells. These authors also
found a bimodal distribution in subclones of the original clones.

    Because of the heterogeneity of the WI-38 cells, it is dif-
ficult at this time to assess the relationship of cell division
patterns to age of the culture. The very nature of the experiments
limits the amount of data that can be gathered and analyzed. As
can be seen in the genealogies in Figures 1, 3 and 4, there is
considerable overlap in interdivision time of progeny cells of
each of these clones. We have not found any direct relationship
between passage level and interdivision time per se. The passage
28 clone was a rapidly dividing culture, exhibiting a synchronous
division pattern with all cells dividing through four generations.
The passage 20 and 53 clones exhibited more variation in interdivision time and less synchrony than the passage 28 clones. Further, within the same culture, the 3 passage 20 clones also exhibited heterogeneity in cell division patterns and interdivision
time. The passage 53 clone appeared to exhibit a more heterogeneous cell division pattern than the passage 20 and 28 clones,
although one clone of the passage 20 culture (cell #3) had a
division pattern similar to that of the passage 53 clone. Heterogeneity in migration activity has also been observed in different
clones. Gradual lengthening of average interdivision time with
successive generations in the clones has been observed in all
genealogies regardless of passage level of the donor culture. The
reasons for this lengthening of the cell cycle are not clear. Depletion of nutrients and space, and accumulation of inhibitory
substances might lead to a lengthening of interdivision time. In
the early generations of the clones, there would be an adequate
supply of space and nutrients; thus lengthening of the cell cycle
possibly involved other factors which do not inhibit mitosis but
do affect interdivision time.

    The duration of the cell cycle could be regulated in a num-
ber of ways: concentration of essential nutrients within a cell,
production of intermediary substances necessary for DNA synthesis
or other essential metabolic steps, or production and elaboration
of either mitotic stimulators, or inhibitors. Any or all of these
regulatory mechanisms could be operative throughout the culture
period. Further, one would not necessarily expect all cells within
any generation in the clone to be affected equally. In the genealogies in Figures 1, 3, and 4, it can be seen that there are variations in interdivision time within a given generation. For
example, in the passage 28 clone one daughter cell of the fifth
generation had an interdivision time of 12.9 hr while its sister
had an interdivision time of 25.2 hr. These relationships are
more striking in the passage 20 and 53 clones. The possible

mechanisms for cell cycle regulation mentioned above may also
account for the shortening of interdivision time of a portion of
cells which was observed in the fifth and sixth generations of
the passage 28 and 53 clones. In these clones, the shortened
interdividion times were evident for a single generation, after
which the cell cycle lengthened once more. In the case of the
passage 20 clones it seems that refeeding stimulated the shortening of interdivision times, since all cells followed exhibited
a decreased interdivision time and the IDT of successive generations
was also low. The average interdivision times of the 2 to 3 generations arising after feeding were comparable to those of the
first few generations. The cell cycle variations seen in our
studies may be expressions of intracellular regulation (or regulation between cells) which are modulated by environmental changes
in the medium. In a recent publication, Burton and Canham (12)
suggested that there might be an exchange of molecules between
cells when cell-cell contact occurs, which could yield either an
inhibitory or stimulatory effect on the division cycle.

     In assessing cell division patterns of human diploid fibro-
blasts in relation to population age, one must consider the occurrence of "non-dividing" cells arising in the clones. Reports in
the literature have suggested that the proportion of non-dividing
cells in the population increases with age, thus accounting for
decline in proliferative capacity of the population (3,4,5).
This concept has recently been challenged by Maciera-Coelho (13)
who suggest that "increased transit time and heterogeneity of cell
cycles"rather than an accumulation of non-cycling cells occurs in
senescent fibroblast cultures. Using incorporation of 3H thymidine
as a marker he found that when human embryonic fibroblast cells
were continuously labelled and repassaged at 11 days, the percentage
of labelled nuclei increases from 66% at 11 days to 92% 14 days
later, thus suggesting that a large proportion of cells not originally labelled were capable of entry into a cycling state. In our
type of experiments it is difficult to determine if a cell has
become a "non-divider" because the clones are not repassaged.
In the passage 53 clones, only 1 daughter cell of the 3rd generation might be called a potential non-divider. Whether this cell
would have divided after feeding or repassaging is not known. Our
data appear to support that of Maciera-Coelho et al.(7) in that
late passage cells show an increase in time for transversing the
cell cycle and in heterogeneity of cell cycles. In the passage
20 culture, however, there was one cell present at the beginning
of filming which did not divide through 12 days in cultures. This
was a large macrophage-like cell similar in description to the
"terminally differentiating" cells in human skin fibroblast clones
reported by Martin et al. (11). These authors suggested that
cells of clones which either did not divide, or were capable of
only a few divisions represented terminally differentiating cells.

These cells were larger and more highly phagocytic than cells of the more highly proliferating clones.

It appears from our data and that of others (6,7,8,10,11) that human diploid fibroblast populations are mixtures of cells of variable morphology, migratory activity and proliferative capacity and this holds true throughout the in vitro lifespan, albeit more variability is noted in late passage cells.

SUMMARY

Genealogies of human diploid embryonic lung fibroblasts, WI-38 were prepare from analysis of filmed sequences of clones at passages 20, 28 and 53. The results indicate heterogeneity in cell division patterns, interdivision time and migration activity. The relationship of the cell division patterns to age of culture is difficult to assess at this time because of the heterogeneity of the clones, however, the late passage culture appeared to be more variable in terms of sister-sister, and mother-daughter relationships. The passage 28 culture was representative of a highly proliferating clone, exhibiting short interdivision times and a synchronous division pattern. The passage 20 and 53 clones exhibited longer interdivision times and a less synchronous division pattern than the passage 28 clone. A gradual lengthening of average interdivision time with successive generations has been observed in all genealogies regardless of passage level of the donor culture. A portion of daughter cells in the fifth and sixth generation exhibited lower interdivision time than the mother cell. The effects of nutrients, space, and mitotic inhibitors or stimulators on interdivision time of the cells within the clones is discussed.

ACKNOWLEDGEMENTS

The authors are grateful to Rosemary Downs, Shelly Henderson, and Dr. Paula Fives-Taylor for assistance collecting film data and establishing the genealogic trees.

This work was supported by National Institutes of Health Contract NICHD-72-2755.

REFERENCES

1. Hayflick, L. and Moorhead, P.S. 1961. The serial cultivation of human diploid cell strains. Exp. Cell Res. 25: 585.

2. Hayflick, L. 1965. The limited in vitro lifetime of human
       diploid cell strains. Exp. Cell Res. 37: 614.

3. Merz, G.S., Jr. and Ross, J.D. 1969. Viability of human dip-
       loid cells as a function of in vitro age. J. Cell. Physiol.
       74: 219.

4. Orgel, L.E. 1973. Ageing of clones of mammalian cells.
       Nature 243: 441.

5. Gelfant, S. and Smith, J.G., Jr. 1972. Aging: noncycling
       cells an explanation. Science 178: 357.

6. Smith, J.R. and Hayflick, L. Variation in the lifespan of
       clones derived from human diploid fibroblast cultures. J.
       Cell Biol., in press.

7. Maciera-Coehlo, A., Ponten, J. and Philipson, L. 1966.
       The division cycle and RNA-synthesis in diploid human cells
       at different passage levels in vitro. Exp. Cell Res. 42:
       673.

8. Cristofalo, V.J. and Sharf, B.B. 1973. Cellular senescence
       and DNA synthesis. Exp. Cell Res. 76: 419.

9. Norrby, K. 1970. Population kinetics of normal, transforming
       and neoplastic cell lines. Acta Path. Microbiol. Scand.
       78, Suppl. 214: 3.

10. Merz, G.S. and Ross, J.D. 1973. Clone size mariation in the
       human diploid cell strain, WI-38. J. Cell. Physiol. 82: 75.

11. Martin, G.M., Sprague, C.A., Norwood, T.H. and Pendergrass, W.R.
       1974. Clonal selection, attenuation and differentiation in
       an in vitro model of hyperplasia. Amer. J. Pathol. 74: 137.

12. Burton, A.C. and Canham, P.B. 1973. The behaviour of coupled
       biochemcial oscillators as a model of contact inhibition
       of cellular division. J. Theor. Biol. 39: 555.

13. Maciera-Coehlo, A. 1974. Are non-dividing cells present in
       ageing cell cultures? Nature 248: 421.

QUESTIONS TO DR. ABSHER:

Dr. Hay: Would it be possible to correlate differences in cell size
with heterogeneity in division times? I am wondering if cells,
within "younger" populations, which are smaller in size, may be shown

to divide more or less rapidly than larger cells of the same age group.

Dr. Absher: We have not found correlations in cell size with division times, except for those very large macrophage-like cells we occasionally see and which do not divide. For example many cells of the passage 28 clone which have short interdivision times were larger than the same cells of the passage 53 clone which had-longer inter- division times were larger than the same cells of the passage 53 clone which had longer interdivision times.

Dr. Martin: How often is there a shift inmmorphology from large to small cells?

Dr. Absher: We have seen this shift from larger to smaller cells but we have not yet quantitatived the frequency of occurence of such shifts in cell size.

CHARACTERISTICS OF PROLIFERATIVE CELLS FROM YOUNG, OLD, AND TRANSFORMED WI 38 CULTURES

Phillip D. Bowman and Charles W. Daniel

Dept. of Biology, Division of Natural Sciences

University of California, Santa Cruz, CA 95064

An important, perhaps central event in the aging of human fibroblasts in vitro is a decline in proliferative capacity exhibited by cell populations as they are carried in serial passage (1-3). In characterizing this aging process at the cell, rather than at the population level, Cristofalo and Sharf (4) have reported that there is a progressive loss by individual cells of the capacity to synthesize DNA and to complete the division cycle. Changes in size of this non-dividing subpopulation may be monitored by auto-radiographic techniques, and these data have predictive value in assessing the amount of potential lifespan completed by a particular cell population during its sojourn in culture.

Other changes are associated with this in vitro aging process, such as increase in cell size (5-8) and ultrastructural alterations (9-13). It is the purpose of this paper to present evidence indicating that these structural changes are mainly associated with the nonreplicating subpopulation of cells, whereas the mitotically active cells, whether from young or old cultures, are phenotypically similar.

RESULTS AND PROCEDURES

Cell Size and DNA Synthetic Ability

Early passage WI 38 and a SV-40 virus transformed variant, VA 13A (hereafter called T/WI 38) were grown from starter cultures obtained from Dr. L. Hayflick. Culture conditions were based upon those detailed by Cristofalo (4), but with modifications for growth

107

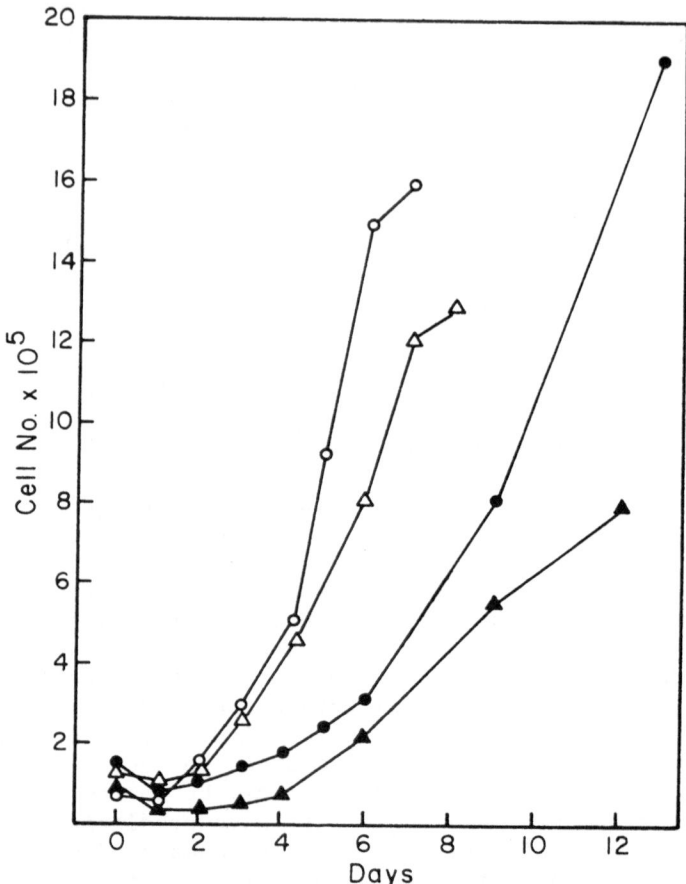

Fig. 1.   Growth curves obtained from cultures grown in 25 cm^2 Falcon flasks.   ● = T/WI 38, o = 10% LSC (P20), △ = 45% LSC (P36), ▲ = 80% LSC (P58).   With permission of Academic Press.

with Hepes buffer in the absence of a 5% CO_2 environment.   Cells were routinely fed every three days, and with this regime the pH never fell below 7.0.   Samples of the cell layer were periodically examined for mycoplasma autoradiographically (14) and by electron microscopy.   These cultural conditions permitted cells to be routinely carried into the mid 60's passage level before entering phase III.   Details of these methods are available elsewhere (15).

Culture age is reported as percent lifespan completed (LSC), determined by the method of Cristofalo and Sharf (4), and accompanied by the conventional passage number.   T/WI 38 cells, which do not age, are characterized by time in culture.   Size distribution

a. 35% LSC (P30): CELL LAYER

WINDOW NO	% CELLS	
1	0	
2	0	
3	0	
4	2.7	XXX
5	3.9	XXXX
6	9.1	XXXXXXXXX
7	13.1	XXXXXXXXXXXX
8	13.1	XXXXXXXXXXXX
9	13.1	XXXXXXXXXXXX
10	13.4	XXXXXXXXXXXXX
11	9.4	XXXXXXXXX
12	3.0	XXX
13	5.8	XXXXXX
14	3.0	XXX
15	2.1	XX
16	2.2	XX
17	1.2	XX
18	1.2	X
19	0.9	X
20	0.6	X
21	0.9	X
22	0.6	X
23	0	
24	0	
25	0	
MEAN	9.7	2440 µ³
STD DEV	3.6	
SKEWNESS	1.1	
KURTOSIS	1.3	

b. 35% LSC (P30): MITOTIC CELLS

WINDOW NO	% CELLS	
1	0	
2	0	
3	0	
4	0	
5	0	
6	0	
7	0	
8	0	
9	0	
10	6.6	XXXXX
11	8.6	XXXXXXX
12	9.9	XXXXXXXX
13	11.3	XXXXXXXXXX
14	8.6	XXXXXXX
15	10.6	XXXXXXXXX
16	10.6	XXXXXXXXX
17	9.3	XXXXXXXX
18	7.9	XXXXXX
19	6.0	XXXXX
20	4.0	XXXX
21	3.3	XXX
22	2.0	XX
23	1.3	X
24	0	
25	0	
MEAN	15.1	3990 µ³
STD DEV	3.3	
SKEWNESS	0.3	
KURTOSIS	-0.7	

Fig. 2. Size distribution data from 35% LSC (P30) WI 38 cells, facsimile of computer printout. Each window corresponds to a volume of 266 µ³. With permission of Academic Press.

a. 80% LSC (P55): CELL LAYER

WINDOW NO	% CELLS	
1	0	
2	0	
3	0	
4	0	
5	12.1	XXXXXXXXXXX
6	6.4	XXXXXX
7	5.5	XXXXX
8	3.2	XXX
9	3.8	XXXX
10	4.3	XXXX
11	4.1	XXXX
12	4.3	XXXX
13	5.1	XXXX
14	4.2	XXXX
15	5.3	XXXXX
16	4.5	XXXXX
17	4.3	XXXX
18	4.1	XXXX
19	4.1	XXXX
20	4.9	XXXXX
21	4.5	XXXXX
22	3.2	XXX
23	3.8	XXXX
24	3.8	XXXX
25	3.2	XXX
MEAN	13.7	
STD DEV	6.2	
SKEWNESS	0.1	
KURTOSIS	-1.2	

5550 μ^3

b. 80% LSC (P55): MITOTIC CELLS

WINDOW NO	% CELLS	
1	0	
2	0	
3	0	
4	0	
5	0	
6	0	
7	0	
8	16.2	XXXXXXXXXXXXXXXX
9	16.2	XXXXXXXXXXXXXXXX
10	2.7	XXX
11	6.8	XXXXXXX
12	12.2	XXXXXXXXXXXX
13	8.8	XXXXXXXXX
14	5.4	XXXXX
15	7.4	XXXXXXX
16	5.4	XXXXX
17	6.1	XXXXXX
18	4.7	XXXXX
19	5.4	XXXXX
20	2.7	XXX
21	0	
22	0	
23	0	
24	0	
25	0	
MEAN	12.6	
STD DEV	3.7	
SKEWNESS	0.4	
KURTOSIS	-1.0	

3900 μ^3

Fig. 3. Size distribution data from 80% LSC (P55) WI 38 cells, facsimile of computer printout. Each window corresponds to a volume of 266 μ^3. With permission of Academic Press.

analysis was performed using the Coulter Counter Model B, with
plotter, and using the 100 μ aperture calibrated with ragweed
pollen.

Growth curves for early, intermediate, and late passage WI 38,
as well as T/WI 38, are presented in Figure 1. Reduced saturation
densities at higher passage levels are perhaps due to the increased
size of individual cells. The very slow initial growth of T/WI 38
results from an extended interdivision time, which we find to be
characteristic of this cell line. T/WI 38 cells are capable,
however, of reaching high saturation densities.

In Figures 2-4, the cell size frequency distribution of both
mitotic and monolayer cells are shown. Mitotic cells were obtained
by a manual shake-off procedure similar to that of Klevecz and Kapp
(16), with which as many as 1.5×10^5 cells were obtained from a
vigorously growing culture in a 690 cm^2 roller bottle. Microscopic
examination of the shake-off cells after fixation and Giemsa
staining indicated that 95% were mitotic. In Figures 2a and 3a it
is apparent that in the cell layer the mean volume increases in
late passage normal cells, and in addition the distribution becomes
conspicuously skewed. By altering the aperture current and ampli-
fication settings, very large cells, up to 15,000 $μ^3$, could be
detected in these late passage cultures (not shown). T/WI 38 cells
display volume characteristics similar to early passage, 35% LSC
normal cells (Fig. 4a). Mitotic cells, whether from young, old, or
transformed cultures, have similar mean volumes (Figs. 2b, 3b and
4b). Mitotic cells from 80% LSC cultures (Fig. 3b) characteristically displayed an irregular frequency distribution for reasons
that are not understood.

Results obtained electronically were confirmed by optical
means. Ocular micrometer measurements indicated a mean cell volume
for 10% LSC (P20) mitotic cells of 4312 $μ^3$, and for 60% LSC (P44)
mitotic cells a mean molume of 3960 $μ^3$.

In addition to making point comparisons between early and late
passage cells, cumulative data was obtained by following cultures
over a six month period. By comparing mean cell volumes of mitotic
cells with monolayer cells at several points in their lifespan
(Fig. 5), it is shown that only in the cell layer is there a shift
toward increased cell volume, that is roughly proportional to percent lifespan completed. Mitotically shaken off cells maintained
a relatively stable volume throughout their lifespan. In the case
of T/WI 38 cells a different relationship between mitotic and
monolayer cells was obtained, in which the volume distributions
were constant and parallel (Fig. 6).

a. T/WI 38: CELL LAYER

WINDOW NO	% CELLS	
1	0	
2	0	
3	0	
4	2.1	XX
5	2.4	XX
6	4.8	XXXXX
7	7.2	XXXXXXX
8	11.0	XXXXXXXXXXX
9	12.0	XXXXXXXXXXXX
10	10.1	XXXXXXXXXX
11	9.6	XXXXXXXXX
12	8.1	XXXXXXXX
13	7.3	XXXXXXX
14	5.4	XXXXX
15	3.7	XXXX
16	4.2	XXXX
17	2.9	XXX
18	1.9	XX
19	2.1	XX
20	1.1	X
21	1.0	X
22	1.3	X
23	0.8	X
24	0.6	X
25	0.5	
MEAN	11.3	
STD DEV	4.3	
SKEWNESS	0.9	
KURTOSIS	0.6	

2843 μ^3

b. T/38: MITOTIC CELLS

WINDOW NO	% CELLS	
1	0	
2	0	
3	0	
4	0	
5	0	
6	0	
7	0	
8	0	
9	15.6	XXXXXXXXXXXXXXX
10	16.1	XXXXXXXXXXXXXXXX
11	10.7	XXXXXXXXXX
12	10.2	XXXXXXXXXX
13	8.8	XXXXXXXX
14	6.8	XXXXXX
15	6.8	XXXXXX
16	6.3	XXXXXX
17	6.3	XXXXXX
18	7.3	XXXXXXX
19	4.9	XXXXX
20	0	
21	0	
22	0	
23	0	
24	0	
25	0	
MEAN	12.9	
STD DEV	3.2	
SKEWNESS	0.5	
KURTOSIS	-1.1	

3221 μ^3

Fig. 4. Size distribution data from T/WI 38 cells, facsimile of computer printout. Each window corresponds to a volume of 266 μ^3. With permission of Academic Press.

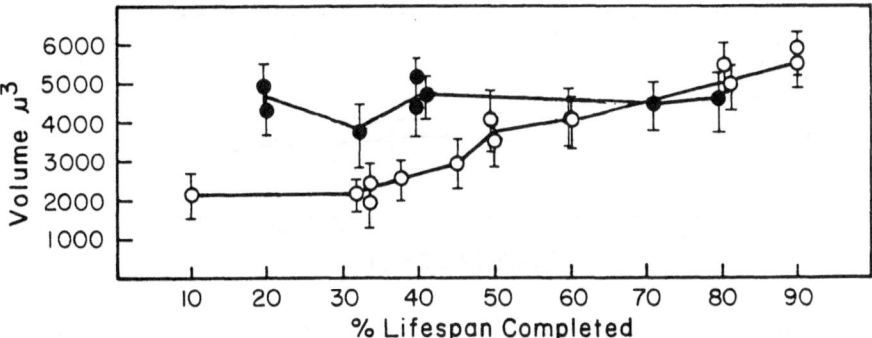

Fig. 5. Mean cell volumes throughout the in vitro lifespan of
WI 38 cells. Measurements made on different substrains are
presented individually, rather than averaged. o = cell layer,
● = mitotic cells. With permission of Academic Press.

These data suggest that mitotic cells, regardless of passage
level, have generally similar mean volumes of approximately 4000
μ^3, and they never exceed 5200 $\mu3$. In contrast, the mean cell
volume in an 80% LSC culture, including the cell layer, is 5500 μ^3
(Figs. 2 and 3). This suggests that increased cell volume associated with cell aging is an event which occurs after cells have
become nondividers. The experimental production of nondividing cells
in young and transformed cultures should, therefore, lead to a detectable increase in mean cell volume.

Nondividing cells were produced by adding the bifunctional
alkylating agent, Mitomycin C, which inhibits DNA synthesis but
has little immediate effect on RNA and protein synthesis (17,18).
Size distributions made five days after treatment indicate that,
under these conditions, a conspicuous increase in size occurs
following cessation of DNA synthesis (Fig. 7).

Ultrastructure of Dividing Cells

Mitotic cells were pelleted, fixed in Karnovsky's fluid, post-
fixed in osmium tetroxide, stained in block with uranyl acetate,
dehydrated, and embedded in Epon-Araldite. Details of these procedures are available (19).

Sections of cells displaying clearly identifiable chromosomes
were photographed, and at least 50 cells from early, late passage,
and transformed cultures were examined. Representative micrographs
of the three classes of mitotic cells are shown in Figures 8 and 9.

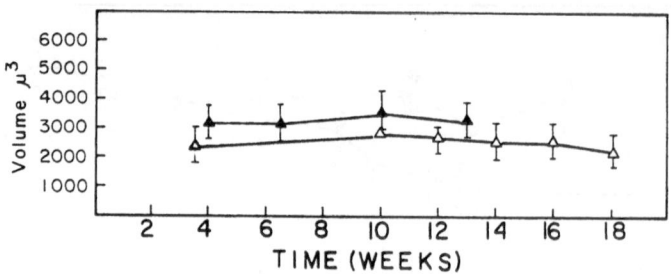

Fig. 6. Mean cell volumes of T/WI 38 cells during serial passage.
Because these cells maintain a stable labeling index, volume is
plotted against time in culture. Δ = cell layer, $\blacktriangle$ = mitotic
cells. With permission of Academic Press.

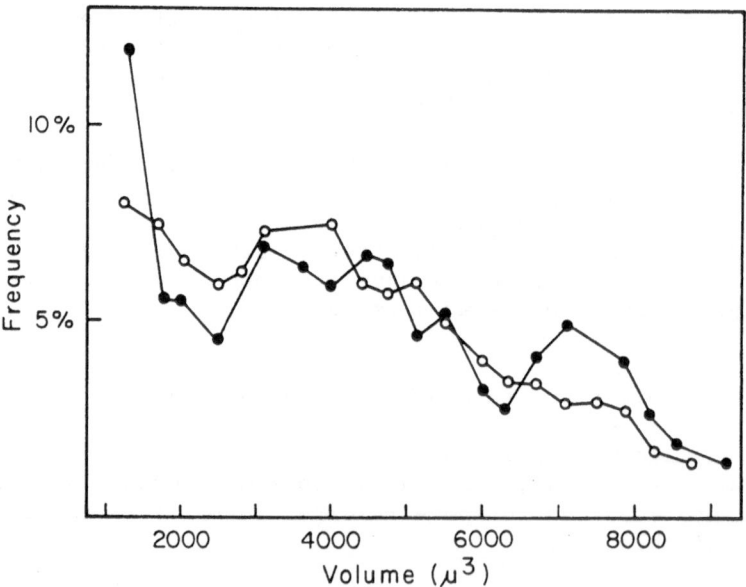

Fig. 7. Volume changes recorded five days after a 30 min exposure
to 25 µg/ml Mitomycin C at 37°C. o = 10% LSC (P20), ● = T/WI 38.
With permission of Academic Press.

The cells appear qualitatively similar in regard to cytoplasmic
texture and distribution of organelles. Because changes in the
number of lysosomes has been reported to increase during in vitro
aging (9,12) the number of lysosomes and mitochondria were quanti-
tated by two independent observers working without knowledge of cell
types.

The results of these counts are given in Table I. In normal
WI 38 cells at various passage levels the average number of mitochondria is very similar, with the exception of 85% LSC, in which
the number is somewhat reduced. This failure to find a consistent
change in mitochondrial numbers during aging is in agreement with
studies of Lipetz and Cristofalo (11), who reported similar numbers
of mitochondria in young and old fibroblasts. Mitochondria were
found with similar frequency in T/WI 38.

Similarly, no consistent pattern was found in the distribution
of lysosomes with increasing passage level. Only in 71% LSC cells
was a significant difference obtained, where the numbers were
reduced. Lysosomes in monolayer cells, in contrast, are reported
to increase in number during in vitro aging (9,11-13). T/WI 38
cells displayed fewer lysosome-like bodies than did normal cells.

TABLE I.   Distribution of Mitochondria and Lysosome-Like Bodies
           In WI 38 and T/WI 38 Cells

Cell Type	No. cells examined	Mean/100 μ^2 ± S.E.	
		Mitochondria	Lysosome-like bodies
20% LSC (P25)	21	21.8 ± 5.6	11.6 ± 6
37% LSC (P31)	32	17.1 ± 8.1	10.8 ± 5.3
71% LSC (P42)	28	18.4 ± 7.0	6.8 ± 5**
85% LSC (P51)	15	14.9 ± 6.3*	10.6 ± 2.8
Total WI 38	96	18.1 ± 7.4	9.1 ± 5.0
T/WI 38	40	19.3 ± 8.3	6.7 ± 4.2†

*   Differs from 20% LSC (P < 0.01) and 77% LSC (P < 0.05).

**   Differs from other WI 38 (P < 0.01).

†   Differs from total WI 38 (P < 0.01).

DISCUSSION

Aging in vitro has generally been described in terms of changes occuring to large populations of cells. Various parameters used to measure this aging process, such as number of population doublings, time to reach confluency, and saturation density, all describe the aggregate behavior of mass cultures. It is apparent from the work of Cristofalo and Sharf (4) that cultures of human fibroblasts consist of two subpopulations, one that loses the ability to incorporate ^{3}H-TdR and one that does not, and there is an age-associated decrease in the latter group--those cells which remain able to replicate DNA and to complete division.

The mitotic shake off technique allows the separation of proliferating cells and provides a means of characterizing that subpopulation. In the present paper we report that mitotic cells do not display the conspicuous aging changes with respect to cell volume and ultrastructure that are associated with Phase III. We therefore suggest that it is useful to consider aging as a two-step process, the first step resulting in the loss of division potential and the second resulting in cell senescence and death. Different mechanisms may be involved in different stages of this aging sequence.

Failure to Initiate DNA Synthesis

Human fibroblasts display a limited replicative lifespan in culture as a consequence of an increasing proportion of noncycling cells. These cells are mainly arrested in the G_1 phase, although a small proportion of G_2 nondividers are also present.

It may be argued that failure to commence DNA synthesis is not an aging process, for G_1 arrests are a basic feature of many metazoan tissues, and cultured cells in confluency are similarly restrained. These noncycling cells associated with aging differ from normally regulated cells in two respects, however. First, complete failure to synthesize DNA as a result of aging is irreversible except by transformation. Second, the incidence of noncycling cells in culture increases with population doublings in human fibroblasts. Nevertheless this aspect of cell aging may be considered as an age-related modification of normal regulatory processes, such that cells become progressively unable to respond to those nutritional or hormonal regulatory factors within the culture vessel

Fig. 8. Electron micrograph of mitotic shake off cells. Darkly staining objects are metaphase chromosomes. a. 20% LSC (P10). b. 71% LSC (42). With permission of Academic Press.

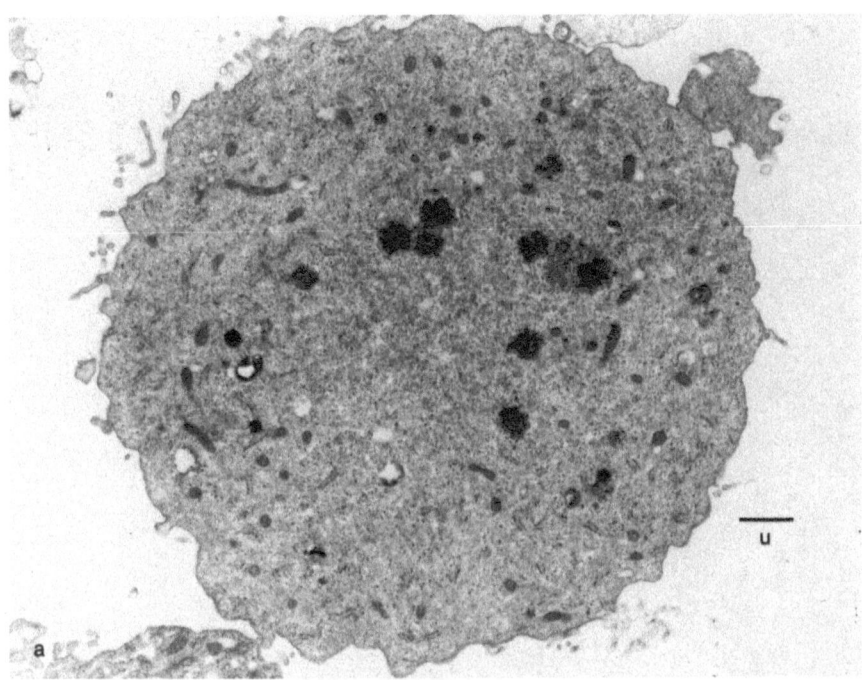

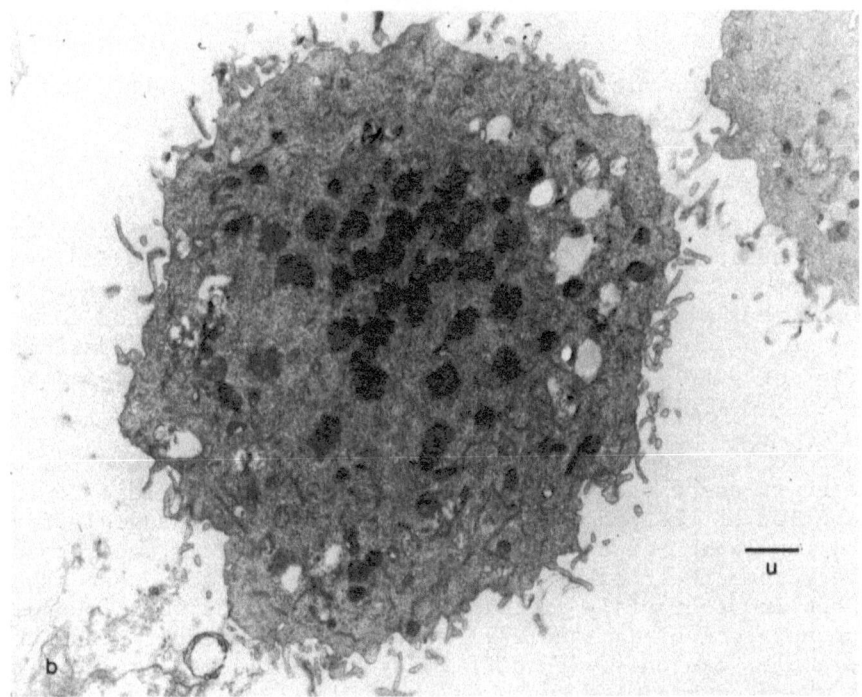

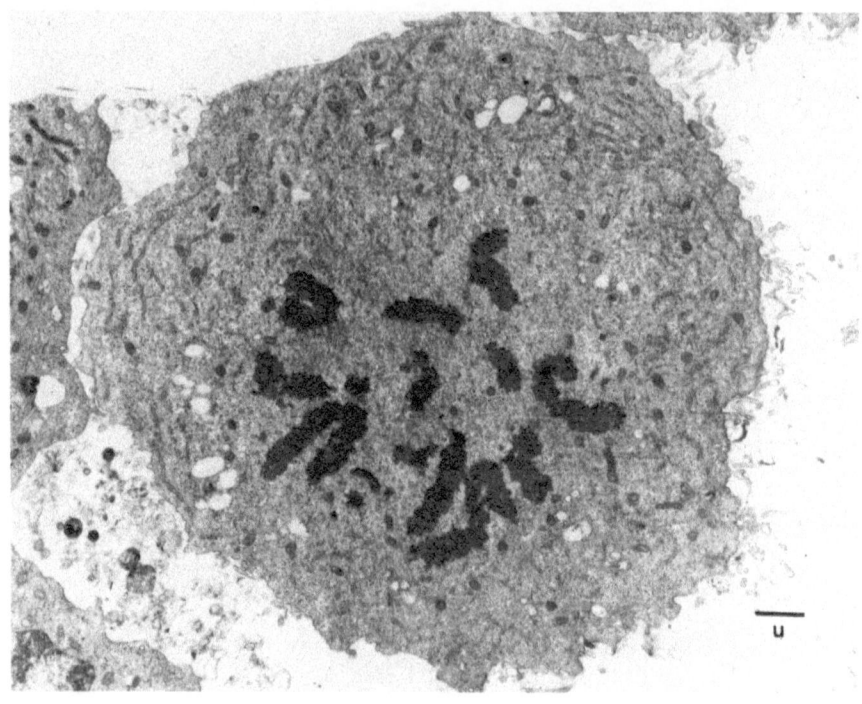

Fig. 9. Electron micrograph of a mitotic T/WI 38 cell. With
permission of Academic Press.

which normally stimulate DNA synthesis. An in vivo analogy is
found in serially transplanted mouse mammary tissue, which progressively loses its ability to synthesize DNA in response to certain
hormones, but which can begin to do so again, if exposed to a different set of endocrine stimuli (15). It would be of considerable
interest to determine if normal, noncycling human fibroblasts could
commence DNA synthesis if implanted in vivo.

     If failure to synthesize DNA is considered as the first obser-
vable event associated with aging it still does not lead to an
explanation of the process, for the mechanisms responsible for
normal regulation of cell divisions are imperfectly understood (20).
However, some theories of aging appear more probable than others
when applied to rapidly dividing cells. Theories which suppose
an accumulation of damaged cell materials as a result of oxidation,
for example, can only with difficulty be applied to proliferating
cells where such damage could be rapidly diluted. Similarly, error
accumulation, either primary or as a result of a positive feedback

loop resulting in error catastrophy (21), is most convincingly
applied to cells after the potential for division has been lost.
Theories which postulate a genetic program, perhaps including
some mechanism which counts the number of cell divisions undergone,
might appear more appropriate. Unfortunately these theories are
too vaguely formulated to provide a useful framework for experimental design.

Cell Senescence

Noncycling Phase III cells in culture may remain alive and
metabolically active for weeks or even months, during which time
a somewhat reduced level of RNA synthesis is maintained and cells
contain normal amounts of protein (20). The generally satisfactory
state of cellular physiology is indicated by experiments of Holland
et al. (22), which show that both old and young WI 38 cells are
capable of producing equal amounts of infective viruses. In the
absence of cell division but in the presence of continued cell metabolism, hypertrophy occurs and cells may attain sizes of 8000 μ^3
or more. This hypertrophy appears to be a direct consequence of
impaired DNA synthesis, and can be produced by any of several
mechanisms which are known to prevent chromosomal replication—aging
during serial passage, treatment with Mitomycin C as discussed herein,
or by irradiation (23).

After maring periods of time cells begin to display evidence
of degenerative senescent changes. Unlike the apparently specific
nature of the changes which impair DNA replication, these are mani-
fold, and include both biochemical alterations and ultrastructural
modifications (9,11-13). This progressive and variable onset of
senescence does not suggest a specific cell program, and mechanisms
involving a clock hypothesis appear unlikely. Instead, the increase
in cell size suggests a state of metabolic imbalance which could be
related to the accumulation of damaged products. The appearance
and continued production of error proteins (21) could also be
expected to lead to generalized disturbances. Such theories would
appear especially tenable if it is shown that a consequence of
these metabolic imbalances is an alteration in the rate of repair
and replacement of intracellular constituents.

SUMMARY

Mitotic cells were obtained by a shake off procedure from cultures of normal WI 38 cells at various passage levels, and from SV-40
virus transformed cells. The size of all mitotic WI 38 cells was
similar regardless of in vitro age, whereas cells from the monolayer
displayed an age-related increase in size. Mitotic transformed

cells were similar to normal in size, but no size changes were observed in transformed monolayer cells during serial passage. Ultrastructural studies of mitotic WI 38 cells revealed no consistent change in the numbers of mitochondria or lysosome-like bodies during aging in culture. Mitotic transformed cells displayed numbers of mitochondria comparable to normal cells, but lysosome-like bodies occurred less frequently.

Size distribution and structural characteristic are presented in relation to the ability of cells to synthesize DNA and to divide. These results support the contention that aging in WI 38 cultures is characterized by a declining fraction of homogeneous, actively dividing cells, and an increasing fraction of heterogenous non-dividers that display senescent changes.

ACKNOWLEDGEMENTS

This work was supported by USPHS Research Grant HD 05875 from the National Institute of Child Health and Human Development. P.B. was supported in part by funds awarded by the Danforth Foundation.

REFERENCES

1. Swim, E. and Parker, R. 1957. Culture characteristics of human fibroblasts propagated serially. American Journal of Hygiene 66: 235.

2. Hayflick, L. and Moorhead, P.S. 1961. The serial cultivation of human diploid cell strains. Exp. Cell Res. 25: 585.

3. Hayflick, L. 1961. The limited in vitro lifetime of human diploid cell strains. Exp. Cell Res. 25: 585.

4. Cristofalo, V.J. and Sharf, B.B. 1973. Cellular senescence and DNA synthesis. Exp. Cell Res. 76: 419.

5. Cristofalo, V.J. and Kritchevsky, D. 1969. Cell size and nucleic acid content in the diploid human cell line WI 38 during aging. Medicina Experimentalis 19(6): 313.

6. Macieira-Coelho, A. and Ponten, J. 1969. Analogy in growth between late passage human embryonic and early passage human adult fibroblasts. J. Cell Biol. 43: 374.

7. Macieira-Coelho, A. 1970. The decreased growth potential in vitro of human fibroblasts of adult origin. In: Aging in Cell and Tissue Culture. E. Holečková and V.J. Cristofalo, eds., Plenum Press, New York, p. 121.

8.   Simons, J. 1969. A theoretical and experimental approach to
     the relationship between cell variability and aging in vitro.
     Aging in Tissue Culture Symposium 19: 25.

9.   Robbins, E., Levine, E. and Eagle, H. 1970. Morphologic
     changes accompanying senescence of cultured human diploid
     cells. J. Exp. Med. 131: 1211.

10. Franks, L.M. and Cooper, T. 1972. The origin of human embryo
     lung cells in culture: A comment on cell differentiation,
     in vitro growth and neoplasia. Inter. J. Cancer 9: 19.

11. Lipetz, J. and Cristofalo, V.J. 1972. Ultrastructural changes
     accompanying the aging of human diploid cells in culture.
     J. Ultrastruc. Res. 39: 43.

12. Brandes, D., Murphy, D., Anton, E. and Barnárd, S. 1972.
     Ultrastructural and cytochemical changes in cultured human
     lung cells. J. Ultrastruc. Res. 39: 465.

13. Brunk, V., Ericsson, J.L.E., Ponten, J. and Westermark, B.
     1973. Residual bodies and "aging" in culture human glia
     cells. Exp. Cell Res. 79: 1.

14. Studzinski, G.P., Gierthy, J.F. and Cholon, J.D. 1973.
     An autoradiographic screening test for mycoplasmal contamina-
     tion of mammalian cell culture. In Vitro 8(6): 466.

15. Bowman, P.D., Meek, R.L. and Daniel, C.W. Aging of human
     fibroblasts in vitro: correlations between DNA synthetic
     ability and cell size. In press.

16. Klevecz, R.R. and Kapp, L.N. 1973. Intermittent DNA synthesis
     and periodic expression of enzyme activity in the cell cycle
     of WI 38. J. Cell Biol. 58: 504.

17. Kuroda, Y. and Furuyama, J. 1963. Physiological and biochem-
     ical studies of effects of mitomycin C on strain HeLa cells
     in culture. J. Cancer Res. 23: 682.

18. Kihlman, B.A. 1966. Actions of Chemicals on Dividing Cells.
     Englewood Cliffs, New Jersey, Prentice Hall, Inc.

19. Bowman, P.D. and Daniel, C.W. The ultrastructure of proli-
     ferating WI 38 cells and its relationship to in vitro aging.
     Exp. Cell Res. (in press).

20. Cristofalo, V.J. 1972. Animal cell cultures as a model system
     for the study of aging. In: Advances in Gerontological
     Research. B.L. Strehler, ed., Academic Press, Vol. 4,
     p. 45.

21. Orgel, L.E. 1973. Ageing of clones of mammalian cells.
     Nature 243: 441.

22. Holland, J.J., Kohne, D. and Doyle, M.V. 1973. Analysis of
     virus replication in aging human fibroblast cultures. Nature
     245: 316.

23. Painter, R.S. 1970. Ultraviolet light on mammalian cells.
     In: Photophysiology V. A.C. Giese, ed.

ALTERATIONS IN CHROMATIN FUNCTIONS DURING AGING IN VITRO

Jon M. Ryan

The Wistar Institute of Anatomy and Biology

Philadelphia, Pennsylvania   19104

Changes in the metabolism of the human diploid cell line WI-38 during the aging process have been extensively characterized. For example, studies have shown age-associated increases in glycogen content (1), lipid synthesis (2), lipid content (3), the number of lysosomes (4,5) and the specific activities of lysosomal enzymes (6-8). In addition, age related decreases in the specific activities of transketolase and 6-phosphogluconate dehydrogenase have also been reported (9). If these age-related changes represent the phenotype of fundamental changes in the expression of the cellular genome, then alterations in chromatin functions could underlie the modulation of the aging process. In this regard, the metabolic alterations mentioned earlier may only reflect functional changes as they occur within the cellular chromatin. However, few reports have been published on age-associated alterations in chromatin functions.

Chromatin that has been isolated from interphase cells has been characterized as being composed of DNA, histone and nonhistone proteins and RNA in the ratio of 0.25:0.71:0.04, respectively (10). Since the histone proteins have been well characterized in regard to their metabolism (11), synthesis (12), post-synthetic chemical modification (13-15) and amino acid composition and sequence (16), they represent well defined proteins in which age-associated changes may be determined and related to other alterations in chromatin functions.

In this report, results are presented showing alterations in the acetylation of the histone proteins, changes in RNA synthesis and the temporal relationship of histone acetylation and RNA synthesis during the aging process of human diploid cells in culture.

MATERIALS AND METHODS

For all experiments described below, cultures of WI-38 cells were grown as previously described (17). Culture age was determined using the autoradiographic technique described by Cristofalo and Sharf (17) as well as by total population doublings. Cultures were checked routinely throughout their lifespan for the presence of mycoplasma using the method described by Levine (18).

To determine the rate of histone acetylation in young and old cells, exponentially growing cultures were pulsed for 48 hr with radioactive sodium acetate under conditions in which the external acetate concentration was not rate limiting (240 µM). The histone protein was isolated using a method similar to that described by Shepherd (19) and incorporation of the labelled acetate into the acetyl form was verified by volatilization of the label at 118 C.

To determine if changes in the rates of histone acetylation and RNA metabolism were closely associated, sister cultures of young cells were grown to confluency and the medium replaced with fresh medium containing 30% serum. To one culture, ^{3}H-sodium acetate at a concentration of 77 µM (spec. act. 20 µCi/µM) was added and the culture allowed to incubate for 3 hr before isolation of the histone protein. To the other culture, uridine-5-^{3}H (spec. act. 25 Ci/mM) was added (2 µCi/ml) and the culture allowed to incubate for 30 min before isolation of the cellular RNA. RNA was isolated according to the method described by Volkin and Cohn (20).

Chromatin was isolated from exponentially growing young and old cells as described by Augenlicht and Baserga (21). The template activity of isolated chromatin using E. coli RNA polymerase (Gen. Biochem., Chagrin Falls, Ohio) was assayed by measuring the rate of ^{14}C-AMP incorporated into RNA (22). Chromatin template activity of exponentially growing young and old cultures using the endogenous RNA polymerase of young and old cells was determined by preparing "nuclear monolayers" and measuring the rate of ^{3}H-UMP incorporated into RNA (23).

Sodium acetate-^{3}H (spec. act. 100 mCi/mM), uridine 5, 6-^{3}H (spec. act. 40 Ci/mM), uridine 5-^{3}H (spec. act. 25 Ci/mM) and the AMP and UMP precursors adenosine 5'-triphosphate 8-^{14}C (spec. act. 50 mCi/mM) and uridine 5'-triphosphate 5, 6-^{3}H (spec. act. 37 Ci/mM) were obtained from New England Nuclear, Boston, Mass.

RESULTS AND DISCUSSION

Our initial studies were concerned with possible age-associated changes in the rate of histone acetylation. The results of this

TABLE I. Incorporation of ^{3}H-acetate into histone of young and old cell populations grown in the presence of 240 μM sodium acetate (24). Histone was isolated as described in Materials and Methods.

Cell Age Population Doublings	N*	^{3}H-acetate Incorporated into Histone (cpm/μg Histone) Mean ± SEM**
19-25	12	21 ± 1.64
45	8	8 ± 1.1

p < .001

*N = number of determinations

**SEM = standard error of the mean

study (Table I) indicated over a twofold reduction in the rate of histone acetylation in old cell populations. To minimize possible effects from differences in the acetate pool size of young and old cells the intracellular acetate pool had 48 hr to equilibrate with the extracellular acetate which was in excess (24).

The reduction in the rate of histone acetylation may be representative of other alterations in chromatin functions (e.g., RNA synthesis). If so, one might expect to observe a temporal relationship between histone acetylation and RNA synthesis as well as a similar reduction in the rate of RNA synthesis. Indeed, studies concerned with alteration in the rate of histone acetylation and RNA metabolism in other systems have shown that these events are closely associated. Pogo and Allfrey (25), for example, have shown that an increase in RNA synthesis in phytohemagglutin-stimulated lymphocytes is preceded by an increase in histone acetylation.

To determine if a change in histone acetylation was closely associated with a change in RNA metabolism in WI-38 cells, confluent nondividing cultures of young cells were refed with fresh medium containing 30% serum and the rate of uridine incorporated into RNA and the rate of histone acetylation were followed over the next 24 hr. The results (Fig. 1) indicate that an increase in the rates of histone acetylation and ^{3}H-uridine incorporated into RNA are temporally related. The increase in the rate of histone acetylation occurred between 6 and 9 hr after addition of fresh medium and increased for an additional 9 hr before leveling off, whereas an increase in the incorporation of uridine into RNA occurred between 9 and 12 hr later and increased throughout the experimental period.

Since a change in the rate of ^{3}H-uridine incorporated into RNA was concomitant with alterations in histone acetylation in young, confluent, nondividing cultures stimulated to synthesize RNA by addition of fresh medium, it was of interest to determine if cellular RNA synthesis in old cultures was reduced similarly to histone acetylation. For these studies, three approaches were used: (1) measurement of the rate of ^{3}H-uridine incorporated into RNA of young and old cells; (2) comparison of the template activity of isolated chromatin from old and young cells using E. coli DNA-dependent RNA polymerase; and (3) measurement of chromatin template activity using the endogenous RNA polymerase of young and old cells.

To measure the rate of incorporation of ^{3}H-uridine into RNA, confluent nondividing young and old cells were stimulated to synthesize RNA by addition of fresh medium containing 30% serum. The

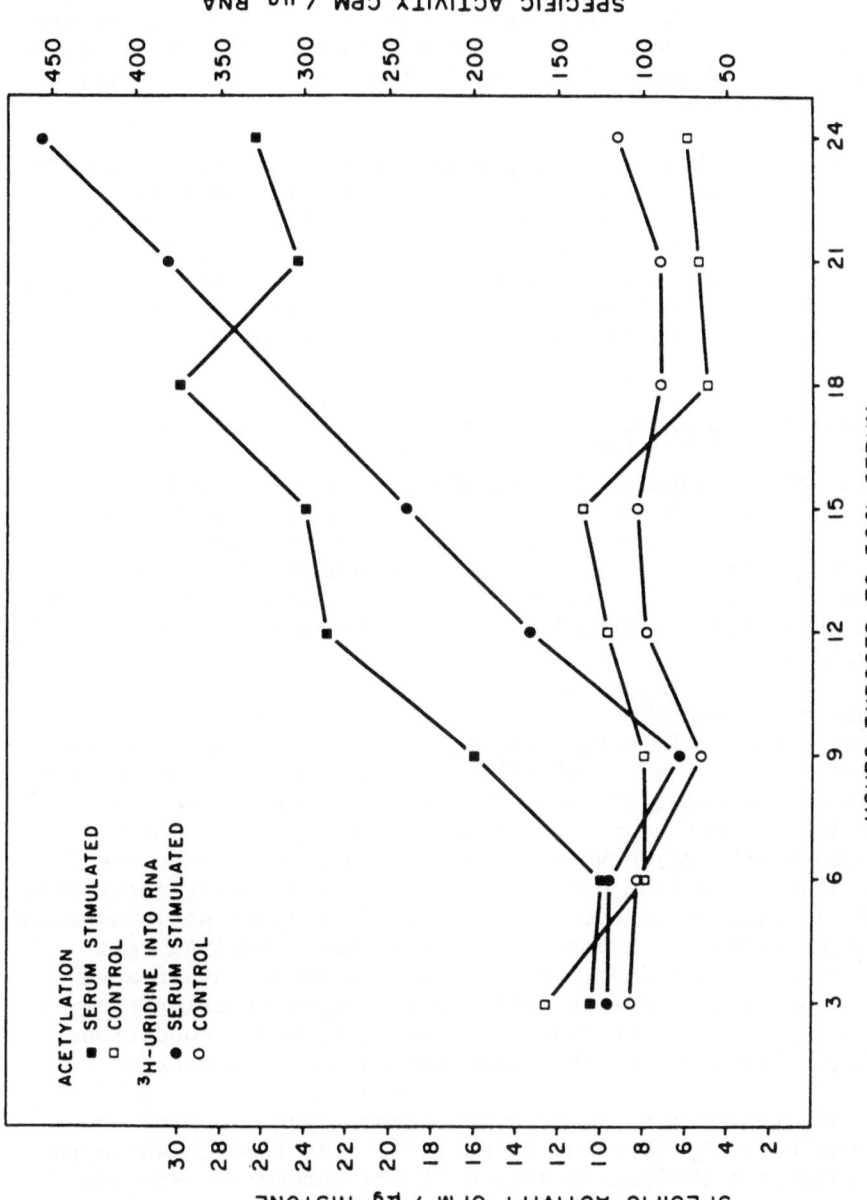

Fig. 1. Incorporation of ^{3}H-uridine into RNA and ^{3}H-acetate into histone of young cell populations following stimulation of confluent, nondividing cultures with fresh medium containing 30% serum. Cellular RNA and histone protein were isolated as described in Materials and Methods.

results (Fig. 2) showed a marked reduction in the rate of uridine incorporated into RNA in old cultures. In addition, since both young and old cultures demonstrated an initial increase in the rate of ^{3}H-uridine incorporated into RNA between 9 and 12 hr after addition of fresh medium, the data would suggest that there is no age related difference in the time necessary to respond to the addition of fresh medium containing 30% serum, but only in the rate in which uridine is incoporated into RNA.

Interpretation of studies comparing the rate of uridine incorporation into the RNA of cells of different ages is complicated however by possible differences in the rates of transport of the labelled precursor as well as possible age-associated differences in the nucleotide pool size. A more direct method of determining if age-associated changes in RNA synthesis do occur is by direct measurement of the chromatin template activity of young and old cells.

The chromatin template activity of aged cells was determined by comparing the rate of ^{14}C-AMP incorporated into RNA using isolated chromatin in the presence of E. coli DNA-dependent RNA polymerase. The results of these experiments (Table II) indicated a threefold reduction in the chromatin template activity of the older cells. These data, which confirmed earlier observations by Srivastava (26), were probably not influenced by differences in RNase activity since it has been shown that no significant age related changes occur in RNase activity. (26).

Although experiments measuring the template activity of isolated chromatin have eliminated possible age-associated differences in nucleotide transport and pool size, a number of reservations have been expressed concerning the use of isolated chromatin to study template activity. For example, Butterworth et al. (27) and Keshgegian and Furth (28) have mentioned the variability in the chromatin preparations from one batch to another and have questioned the biological validity of using bacterial RNA polymerase in assaying the template activity of chromatin isolated from mammalian cells. It was of interest, therefore, to determine if an age-associated reduction in chromatin template activity was apparent using the endogenous RNA polymerase of young and old cells under conditions in which direct isolation of the chromatin was not necessary.

For these experiments, nuclear monolayers were prepared by the method described by Tsai and Green (23). The results shown in Table III indicate a twofold reduction in the chromatin template activity of old cells measured in this way essentially paralleling the data found for isolated chromatin.

The above results give additional support for the validity of an age-associated reduction in the chromatin template activity

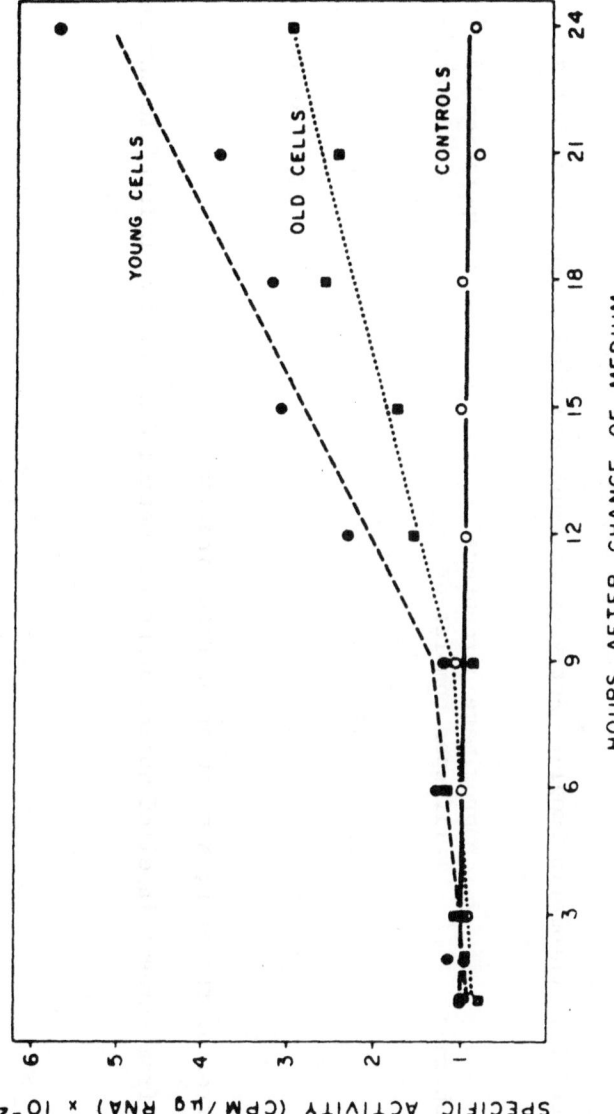

Fig. 2. Incorporation of ^{3}H-uridine into RNA of young and old cells following stimulation of RNA synthesis in confluent monolayers.

TABLE II. Chromatin Template Activity in Young and Middle Aged WI-38 Cells

Cell age Population Doublings	Approximate percentage lifespan completed*	Incorporation of ^{14}C-AMP into RNA cpm/7.49 μg DNA
24	45	686 ± 97 (N = 3)
34	65	235 ± 3 (N = 2)

± Standard Error of the mean; N = number of determinations.

* Determined from percent labelled nuclei using the method described by Cristofalo and Sharf (17).

of WI-38 cells. Use of this "nuclear monolayer" technique (23) cir-
cumvents the objections of possible differences in the integrity of
isolated chromatin from young and old cultures and the use of an
exogenous bacterial RNA polymerase. Of course, as in most experimental procedures, new artifacts are introduced, but these artifacts
are substantially different from those used for the isolation and
purification of chromatin. The consistency between these two methods
suggests that the reduction in chromatin template activity is age-
associated. However, the above evidence is based on the isolation
of chromatin from mass cultures and does not distinguish between
the possible reduced ability to synthesize RNA of the entire culture
(nondividing and dividing cells) or a selective decrease in RNA
synthesis that is confined to a subpopulation of cells (e.g., the
nondividing population). Experiments are currently underway to
distinguish between these two possibilities.

     The above results showing age-associated alterations in histone
acetylation and chromatin template activity suggest that the metabolic alterations which were mentioned earlier (1-9) as partially
characterizing the aging of WI-38 cells in culture may be related
to dynamic changes in the activity of the chromatin components in
the aging population. The nature of the relationship between
reduced functional capacities and the reduced rates of histone
acetylation and RNA synthesis in old cultures may be explained in
several ways. For example, an inability to initiate DNA synthesis
and cell division by some mechanism independent of chromatin
metabolism (e.g., membrane dysfunction) could result in a gradual
decline in the rates of RNA synthesis and histone acetylation.
On the other hand, the decrease in the activity of RNA synthesis
and histone acetylation may prevent initiation of DNA synthesis and
cell division or finally, the gradual reduction in RNA synthesis or
histone acetylation may in turn cause a gradual decline in the
activity of some other chromatin functions which in turn results in
reduced initiation of DNA synthesis and cell division. For example,
although the dramatic decrease in the chromatin template activity of
old cells probably reflects the noncycling state of the cell population (29-31), it is possible that the noncycling state is the result
of the cells' inability to synthesize specific RNA molecules (and
perhaps therefore, specific proteins) needed for the cell to continue
through the cell cycle. This would result in a gradual accumulation
of cells in the early G_1 phase of the cell cycle that would be reflected in changes in cellular metabolism characteristic of G_1
arrested populations: i.e., gradual reduction in the rates of histone acetylation (24), and the incorporation of thymidine into DNA
and a decline in the number of cells undergoing cell division (17).
Resolution of these possible explanations is currently underway.

TABLE III. Chromatin Template Activity in Young and Old Nuclear Monolayers of WI-38 Cells

Cell age Population Doublings	Approximate percentage lifespan completed*	Incorporation of ^{3}H-UMP into DNA cpm/7.94 µg DNA
20-21	45	1435 ± 256 (N = 6)
41-51	95	701 ± 120 (N = 10)

$\pm$ Standard error of the mean; N = number of determinations.

* Determined from percent labelled nuclei using the method described by Cristofalo and Sharf (17).

$p < 0.02$

SUMMARY

Age-associated alterations in the chromatin functions of human
diploid cells have been observed. These alterations include: (1) a
decline in the rate of histone acetylation; (2) a reduction in the
rate of RNA synthesis as measured by (a) the rate of ^{3}H-uridine
incorporated into the RNA of young and old cells; (b) comparison
of the template activity of isolated chromatin from young and old
cells using E. coli RNA polymerase and (c) measurement of chromatin
template activity using the endogenous RNA polymerase of young and
old cells.

It is suggested that the nondividing state of old cells may be
the result of the inability to synthesize specific RNA molecules
(and perhaps specific proteins) necessary for the cell to continue
through the cell cycle.

ACKNOWLEDGEMENTS

The author is grateful to Dr. V.J. Cristofalo for his helpful
suggestions and discussions during this work. The work reported
herein was supported by U.S. Public Service Research Grants HD02721
and HD06323 from the National Institute of Child Health and Human
Development and GM-0042 from General Medical Sciences.

REFERENCES

1. Cristofalo, V.J., Howard, B.V. and Kritchevsky, D. 1970. The
    biochemistry of human cells in culture. In: Organic Biologi-
    cal and Medicinal Chemistry. U. Gallo and L. Santamaria, eds.,
    North Holland Publ., Amsterdam, 2: 95.

2. Rothblat, G., Boyd, R. and Deal, C. 1971. Cholesterol biosyn-
    thesis in WI-38 and WI-38VA13A tissue culture cells. Exp.
    Cell Res. 67: 436.

3. Kritchevsky, D. and Howard, B.V. 1966. The lipids of human
    diploid cell strain WI-38. Ann. Med. Exp. Biol. Fenn 44: 343.

4. Lipetz, J. and Cristofalo, V.J. 1972. Ultrastructural changes
    accompanying the aging of human diploid cells in culture. J.
    Ultrastr. Res. 39: 43.

5. Robbins, E., Levine, E.M. and Eagle, H. 1970. Morphologic
    changes accompanying senescence of cultured human diploid
    cells. J. Exp. Med. 131: 1211.

6.   Cristofalo, V.J., Parris, N. and Kritchevsky, D. 1967. Enzyme
     activity during the growth and aging of human cells in vitro.
     J. Cell Physiol. 69: 263.

7.   Wang, K.M., Rose, N.R., Bartholomew, E.A., Balzer, M., Berde,
     K. and Foldvary, M. 1970. Changes of enzyme activities in
     human diploid cell line WI-38 at various passages. Exp. Cell
     Res. 61: 357.

8.   Milisauskas, V. and Rose, N.R. 1973. Immunochemical quanti-
     tation of enzymes in human diploid cell line WI-38. Exp.
     Cell Res. 81: 279.

9.   Cristofalo, V.J. 1970. Metabolic aspects of aging in diploid
     human cells. In: Aging in Cell and Tissue Culture. E.
     Holeckova and V.J. Cristofalo, eds., Plenum Press, New York,
     p. 83.

10. Comings, D.E. 1972. The structure and function of chromatin.
     Advances in Human Genetics 3: 237.

11. Gurley, L.R. and Hardin, J.M. 1968. The metabolism of histone
     fractions. I. Synthesis of histone fractions during the life
     cycle of mammalian cells. Arch. Biochem. Biophys. 128: 285.

12. Butler, W.B. and Mueller, G.C. 1973. Control of histone syn-
     thesis in HeLa cells. Biochim. Biophys. Acta 294: 481.

13. Shepherd, G.R., Noland, B.J. and Hardin, J.M. 1971. Histone
     phosphorylation in synchronized mammalian cell cultures. Arch.
     Biochem. Biophys. 142: 299.

14. Borun, T.W., Pearson, D. and Paik, W.K. 1972. Studies of
     histone methylation during the HeLa S-3 cell cycle. J. Biol.
     Chem. 247: 4288.

15. Shepherd, G.R., Noland, B.J. and Hardin, J.M. 1971. Histone
     acetylation in synchronized mammalian cell cultures. Biochim.
     Biophys. Acta 228: 544.

16. Rall, S.C. and Cole, R.D. 1971. Amino acid sequence and
     sequence variability of the amino-terminal regions of lysine-
     rich histones. J. Biol. Chem. 246: 7175.

17. Cristofalo, V.J. and Sharf, B.B. 1973. Cellular senescence
     and DNA synthesis: Thymidine incorporation as a measure of
     population age in human diploid cells. Exp. Cell Res. 76:
     419.

18. Levine, E.M. 1972. Mycoplasma contamination of animal cell
    cultures: A simple, rapid detection method. Exp. Cell Res.
    74: 99.

19. Shepherd, G.R., Noland, B.J. and Roberts, C.N. 1970. Phos-
    phorus in histones. Biochim. Biophys. Acta 199: 265.

20. Volkin, E. and Cohn, W.E. 1954. Estimation of nucleic acids.
    In: Methods of Biochemical Analysis 1. D. Glick, ed., Inter-
    science Publishers, p. 287.

21. Augenlicht, L.H. and Baserga, R. 1973. Preparation and
    partial fractionation of nonhistone chromosomal proteins
    from human diploid fibroblasts. Arch. Biochem. Biophys. 158:
    89.

22. Rovera, G., Farber, J. and Baserga, R. 1971. Gene activation
    in WI-38 fibroblasts stimulated to proliferate. Requirement
    for protein synthesis. Proc. Nat. Acad. Sci. 68: 1725.

23. Tsai, R.L. and Green, H. 1973. Rate of RNA synthesis in
    ghost monolayers obtained from fibroblasts preparing for
    division. Nature New Biol. 243: 168.

24. Ryan, J.M. and Cristofalo, V.J. 1972. Histone acetylation
    during aging of human cells in culture. Biochem. Biophys.
    Res. Comm. 48: 735.

25. Pogo, B.G.T., Allfrey, V.G. and Mirsky, A.E. 1966. RNA syn-
    thesis and histone acetylation during the course of gene acti-
    vation in lymphocytes. Proc. Nat. Acad. Sci. 55: 805.

26. Srivastava, B.I.S. 1973. Changes in enzymic activity during
    cultivation of human cells in vitro. Exp. Cell Res. 80: 305.

27. Butterworth, P.H., Cox, R.F. and Chesterton, C.J. 1971.
    Transcription of mammalian chromatin by mammalian DNA-dependent
    RNA polymerase. Europ. J. Biochem. 23: 229.

28. Keshgegian, A.A. and Furth, J.J. 1972. Comparison of tran-
    scription of chromatin by calf thymus and E. coli RNA poly-
    merases. Biochem. Biophys. Res. Comm. 48: 757.

29. Farber, J., Rovera, G. and Baserga, R. 1971. Template
    activity of chromatin during stimulation of cellular pro-
    liferation in human diploid fibroblasts. Biochem. J. 122:
    189.

30. Costlow, M. and Baserga, R. 1973. Changes in membrane
     transport function in G_0 and G_1 cells. J. Cell Physiol.
     82: 411.

31. Yanishevsky, R., Mendelsohn, M.L., Mayall, B.H. and Cristofalo,
     V.J. 1974. Proliferative capacity and DNA content of aging
     human diploid cells in culture: a cytophotometric analysis.
     J. Cell. Physiol., in press.

QUESTIONS TO DR. RYAN

Dr. Courtois: Don't you think that as cells age, their cell cycle
changes and, then chromatin prepared from an old population would
be different and reflect this phenomenon? It has been well
established that chromatin activity changes during the cell cycle.

Dr. Ryan: Yes, as WI-38 cells undergo the aging process, the
majority of cells accumulate primarily in the G_1 stage of the cell
cycle while a small fraction is in the G_2 stage. If the cells are
not traversing the G_1 phase, then the reduced chromatin activity
of the old cells could reflect this arrested state.

Dr. Courtois: I wonder if you could not make the same observations for your experiments on serum stimulation: In short, do
you know if young cells and old cells are stimulated the same
way and at the same rate by the addition of serum?

Dr. Ryan: If confluent cultures of young and old cells are
stimulated to divide by addition of fresh medium containing 30%
serum they respond in a similar way in that the time required to
illicite an increase in the incorporation of ^{3}H—uridine into RNA
is the same in both young and old cultures. However, the rate
at which they respond is different. Young cells respond by
synthesizing a greater amount of RNA than old cells, which reflects
the reduced capacity to synthesize RNA in old cells.

GROWTH-PROMOTING ALPHA-GLOBULIN AND AGEING

Jiří Michl, Mirosalv Tolar, Věra Spurná and
Dagmar Řezáčová

Institute of Physiology, Praha, Institute of Biophysics,
Brno, Czecholslovak Academy of Sciences, and Institute
of Sera and Vaccines, Czechoslovakia

In addition to growth factors of low molecular weight, most mammalian cells in continuous culture require certain proteins that are supplied in the form of dialyzed or undialyzed serum. Although some progress in identifying growth-promoting factors in serum has been made, the function of serum proteins in supporting the growth of cell cultures is not completely understood. In general, the growth-promoting properties of serum have been attributed to the alpha-globulins; we have reported the isolation of a protein from calf serum which induced mitotic activity in metazoan cells in vitro (1). The effect of the growth-promoting alpha-globulin (GPAG) is demonstrable both in primary cultures and in cell lines (2,3). It is well known that human diploid fibroblasts obtained from normal organs will multiply rapidly for several generations in vitro but will enter a stage of declining growth terminating in cell death (4). The basis for this limitation in proliferative capacity is unknown; it has been suggested that it reflects an intrinsic cell property connected with the ageing process (5).

In our previous study it was demonstrated that young diploid cells were able to proliferate in the same pool of medium supplemented with the growth-promoting alpha-globulin in which older cells were entering the degenerate phase (3). The possibility exists that the utilization of GPAG became impaired in senescent cells. For this reason we have investigated the physiological effect of GPAG in mammalian cells.

RESULTS AND DISCUSSION

GPAG as isolated from calf serum has a strong binding capacity for several precursors; the precursors adsorbed to the protein could not be readily removed by a repeated dialysis or by gel filtration. In our previous experiments considerable binding of $^{32}PO_4^=$ (6), ^3H-thymidine, and ^3H-uridine (7) to GPAG was found. As determined by autoradiography, the incorporation of labelled precursors bound to GPAG by the incubation at 37°C into RNA and DNA is significantly higher than the incorporation of the same precursors in the presence of unlabelled GPAG.

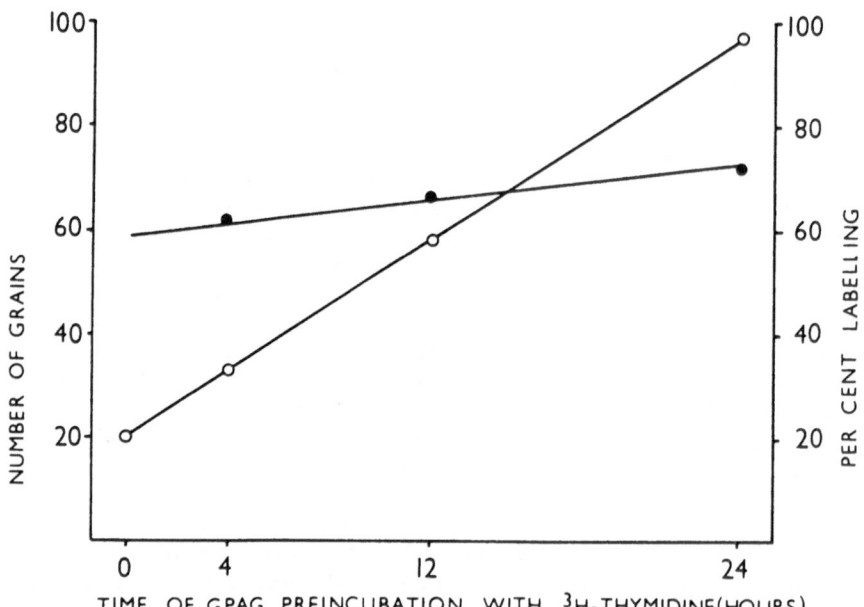

Fig. 1. Effect of GPAG on labelling frequency in nuclei of L cells incubated with ^3H-thymidine. The cells were treated for 1 hr with the labelled precursor. The number of grains in the cultures without GPAG was 21; this value was subtracted from the values found in the cultures with GPAG to give the values plotted in the figure. o——o, number of grains; ●——●, percent labelling.

Figure 1 demonstrates the influence of GPAG on DNA synthesis in L cells; in a protein-free medium the number of grains and labelling frequency are significantly lower than in GPAG—supplemented medium. At the same time an increase in the incorporation of ^{3}H-thymidine into DNA parallel to the preincubation of GPAG with the precursor was observed.

The autoradiographic analysis also showed profound differences in the grain distribution in cells incubated in the presence of ^{3}H-thymidine bound to GPAG and in the presence of free ^{3}H-thymidine. Silver grains were distributed in the nucleus when free ^{3}H-thymidine served as a source of radioactivity in the medium. In the presence of ^{3}H-thymidine bound to GPAG, as prepared by gel filtration (7), the label was found over a relatively small region in cytoplasm in a site next to the nucleus, probably corresponding to the location of the GERL area (Fig. 2). These studies demonstrate that GPAG may enter the cells by pinocytosis.

To get more information as to the process of pinocytosis we observed the cells by phase-contrast microscopy. Figure 3 shows that pinocytosis is easily seen in the cells incubated in Hank's solution supplemented with GPAG.

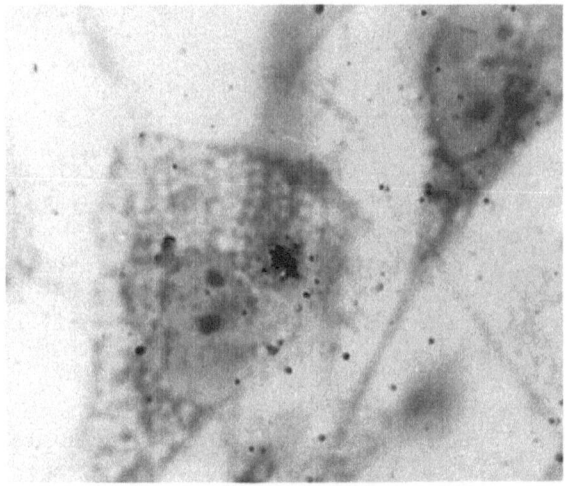

Fig. 2. Autoradiograph of human diploid cells, line LEP 19, incubated with ^{3}H-thymidine bound to GPAG (2 hr, 37°C, 0.1 μCi of ^{3}H-thymidine per ml).

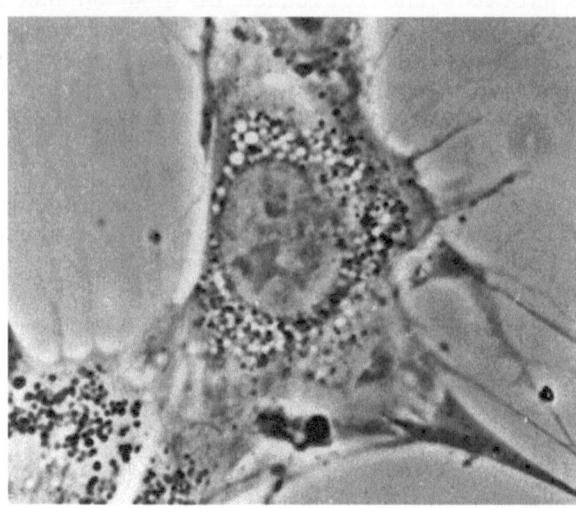

Fig. 3.- Living human diploid cells, line LEP 19; phase contrast, 15
min after the addition of GPAG (1 mg per ml) in Hank s solution
without serum.

     GPAG is needed for an actively proliferating culture of human
diploid cells in vitro and it may be concluded that pinocytosis of
GPAG, as the first step in the biosynthesis of macromolecules in
amounts sufficient for growth, represents the controlling factor
in deciding whether a cell will continue to grow or will go into
the resting state. The present studies are supported by results
showing that in medium supplemented with GPAG the maximum cell
population density of human diploid cells is directly proportional
to protein content (Fig. 4).

     For this reason we have compared human diploid cells, line
LEP 19, in 17th and 52nd passages as to the pinocytosis of ^{3}H-
thymidine bound to GPAG. Figure 5 shows that young diploid cells
incorporate significantly higher amounts of radioactivity than old
cells. At low cell population densities younger cells exert high
pinocytic activity that is inhibited at higher population densities
but pinocytosis is sharply reduced in old cells at low population
densities. From these results it was concluded that the limitation
in proliferative capacity in senescent diploid cells reflects the
inhibition of pinocytic uptake of GPAG.

     As shown in Figure 5, young diploid cells are characterized

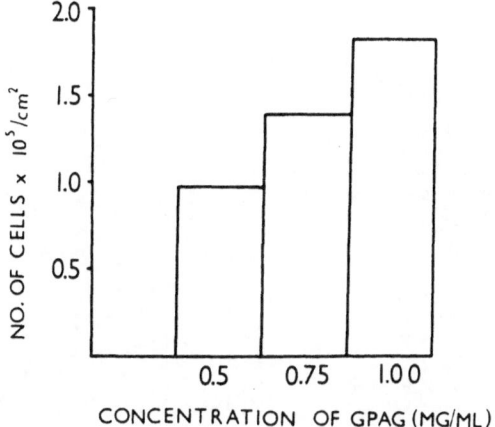

Fig. 4. Cell population density of human diploid cells grown in a
medium supplemented with different amounts of GPAG. The cells were
grown 7 days without medium change. Inoculum: 2.0 X 10^4 per cm^2.
Ordinate: cell population density.

by low pinocytosis of GPAG at higher population densities; it means
that the cells in this stage are nutritionally limited and some
changes in life processes may be expected. In continuous culture
these changes may accumulate and ultimately may result in irreversible damage of cell structures needed for engulfing GPAG.

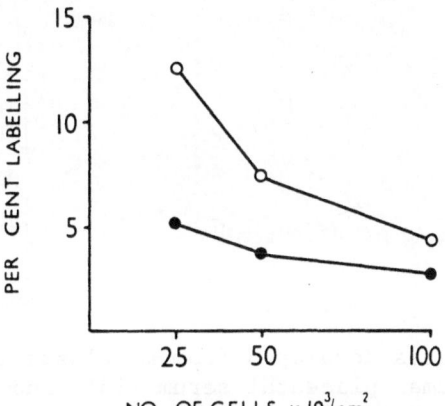

Fig. 5. Percent labelling in human diploid cells, line LEP 19, incubated for 2 hr with ^{3}H-thymidine bound to GPAG. o——o, young cells;
●——●, old cells. Abscissa: cell population density.

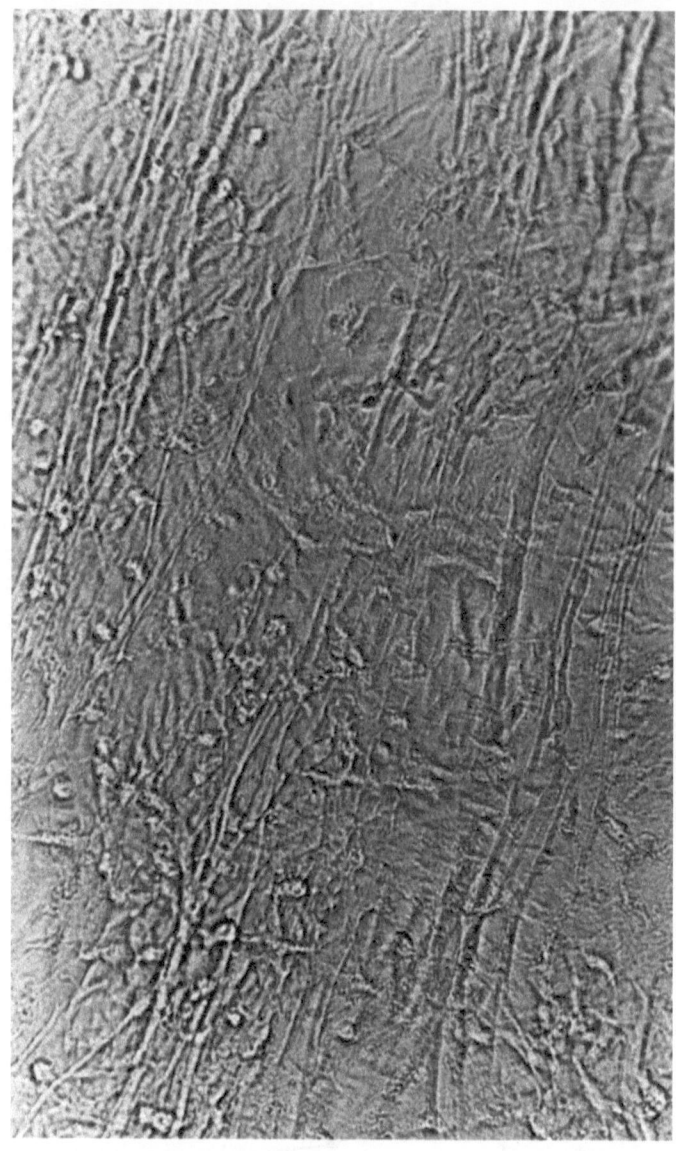

Fig. 6. Chick myotubes developed from myoblasts in Eagle's MEM
supplemented with human placental serum (15%) and chick embryo
extract (5%); 14 days in vitro, living.

Fig. 7. Chick myotubes developed from myoblasts in Eagle's MEM
supplemented with GPAG (3 mg per ml) and chick embryo extract (5%);
14 days in vitro, living.

It is reasonable to suppose that the inhibition of GPAG pinocytosis in diploid cells at higher population densities is connected with mutual cell-to-cell contact. For this reason ageing
is clearly demonstrable in metazoan cells and organisms but is only
proposed and discussed in unicellular organisms.

In further experiments we decided to work with differentiated
tissue to examine the role of GPAG in non-dividing cells. It is
well documented that some differentiated cells are characterized
by a specific inhibition of DNA synthesis. One of these systems is
represented by chick embryo myoblasts that fuse _in vitro_ in post-
mitotic stage in myotubes and ultimately form cross-striated fibers
(8). Figures 6 and 7 demonstrate that differentiation of chick
embryo muscle tissue is potentiated in Eagle's MEM with GPAG in
comparison to serum-supplemented medium. The effect of GPAG is
specific as the factor can not be substituted by other serum proteins. These results strongly indicate that GPAG is not only a
growth-promoting factor but it represents also a specific serum
protein essential for macromolecule biosynthesis in differentiated
tissue and may be involved in the ageing of non-growing cells.

The last, but central question is whether GPAG influences
macromolecule biosynthesis _in vivo_. The results of our experiments
with nuclei isolated from freshly explanted rat liver (9) are
shown in Table I. The relative incorporation of $^{32}PO_4^=$ bound to
GPAG in macromolecule is higher than the incorporation of free
$^{32}PO_4^=$ in the presence of unlabelled GPAG. Furthermore, the
incorporation of $^{32}PO_4^=$ is influenced even in freshly isolated
rat liver nuclei by the presence of GPAG: in comparison with protein-free incubation medium, the incorporation of $^{32}PO_4^=$ in RNA
and DNA is significantly increased in the presence of GPAG.

In conclusion, one may state that primary ageing of metazoan
cells is not a programmed event at the cellular level. Most living
cells programme life, but not their own death. The only exception
by be programmed death described during embryonal development. In
unicellular organisms a programme of continual cell replacement
leads to a system which can apparently remain in a functional stage
indefinitely. Metazoa have a programme of continual renewal of their
germ plasm. However, the capacity for cell replacement in organs
may be lost during the developmental process. To fulfil the
functional tasks the cells in specialized organs live in mutual
cell-to-cell contact. For the biosynthesis of macromolecules in
amounts sufficient for indefinite growth or survival in dividing
as well as non-dividing metazoan cells a specific serum factor,
growth-promoting alpha-globulin, is required. As pinocytic uptake
of GPAG is inhibited at higher population densities _in vitro_ it
is suggested that all cells in organs may be also nutritionally
limited. Specific nutritional deficiency results ultimately in

Table I. $^{32}PO_4^{\equiv}$ incorporation in freshly isolated rat liver nuclei incubated in medium supplemented with $^{32}PO_4^{\equiv}$-labelled GPAG, with a mixture of unlabelled GPAG and $Na_2H^{32}PO_4$ and with $Na_2H^{32}PO_4$. All incubation media contained equal amounts of radioactivity. One hr; 36°C.

	^{32}P - GPAG	GPAG + $Na_2H^{32}PO_4$	$Na_2H^{32}PO_4$
TOTAL RADIOACTIVITY (c.p.m. per 10g wet tissue)	232.094	509.465	258.856
HISTONES	16.4 ± 0.4^x %	36.6 ± 0.9 %	53.3 ± 0.6 %
inorganic phosphate	6.19 ± 0.09 %	10.26 ± 0.3 %	—
phosphoserine + phosphothreonine	0.77 ± 0.1 %	0.46 ± 0.04 %	—
TCA EXTRACT	1.4 ± 0.2 %	11.2 ± 0.4 %	15.2 ± 0.5 %
LIPIDIC EXTRACT	7.4 ± 0.3 %	10.7 ± 0.5 %	18.5 ± 0.6 %
RNA	63.5 ± 0.6 %	35.2 ± 1.0 %	11.5 ± 0.2 %
DNA	7.0 ± 0.4 %	4.0 ± 0.3 %	0.9 ± 0.1 %
RESIDUAL PROTEIN	3.7 ± 0.2 %	1.9 ± 0.09 %	0.3 ± 0.04 %

x standard error

irreversible damage of cell structures important for pinocytosis of GPAG so that the biosynthesis of DNA and RNA continually decreases or stops.

It is highly probable that metazoa age because they are metazoa.

REFERENCES

1. Michl, J. 1961. Metabolism of cells in tissue culture _in vitro_. I. The influence of serum protein fractions on the growth of normal and neoplastic cells. Exp. Cell Res. 23: 324.

2. Michl, J. 1962. Metabolism of cells in tissue culture _in vitro_. II. Long-term cultivation of cell strains and cells isolated directly from animals in a stationary culture. Exp. Cell Res. 26: 129.

3. Macek, M. and Michl, J. 1964. Příspěvek ke kultivaci lidských diploidních buněk. Acta Univ. Carol. Med. 10: 519.

4. Hayflick, L. and Moorhead, P.S. 1961. The serial cultivation of human diploid cell strains. Exp. Cell Res. 25: 585.

5. Hayflick, L. 1965. The limited in vitro lifetime of human diploid cell strains. Exp. Cell Res. 37: 614.

6. Michl, J., Svovodová, J. and Řezáčová, D. 1973. Growth regulatory effect of the growth-promoting alpha-globulin in human diploid and heteroploid cells. Physiol. Bohemoslov. 22: 487.

7. Michl, J. and Spurná, V. 1974. Growth-promoting alpha-globulin. A pleiotypic activator. Exp. Cell Res. 84: 56.

8. Tolar, M. 1973. Vývoj nervosvalového systému in vitro. Avicenum, SZdN, Babákova sbirka, Praha.

9. Michl, J. and Svobodová, J. 1969. Primary function of the growth-promoting alpha-globulin in cell culture. Exp. Cell Res. 58: 174.

QUESTIONS TO DR. MICHL

Dr. Gutmann: Is there any evidence for direct effects of GPAG on membrane transport? Incorporation studies have to cope with the difficulties of definitions and calculations. If the growth promoting effect affects differentiation could this be due also to changes of membrane properties due to GPAG, which could secondarily affect permeability and entry of other agents into the cell?

Dr. Michl: We have evidence for a direct effect of GPAG on transmembrane transport, because GPAG-bound precursors enter the cells by pinocytosis. It is not excluded that pinocytotic uptake of GPAG influences the precursor pool size in some separate area in the cells.

CELL SURFACE ALTERATIONS AND "IN VITRO" AGING OF ANIMAL CELLS

R. Azencott, C. Hughes*, and Y. Courtois

Unité de Recherches Gérontologiques INSERM U. 118, 29
Rue Wilhem, 75016 Paris, France, and the National
Institute for Medical Research *, Mill Hill, London
NW7 IAA, England

It is a well established phenomenon that animal cells
have a finite lifespan in vitro, provided they either are not
infected by a virus or keep a normal karyotype. Conversely, abnormal
cells survive indefinately (1). Several biochemical parameters
have been studied in the past 10 years though very few of those
studied have dealt with cell-surface integrity (2). However it
is feasible to conceive that modifications of cell surface components
would impair cell division since such events have been described
for cell differentiation, cell transformation and even for the
different phases of the cell cycle (3-5).

The most usual way to determine the total doubling-potential
is to trypsinize the given cells when they reach confluency and then
to plate them with a split ratio of 1:2. Several researchers have
shown that if cells are plated with a higher split ratio (1:4 to
1:10) then the number of divisions increased (6).

It is well established furthermore that, as cells age, their
size increases while their density at saturation decreases (7).
Moreover, while young cells are able to form multilayers, old cells
cannot, which inability could be intepreted to mean that they are
more contact-inhibited (8,9).

A recent report has described an increase in cell loss as a
result of detachment in ageing culture (10). Lastly, photometric
as well as electron microscopic observations have shown there to
be alterations at the cell surface (11,12).

In the present study, we have analyzed the cell surface com-

147

ponents (13) and the adhesiveness of chick embryo fibroblasts
during in vitro ageing (14). We have shown that alterations of
cell surface could be responsible for most of the age-related phenomenon previously noted.

MATERIALS AND METHODS

Embryo chick fibroblasts were obtained from leg muscles of
12-day-old chick embryos in the manner described by Hay and
Strehler (6). Cell cultures and subcultures were carried out in
Roller bottles as described elsewhere (11). We used Eagle's
essential medium supplemented with 10% fetal calf serum. After 19
to 23 population doublings, cells reached Hayflick's phase III in
accordance with several authors (6,8).

Membrane biochemistry was carried out by labeling the cell
surface with ^{3}H-glucosamine (0.25 ml, 50 μCi, 12 Ci/mmole) or
^{14}C-glucosamine (0.125 ml, 50 μCi, 12 Ci/mmole) in Roller bottles
filled with 50 ml of medium. A double labeling process permitted
us to analyze material which could be removed from the outside of
the cells via a 10 min treatment in 0.25% trypsin. This fraction was
called the trypsinate. We also analyzed a smooth membrane fraction
prepared according to a modified Friedman's method (13).

In a second set of experiments cell adhesiveness was studied
as a function of the culture age (14). The technique used was
similar to Roth's method (15). The first step was to prepare
sizeable aggregates from dispersed young cells. In a second step,
we labelled young and old cells by ^{3}H-thymidine for 24 to 48 hr,
afterwhich these cells were suspended via an initial 0.25% trypsin
treatment. The last step was to mix the labelled single cells with
few aggregates and to count the number of cells which stuck to the
aggregates in 4 hr at 37°C.

RESULTS

Glucosamine is relatively specific to the different components of the cell surface: N-acetylglucosamine, N-acetylgalacto-
samine in glycoproteins and in sialic acids (3).

We first checked that the radioactive precursors were incorporated linearly during 7 to 8 hr. The material, mostly non-
dialysable, removed from the cells by the trypsin treatment, contained the highest proportion of radioactivity. When analysed on
sepharose 6 B in SDS buffer, the trypsinate profile, prepared from
glucosamine-labelled cells (^{3}H for the young cells and ^{14}C for the
old cells) were similar. However, as shown on Fig. 1 by determining

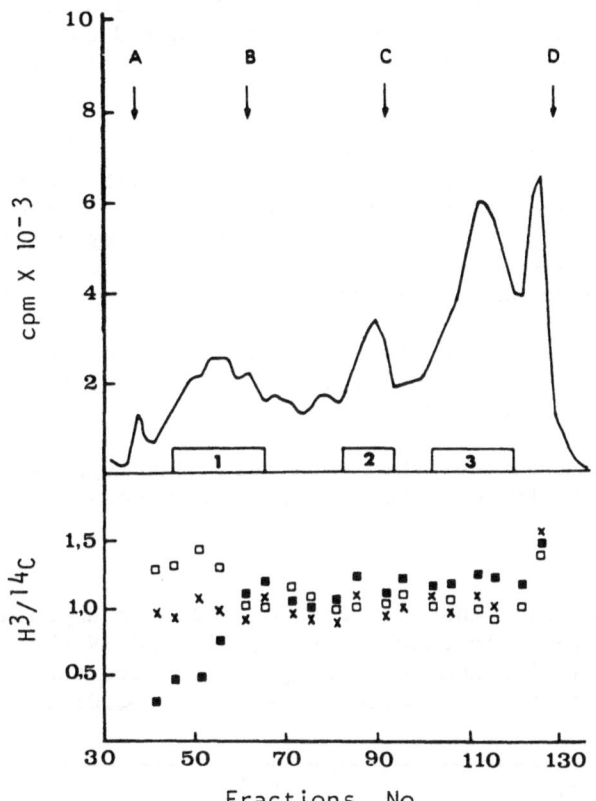

Fig. 1.   Chromatography on Sepharose 6B of mixed trypsinates from
early or late phase cells labelled with radioactive glucosamine.
Top: The samples (2 ml) and column were in 1% sodium dodecyl sulphate -- 10 mM Tris-HCl buffer, pH 7.4.   The column (2.2 cm X 150 cm)
was eluted with the same buffer.   Fractions (3.3 ml) were collected
at a flow rate of 20 ml per hr.   ABC and D indicate the elution
positions respectively of Blue dextran (main peak) immunoglobulin G
(M.W. 150,000), ^{125}I-labelled adenovirus type 5 fibre protein (M.W.
62,000), and ^{14}C-tryptophan markers run separately.   Bottom: ^{3}H/^{14}C
ratios obtained in the co-chromatography of trypsinates from early
passage (5) cells labelled with ^{14}C-glucosamine and late passage (17)
cells labelled with ^{3}H-glucosamine (□); ^{3}H/^{14}C ratios obtained
in an identical experiment in which the labels were reversed (■);
^{3}H/^{14}C ratios obtained in a control experiment in which trypsinates

prepared from early passage (4) cells labelled either with ^{3}H-glucosamine or ^{14}C-glucosamine were co-chromatographed (Δ). The curves were plotted after normalizing the ^{3}H/^{14}C ratios of the unfractionated trypsinates to unity.

the ratio ^{3}H:^{14}C, we were able to see that there was a difference between young and old cells. This difference was related to a ghigh molecular weight material that we analysed further after enzymic digestion (pronase and Hyaluronidases), by chromatography on Sephadex G 50, These treatments gave some high molecular materials which have again striking different labelling patterns.

It is very interesting to note that the smooth membrane fractions of young and old cells gave similar patterns as far as radioactivity was concerned.

In the second part of this work cell adhesiveness was studied (14). As shown in Table I, young cells in suspension are much

TABLE I. Differential Adhesion For Confluent or Exponential Cells at Several Passages. In experiments I, II, and IV where the number of samples were numerous enough to utilize the Mann Whitney's test we calculated for U a value of zero and a $<$.01 for each (from ref. 14).

Experiment number	Passage	Mean number of cells per aggregate
	confluent cells	
I	4	1.880
	22	500
II	1	1.590
	21	90
	exponential cells	
III	1	519
	8	231
IV	3	272
	16	154

more adhesive toward young cell aggregates than are old cells.
This phenomenon as noted immediately after a few cell doublings,
and further increased as cells aged.

Further evidence that cell adhesiveness decreases with age
comes from the impossibility to obtain good size aggregates from
old cells in suspension, even after several days of rotation. As
the assay lasted 4 hr it cannot be ruled out that young and old cells
were repairing at a different rate the damages provoked by the
trypsine treatment. When we measured the amount of non-dialysable
material which incorporated ^{3}H-glucosamine during this period, we
could not detect significant differences between young and old
cells.

While less material was synthesized by cells in suspension
than by cells attached to the bottom of the petri dish, it is likely
that we were unable to detect some slight differences which might be
of the same order as those described in the first part of this
paper.

If this is the case, a particular material, synthesized in
very small amounts, would be different in young and in old cells,
and would be of major importance in explaining senescence pheno-
menona.

These results are suggestive of a hypothesis in which in vitro
ageing of cells would be the consequence of membrane alterations,
capable of impairing the division capacity of cells. These modifications would cumulate during serial population doublings and ultimately cause the cell death. The unbalance with age in synthesis
of some glycoproteins or mucopolysaccharides shown in the first part
of this work would be responsible for this phenomenon.

Does the trypsin treatment damage irreversibly the cell surface
and then accerlerate the rate of in vitro ageing? This may be the
case since chick fibroblasts cultured on a collagen layer and
mechanically subcultured could divide for years (16), instead of a
few months as with trypsinized cells. Similarly a split ratio higher
than 1:2 will increase the lifespan of cells (6).

However this hypothesis would not take into account the fact
that transformed cells (like 3T3) will divide indefinitely even
after many subcultures with trypsin. Conversely, from other
results (4), it may be inferred that transformation impaired the
arrangement of some glycoproteins at the cell surface and made them
more resistant to trypsin damage.

These experiments indicate strongly that one should focus one's
attention on the relationship between cell surface and ageing,

keeping in mind that it is worthwhile, but difficult, to look for similar events in normal in vivo ageing cells.

SUMMARY

The pattern of glucosamine labelling of cell surface components was shown to be different in young and in old chick fibroblasts. Similarly, cell adhesiveness was shown to decrease with age.

From these results, we present an hypothesis in which we assumed that the integrity of cell surface is of major importance to explain most of the phenomena occurring during in vitro ageing of animal cells.

ACKNOWLEDGEMENT

The authors are much indebted to Miss J. Tassin for her excellent technical assistance.

REFERENCES

1. Hayflick, L. 1970. Aging under glass. Exp. Geront. 5: 291.

2. Cristofalo, V.J. 1972. Animal cell cultures as a model system for the study of aging. Adv. Gerontol. Res. 4: 45.

3. Hughes, R.C. 1973. IN: Progress in Biophysics and Molecular Biology (Buttler and B. Noble, editors), Pergamon Press, 26: 189.

4. Nonnan, K.D., Levine, A.J., and Burger, M.M. 1973. Cell cycle-dependent changes in the surface membrane as detected with ^{3}H-concanavalin A. J. Cell. Biol. 58: 491.

5. Shoham, J., Sachs, L. 1974. Different cyclic changes in the surface membrane of normal and malignant transformed cells. Exp. Cell. Res. 85: 8, 1974.

6. Hay, R.J., and Strehler, S.L. The limited growth span of cell strains isolated from the chick embryo. Exp. Geront. 2: 123.

7. Simons, J.W.I. 1969. A theorical and experimental approach to the relationship between cell variability and aging in vitro. IN: Aging in Cell and Tissue Culture (E. Holeckova and V.J. Cristofalo), Plenum Press, New York, p. 25.

Macieira-Coelho, A. 1969. The decreased growth potential
in vitro of human fibroblasts of adult origin, ibid: 121.

8. Macieira-Coelho, A., Ponten, J., and Philipson, L. 1966.
Inhibition of the division cycle in confluent cultures of
human fibroblasts in vitro. Exp. Cell Res. 43: 20.

9. Courtois, Y. 1973. Aging mechanisms "in vitro" for chick
fibroblasts: control at the gene level or damage at the cell
surface. IN: Molecular and cellular mechanisms of aging.
Symposium. I.N.S.E.R.M., Paris, vol 27.

10. Good, P., and Watson, D. 1973. Pairwise loss of mitotic
ability by human diploid fibroblasts. Exp. Geront. 8: 147.

11. Courtois, Y. Chromatin modifications of chick embryo fibroblasts
during "in vitro" ageing. Scand. J. Cl. Lab. Invest. 2: 9,
1972; Mechan. Ageing Devel. 3: 51, 1974

12. Robbins, E., Levine, E.M., and Eagle, H. 1970. Morphologic
changes accompanying senescence of cultured human diploid
cells. J. Exp. Med. 131: 1211.

13. Courtois, Y., and Hughes, C. 1974. The synthesis of glycoproteins at the surface of chick embryo cells during in vitro
aging. Eur. J. Biochem. 44: 131.

14. Azencott, R., and Courtois, Y. 1974. Age related differences
in intercellular adhesion for chick fibroblasts cultured
in vitro. Exp. Cell Res., in press.

15. Roth, S.A., McGuire, E.J., and Roseman, S. 1971. An assay
for intercellular adhesive specificity. J. Cell. Biol. 51:
525.

16. Gey, G.O., Svotelis, M., Foard, M., and Bang, F.B. 1974. Long
term growth of chicken fibroblasts on a collagen substrate.
Exp. Cell Res., 84: 63, 1974.

QUESTION TO DR. COURTOIS

Dr. Hay: Did you try to look at agglutinability of young and old
cells with concanavalin A?

Dr. Courtois: I discussed this point recently with Dr. M. Burger
from Basel and I plan to do it in the near future, with different
lectins.

CHANGES IN LYSOSOMES DURING AGEING OF PARENCHYMAL AND NON-

PARENCHYMAL LIVER CELLS

D. L. Knook, E. C. Sleyster and M. J. van Noord

Institute for Experimental Gerontology TNO Rijswijk ZH,
The Netherlands

Previous studies at our Institute revealed a decrease in the
functional reserve capacity of the rat liver during ageing (1,2).
This reduction in function and other changes occurring in the liver
with age (1) will be reflected in metabolic changes at the cellular
level. Since the liver consists of different cell types, a particular ageing phenomenon observed for the intact organ or measured
with a liver homogenate may be an expression of a general pattern
in all liver cells; alternatively, different cell types may show
different patterns.

The various cell types in the mammalian liver can be divided
into two large groups: parenchymal cells (hepatocytes and non-
parenchymal cells. The latter cell population consists mainly of
endothelial and Kupffer cells. Parenchymal cells, representing
about 60-65% of the total number of liver cells and more than 90%
of the liver mass, form the major cell type. Procedures have been
recently developed for the isolation of parenchymal cells (3) and
non-parenchymal cells (4). The availability of populations of
the two liver cell classes is essential for studies on the question
as to whether the reduction in functional capacity of the liver can
be attributed to the decreased metabolic activities in individual
cells as well as to define the contribution of each cell class to
the phenomena of liver ageing.

Of course, there may be many mechanisms which can account for
metabolic changes in ageing liver cells. For several reasons,
the possible role of lysosomes deserves careful attention. One
main reason is based on the following consideration. Cell damage
is widely believed to be a major cause of ageing. Lysosomal

enzymes are concerned in processes such as protein degradation and turnover and in the breakdown in the regulation of lysosomal activity in cells may account for multiple forms of damage to the cellular machinery.

Another indication for a role of lysosomes in cellular ageing is the increase in activity of some lysosomal enzymes that has been reported to occur during ageing of cultured human cells (5). In recent review papers by Finch (6) and Wilson (7), literature data have been summarized on alterations in the level of a large number of enzymes, including lysosomal enzymes, in various mammalian tissues. For the rat liver both age-related increases in the activities of some lysosomal enzymes as well as decreases for the same enzymes or even the absence of changes in the enzyme levels have been reported. Since the values for the rat liver have been determined with tissue homogenates, it is clear that one main reason for the controversial results will be a large variation in the contribution of the various liver cell populations to the total enzyme activities. Various other biological factors which contribute to the divergence in the enzyme activities have been discussed by Finch (6).

Up to now, no data on age-related enzyme patterns in distinct liver cell classes have been available. In this study, various lysosomal enzyme activities as well as morphological and morphometric changes in lysosomes have been investigated not only in intact liver tissue but also in isolated parenchymal and non-parenchymal liver cell populations. The results obtained for young and old rats were compared. In this way, the determination of the contribution of each cell class to the age-related changes in levels of liver enzymes can contribute to the solution of problems associated with liver cell heterogeneity and enzyme distribution.

Female WAG/Rij and BN/BiRij rats aged 3 months (young) and 30-35 months (old) were used in the experiments. The mean lifespan of females of these strains is 31 months (8,9). It is interesting to note that there is no significant difference in the lifespan of male and female BN/BiRij rats (9), whereas female WAG/Rij rats live much longer than do male rats of this strain (8).

Parenchymal cells were isolated from the livers of young and old rats by use of the enzymes collagenase and hyaluronidase (10). At least 85-90% of the isolated cells were viable as judged from trypan blue exclusion, electron microscopic observation, and characteristics of endogenous respiration (10). Liver cells of other types were absent in the parenchymal cell preparation. An electron micrograph of a parenchymal cell isolated from a 30-month-old rat is shown in Figure 1. The cell has a relatively well-preserved ultrastructure and is surrounded by a continuous plasma membrane. The cytoplasmic organelles are regularly distributed

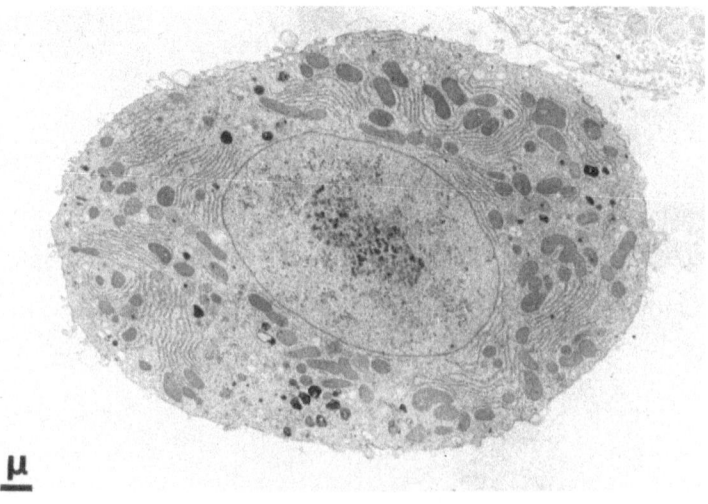

Fig. 1. Isolated parenchymal cell from the liver of a 30-month-old female WAG/Rij rat. Magn. 4000 X.

over the cell and several lysosome-derived structures are present.

Non-parenchymal liver cell suspensions were obtained according to a modification of an existing method (11). The liver was perfused in situ with Gey's balanced salt solution containing pronase, followed by incubation of small liver fragments in the same solution. The resulting suspension consisted of several cell types. Judged by electron microscopic observation, approximately half of the cells in the suspension could be considered as endothelial cells, whereas one-fourth were Kupffer cells. Kupffer cells could also be identified cytochemically by their peroxidase activity (12). The other cell types present in the final suspension were mainly fat-storing cells and blood cells. Parenchymal cells appeared to be nearly completely absent and could be removed by centrifugation at low g values. All Kupffer and endothelial cells, from both young and old rats, excluded trypan blue and showed a well-preserved ultra-structure. An electron micrograph of a representative Kupffer cell isolated from the liver of an old rat is shown in Figure 2. The cell is surrounded by a continuous plasma membrane and shows a number of pseudopods. The mitochondria are well-preserved and are much smaller in number and size than those of parenchymal cells. The cytoplasm contains a great variety of lysosomal structures differing in size and density. Because of the phagocytic activity of Kupffer cells, a number of lysosomal structures contain various

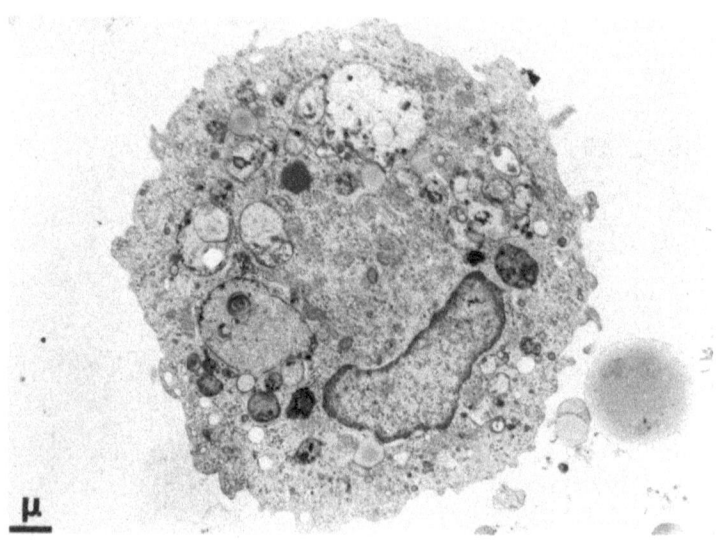

Fig. 2. Kupffer cell isolated from the liver of a 35-month-old
female BN/BiRij rat. Magn. 5800 X.

materials. Among them may be fragments of parenchymal cells,
possibly released during the pronase treatment.

     The lysosomal structures in parenchymal, Kupffer, and endo-
thelial cells isolated from old animals are shown in more detail
in Figure 3 and 4. As mentioned above, the cytoplasm of parenchymal
cells of old rats shows relatively large quantities of lysosome-
derived structures, including lipofuscin (Fig. 3). The accumulation
of this pigment in various tissues is one of the most consistent
cytological changes associated with ageing. The chemical compositon
and the function of lipofuscin are not yet completely understood (13).
The lysosomal structures in Kupffer cells show considerable poly-
morphism as a result of their association with the various materials
that are phagocytized by these cells (Fig. 4A). In one of the
larger structures in the cell in Figure 4A, the remnants of a
mitochondrion can be seen. Figure 4B shows that the lysosomes are
very sparse in endothelial cells. They have a rounded shape and a
diameter of approximately 0.6 . In view of the small number of
lysosomes and the restricted endocytotic capacity of endothelial
cells (14), the lysosomal activities observed for non-parenchymal

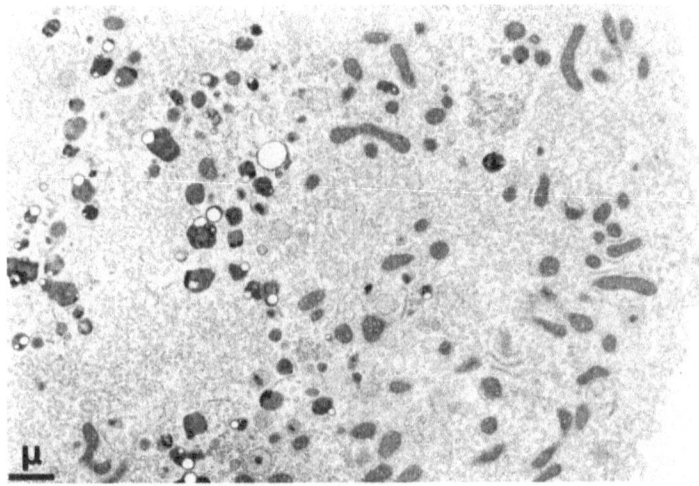

Fig. 3. Lysosome-derived structures in a parenchymal liver cell
isolated from a 30-month-old female WAG/Rij rat. Magn. 6000 X.

cell suspensions can be attributed mainly to the numberous lysosomal structures in Kupffer cells.

Electron microscopic observations had indicated age-related
increases in lysosomal structures such as sencondary lysosomes and
lipofuscin in parenchymal cells. In order to obtain more quantitative data, a morhpometric study of the lysosome-derived structures
was performed on parenchymal cells in situ. Sterological methods
for morphometrical cytology (15) were applied to micrographs which
were studied at the level of both light microscopy and electron
microscopy. The volume fractions occupied by lysosomal structures
in parenchymal cells from young and old rats are summarized in Table
I. Although there is a large increase in the total and cytoplasmic
volume of hepatocytes with age, which is caused by a change to
polyploid cells (10), there is a still greater increase in the
percentage of the cytoplasmic volume occupied by lysosomal
structures. Up to now, the morphometric studies have been confined to parenchymal cells. It appears to be very difficult to
perform morphometric measurements on sinusoidal cells, such as
Kupffer and endothelial cells, in situ. However, the retained
structural integrity of the Kupffer and endothelial cells after
isolation now offers the possibility to perform morphometric

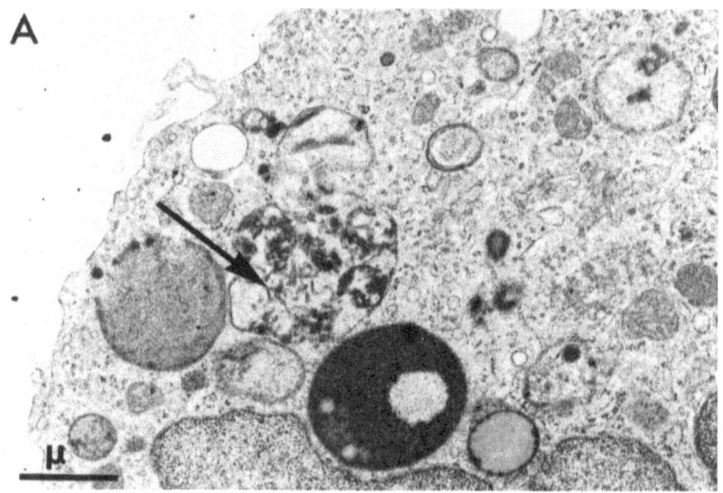

Fig. 4A. Heterogeneous lysosomal structures in a Kupffer cell isolated from the liver of a 35-month-old female BN/BiRij rat. The
arrow indicates the remnants of a mitochondrion in one of these
lysosomes. Magn. 13,500 X.

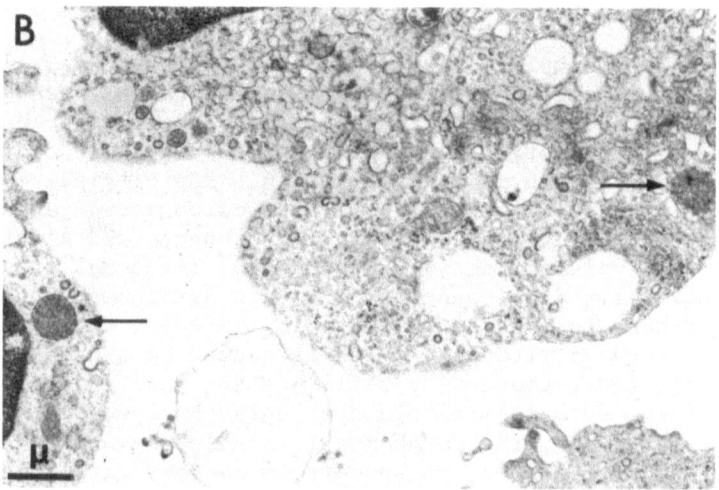

Fig. 4B. Details of two isolated endothelial cells from the liver
of a 35-month-old female BN/BiRij rat. Arrows indicate the small
lysosomes present in these cells. Magn. 8,300 X.

Table I. Cytoplasmic Volume Fraction of Lysosomal Structures in Parenchymal Liver Tissue from Young (5 Months) and Old (35 Months) WAG/Rij Rats

	Cell Type					
	Young			Old		
	Central	Midzonal	Peripheral	Central	Midzonal	Peripheral
Volume (μ³)*						
Hepatocytes	4230 ± 160	4100 ± 160	2710 ± 90	6970 ± 320	5930 ± 210	4190 ± 200
Nucleus	223 ± 10	183 ± 9	140 ± 8	425 ± 21	290 ± 13	245 ± 16
Cytoplasm	4010 ± 160	3920 ± 160	2570 ± 90	6545 ± 320	5640 ± 210	3955 ± 200
Ratio nucleus:cell	0.053	0.045	0.052	0.060	0.049	0.058
% Cytoplasmic volume** Lysosomal structures	0.96 ± 0.16	0.56 ± 0.12	0.45 ± 0.11	1.33 ± 0.19	1.65 ± 0.22	0.23 ± 0.08

*Relative volumes based on diameters of cells and nuclei measured for 0.5μ sections of cells in parenchymal liver tissue using a Zeiss Particle Size Analyzer (TGZ 3). Values represent the mean value ± the standard error determined on 290 micrographs.

**Values represent the mean ± standard deviation from the Poisson distribution.

studies on these isolated cells.

The morphometrically measured increase for lysosome-derived structures in hepatocytes from old livers suggests a role of lysosomes in the cellular ageing processes in the rat liver. To correlate this increase in number and size of the lysosome-derived structures with possible alterations in the lysosomal activities, various lysosomal enzymes were determined in suspensions of parenchymal and non-parenchymal liver cells. The enzymes acid phosphatase (EC 3.1.3.2) and β-galactosidase (EC 3.2.1.23) were assayed fluorimetrically (16,17); cathepsin D (EC 3.4.4.23), acid deoxyribonuclease (EC 3.1.4.6), and arylsulphatase B (EC 3.1.6.1) were assayed by spectrophotometric methods (18). Prior to enzyme determinations, the cells were treated with 0.05% Triton X-100 and the measured values represent the total enzyme activities.

In Table II, the values which were obtained for the lysosomal enzyme activities in parenchymal and non-parenchymal cells from young rats are compared. The activities are expressed on both a protein basis and per number of cells. All enzymes are present in both parenchymal and non-parenchymal cells, but there are great differences in the distribution of the various enzyme activities over the two cell populations. All activities per mg of protein are much higher in non-parenchymal cells than in parenchymal cells. The most stiking differences are observed for cathepsin D, β-galactosidase, and arylsulphatase B. The specific activity of cathepsin D--estimated on a protein basis--is about 13 times higher in non-parenchymal cells than in parenchymal cells. "Per cell", non-parenchymal cells possess about 2 times as much cathepsin as do parenchymal cells. The activities per mg of protein of the two β-galactosidase isoenzymes and of arylsulphatase B are about 4 times higher in non-parenchymal cells than in parenchymal cells. Another study has also indicated a higher specific arylsulphatase activity in Kupffer cells (19). These data are in good accord with the electron microscopic demonstration of the presence of numerous large lysosomes in Kupffer cells. The results demonstrate that among the lysosomal enzymes investigated, cathepsin D, β-galactosidase, and arylsulphatase B are selectively concentrated in non-parenchymal cells. It is clear that lysosomes in parenchymal and non-parenchymal cells differ in terms of their enzyme composition. The reasons for the high activities of cathepsin D, arylsulphatase B, and the β-galactosidases in non-parenchymal cells are not yet known, but they are related to the high endocytotic activity of these cells.

The influence of age on the activities of acid phosphatase, cathepsin D, and acid DNAse in parenchymal and non-parenchymal cells is shown in Figure 5. The enzyme activities were estimated on a protein basis and per 10^6 cells. The three enzymes measured

Table II. Distribution of Lysosomal Enzyme Activities Over Parenchymal and Non-parenchymal Liver Cells Isolated From Young BN/BiRij rats

Enzymes measured	Act. */mg protein		Act. */10^6 cells	
	Parenchymal cells	Non-parenchymal cells	Parenchymal cells	Non-parenchymal cells
Acid phosphatase	33.88 ± 3.94 (13)	53.70 ± 4.17 (7)	64.04 ± 5.13 (13)	12.87 ± 1.60 (7)
β-Glactosidase pH 3.6	2.62 ± 0.28 (8)	10.97 ± 1.43 (5)	5.12 ± 0.60 (8)	2.53 ± 0.40 (5)
pH 4.5	1.97 ± 0.39 (4)	7.55 ± 1.02 (5)	3.63 ± 0.87 (4)	1.69 ± 0.21 (5)
Cathepsin D	8.23 ± 1.13 (6)	110.79 ± 16.48 (4)	16.27 ± 2.46 (6)	35.70 ± 6.62 (4)
Acid DNAse	11.30 ± 1.64 (7)	27.38 ± 7.91 (5)	21.05 ± 1.87 (7)	7.32 ± 1.40 (5)
Arylsulphatase B	6.63 ± 0.26 (5)	20.73 ± 3.65 (5)	12.96 ± 0.56 (5)	3.91 ± 0.31 (5)

*Enzyme activities are expressed as nmoles 4-methylumbelliferone (acid phosphatase, β-galactosidase), nmoles tyrosine (cathepsin D), nmoles mononucleotide equivalents (acid DNAse) or nmoles nitrocatechol (arylsulphatase B) released per min at 37 C. The values represent the mean ± the standard error with the number of determinations in parentheses.

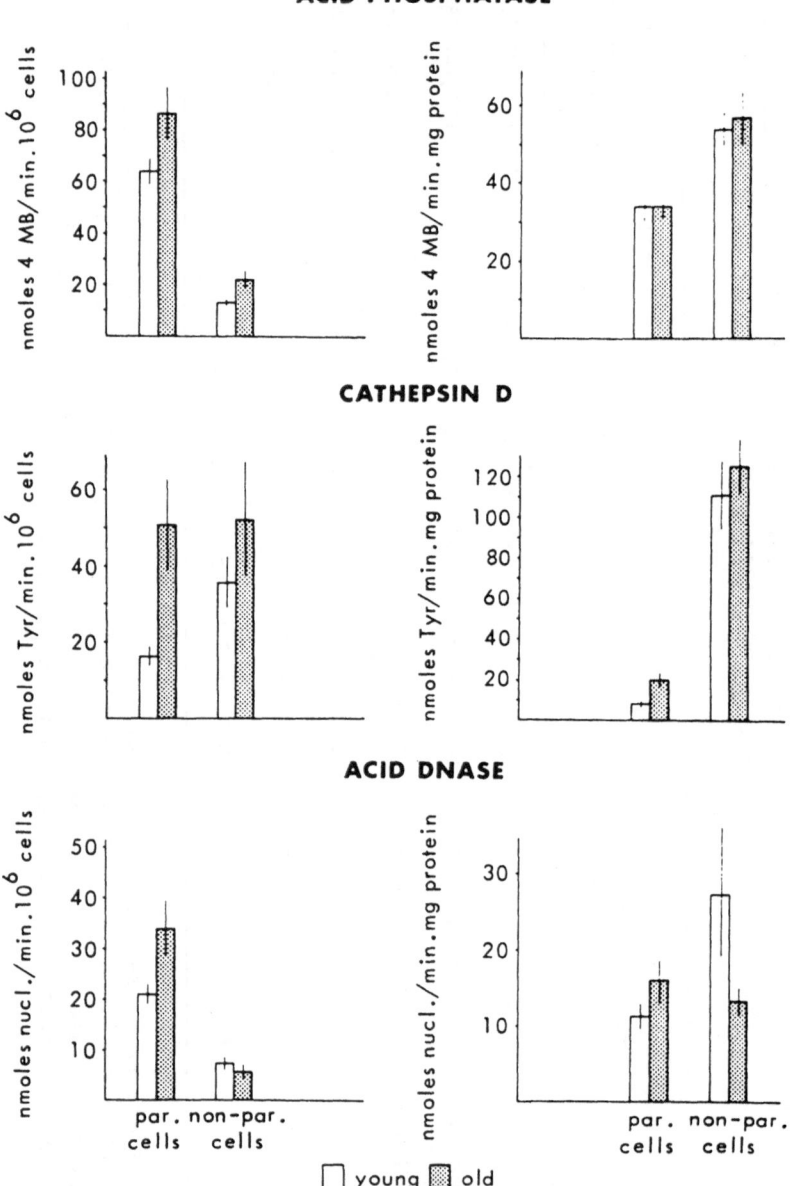

Fig. 5. Lysosomal enzyme activities in parenchymal and non-parenchymal liver cells isolated from young (aged 3-months) and old (aged 30-35-months) female BN/BiRij rats. The enzymatic activities are expressed in nmoles per min per mg of protein and in nmoles per min per 10^6 cells as described under Table II.

exhibit different patterns during ageing. The specific activity
per milligram of protein of the acid phosphatase does not significantly increase in either parenchymal or non-parenchymal cells.
In contrast, the specific cathepsin activity increases more than
2-fold in parenchymal cells, but remains nearly constant in non-
parenchymal cells. The specific activity of acid DNAse in parenchymal
cells from old rats is somewhat higher than in parenchymal cells
from young animals, but a strong decrease with ageing is found for
the acid DNAse activity in non-parenchymal cells.

The activities of the lysosomal enzymes, expressed on a cellular basis, increases in the two cell classes. The higher activities
"per cell" in the parenchymal cells are in agreement with the shift
to polyploidy resulting in larger hepatocytes which occurs during
ageing of the rat liver (2). The changes in polyploid characteristics of the liver are also reflected in the protein content of the
parenchymal cells which increases from 2.09 ± 0.09 mg to 3.04 ± 0.28
mg per 10^6 cells. Non-parenchymal cells have a completely different life cycle as compared with parenchymal cells and show no
shift to polyploid cells. In sipite of this, their protein content
does increase from 0.24 ± 0.03 to 0.39 ± 0.08 mg per 10^6 cells.

As already mentioned, the data of Figure 5 demonstrate that
there is an age-related increase of lysosomal enzyme activities
"per cell" for parenchymal and non-parenchymal cells. The results
obtained for the activities per mg of protein for the lysosomal
enzymes mentioned above, and preliminary results obtained with
 -galactosidase and arylsulphatase B, suggest that, in general,
there is an increase in the specific lysosomal enzyme activities
in parenchymal cells with ageing. The great differences in the
degree of increase for the various enzymes suggest that lysosomes
in parenchymal cells are heterogeneous with regard to their enzyme
content and that this heterogeneity becomes more pronounced during
ageing of the cells. In contrast to the situation in parenchymal
cells, the specific activities (estimated on a protein basis) of
most lysosomal enzymes in non-parenchymal cells remained constant
with ageing or even decreased.

These results also illustrate that the data reported in the
literature (6,7) with regard to age-related changes in enzyme
activities of tissue homogenates represent a mean value which will
not reflect the actual changes in the different cell populations
of the liver. Further studies on distinct liver cell populations
from young and old rats may reveal the contribution of the different cell types to the decrease in the functional capacity of
the liver with age and provide an understanding of the changes in
cellular metabolism with age.

SUMMARY

Intact and viable parenchymal and non-parenchymal liver cell preparations were isolated by enzyme perfusion techniques from young and old rats. The distribution of the lysosomal enzymes acid phosphatase, β-galactosidase, cathepsin D, acid DNAse, and arylsulphatase B over parenchymal and non-parenchymal cells was determined. In addition, morphological and morphometric changes which occur in parenchymal cells with age were investigated.

All lysosomal enzymes studied are present in both cell classes, but non-parenchymal cells possess much higher activities per mg protein than do parenchymal cells. This phenomenon is most pronounced for cathepsin D with a 13-times higher specific activity in non-parenchymal cells. Electron microscopic observations demonstrated that the lysosomal activities in non-parenchymal cells can be attributed mainly to the large and numerous lysosomal structures in Kupffer cells.

Parenchymal cells from old rats have higher lysosomal enzyme activities per mg protein than do hepatocytes from young rats. This observation is in agreement with the general increase with age in the cytoplasmic volume fraction occupied by lysosomal structures in parenchymal cells. In general, non-parenchymal cells show no increase in specific enzyme activities with age. The results obtained suggest an increase in the heterogeneity--in both appearance and enzyme content--of the lysosomal structures in parenchymal cells with age.

REFERENCES

1. de Leeuw-Israel, F.R. 1971. Aging changes in the rat liver. An experimental study of hepato-cellular function and morphology. Thesis, University of Leiden.

2. van Bezooijen, C.F.A., de Leeuw-Israel, F.R. and Hollander, C.F. 1972-73. On the role of hepatic cell ploidy in changes in liver function with age and following partial hepatectomy. Mech. Age. Dev. 1: 351.

3. Muller, M., Schreiber, M., Kartenbeck, J. and Schreiber, G. 1972. Preparation of single-cell suspensions from normal liver, regenerating liver, and Morris hepatomas 9121 and 5123 tc. Cancer Res. 32: 2568.

4. Mills, D.M. and Zucker-Franklin, D. 1969. Electron microscopic study of isolated Kupffer cells. Am J. Pathol. 54: 147.

5. Cristofalo, V.J. 1970. Metabolic aspects of aging in diploid
    human cells, IN Aging in cell and Tissue Culture (E. Holeckova
    and V.J. Cristofalo, editors), New York-London, Plenum Press,
    p. 83.

6. Finch, C.E. 1972. Enzyme activities, gene function and
    ageing in mammals (Review). Exp. Geront. 7: 53.

7. Wilson, P. 1973. Enzyme changes in ageing mammals. Geron-
    tologia 19: 79.

8. Boorman, G.A. and Hollander, C.F. 1973. Spontaneous lesions
    in the female WAG/Rij (Wistar) rat. J. Gernotol. 28: 152.

9. Boorman, G.A. Personal communication.

10. van Bezooijen, van Noord, M.J. and Knook, D.L. The viability
    of parenchymal liver cells isolated from young and old
    rats. Mech. Age. Dev., in press.

11. Mills, D.H. and Zucker-Franklin, D. 1969. Electron micro-
    scopic study of isolated Kupffer cells. Am. J. Pathol. 54:
    147.

12. Wisse, E., Roels, F., de Prest, D, van der Meulen, J., Emeis, J.J.
    and Daems, W. Th. 1973. Peroxidatic reaction of Kupffer
    and parenchymal cells compred in rat liver. In Electron
    Microscopy and Cytochemistry (E. Wisse, W. Th. Daems,
    I. Molenaar and P. van Duijn, editors), Amsterdam, North-
    Holland Publishing Company, p. 119.

13. Hasan, M. and Glees, P. 1972. Genesis and possible dis-
    solution of neuronal lipofuscin. Gerontologia 18: 217.

14. Wisse, E. Observations on the fine structure and peroxidase
    cytochemistry of normal rat liver Kupffer cells. J.
    Ultrastruct. Res., in press.

15. Loud, A.V. 1968. A quantitative sterological description of
    the ultrastructure of normal rat liver parenchymal cells.
    J. Cell Biol. 37: 27.

16. Robinson, D. and Willcox, P. 1969. 4-Methylumbelliferylphos-
    phate as a substrate for lysosomal acid phosphatase. Biochim.
    Biophys. Acta 191: 181.

17. Furth, A.J. and Robinson, D. 1965. Specificity and multiple
    forms of β-galactosidase in the rat. Biochem. J. 97: 59.

18. Barrett, A.J. 1972. Lysosomal enzymes. IN Lysosomes-
     a Laboratory Handbook (J.T. Dingle, editor), Amsterdam-
     London, North-Holland Publishing Company, p. 46.

19. Arborgh, B., Berg, T. and Ericsson, J.L.E. 1973. FEBS Letters
     35: 51.

                          QUESTIONS TO DR. KNOOK

Dr. Martin: Is the choice of the 5 month-old rat an appropriate
"young" control, since these animals may still be undergoing appreciable growth? In aging research generally, it is best to choose
as ones control a mature animal not undergoing active growth?

Dr. Knook: We have decided to compare rats of 3 age groups, WI rats
3-months of age or young animals. Twelve month-old rats or adult
animals and 30-35-month-old rats or old animals. In my presentation
I have compared the results obtained for young and old rats and we
will investigate adult rats in the near future.

Dr. Leuenberger: Did you specify the observed decrease of certain
acid hydrolases of ageing non-parenchymal cells within different
cell fractions (cytoplasmic, lysosomal nuclear)? Have you any idea
about an uneven distribution of lysosomal enzymes within different
types of lysosomes?

Dr. Knook: Concerning the first part of your question, we didn't
investigate the distribution of lysosomal enzymes in various subcellular fractions. It is possible to distinguish between the
"free" activity of the lysosomal enzymes, which means the activity
present in the cytoplasm outside the lysosomes and the "bound"
activity representing the enzyme activities within the lysosomes.
However, both the distributions of the enzymes over the various
fractions and the ratio of free and bound activity will be influenced by the homogenization technique used. Since Kupffer cells,
as was shown in the electron micrographs contain some very large
and probably unstable lysosomal structures. These structures would
be easily disrupted upon homogenization of the Kupffer cells. I am
afraid that the data obtained after homogenization don't represent
the original localization of the lysosomal enzymes in the Kupffer
cells. With regard to your second question I have to say that indeed
a great variety of the lysosomal structures in Kupffer cells suggests
an uneven distribution of the lysosomal enzymes, but we have no
data yet to support this idea.

Dr. Kasten: Have you considered examining the functional activities
of lysosomes in Kupffer cells by challenging the animals with India
ink and looking at the uptake in young and old animals?

Dr. Knook: From the literature it is known that isolated Kupffer cells activity phagocytize various materials, e.g., latex particles. In my opinion the best approach for the study of age related changes in functional activities of lysosomes in Kupffer cells will be the examination of the uptake of latex particles or labelled compounds by Kupffer cells isolated from young and old rats.

Dr. Maceiera-Coelho: Is there any relationship between the increase of lysosomal enzyme activities and the levels of protein synthesis?

Dr. Knook: Enhanced enzyme levels can be explained in 2 ways: both by an increase in protein synthesis and by a decrease in the degradation of active enzyme molecules. It is rather difficult to discriminate between these 2 possibilities.

Dr. Cristofalo: I wonder if you would comment on your view of the significance of the increased lysosomal size and its relationship to the functional impairment observed in liver from older animals.

Dr. Knook: The increased lysosomal size can be considered as an indication of increased lysosomal activities. Concerning the reported increase of lysosomal enzymes I should like to point out that an important question is the localization of lysosomal activities in the cells. If a considerable amount of enzymes is present outside the lysosomes, then it is easy to understand that then extra-lysosomal activity may cause great damage to the cellular metabolism. We have planned some experiments to determine the amount of lysosome activity. Another point is that the increase in lysosomal structures in parenchymal cells is also correlated with the increase in the amount of lipofuscin. We have already performed some experiments on the possible effects of lipofuscin on mitochondrial and lysosomal activities, but that is another story.

THE EFFECT OF AGE ON MITOCHONDRIAL ULTRASTRUCTURE AND ENZYMES

P. D. Wilson and L. M. Franks

Imperial Cancer Research Fund, 44, Lincoln's Inn Fields
London WC2A 3PX, England

SUMMARY

The ultrastructure of perfused livers and of mitochondrial fractions from 6 months and 30 month-old C57/BL mice were studied. In old mice the liver cell mitochondria were enlarged and rounded with a light "foamy", vacuolated matrix, short cristae and a loss of dense granules. Quantitative studies showed a 60% increase in the mean size and an increased proportion of larger mitochondria in intact 30 month-old perfused livers. Endothelial and Kupffer cell mitochondria were smaller than those of the parenchymal cells.

Mitochondria in pellets prepared from 6 and 30 month-old livers were rounded and condensed although there were a few larger and "foamy" mitochondria in the preparations from old mice. Up to 47% of large mitochondria in the old livers were lost during cell fractionation. The levels of cytochrome oxidase and malate dehydrogenase were slightly decreased with age but their cytochemical localization was unchanged.

INTRODUCTION

There is considerable evidence to suggest that there are mitochondrial changes in ageing and in tumours (1-9) but there are few detailed studies. In this preliminary paper we review the problem and describe age-associated changes in mitochondria in the mouse liver in our colony of ageing mice. We also attempt to make a correlation between structural and functional changes in mitochondria.

MATERIALS AND METHODS

The livers from 6 month-old and 30 month-old male C57BL/ICRF/a[t] mice were compared. For morphology the livers were perfused with Waymouth's medium and 2.5% glutaraldehyde in 0.2 M cacodylate buffer. Mitochondrial fractions were prepared from liver homogenates by a modification of Schneider's method (10). The pellets were then fixed in 2.5% glutaraldehyde or used for biochemistry. All specimens for morphology were post-fixed in Palade's fluid, dehydrated in graded alcohols and embedded in Araldite using expoxy propane as transitional solvent. Uranyl acetate and lead citrate stained sections were viewed under an Hitachi HS7S or Siemens 1 electron microscope. For cytochemistry, the livers were fixed for 1 hr in 0.1 M phosphate buffered 4% formaldehyde, or for 30 min in 1.5% glutaraldehyde in 0.2 M cacodylate buffer. Tissues were then rinsed in the appropriate buffer and chopped into 50 to 100 slices, and incubated in media for the demonstration of cytochrome oxidase (11) or malate dehydrogenase (12) for 45 min at 37°C. After rinsing OSO_4 post-fixation and rapid alcohol dehydration, the tissues were embedded in Spurr low viscosity embedding medium (13). Cytochrome oxidase and malate dehydrogenase activities were assayed spectrophotometrically in total liver homogenates and in mitochondrial preparations (14,15). The number and size of mitochondria were measured using a modification of the techniques described by Weibel (16) and Berger (17).

RESULTS

Morphology of Perfused Young and Old Liver Mitochondria

There were obvious morphological differences between the mitochondria of young (6 month) and old (30 month) mouse livers which had been similarly "well" fixed by perfusion with glutaraldehyde and postfixing with OSO_4.

In the parenchymal cells the mitochondria of young mice appeared to be smaller and perhaps more plentiful in number. Although there was some variation in shape, the majority of mitochondria in young liver were elongated and rod-lie (the "classical" mitochondrial structure) while old liver mitochondria were mostly rounded and enlarged. The degree of enlargement was not uniform but depended on the region of the lobule. The largest mitochondria were found in the central region of the lobules. The arrangement and number of cristae varied but the cristae in young liver mitochondria tended to extend further into the matrix and showed a more regular arrangement than those in the old mitochondria. In the old mitochondria cristae tended to be short and irregularly spaced. An obvious difference was found in matrix density. The matrix in young mitochondria was dense and homogeneous while in the old mitochondria the matrix was light often had a "foamy" vacuolated appearance. Occasionally foamy mitochondria were found in the 6 month-old liver but these were very rare. The most swollen areas of the mitochondria showed the greatest vacuolation. Although normal mitochondrial dense granules were common in young mitochondria they were rarely seen in the old mitochondria. In some mitochondria a separation of the inner and outer membranes was noted but his seemed to occur in both young and old mitochondria.

There was also a considerable difference between the morphology of mitochondria of different cell types in the liver within one age group. The mitochondria of the Kupffer and endothelial cells were much smaller than those of the parenchymal cells, although sometimes elongated. The cristal pattern was characteristic. In young liver the cristae often extended almost the complete width of the mitochondria, sometimes with a wavy outline. Occasionally, the cristae were oblique or longitudinal within the mitochondria. In the old liver the Kupffer and endothelial cell mitochondria were often increased in size, while the matrix density decreased and showed patchy vacuolation. Some endothelial cell cristae showed localized swelling.

The age-associated changes did not affect all cells or all mitochondria in a single cell to the same extent. In some cells apparently severely affected and apparently normal mitochondria could be found side by side.

In the old mouse liver there was some evidence of fibrosis in the subendothelial space of Disse, which contained an increased amount of collagen and interstitial material. Lipofuchsin granules were present in many cells.

Morphology of Isolated Mitochondrial Fractions

The fractions obtained were relatively pure consisting of mitochondria with some glycogen contamination. There was little if any difference in the structure of mitochondria isolated from young or old mice. They were rounded and showed the "condensed" configuration with dense matrix and swollen cristae characteristic of active respiration. The mitochondria from old liver tended to be larger and occasionally there were distended mitochondria with a light "foamy" matrix. These were rarely found in young preparations.

Quantitation

The total profile area in a plane through random particles
in space is equal to the total volume occupied by the particles in
that space (17). The principle was employed by Weibel (16) in his
development of stereological techniques. The method was employed on
electron micrographs of young and old perfused liver and of isolated
mitochondria to give a comparison of size. A transparent sheet
of equally spaced dots was superimposed over electron micrographs
of the same magnification and the numbers of dots touching any
mitochondrion was recorded. An estimate of relative number of
mitochondria was made by counting mitochondria per unit area.

Perfused Mouse Liver

Table I shows the results of the determination of the mean
size of mitochondria in perfused liver from 6 month and 30 month-
old mice. This was measured as the mean number of dots per mitochondrion. There was a significant increase of approximately 60%
in the mean size of mitochondria in th 30 month-old liver as compared
with the 6 month-old. The standard deviation (S.D.) was larger in
the 30 month-old livers indicating that there was a wider range
of variation in mitochondrial size in the old animals.

TABLE I

Mean Size of Mitochondria (number of dots per mitochondrion at
standard magnification)

	6 Month	30 Month
Perfused liver	3.66	5.77
	S.D. 0.38	S.D. 1.03
Isolated mitochondria	2.47	2.96
	S.D. 0.32	S.D. 0.26

The size distribution of mitochondria was different in the
6 month and 30 month-old liver (Fig. 1a). The 30 month-old
tissue contained a lower proportion of small mitochondria and a
higher proportion of larger mitochondria than the livers of 6 month-
old animals.

Table II shows that there was a decrease in the total number of mitochondria per unit area of tissue in the 30 month-old liver, but the difference was not statistically significant.

Isolated Mitochondria

There was a slight increase in the mean size of mitochondria isolated from 30 month-old livers compared with those from 6 month-old animals, but this increase was much less than observed for the perfused livers (Table I). Figure 1b shows that the size distributions for mitochondria isolated from 6 month-old and 30 month-old livers were different. Again, the 30 month-old mitochondrial fractions contained a lower proportion of small mitochondria and a higher percentage of large mitochondria than those from 6 month-old livers. None of the very large mitochondria was present in either fraction (over 12 dots) and there was a loss of a proportion of all large mitochondria (4 dots and over) compared with the intact livers. This amounted to 28% of the total in the 6 month-old and 47% in the 30 month-old animals.

TABLE II

Total Number of Mitochondria (per 10 X 8 field of perfused liver at standard magnification)

6 Month	30 Month
182	145
S.D. 41.5	S.D. 31.1

Cytochemistry

Cytochrome oxidase was present on the outer surface of the inner mitochondrial membrane in all mitochondria in 6 month-old livers. Reaction product usually extended into the space between the inner and outer limiting membranes and into the intracristal space. There was some local variation in intensity of reaction product within the cristae of individual mitochondria. Cytochrome oxidase was similarly present on the internal mitochondrial membrane in 30 month-old mouse livers but occasionally mitochondria were seen containing little or no reaction product. This was local variation in intensity of the product within the cristae.

Malate dehydrogenase was localized in the matrix of the mitochondria in 6 month-old and 30 month-old livers. There was no obvious change in the localization with age.

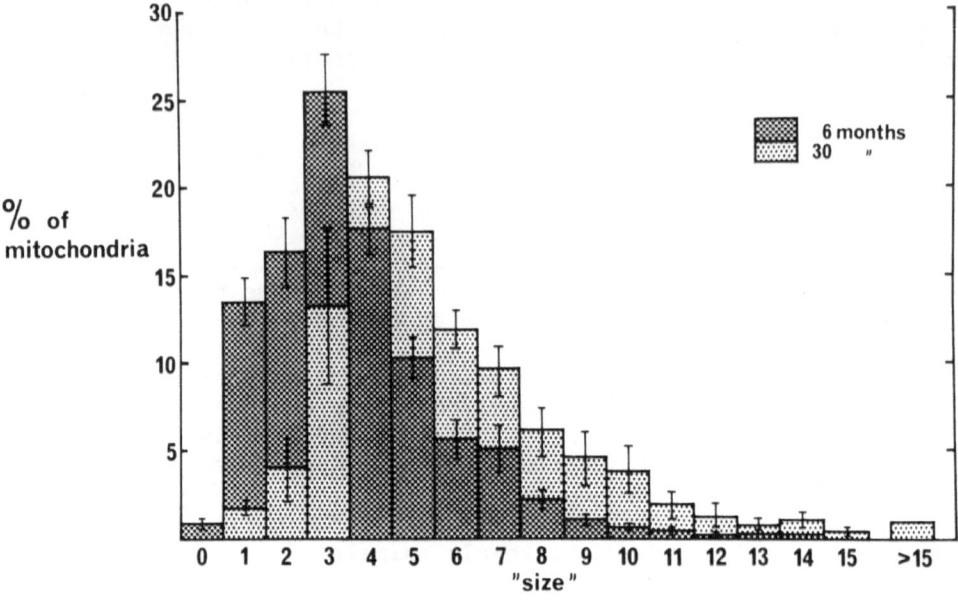

Fig. 1a.

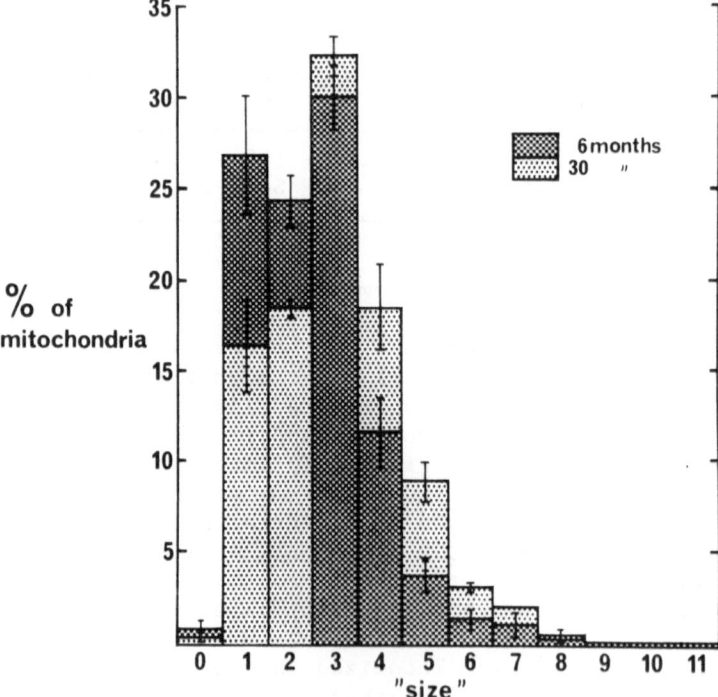

Fig. 1b.

Fig. 1. Size distribution of mitochondria in 6 month and 30 month-old mouse livers. The percentage of mitochondria was plotted against the "size" as estimated by the number of dots contained. Eight mice in each age group. (a) Perfused mouse liver. (b) Isolated mitochondria.

Biochemistry

The biochemical assays of cytochrome oxidase and malate dehydrogenase activities showed that there was a tendency for both of these enzymes to be decreased in the whole tissue homogenates and in the mitochondrial suspensions from the 30 month-old mouse livers. The decrease was small at approximately 10% and may not have been statistically significant (Table III).

TABLE III

Enzyme Activities (units per mg protein)

	6 Month	30 Month	% Decrease
Malate Dehydrogenase			
Whole homogenate	620	581	6
Mitochondria	271	219	19
Cytochrome Oxidase			
Whole homogenate	19.2	15.0	22
Mitochondria	34.9	30.8	12

(1 unit causes a decrease in O.D. of 0.01 per min. Mean values from 4 mice of each age-group.)

DISCUSSION

The morphological changes we have found in old liver cell mitochondria strongly suggest that there may also be functional changes
but the nature of these changes is difficult to define, since
there are many problems involved in any attempt to measure these
changes biochemically. Measurement of total enzyme activity in whole
cell homogenates may not reflect the true functional capacity of
the cells since results obtained with homogenates measure maximal
activity under optimal conditions and take no account of local
intracellular factors. Methods involving cell fractionation and
preparation of isolated mitochondria also have their drawbacks since
our results show that the fractionation procedure leads to the loss
of abnormal mitochondria. Comparative studies using morphological
and biochemical methods should help to overcome some of the difficulties. A particular virtue of the morphological studies is that
they have demonstrated clearly that the changes do not affect all
mitochondria to the same extent. In some instances grossly altered
and normal mitochondria could be seen side by side in the same cell.

The main age-associated structural changes in mitochondria are
a greater variation in mitochondrial size, with a considerable increase in the number of larger mitochondria, decreased matrix density
with vacuolation and loss of dense granules. Some but not all the
differences in morphology resemble those produced by imperfect
fixation. Although this seems unlikely since young and old tissues
were treated in the same way it is possible that the "old" mitochondria may be more fragile and therefore more likely to be damaged
during fixation. Many of the changes have also been described
following a wide range of treatments with toxic agents, including
hydrocortisone (18), chloramphenicol and ethidium bromide (19), etc.
Variation in matrix density is thought to be characteristic of oxygen
deficiency (20).

The occurrence of large mitochondria has been induced by many
agents and in abnormal metabolic conditions, e.g., ACTH (21), cortisone (22), chloramphenicol, and ethidium bromide (19), cuprizone (23),
riboflavin deficiency (24), water deprivation (25), freezing and
thawing (26), and x-irradiation (27). Landis (28) described progressive mitochondrial degeneration in the Purkinje cells of mutant
mice. The features he described were all found in our aged liver
mitochondria. In abnormally large liver mitochondria induced by
cortisone administration (22) there were multiple defects in the
respiratory chain. This has led us to believe that the same may be
true in the aged liver. Preliminary experiments using a Clarke
oxygen electrode (unpublished observations) suggests that there
is a decreased rate of respiration and a resistance to the inhibition
of respiration by antimycin in mitochondria isolated from a 30
month-old liver compared with those from a 6 month-old.

The increase in size in old mouse liver mitochondria varied with the region of the liver love. This agreed with the reported gradient of size of normal liver mitochondria decreasing from the central arteriole towards the edge of the lobule (17). Kimberg and Loeb (22) also found that the increase in size of liver mitochondria induced by cortisone varied from 35% in the centrilobular cells to 400% in the peripheral cells. The increase in size of the mitochondria in old liver was apparently possibly accompanied by a decrease in number. This could be due to the loss of some mitochondria and the swelling of others or the fusion of mitochondria.

The results of the enzyme cytochemical studies in vivo suggest that there is little change in the distribution of the marker enzymes for the inner membrane (cytochrome oxidase) and the matrix (malate dehydrogenase).

Although there were biochemical changes in the levels of these enzymes their significance is debatable. In these experiments the decreased between 6 and 30 months were small, approximately 10% which may not be significant. In an earlier survey of 13 different enzyme activities in male and female mouse kidneys, livers and lungs (29,30) the only enzymes which showed relatively similar patterns of change with age in these three tissues were the mitochondrial enzymes succinic dehydrogenase and cytochrome oxidase. Characteristically, these enzymes were slightly decreased at 30 months when compared with their levels at 6 months but there was a high peak of activity at 18 months. When others reported enzyme changes in the ageing of other tissues and species were taken into consideration it was apparent that an extremely complex situation existed concerning enzymes and ageing. Any particular enzyme could decrease, remain constant, or increase with age depending on its location. However, it was found that decreased mitochondrial enzyme activity was a common effect of tissue ageing (31). This has been reported more frequently for enzymes of the inner mitochondrial membrane than for those of the outer membrane. The matrix enzymes have not been studied in much detail.

In conclusion, our results confirm the earlier suggestion that there is an increase in mitochondrial size with age and that this is probably accompanied by functional changes in the mitochondria. Morphological studies suggest that the major changes occur in the matrix though these maybe a function of changes in membrane permeability. Morphological studies also show that the mitochondria of different cell types of the complex organ may be affected in different ways, and this type of alternation may be masked using standard biochemical methods of homogenization and fractionation.

REFERENCES

1. Tauchi, H., Sato, T., Hashno, M., Kobayarhi, H., Ada Chi, F.,
    Aoki, J., and Masuko, T. 1964. Studies on correlation be-
    tween ultrastructure and enzymatic activities on the parenchy-
    mal cells in senescence. IN: Age with a Future (Hansen, P.F.,
    editor), Munksgarrd, Copenhagen.

2. Chen, J.C., Warshaw, J.B., and Sanadi, D.R. 1972. Regulation
    of mitochondrial respiration in senescence. J. Cell. Phys. 80:
    141.

3. Menzies, R.A., and Gold, P.H. 1972. The apparent turnover of
    mitochondria, ribosomes and sRNA of the brain in young adult
    and aged rats. J. Neurochem. 19: 1671.

4. Tribe, M.A., and Ashurst, D.E. 1972. Biochemical and structural
    variation in the flight muscle mitochondria of ageing blow-
    flies, Calliphora erythrocephala. J. Cell Sci. 10: 443.

5. Siliprandi, D., Siliprandi, M., Scutari, G., and Zoccarato, F.
    1973. Restoration of some energy linked processes lost during
    the ageing of rat liver mitochondria. Biochem. Biophys. Res.
    Comm. 55: 563.

6. Bernhard, W. 1969. Ultrastructure of the cancer cell. IN:
    Handbook of Molecular Cytology (A. Lima de Faria, editor),
    North Holland.

7. Frolkis, W.V., and Bogatskaya, L.M. 1968. The energy metabolism
    of myocardium and its regulation in animals of various ages.
    Exper. Geront. 3: 199.

8. Glew, R.H., Zatzkin, J.B., and Kayman, S.C. 1973. A comparative
    study of the interaction between concanavalin A and mitochondria
    from normal and malignant cells. Cancer Res. 33: 2135.

9. Martin, A.P., Cornbleet, P.J., Lucas, F.V., Morris, H.P., and
    Vorbeck, M.L. 1974. Mitochondrial membrane associated
    properties of Morris hepatomas. Cancer Res. 34: 850.

10. Falcone, A.B., and Hadler, H.I. 1968. Action of granucidin
    on mitochondria. I. Ion-dependent mitochondrial volume changes
    energized by adenosine 5'-triphosphate. Arch. Biochem.
    Biophys. 124: 91.

11. Seligman, A.H., Marnorsky, M.J., Wasserkrug, H.L., and Hanker,
    J.S. 1968. Nondroplet ultrastructural demonstration of cyto-
    chrome oxidase activity with a polymerizing osmophilic reagent,

diamino benzidine (DAB). J. Cell Biol. 38: 1.

12. Pearse, A.G.E. 1968. Histochemistry: Theoretical and applied.
     Volume II, 3rd Edition, Churchill.

13. Spurr, A.R. 1969. A low viscosity epoxy resin embedding medium
     for electron microscopy. J. Ultrastruct. Res. 26: 31.

14. Cooperstein, S.J., and Lazarow, A. 1951. A microspectrophoto-
     metric method for the determination of cytochrome oxidase.
     J. Biol. Chem. 189: 665.

15. Ochoa, S. 1955. Methods in Enzymology (Colowick and Kaplan,
     editors), volume 1, p. 735, Academic Press, Inc.

16. Weibel, E.R., Kistler, G.S., and Scherele, W.F. 1966. Practical
     sterological methods for morphometric cytology. J. Cell Biol.
     30: 23.

17. Berger, E.R. 1973. The morphologically different mitochondrial
     populations in the rat hepatocyte as determined by quantitative
     three-dimensional electron microscopy. J. Ultrastruct. Res.
     45: 303.

18. Kodama, M., Sugiura, T., and Kodama, T. 1972. Ultrastructural
     changes of mitochondria of the neoplastic cells following the
     administration of corticosteroids. Cancer Res. 32: 215.

19. King, M., Godman, G., and King, D. 1972. Respiratory enzymes
     and mitochondrial morphology of HeLa and L cells treated with
     chloramphemicol and ethidium bromide. J. Cell Biol. 53: 127.

20. King, M., and King, D. 1971. Respiratory enzyme activity and
     mitochondrial morphology of L cells under prolonged oxygen
     deprivation. Lab. Invest. 25: 374.

21. Kahri, A. Effect of actinomycin D and puromycin on the ACTH
     induced ultrastructural transformation of mitochondria of cor-
     tical cells of rat adrenals in tissue culture. J. Cell Biol.
     36: 181.

22. Kimberg, D.V., and Loeb, J.N. 1972. Effect of cortisone ad-
     ministration on rat liver mitochondria. Support for the con-
     cept of mitochondrial fusion. J. Cell. Biol. 55: 635.

23. Tandler, B., and Hoppel, C.L. 1973. Division of giant mito-
     chondria during recovery from cuprizone intoxication. J.
     Cell Biol. 56: 266.

24. Tandler, B., Erlandson, R.A., Smith, A.L., and Wynder, E.L.
     1969. Riboflavin and mouse hepatic cell structure and function.
     II. Division of mitochondria during recovery from simple
     deficiency. J. Cell. Biol. 41: 477.

25. Bartok, I., Viragh, S., and Henyhart, J. 1973. Prompt divisions
     and peculiar transformation of cristae in liver mitochondria
     of rats rehydrated after prolonged water deprivation. J.
     Ultrastruct. Res. 44: 49.

26. Sherman, J.K. 1971. Correlation in ultrastructural cryo-
     injury of mitochondria with aspects of their respiratory
     function. Exp. Cell Res. 66: 378.

27. Jordan, S.W., Dean, P.H., and Ahlquist, J. 1972. Early
     ultrastructural effects of ionizing radiation. I. Mitochondrial
     and nuclear changes. Lab. Invest. 27: 538.

28. Landis, S.C. 1973. Ultrastructural changes in the mitochondria
     of cerebellar Purkinje cells of nervous mutant mice. J. Cell
     Biol. 57: 782.

29. Wilson, P.D., and Franks, L.M. 1971. Enzyme patterns in
     young and old mouse kidneys. Gerontologia 17: 16.

30. Wilson, P.D. 1972. Enzyme patterns in young and old mouse
     livers and lungs. Gerontologia 18: 36.

31. Wilson, P.D. 1973. Enzyme changes in ageing mammals. Geron-
     tologia 19: 79.

QUESTIONS TO DR. WILSON

Dr. Leuenbereger: The observed differences in shape and volume
between in situ perfusion fixed and isolated mitochondria are
striking and might indicate an increased sensitivity of mitochondrial membranes against glutaraldehyde fixation. Did you try
straight osmium fixation?

Dr. Wilson: We have not used osmium fixation alone but we have
had experience with immersion in glutaraldehyde and with a modified
Karnovsky fixative which is a combination of glutaraldehyde and
formaldehyde which penetrates tissues rapidly and is reputed to be
an optimal fixative for mitochondrial ultrastructure. In all cases
we have found similar differences in the structure of isolated
mitochondria and those in intact tissue.

Dr. Knook: My question also concerns the dense ultrastructure of your mitochondria after isolation. In isolated parenchymal cells we often observe mitochondria with a dense ultrastructure and after careful isolation of these mitochondria from the isolated parenchymal cells, the mitochondria appear to possess an uncoupled respiration. Mitochondria isolated from liver tissue do show, of course, a coupled respiration. Since Buffa et al. in 1970 (Nature 226: 272, 1970) suggested that there might be a relationship between the dense ultrastructural appearance of mitochondria and uncoupling of phosphorylation and respiration, I wonder whether you performed biochemical measurements on the respiration and phosphorylation of the isolated mitochondria?

Dr. Wilson: We have just started to look at this problem using a Clarke oxygen electrode. It has been suggested that condensed configurations are characteristic of active oxidative phosphorylation and a high energy state. Hackenknock and others have noticed changes from the orthodox to the condensed configurations by adding 2-deoxyglucose to generate ADP.

Dr. Knook: You will know that some years ago Tanchi reported the presence of large and irregularly formed mitochondria in old hepatocytes. We have also observed the presence of this kind of mitochondria, not only in parenchymal line from old rats, after isolation and in situ, but also in parenchymal cells from young rats. Performing morphometric studies among comparable lines as you did, we have observed the absence of significant increases in the cytoplasmic volume occupied by mitochondrial structures in parenchymal cells in situ. These results are in contradiction to your results for the ageing mouse liver.

Dr. Wilson: Yes, that is so. However, I would not think that mitochrondria were causal in ageing but rather in our particular system, they represent an adverse reaction to stress in the old liver, not apparent in the young.

Dr. Gutmann: Do you have any data on the number of mitochondria based on cell number or cell surface. As cell size probably decreases, this might affect functional significance by alteration of percent ratio of number to cell size.

Dr. Wilson: No, we have not been able to get accurate counts of mitochondrial number per cell because of the nature of the liver. The cell outlines are frequently indistinct in liver. However, I am interested in the prospect of being able to isolate parenchymal cells from non-parenchymal cells as Dr. Knook has described.

CREATINEPHOSPHOKINASE IN HUMAN DIPLOID CELL LINES

M. Macek, H. Tomášová, J. Hurych, D. Řezáčová

Institute for the Child Development Research, Faculty
of Pediatrics, Charles University, Institute of Hygiene
and Epidemiology and Institute of Sera and Vaccines,
Prague, Czechoslovakia

The creatinephosphokinase (CPK) has been intensively studied in
human serum of normal individuals and in different pathological
conditions, especially in various neuromuscular diseases. In some
of them, as in Duchenne myopathy, its activity is significantly
increased (1). In human tissues, the highest activity was found in
muscle, heart and brain. Very low activity was detected in uterus,
gall bladder, pylorus, adrenal and thyroid glands (2) and lung.

Thus far, CPK has not been examined in human diploid cell
strains. The activity of CPK and its isoenzymes has been investigated only in tissue cultures of chick and rat muscle (3-7).
Therefore, the aim of our study was to determine the CPK activity
in long term cultivated human diploid strains from different tissues
in order to present some new information about this enzyme which is
involved in the energy metabolism of cells.

MATERIAL AND METHODS

The cultures examined were derived from skin, thymus, spleen,
lung, muscle, pericardium, umbilical cord tissue fragments (8) and
amniotic fluid cells (9) of one embryo, 13 foetuses, and from 3
children of both sex, started 2-3 days after their death.

The amniotic fluid cells and all children tissues were cultivated in medium EPL (10), enriched by vitamins of Eagle's MEM medium
and 5-10% of calf serum. In the second phase of cultivation, these
cultures were propagated without serum, similarily as embryonal and
other foetal cultures, which were intiated in this medium without
serum.

185

The cultures were investigated 3-25 weeks after their initiation in the phase of active growth. They were harvested 6-7 days
after the subcultivation, when confluent cell monolayers were
formed. The homogenized cell suspension was centrifuged (20',
19,000 g, 4°C). The activity of CPK and nitrogen (N) were determined in the supernate immediately. The activity of CPK was measured
by the optical UV test and expressed as a ratio of i.u. CPK/mg N.
The specificity of CPK determination was confirmed by the inhibition of the reaction by monoiodoacetic acid.

Simultaneously, the concentration of lactic acid and pH of the
culture medium were determined in order to compare the enzymatic
and metabolic activity of cultivated cells. All data are given by
arithmetic mean and its 95% upper and lower confidence limits $(ts_{\bar{x}})$.

RESULTS

All strains were characterized by the uniform fibroblastoid
morphology and by normal chromosomal constitution.

No consistent changes of CPK activity were detected in samples
analyzed 3-12 days after subcultivation as well as during 3-25 weeks

TABLE I. Comparison of Creatinephosphokinase Activity in Amniotic
Fluid Cell Cultures and Other Human Diploid Cell Lines

CULTURE	N	CPK i.u./mgN		LACTIC ACID mg / 100 ml		pH	
		$\bar{X}$	$\pm ts\,\bar{X}$	$\bar{X}$	$\pm ts\,\bar{X}$	$\bar{X}$	$\pm ts\,\bar{X}$
AMNIOTIC FLUID CELLS	31	299	143	26.5	7.4	7.05	0.09
SKIN	15	136.3	47.5	24.5	8.1	7.04	0.11
THYMUS	5	107.24	55	37.3	ND	6.80	ND
SPLEEN	2	166	ND	39.3	ND	6.73	ND
MUSCLE	6	415.3	295.5	28	ND	6.97	0.14
PERICARDIUM	8	152.7	68.9	32	11.5	6.88	0.10

N - number of cultures analyzed
ND - not done

of cultivation.   Therefore, it was possible to calculate the average values of CPK activity for each strain.

The CPK activity was present, not only in strains derived from children muscles (Table I), but also in long term cultivated strains of amniotic fluid cells, skin, thymus, spleen, pericardium and umbilical cord.   There were no statistically significant differences between cells derived from different tissues in our sample under investigation.   Higher individual variability was disclosed in some amniotic fluid cell and muscle strains.

The activity of CPK in amniotic fluid cell cultures does not differ from activities of other foetal cultures (skin, thymus, spleen, umbilical cord) derived from foetuses of the same period in pregnancy (15-24 weeks).   The activity of CPK of the whole sample under investigation is 224.8 $\pm$ 68.3.

The comparison of the average values of CPK of embryonal, foetal and children tissues did not disclose essential differences (Fig. 1).   The concentrations of lactic acid did not differ statistically nor did the pH of investigated culture mediums of our strains (Table I, Fig. 1).

In some of amniotic fluid cell cultures, the collagen synthesis was examined in parallel (Fig. 2).   In cultures E 12 and E 26, the identical activity correlates with the similar rate of collagen

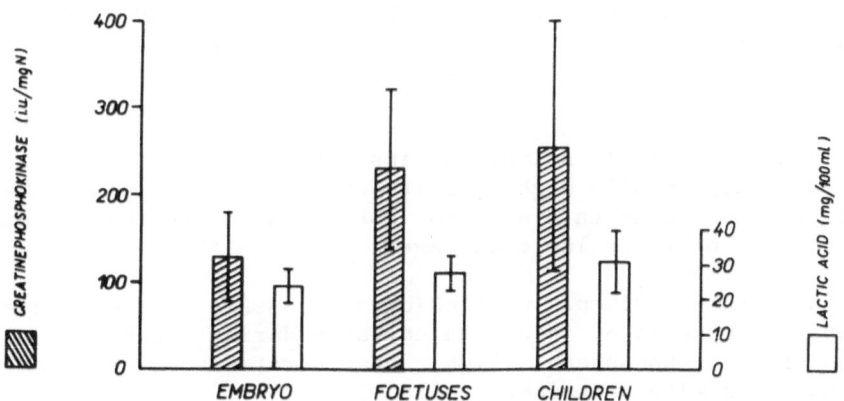

Fig. 1.   Comparison of creatinephosphokinase in embryonal, foetal and children tissue cultures.   Embryonal tissue: lung.   Foetal tissues: amniotic fluid cells, thymus, spleen, skin, umbilical cord.   Children tissues: muscle, thymus, pericardium (post mortem cultures).

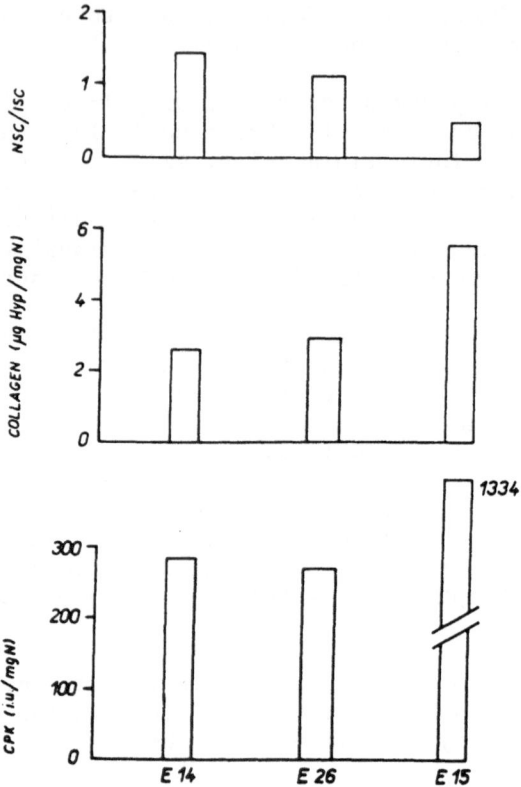

Fig. 2. Creatinephosphokinase activity and collagen synthesis in amniotic cell cultures. NSC = soluble collagen. ISC = insoluble collagen.

synthesis and ratio of soluble to insoluble collagen. In culture E 15 the exceptionally high activity of CPK was accompanied by a higher rate of collagen synthesis, corresponding to the increase in the proportion of insoluble collagen.

Separate experiments with chick embryo skin slices suggest increased synthesis of insoluble collagen during incubation with 0.5 mM creatine phosphate. Higher concentrations of creatine phosphate are without effect (Table II).

DISCUSSION

Our results proved that CPK is present not only in cultures derived from tissues with the high CPK activity in vivo, as could

TABLE II.   Influence of Creatine Phosphate on the Synthesis
            of Collagen In Vitro

Agent	Concentration /mM/	^{14}C–hydroxyproline /counts/min/μmole/	% of control
Control	--	7260[a]	100.0
Creatine phosphate	0.5	8150[b]	112.3
	2.0	7165[b]	98.7
	5.0	7289[b]	100.4

Chick embryo skin slices/2g/ were incubated with ^{14}C–proline/4.63/
μCi/ and creatine phosphate/sodium salt, Koch–Light England/ for 2h
at 37° . Hydroxyproline represents collagen left after the extraction
with 0.5 M NaCL pH 7.4/ insoluble collagen/. Hydroxyproline was
separated after acid hydrolysis on a Dowex 50X8 column.
Efficiency of scintillation counting was 56.7%
a/ mean of 3 parallel samples and 2 hydroxyproline separations
b/ mean of 2 hydroxyproline separations

be expected for cultures derived from muscle. CPK was also found
in cultures from tissues with very low activity in vivo, as for
example in lung. Moreover, the CPK activity was detected in cultures
from tissues, in which it was never investigated in vivo, such as
umbilical cord, thymus, spleen, pericardium, amniotic fluid cells.

On the basis of our observations, we can presume, that CPK
activity does not differ significantly in cultivated cells from different tissues, despite the striking differences in activity in
different tissues in vivo. However, further investigations are
necessary to elucidate the proglems of the greater variability
of CPK activity in some of amniotic fluid cell and muscle cultures.
This might reflect individual and tissue differences which are at
least partially genetically determined, and which influence the

CPK activity of cultivated cells in vitro.

The similarity of the activity of CPK in our samples might be due to the proliferation in vitro of cells with similar biological properties, as suggested by their fibroblast morphology and their collagen synthesis, as documented on parallel investigations of collagen in some amniotic fluid cell cultures.

The comparison of CPK activity and collagen synthesis in particular cultures and in cultures in which we reported collagen synthesis earlier (11-13), indicate that CPK might belong to the systems which provide energy for collagen synthesis. The source of it is so far not well recognized. This hypothesis is supported by our recent findings from experiments with chick embryo skin slices in which creatine phosphate in low concentration stimulated the synthesis of collagen.

The concentration of lactic acid and pH measurements of cultivation medium confirmed that the CPK investigation was carried out under comparable cultivation conditions and similar metabolic activity of the cultivated cells.

It is interesting that the activity of CPK was not altered by postmortem autolysis. It documents the activity in cultures of muscle, thymus and pericardium, started 2-3 days after the death of children. The activity in post mortem thymus cultures was identical as in foetal thymus, where the culture was started immediately after the sectio minor.

Our results demonstrate that it is possible to investigate the pathophysiology of CPK not only in human tissues and extracellular fluids in vivo, but also in fibroblast cultures from different tissues, initiated even post mortem and in the prenatal period of life in long term cultivated amniotic fluid cells.

ACKNOWLEDGEMENT

The excellent technical assistance of H. Ducholkvá, H. Zemanová, and S. Jeník is highly appreciated. The statistical analysis was carried out by L. Pellar. Sectiones minores were performed at the Clinic of Gynaecology and Obstetrics, Faculty of Pediatrics, Charles University, Prague.

SUMMARY

The activity of CPK was proved in cultivated cells from amniotic fluid cells, thymus, spleen, pericardium, umbilical cord, skin

and muscle. The activity in these strains did not significantly
differ. The average value is 224 ± 68 i.u. CPK/mg N. No differences of activity were disclosed between embryonal, foetal and
children cultures. The activity of CPK in post mortem cultures was
not altered. The relationship between collagen synthesis and CPK
activity is suggested. The evidence of CPK activity in amniotic
fluid cells corresponding to the activity in other foetal cultures
and the whole sample under investigation opens the way to the prenatal investigation of pathophysiology and inborn errors of the
metabolism of this enzyme.

REFERENCES

1. Pearce, J.M.S., Pennington, R.J. and Walton, J.N. 1964. Serum
    enzyme studies in muscle disease. II. Serum kinase activity
    in normal individuals. J. Neurol. Neurosurg. Psychiat. 27: 961.

2. Hess, J.W., MacDonald, R.P., Frederick, R.J., Jones, R.N. and
    Nelly, J. 1964. Serum creatine phosphokinase (CPK) activity
    in disorders of heart and skeletal muscle. Ann. Int. Med. 61:
    1015.

3. Reporter, M.C., Konigsberg, I.R. and Strehler, B.L. 1963.
    Kinetics of accumulation of creatine phosphokinase activity
    in developing embryonic skeletal muscle in vivo and in mono-
    layer culture. Exp. Cell Res. 30: 410.

4. Coleman, G.R. and Coleman, A.W. 1968. Muscle differentiation
    and macromolecular synthesis. J. Cell Physiol. 72: 19.

5. Shainberg, A., Jagil, G. and Jaffe, D. 1971. Alteration of
    enzymatic activities during muscle differentiation in vitro.
    Develop. Biol. 25: 1.

6. Morris, G.E., Cooke, A. and Cole, R.J. 1972. Isoenzymes of
    creatinephosphokinase during myogenesis in vitro. Exp. Cell
    Res. 74: 582.

7. Zalin, R. 1972. Creatine kinase activity in cultures of dif-
    ferentiating myoblasts. Biochem. J. 130: 79.

8. Macek, M. and Michl, J. 1964. Contribution to the cultivation
    of human diploid cells. Acta Univ. Car. Med. (Prague) 10: 518.

9. Macek, M., Řezáčová, D. and Kotásek, A. 1972. Simple smear
    method for the sampling and culturing of embryonal and fetal
    tissues. Humangenetik 16: 245.

10. Michl, J. 1961. Metabolism of cells in tissue cultures in vitro. Exp. Cell Res. 23: 324.

11. Macek, M., Hurych, J. and Chvapil, M. 1967. Total hydroxyproline in post-mortem tissue cultures. Cytologia 32: 308.

12. Macek, M., Hurych, J. and Chvapil, M. 1967. The collagen protein formation in tissue cultures of human diploid strains. Cytologia 32: 426.

13. Macek, M., Hurych, J. and Řezáčová, D. 1973. Collagen synthesis in long term amniotic fluid cell cultures. Nature 243: 289.

QUESTION TO DR. MACEK

Dr. Rotzsch: I would like to ask Dr. Macek for the method used for measuring the activity of the CPK? Have you measured the total activity or only the specific one?

Dr. Macek: We have used the UV test (Boehringer) and measured the total activity.

CYTOCHEMICAL AND CYTOGENETIC FINDINGS IN FIVE HUMAN LEUKOCYTE

LONG-TERM CULTURES (LAHL) OF DIFFERENT ORIGIN

V. Pössnerová, M. Macek, F. Heřmanský, J. Fortýnová,
S. Jeník, J. Holý, M. Křeček

Research Laboratory for Blood and Liver Diseases,
Charles University, Institute for the Child Development
Research, Hematology and Blood Transfusion Institute,
Prague, Czechoslovakia

The origin of lymphoblastoid cells in long-term leukocyte suspension cultures remains unclear despite the fact that since the
first report of Iwakata et al. (1) and Moore et al. (2) a great
number of lymphoblastoid lines have been derived. With the aim of
contributing to the clarification of this problem a study was
conducted of five long-term suspension cultures of leukocytes
established in our laboratory from different material. These
lines are designated LAHL 1, 3, 4, 5, 6 and have been examined by
means of cytochemical methods used in hematology, by cytogenetic
methods and by ^{3}H-thymidine autoradiography.

MATERIALS

The donors were patients 36–51 years old. The culture LAHL 1
was obtained from the peripheral blood of a patient with chronic
myelosis in a blastic crisis (Fig. 1). Culture No. 3 was derived
from the spleen of a patient with icterus hemolyticus, culture No. 4
from the peripheral blood of a patient with acute myelomonocytic
leukemia, culture No. 5 from the spleen of a patient with panmyelopathy, and culture No. 6 from the peripheral blood of a patient with
paramyeloblastic leukemia (Table I).

Both at the beginning and after more than 100 days of cultivation, comparative cytochemical examinations were performed. After
100 days all other cells except proliferating blasts disappeared.
Simultaneously the karyotype and the incorporation of ^{3}H-thymidine
by autoradiographic analysis were carried out (1 µCi/ml; 27 Ci/m
mole; 60 min). The doubling time varied between 20–70 hr. The
techniques are described elsewhere (3–6). Three of these five

TABLE I.

lahl cultures	1	3	4	5	6
source of cells	peripheral blood from CML in blastic crisis	spleen from acquired hemolyt. anemia	periph. blood from acute monocytic leukaemia	spleen from pancyto-penia	periph. blood from myeloblastic leukaemia
days of cultivation	300 (end)	164 (end)	180 (end)	364 frozen for two years continued	186 continued
3H- thymidine incorporation % of positive cells >40 grains	19.7	17.6	27.2	18.8	39.0

cultures were lost because of bacterial contamination on days 300, 164 and 180. LAHL No. 5 was frozen at 364 days and thawed after two years. This line LAHL No. 5 and LAIIL No. 6 are still cultivated.

RESULTS

All cells in the cultures were grown entirely in suspension. If not shaken, they fell in clumps to the bottom of the culture flask. Microscopic examination revealed that the transient phenomenon of pseudogranules in some cultures rarely occurred and it lasted a short time only (Fig. 9). In the initial phase of the culture LAHL 3 and LAHL 5, which originated from human spleens, macrophages with a strongly positive Perl's reaction for iron were observed (Fig. 8).

However, after 100 days of cultivation all cells resembled dedifferentiated blastoid elements with basophilic cytoplasm and varying distinctly in their size. The basophilic cytoplasm of these cells was occasionally vacuolated without specific granules; sometimes it contained azurophilic "dust". Signs of phagocytosis or even canibalism were rare. At this time all the nuclei displayed a fine chromatin structure and the staining of ribonucleoprotein structure in the cytoplasm and in nucleoli was strongly positive (Fig. 6). The PAS reaction was positive with certain variations

of intensity. When stained with Sudan Black B or with Oil-red O
for neutral fats, the reaction was negative except for some of the
vacuoles which had blue-grey and pinkish shades, respectively (Fig.
5).   In this advanced time, the Perl's reaction was negative (Table
I).   Lactate dehydrogenase and acid phosphatase showed slightly to
moderately increased positivity (Figs. 4, 7).   Naphtol AS-D
chloroacetate esterase was occasionally moderately positive in the
form of granules dispersed irregularly in the cytoplasm and in the
vacuoles. The positivity of nonspecific esterase was slightly
increased (Fig. 3). The strong positivity, typical in blood smears
for macrophages, was found neither in culture 3 with phagocytosis
nor in culture 4, which was derived from myelo-monocytic leukemia.
Blocking of the nonspecific esterase with fluoride diminished the
positivity of the cytoplasm.   However, the intensity of the reaction
in vacuoles remained unchanged (Table II).   The autoradiographic
examination of the cultures revealed 17.8-39% of cells with more
than 40 granules (Fig. 2).   Incorporation of this nucleotide, PAS
positivity and activities of enzymes fluctuated during the growth
to a certain extent, which could be explained in part by the
inevitable changes in the culture conditions.

TABLE II.   LAHL Cultures After 100 Days

METHOD	1	3	4	5	6	
RNA	+ +	+ +	+ +	+ +	+ +	nucleoli active
Feulgen (DNA)	fine chromatin					
sudan black B	V	V	V	V	V	
phospho-lipids	V	V	V	V	V	
Fe	0	0	0	0	0	
PAS	+	+	+	+	±	
oil red-O	V		V	V	V	

V = some vacuoles only positive

V. POSSNEROVA ET AL.

TABLE III.   LAHL Long-Term Cultures After 100 Days

METHOD	1	3	4	5	6
peroxydase	0	0	0	0	0
alcaline phosphatase	0	0	0	0	0
acid phosphatase	+	+	+	+ +	+ +
non specific esterase	+	+	+	+ +	+ +
naphtol-AS-D chloro-acetate esterase	V	V	V	V	V
lactic acid dehydrogenase	+ +	+ +	+ +	+ +	+ +

V = some vacuoles moderatly positive

Chromosomal investigation of lines LAHL No. 1 and LAHL No. 3 reported earlier brought evidence that their main cell population was diploid with a low number of heteroploid cells and frequency of aberrations of chromosome structure.

The analysis of two other lines, LAHL 5 and 6, revealed that line LAHL 6 is diploid. Line LAHL 5 is hypodiploid, 66% of the cells had 45 chromosomes. Similarly as in lines LAHL 1 and LAHL 3, the number of hyperdiploid, hypo- and hypertetraploid mitoses was not increased. The frequency of unstable aberrations was not elevated either. The marker chromosomes m 2 with the centromere in the submedian region were also found in low (1%) frequency. The subterminal constriction on the long arm of C 9 (Cqh + 1) chromosome was detected in 1-3% of cells as common in other lymphoblastoid cell lines. Both lines were EBV positive, as shown by the immunofluorescent examinations (Table IV).

The diploid line LAHL 6 derived from a person affected with acute myeloblastoid leukemia was characterized by normal karyotype

TABLE IV.   Chromosomal Examination of Lymphoblastoid Lines LAHL

LINE	N	NUMBER OF CHROMOSOMES (%)						STRUCTURAL ABERRATIONS(%)			EBV
		◂45	45	46	47	48-69	↓4x↑	US	SA	Cgh+	
LAHL 5[†]	100	19	66	12	4	-	-	-	1m₂	1	+
LAHL 6[††]	100	-	11	86	3	-	-	3	1m_T	3	±

US - UNSTABLE STRUCTURAL ABERRATIONS

SA - STABLE STRUCTURAL ABERRATIONS

↓4x↑ - MITOSES WITH HYPO- AND HYPERTETRAPLOID NUMBERS OF CHROMOSOMES

MARKER m_T - CENTROMERE IN TERMINAL REGION

MARKER m₂ - CENTROMERE IN SUBMEDIAN REGION

N - NUMBER OF INVESTIGATED MITOSES

† - PANCYTOPENIA

†† - ACUTE LYMPHOBLASTOID LEUKEMIA

(Table V) 46 XY. It was interesting that 27% of karyotypes exhibited trisomy of C group chromosomes in different karyotype combinations. Monosomy D was evident in 16% of mitoses. The line LAHL 5, derived from a patient with pancytopenia was represented by a population of hypodiploid cells 45 XX C- and cells with normal karyotype 46 XX.

The monosomy C was present in 53% of karyotypes.

DISCUSSION

Long-term leukocyte cultures have attracted a great deal of interest because of their unlimited lifespan (7). Unfortunately, their EB virus positivity diminishes the value of this experimental model. The precise cellular origin of those leukocyte cell lines could not be established with certainty on the basis of morphological characteristics (8-10). We used a spectrum of cytochemical methods, but the results were also inconclusive.

The transient phenomenon of pseudogranula, reported by Clarkson (11) was described by us elsewhere (12) as an artificial phenomenon in samples where residua of Heparin and Neomycin were present. In

V. POSSNEROVA ET AL.

TABLE V.   Results of Karyotype Analysis of Lymphoblastoid Lines LAHL

LINE	N	TRISOMY (%)					MONOSOMY (%)					KARYOTYPE
		C	D	E	F	G	C	D	E	F	G	
LAHL 5	31	3	3	3	7	0	53	3	0	0	0	45,XX,C-/46,XX
LAHL 6	29	27	13	0	0	3	7	16	7	13	7	46,XY

N - NUMBER OF KARYOTYPES INVESTIGATED

the initial stage the iron in macrophages present in the two cultures originating from spleens was derived from phagocytosed erythrocytes. The results of cytochemical examination after 100 days were similar in all five long-term cultures of different origin and did not differ from those which were observed in lymphocytes stimulated by phytomitogens. That is in agreement with the findings of Trujillo (9). At present it is difficult to explain the positivity of some vacuoles of cultivated blasts when stained by Sudan Black B and naphtol-AS-D chloroacetate esterase in the course of cultivation. The positivity of both staining reactions is characteristic for the neutrophilic series only and is used for its identification in hematology. In the long-term cultures the positivity seems to be unspecific, perhaps due to some degenerative changes.

In our opinion the striking similarity of all these 5 long-term suspension cell cultures irrespective of their origin suggest that they originate from lymphoid cells.

The chromosome analysis of our lines LAHL 1 and LAHL 3 was reported earlier (4). Chromosome investigation of 3 from 4 of our lines reveals their diploid model number of chromosomes and their genetic stability upon continuous cultivation (13, 14). The EBV positivity was detected in both lines 5 and 6. In the line LAHL 6 the immunofluorescence was weaker. The subterminal constriction on long arms of C 9 chromosomes was also evident in a low percentage of cells as common to the majority of lymphoblastoid lines. It is remarkable that in lines LAHL 5 and LAHL 6 the karyotype anomalies of C group chromosomes were proved in high proportion of cells. This corresponds to previous findings (13-15). The occurence of similar aberrations described in Burkitt's lymphoma (16, 17), acute leukemia (18), cervical carcinoma (19), suggest the possible similar pathway of clonal evolution of lymphoblastoid lines in vitro and during proliferation of tumor cells in vivo (12).

SUMMARY

     Five human leukocyte long-term suspension cultures were inves-
tigated by means of cytochemical and cytogenetic methods. A striking
resemblance in the morphology of these cells originating either from
peripheral blood or from spleen of patients with or without hemato-
logical disorders was found. Lymphocytic origin of all five
cultures is suspected. In the three cultures a diploid karyotype
was found with some aberrations. In one culture, derived from the
spleen of a patient with panmyelopathy the mosaic 45 XX C -/46 XX
was detected. Anomalies of C group chromosomes were the most consistent type of chromosome aberrations in two lines. Both were EBV
positive.

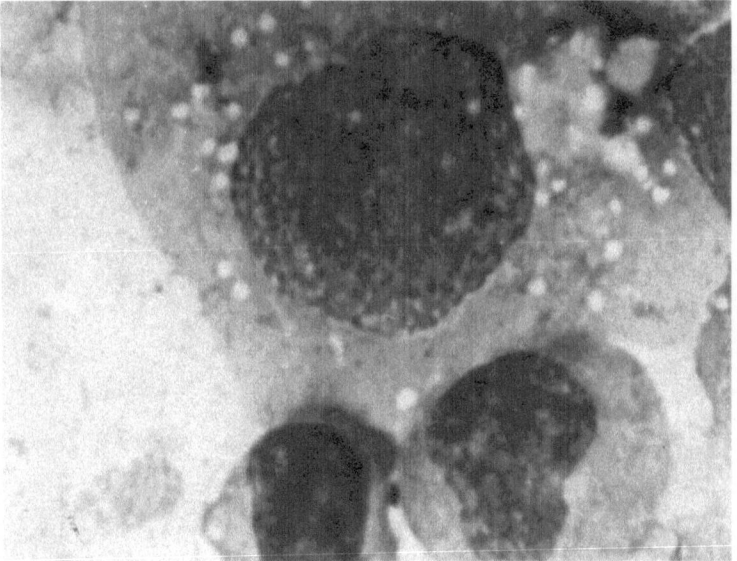

Fig. 1. LAHL 1 culture after 100 days. Giemsa 1000X.

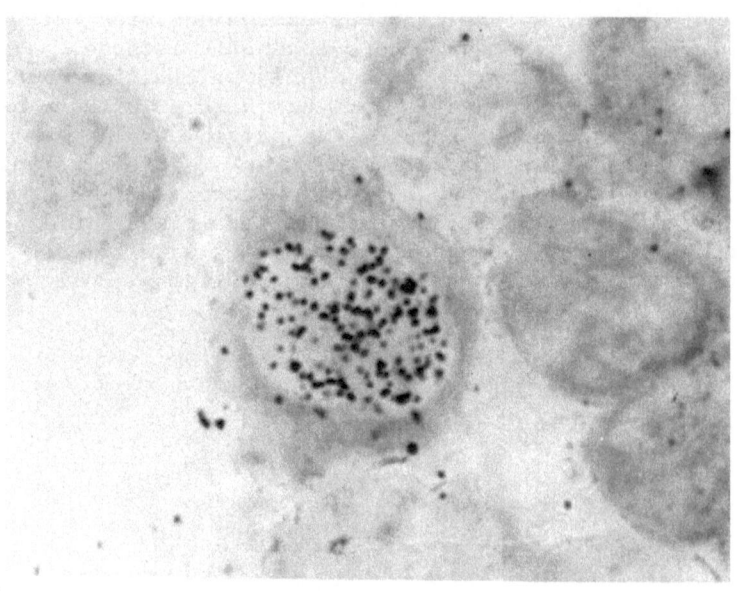

Fig. 2. LAHL 1 Autoradiography by ^{3}H-thymidine on the 189th day.

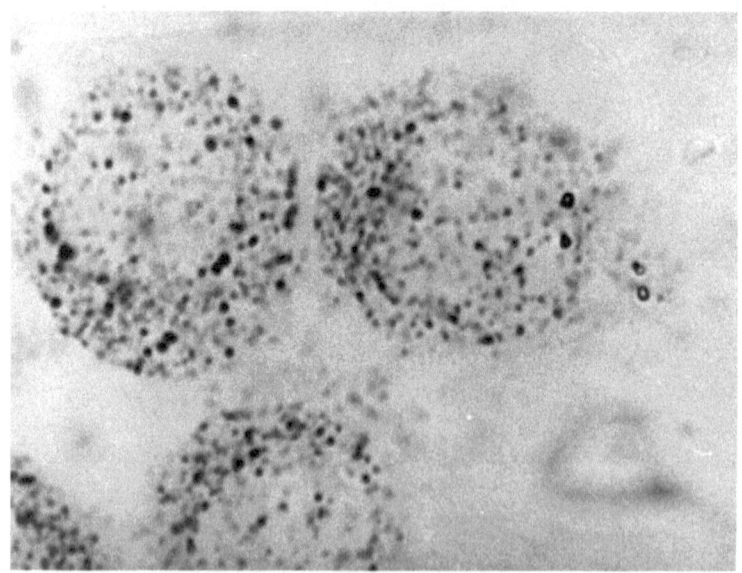

Fig. 3. LAHL 1 Reaction for non-specific esterase on the 252nd day.

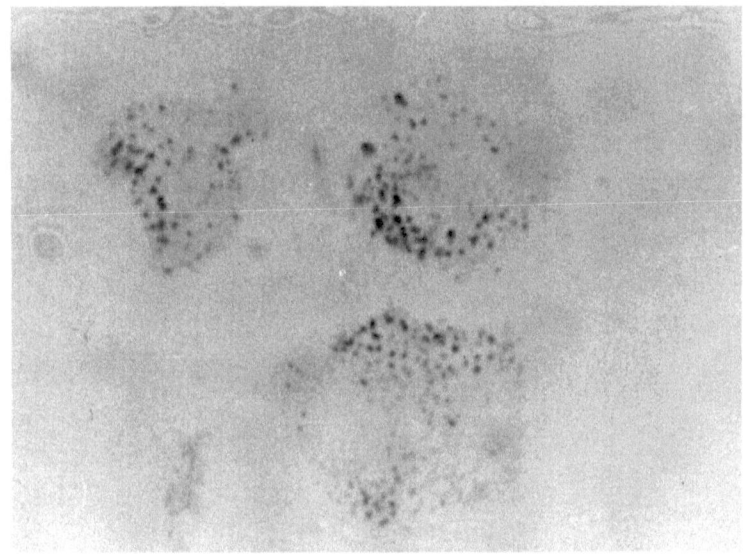

Fig. 4.   LAHL 1   Lactic acid dehydrogenase reaction on the 185th day.

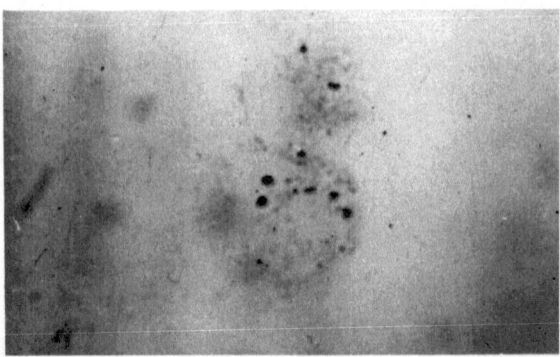

Fig. 5.   LAHL 1 culture 120 days.   Positive staining of some vacuoles with Sudan-Black B.

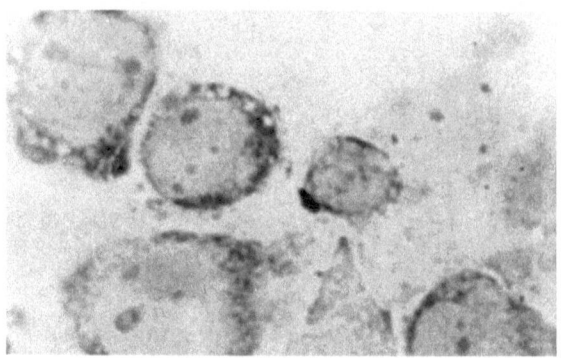

Fig. 6. LAHL 1 culture after 148 days. Staining of structures containing RNA by buffered toluidine blue.

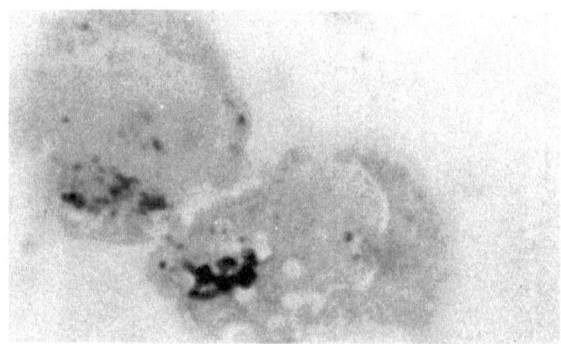

Fig. 7. LAHL 1 culture after 153 days. Reaction for acid phosphatase.

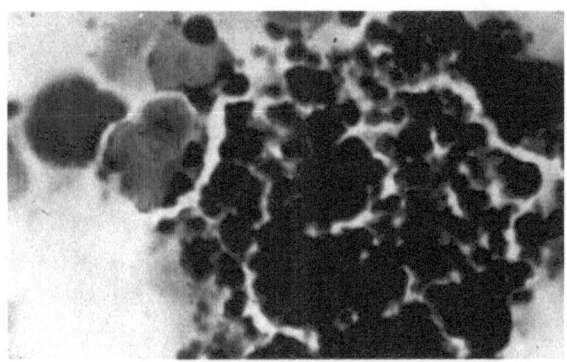

Fig. 8. LAHL 3. Perl's reaction for iron in monocyte on day
36.

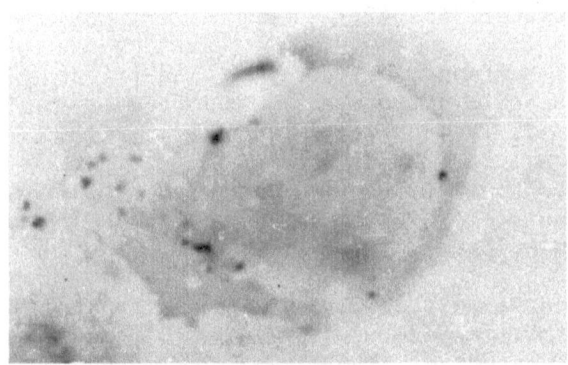

Fig. 9. LAHL 3. Blast with pseudogranules.

REFERENCES

1. Iwakata, S. and Grace, J.T.   1964.   Cultivation in vitro of
    myeloblasts from human leukemia.   New York State J. Med. 64:
    2279.

2. Moore, G.E., Ito, M.D., Ulrich, K. and Sandberg, A. A.   1966.
    Culture of human leukemia cells.   Cancer 19: 713.

3. Possnerová, V.   1967.   Some cytochemical investigations on
    leukocytes from the peripheral blood cultivated in vitro.
    Fol. Morphol. 15: 160.

4. Possnerová, V., Hermanský, F., Smetana, K., Křeček, M. and
    Fortýnová, J.   1973.   Long-term in vitro culture of leukocyte
    lines LAHL 1,2,3.   Neoplasma 20: 395.

5. Possnerová, V., Smetana, K., Křeček, M., Hermanský, F. and
    Fortýnová, J.   1970.   Long-term culture of human leukaemic
    leukocytes.   Neoplasma 17: 513.

6. Macek, M.   1972.   Chromosomal analysis of lymphoblastoid cell
    lines.   Neoplasma 19: 51.

7. Christofinis, G.J.   1969.   Chromosome and transplantation
    results of a human leukocyte cell line derived from a healthy
    individual.   Cancer 24: 649.

8. Moore, G.E. and McLimans, W.F.   1968.   The lifespan of the
    cultured normal cell: Concept derived from studies of human
    lymphoblasts.   J. Theoret. Biol. 20: 217.

9. Trufillo, J.M., Drewinko, B. and Ahearn, M.J.   1972.   The
    ability of tumor cells of the lymphoreticular system to grow
    in vitro.   Cancer Res. 32: 1057.

10. Belpomme, D., de Kaouel, Ch.Le Borgne, Paintrasid, M., Grandjou,
    D., Vennat, A.M., de Nava, Ch., Thoyer, C., de Grouchy, J.,
    de Thé, G. and Mathé, G.   1969.   Aspects morphologiques de
    plusieurs lignées permanents (ICI) établies a partir de sang
    leucémique humain.   Rev. franc. Et. clin. Biol. 14: 848.

11. Clarkson, B. Strife, A. and de Harven, E.   1967.   Continuous
    culture of seven new cell lines (SK 1-7) from patients with
    acute leukaemia.   Cancer 20: 926.

12. Possnerova, V.   1969.   Experimental reproduction of intra-
    cellular granules from heparin and neomycin in cultivated
    normal and leukaemic leucocytes.   Neoplasma 16: 573.

13. Huang, C.C. and Moore, G.E. 1969. Chromosomes of 14 hemato-
     poietic cell lines derived from peripheral blood of persons
     with and without chromosomal anomalies. J. Nat Cancer Inst.
     43: 1119.

14. Macek, M., Seidel, E.H., Lewis, R.T., Brunschwig, J.P. and
     Wimberly, I. 1971. Cytogenetic studies of E.B. virus-
     positive and EB virus-negative lymphoblastoid cell lines.
     Cancer Res. 31: 308.

15. Kohn, G., Diehl, V., Mellman, W.J., Henle, W. and Henle, G.
     1968. C-group chromosome marker in long-term leukocyte
     cultures. J. Nat. Cancer Inst. 41: 795.

16. Chu, E.W., Whang, J.J.K. and Rabson, A.S. 1966. Cytogenetic
     studies of lymphoma cells from an American patient with a
     tumor similar to Burkitt's tumors in African children. J.
     Nat. Cancer Inst. 37: 885.

17. Jacobs, P.A., Tough, I.M. and Wright, D.H. 1963. Cytogenetic
     studies in Burkitt's lymphoma. Lancet 2: 1144.

18. Sandberg, A.A., Takagi, N., Sofuni, T. and Crosswhite, L.H.
     1968. Chromosomes and causation of human cancer and leukemia.
     V: Karyotypic aspects of acute leukaemia. Cancer 22: 1268.

19. Lin, C.C., Dent, P.B., Ward, E.J. and Culloch, P.B. 1973.
     Extra chromosome 7 in cell cultures from cervical carcinoma.
     J. Nat. Cancer Inst. 50: 1399.

QUESTION TO DR. PÖSSNEROVÁ

Dr. Macieira-Coelho: Do you think that the established cell is
the same, when obtained from either normal or leukemic blood?

Dr. Pössnerová: The results of cytochemical examinations were
the same in established cultures from peripheral blood or spleens
of patients either with or without leukemia.

ENZYMATIC DIFFERENCES BETWEEN SHORT-LIVED AND LONG-LIVED DROSOPHILA STRAINS

Jaroslava Skřivanová, Františka A. Gadirová and
Emma Holečková

Department of General Biology, Medical Faculty
Charles University, and Institute of Physiology,
Czechoslovak Academy of Sciences, Prague

One of the phenomena characterizing age ing of the organism
as well as of diploid cells in culture is an increase of the
activity of lysosomal enzymes, among others of the acid phosphatase EC 3.1.3.2. On the other hand the activity of alkaline
phosphatase EC 3.1.3.1. does not change in ageing diploid cells
(1), although its activity increased in hydrocortisone-treated
diploid cells with prolonged lifespan (2). However, the activity
of this enzyme considerably alters during ontogenesis. The
alkaline phosphatase is active in all cells in the course of
early stages of embryogenesis and it is unde tectable during
intermediate period in most tissues. The alkaline phosphatase
again becomes active at later stages in certain tissues (3).
Remarkable alkaline phosphatase activity was also found at early
states of wound healing in fibroblastic cells, but the activity
level of normal fibroblasts was very low (4).

Long- and short-lived banana fly (Drosophila melanogaster)
strains have been used as an experimental model for studies of
metabolic differences between organis ms w ith short and long life-
spans. Short-lived vestigial Drosophilas had a lower esterolytic
activity (5) and a lower intensity of oxidations (6) than the long-
lived wild Drosophilas.

Because both acid and alkaline phosphatase activities stand
in relation to development and ageing, we compared these two
enzymes in short-lived vestigial and long-lived Drosophila
strains.

MATERIALS AND METHODS

Both Drosophila strains used were cultivated on standard glucose-Faex siccata-agar medium at 20-22°C. Larvae, pupae and 1-5 day old flies and 20-25 day old flies of the vestigial and white strain were used for experiments. About 100-150 individuals were collected as the representatives of every ontogenetic stage or population. Larvae and pupae were handled without any treatment, flies under light either narcosis.

Enzyme Preparation

The homogenates of whole larvae, pupae or flies were used as a source of the enzyme. The collected individuals were weighted and homogenized for a short time in 1% sodium desoxycholate solution. After the dilution to final concentration (100 mg of fresh material in 5 ml of desoxycholate solution) the suspension was vigorously stirred for 20 min at room temperature. After this operation the homogenates were diluted with equal volume of 0.9% NaCl. The enzyme preparations were always stored at -20°C before the use (7).

Determination of Enzyme Activity

Alkaline phosphatase activity was measured by hydrolysis of p-nitrophenylphosphate sodium salt in 0.05 M glycine buffer at pH 10.0 and 37°C. After 30 min of the incubation the reaction was stopped by adding 0.25 N NaOH. The intensity of colour obtained was measured at 410 nm. The activity of enzyme was expressed in micrograms of phenol liberated per 1 mg of protein per hr.

Acid phosphatase activity was measured in 0.05 M citrate buffer, pH 5.0 with p-nitrophenylphosphate sodium salt as a substrate. The assay procedure and the determination of colour intensity was the same as with alkaline phosphatase activity. The activity of enzyme was expressed as micrograms of phenol liberated per 1 mg proteins per hr.

Determination of protein content was carried out after the method of Lowry (8).

Each point on the curves, presented in following figures, represents the average of two values obtained from two separately cultivated Drosophila populations.

RESULTS

Figure 1 represents the changes in the activity of acid phosphatase per 1 mg of protein in larvae, pupae and flies. In the wild strain, whose lifespan is twice as long as that of the vestigial strain, the activity of acid phosphatase increases during development from larvae to 1-5 day old imagos. In 20-25 day old imagos, the activity remains unchanged. In flies of the vestigial strain, the changes in enzyme activity of larvae and pupae do not differ from those found in the wild strain. Marked differences appear in imagos: in flies 1-5 days of age, the activity of acid phosphatase is higher in vestigial than in the wild flies of corresponding age, and in the older (20-25 days) flies, the difference still increases.

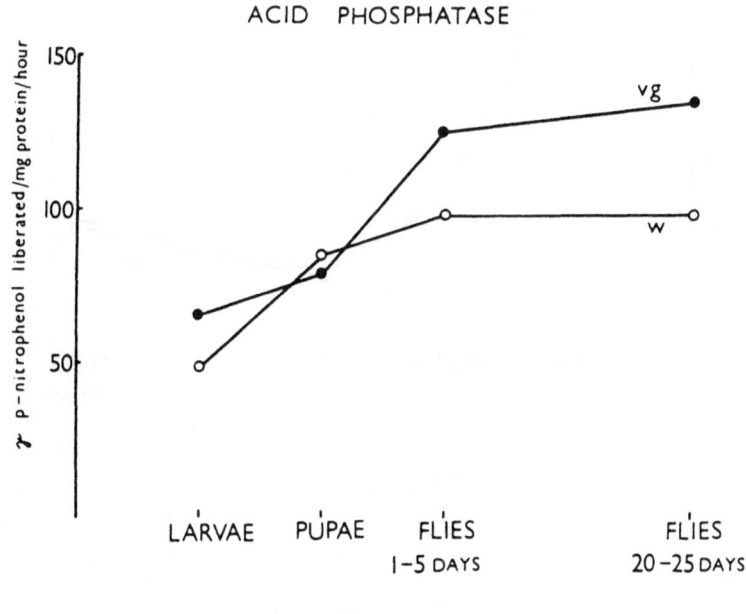

Fig. 1

Figure 2 shows changes in the activity of alkaline phosphatase
in the same material. Here, the activity falls rapidly from the
larval stage to 1-5 day old imagos. Then, it again slightly increases in flies 20-25 days old. No difference between the short-
lived vestigial and the long-lived white flies was found.

The protein content was determined in larvae, pupae and flies
of both Drosophila strains used. No difference was found between
the ontogenetic stages of one strain or between these two strains.

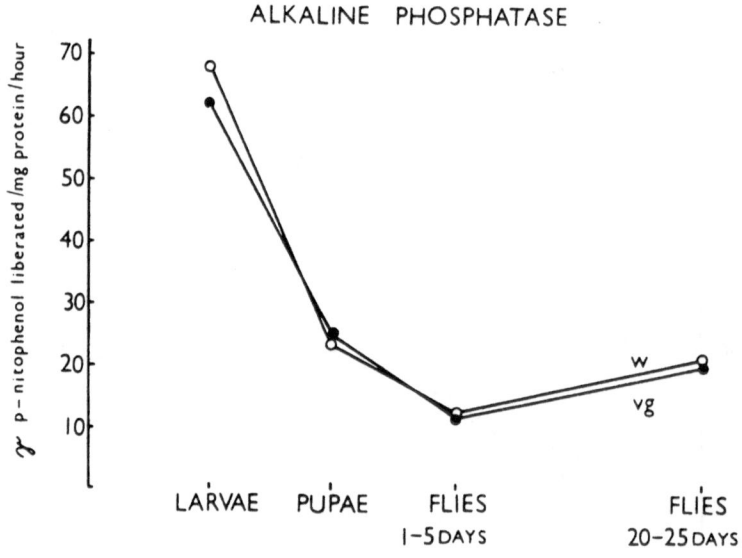

Fig. 2

DISCUSSION

The mean lifespan of the wild Drosophila melanogaster strains is 46 days, that of the vestigial strain is 20 days; shorter life is invariably associated with the character of vestigial wings (9). The white strain which we used in our experiments has normal length wings and can be considered as long living as the wild Drosophilas. Therefore the vestigial flies aged 20-25 days approach the beginning of old age, while the white flies are "adult" but not yet old.

In both strains of flies we have studied, the activity of acid phosphatase increased with age. In the vestigial strain this increase was more pronounced than in the wild strain, which ages at a slower rate. This is in good agreement with the results of Cristofalo (1,2) in ageing diploid cells, where the activity of acid phosphatase increased simultaneously with the prolongation of their in vitro cultivation. It is worthwhile to note that the changes in the activity of this enzyme have practically the same trend in phylogenetically so distant objects.

As for alkaline phosphatase, the decrease of its activity during ontogenesis agrees with the results of some other authors. But also, the slight increase of this enzyme activity, which we can see in older flies, is in conformity with the higher level of several dephosphorylating enzymes in old animals (10).

The metabolic importance of both phosphatases, and especially of the alkaline phosphatase is not quite clear. But they take part in the degradation of many metabolically very important organic phosphates. On the other hand, they influence simultaneously the level of organic phosphates in cells. The metabolic consequence of these changes may play a significant role in the processes maintaining the vitality of the cell.

SUMMARY

The activities of acid and alkaline phosphatase were measured. The activity of acid phosphatase increased during the whole ontogenesis of both Drosophila strains studied--the short-lived vestigial and the long-lived white. This increase was more marked in the more rapidly ageing vestigial flies. The activity of alkaline phosphatase decreased in both strains of flies simultaneously from the larval stage to young imagos. A slight increase of the activity as in old vestigial as in adult white flies appeared.

ACKNOWLEDGEMENTS

The authors are indebted to Dr. M. Ondrej from the Institute of Experimental Botany, Czechoslovak Academy of Sciences, Prague, for his gift of Drosophila strains.

REFERENCES

1. Cristofalo, V.J., Parris, N. and Kritchevsky, D. 1967.
    Enzyme activity during the growth and aging of human cells
    in vitro. J. Cell. Physiol. 69: 263.

2. Cristofalo, V.J. 1970. Metabolic aspects of aging in diploid
    human cells, IN Aging in Cell and Tissue Culture (E. Holeckova
    and V.J. Cristofalo, eds.) Plenum Press, New York-London,
    p. 83.

3. Moog, F. 1959. Cell organism and millieu (D. Rudnick, ed.)
    Ronald Press, New York, p. 121.

4. Fell, H.B. and Danielli, J. 1943. The enzymes of healing
    wounds. Brit. J. Exp. Pathol. 24: 196.

5. Sekla, B. 1928. Esterolytic processes and duration of life
    in Drosophila melanogaster. Brit. J. Exptl. Biol. 4: 161.
6. Fourche, J. 1959. La
6. Fourche, J. 1959. La respiration embryonnaire chez Drosophila
    melanogaster. Comptes Rendus Acad. Sci. 249: 1148.

7. Cox, R.P. and MacLeod, C.M. 1962. Alkaline phosphatase con-
    tent and the effect of prednisolone on mammalian cells in
    culture. J. Gen. Physiol. 45: 439.

8. Lowry, H.O., Rosenbrough, N.J., Farr, A.L. and Randall, R.J.
    1951. Protein measurement with the Folin phenol reagent.
    J. Biol. Chem. 193: 265.

9. Pearl, R. 1921. Experimental studies on the duration of life.
    I. Introductory discussions on the duration of life in
    Drosophila melanogaster. Amer. Naturalist 55: 481.

10. Bourne, G.H. 1960. General aspects of aging in cells from a
    physiological point of view, IN The Biology of Aging
    (B. L. Strehler et al., eds.), Amer. Inst. Biol. Sci.,
    Washington D.C., No. 6, p. 133.

QUESTIONS TO ING. SKRIVANOVA

Dr. Franks: What proportion of the alkaline phosphatase changes ocurred in the intestinal tract?

Ing. Skřivanová: We do not know anything about the origin of the alkaline phosphatase. In this work we studied homogenates of whole organisms only.

Dr. Kasten: I believe that vestigial is due to a recessive gene. Did you consider looking at the heterozygotes to see if there might be a biochemical expression of acid phosphatase?

Ing. Skřivanová: In all experiments presented we studied the homozygotes for vestigial gene only. We plan some experiments with heterozygotes for this gene in the future.

Dr. Gutmann: It might be interesting to consider the possible relation of changes in acid phosphatase to changes in developmental degradation of proteins (histolysis) connected with the increase of lysosomal enzymes.

CELL AGEING IN THE INTESTINAL TRACT

C. Rowlatt

Imperial Cancer Research Fund, 44, Lincoln's Inn
Fields, London WC2A 3PX, England

There has been much discussion whether ageing in a fixed post-
mitotic cell is the same as ageing in a constantly renewing stem cell
population. This can be tested by comparing cells from the crypts
and different levels on the villus in the small intestine. We know
from the work of Lesher's group (1) that the transit time from crypt
base to villus tip in the mouse jejunum is just over 50 hr in both
1-year-old and 2 1/2-year-old animals, of which about 6 hr is spent
in the crypt. There is therefore no difference in the post-mitotic
age of cells at a similar level in the villus in adult and aged mice,
although they are derived from stem cells of different ages.

The digestive tract has been extensively studied (see ref. 2
for morphological review), for example the morphology of the maturation process in the embryo, new-born and adult (3), the glycoprotein
incorporation (4), etc. There have been few morphological studies
on age changes (5) particularly at the ultrastructural level.

The work which I am reporting is the result of a study of
C57BL mice from our ageing colony. Carcinoma of the small bowel
occurs in these animals (2%) (6) and jejunum was used as there is
less amyloid than further down the small intestine. Seven adult
(3-9 months), 2 germ-free (5 weeks and 5 months) and 7 aged (30
months) animals were used. The tissues were glutaraldehyde and
osmic acid fixed, embedded in Araldite, and sections post-stained in
uranium and lead solutions.

The normal appearances of the epithelium in the crypt and low and
high on the villus in an adult mouse are quite characteristic. The
cells in the crypt are typical relatively undifferentiated epithelial

cells. The villus cells are differentiated although there is a
gradient from low to high in the villus. In the nucleus there is
increasing condensation of peripheral chromatin. The mitochondria,
which are scanty and evenly distributed in the crypt, have, by the
villus tip, become collected mainly in the subnuclear zone. The
microvilli are short and irregular in the crypt and become well
ordered on the villus. They increase in height between the low
villus and high villus, usually with a more pronounced fuzzy coat
on the high villus. Definite changes occur at the extrusion zone.
Before the cells are shed at the villus tip the nuclei become pykno-
tic, the cytoplasmic structure degenerates, and at the apical margin,
the microvilli break down becoming short and degenerate. The apical
junctional complex is still intact. The cells die before they are
shed.

But here we came up against the standard problem in studying
cell ageing in animals, that is the problem of the different cellular environment in the aged animal. Mrs. Defries in our group
pointed out that while the thymidine uptake in the adult was severely
depressed by starvation, in the aged animals it was much less affected. We have therefore used fed animals in spite of the dis-
advantages. Feeding produces a number of changes in cytology, no-
tably the alteration in the balance of cytoplasmic components and the
presence of extensive lateral blebbing of the villus cells in their
basal regions. The nuclear pattern is unchanged. The cell pattern
is similar in germ-free and conventional animals.

In the 30 month animals, the sequence of changes from crypt to
villus tip is similar to the adult. The villus cells in these fed
animals are tall and active in both low and high regions, with long
microvilli and patchy distribution of cytoplasmic contents. Protrusions of epithelial cytoplasm through the basal lamina are much
more marked in old animals. There are many membrane-bound blebs
of epithelial cytoplasm filling the stromal space in the villus. A
comparison was made of the basal lamina at the various heights in
the villus in 7 month and 30 month animals. The basal lamina is
not thickened in any of these sites, but high in the aged villus
there is a suggestion that it is a less dense structure than in the
adult. The more extensive protrusions suggest possible weakening
of the basal lamina.

Another age-associated change affects the mitochondria. Although some are normal, others are irregularly enlarged with alterations in the homogeneity of the mitochondrial matrix. These
changes are very similar to those described by Wilson and Franks
(7). Preliminary results show that the mitochondrial changes seem
to affect the majority of cells in a given crypt, although other
crypts may be normal. This suggests that this age-associated change
may affect and be transmitted by stem cells, but we cannot yet ex-

clude the possility that the changes may be due to local conditions such as amyloid.

We are continuing investigations into other features such as the surface staining properties of these cells, and are also extending the work to include the large bowel and its tumors.

REFERENCES

1. Fry, R.J.M., Lesher, S. and Kohn, H.I. 1961. Age effect on cell-transit time in mouse jejunal epithelium. Amer. J. Physiol. 201: 213.

2. Toner, P.G., Carr, K.E. and Wyburn, G.M. 1971. The digestive system: an ultrastructural atlas and review. Butterworths.

3. Merrill, T.G., Sprinz, H. and Tousimis, A.J. 1967. Changes of intestinal absorptive cells during maturation: an electron microscopic study of prenatal, postnatal adult Guinea pig ileum. J. Ultrastruct. Res. 19: 304.

4. Weiser, M.M. 1973. Intestinal epithelial cell surface membrane glycoprotein synthesis: I. An indicator of cellular differentiation. J. Biol. Chem. 248: 2536.

5. Andrew, W. 1971. The anatomy of aging in man and animals. Heineman Medical.

6. Rowlatt, C., Franks, L.M., Sheriff, M.U. and Chesterman, F.C. 1969. Naturally occurring tumors and other lesions of the digestive tract in untreated C58BL mice. J. Natl. Cancer Inst. 43: 1353.

7. Wilson, P.D. and Franks, L.M. The effect of age on mitochonrial ultrastructure and enzymes. This meeting.

QUESTIONS TO DR. ROWLATT

Dr. Rother: The aging changes in epithelial cells of the intestine are influenced by diurnal rhythms. We measured, for instance, the nucleus - cytoplasmic relationship of crypt cells and found a significant change of this parameter from 8 a.m. to 2 p.m. Did you also observe such an influence of diurnal rhythms on your findings?

Dr. Rowlatt: Thank you. No studies on the diurnal rhythms were done.

STUDIES ON THE PROLIFERATIVE CAPACITY OF MOUSE SPLEEN CELLS IN

SERIAL TRANSPLANTATION

J.W.I.M. Simons and C. van den Broek

Department of Radiation Genetics and Chemical
Mutagenesis, University of Leiden
The Netherlands

INTRODUCTION

There are many aspects of cellular ageing which are not yet
solved. Two important problems are whether the phenomenon of
limited lifespan in vitro reflects true ageing and whether the
limited proliferative capacity of normal somatic cells in vitro and
in vivo is an intrinsic and inevitable process.

According to the hypothesis of Hayflick and Moorhead the limited
lifespan of diploid cells in vitro could be due to senescence at the
cellular level (1). Data are available which support the idea that
the ageing phenomena in vitro and in vivo are similar at the cellular
level. For instance, the duration of the lifespan of human diploid
fibroblasts in vitro depends on the age of the donor tissue: the
older the donor tissue, the shorter the life expectancy of cells
in vitro (2,3,4). Furthermore the lifespan in vitro of cells
derived from individuals with premature ageing is shorter than the
lifespan of cells from normal individuals (5). Another similarity
is the prolongation of the G2 period with ageing which occurs both
in vivo and in vitro (6,7).

A further correlation between ageing in vivo and ageing in
vitro appears to be the increase in cell size with increasing age.
In vitro this has been observed for human skin fibroblasts (8),
human fetal lung fibroblasts (9,10), embryonic heart cells of the
marsupial Potorous tridactylis (11) and chick embryo fibroblasts
(12). In vivo an increase in cell size with advancing age has been
found for rat parietal stomach cells (13), rat kidney cells (14),
liver cells of seven domesticated mammalian species (15), cells of
the superior cervical sympathetic ganglion from the rat (16,17) and

human circulating lymphocytes (18). The reason for this increase
remains obscure, but it may be due to the accumulation of defective
protein which also occurs both in vivo and in vitro (19).

In this paper we show that the increase in cell size also
occurs during ageing of mouse spleen cells.

The limited lifespan of cells appears to occur in vivo also.
Limited lifespan phenomena in vivo have been reported for mouse skin
(20), mouse mammary gland (21) and mouse bone marrow cells (22-25).
This limited lifespan of normal cells contrasts markedly with the
unlimited lifespan of transformed cells. It is still a question
whether this limited lifespan is due to ageing as the technique of
transplantation may influence the outcome of the experiment. For
mouse spleen and bone marrow the number of serial transplantations
is four (22-25) or five (23), but the length of the time interval
between transfers is of considerable importance as was shown by
van Bekkum and Weyzen (26). These authors even achieved 13 passages
without losing the transfer line, but their irradiation dose maybe
did not prevent endogenous colony formation. However, their data
agreed with the assumption that the proliferation rate of the stem
cells is higher than that of the more mature ones leading to a
gradual shift in favour of the former category when serial transfer
is performed at short intervals. As survival of lethally irradiated
mice seems primarily determined by the injection of a sufficiently
large number of relatively mature hematopoietic cells longer intervals would permit more time for maturation and more serial transplantations. This was investigated in the experiments described in
this paper. The determination of the volume distribution of transplanted spleen cells served as an indicator for ageing phenomena
under these conditions.

MATERIALS AND METHODS

The mice used in the present investigation were CBA-Rij, an
inbred line from the Dutch Radiobiological Institute at Rijswijk.
The donors for the primary transplantation were 8-10-week-old males,
as were the recipients for each passage; some females were also
used as recipients. Whole-body irradiation was administered at a
dosage of 800 rad using a Philips-Muller roentgen apparatus (300 kV;
10 mA, 2mm Cu; exposure rate of 60 rad/min). For each transplantation three donor mice were sacrificed. The spleens were pressed
through a nylon filter six layers thick, the femoral marrow was
aspirated into RPMI or Ham's F-10 solution and the cell suspension
was also passed through the nylon filter. The number of cells was
counted in a hemocytometer. For each transplantation 5×10^6 spleen
cells or 2×10^6 bone marrow cells were injected into recipient mice
via the lateral tail vein within 5-8 hrs after irradiation.

The frequency distribution of cell volumes was determined using a Coulter counter Model B connected to a Coulter size-distribution plotter. The cells were diluted to 20,000 cells per 0.5 ml CSS (counter salt solution); 5 size distributions were plotted per sample. The data from the plotter were fed into an IBM computer; after logarithmic transformation of the cell volumes, to obtain a better approximation of the normal distribution, the mean and the standard deviation were estimated. A more detailed description of the technique for measuring cell volumes is found elsewhere (27).

Spleen cells to be measured by the Coulter counter are purified as follows: a 25 ml Erlemeyer flask containing 150 mg carbonyl iron powder (G.A.F. Delft, Holland, grade 341), 10 siliconized glass beads and 10 ml cell suspension is agitated in a water bath at 37°C for 15 min. The phagocytizing cells absorb the iron and become heavy. Eight ml of this suspension is mixed with 2 ml dextran (Pharmacia, Uppsala, Sweden) and left at 37°C for 45 min, during which time the erythrocytes agglutinate and settle. The cell suspension is poured over a mixture of isopague (Natrii N-methyl-3,5-diacetamido-2.4.6-tryod-benzoas; Nijegaard e Co., Oslo, Norway) and ficoll (Pharmacia, Uppsala, Sweden) with a density of 1.077 g/ml and centrifuged for 15 min at 1,000 g. The iron-loaded phagocytizing cells spin down and the purified spleen cells appear at the interphase. They are washed twice in Hanks BSS without cA and Mg and suspended in 10 ml of medium consisting of 70% RPMI or Ham's F-10 and 30% nonfetal calf serum.

RESULTS

Parameters of the Cell Volume Distribution With Respect to Age

In Tables I, II, III and IV the means and standard deviations of the log volume distributions of erythrocytes and spleen cells are given for males and females of three different age groups: Young animals 8-10 weeks of age, a middle-aged group varying in age from 26-52 weeks and an old age group varying in age from 116-131 weeks. Significant changes in relation to age were found. The mean log volume of spleen cells increases significantly with age as tested according to the method of Kruskall and Wallis (28) ($p < 0.001$). The standard deviation of these cells is increased in the old age group. The data for males and females were pooled as no differences were observed between males and females within each age group with Wilcoxon's test. The data for the parameters for the erythrocytes are less clear-cut: the standard deviations for males and females cannot be pooled since a very significant difference is found in the middle-age group. This difference can be due to sex as well as sampling errors. A significant difference between age groups

TABLE I. Means of Cell Volume Distributions in Cubic μ (After Logarithmic Transformation) For Spleen Cells From Mice of Three Different Age-Groups

Males			Females		
8-10 weeks	26-52 weeks	116-131 weeks	8-10 weeks	26-52 weeks	116-131 weeks
1.9687	2.0188	1.9964	1.9684	2.0078	2.0333
1.9674	1.9780	2.0192	1.9646	1.9610	2.0188
1.9665	1.9575	2.0012	1.9801	2.0152	2.0041
1.9720	2.0094	2.0308	1.9619	1.9979	2.0368
1.9734	2.0049	2.0132	1.9529	2.0086	2.0125
1.9779	2.0019	2.0350	1.9638	2.0170	2.0493
1.9861	1.9888	2.0124	1.9482	2.0005	2.0453
2.0037	1.9744	2.0148	1.9804	2.0230	2.0204
1.9939	1.9815	2.0077	1.9874	2.0233	2.0386
1.9919	1.9977		1.9736	2.0124	2.0249

mean value:

Males			Females		
1.9801	1.9913	2.0145	1.9681	2.0067	2.0248

chi-square = 14.33   d.f. = 2   P<0.001   chi-square = 20.31   d.f. = 2   P<0.001

chi-square = 34.66   d.f. = 2   P<0.001

TABLE II. Standard Deviations of Cell Volume Distributions (After Logarithmic Transformation) For Spleen Cells From Mice of Three Different Age-Groups

	males			females		
	8-10 weeks	26-52 weeks	116-131 weeks	8-10 weeks	26-52 weeks	116-131 weeks
	0.1002	0.0947	0.1050	0.0893	0.1032	0.0934
	0.0986	0.0984	0.1047	0.0933	0.1537	0.1047
	0.0926	0.1059	0.0996	0.0998	0.0827	0.1023
	0.0938	0.0831	0.1097	0.0832	0.0940	0.1136
	0.0946	0.0811	0.1062	0.0777	0.0889	0.1196
	0.0968	0.0895	0.1014	0.0861	0.0830	0.1046
	0.1040	0.0941	0.1030	0.1008	0.1015	0.0996
	0.0955	0.0903	0.0906	0.1037	0.0916	0.1273
	0.1023	0.0921	0.0904	0.1049	0.0914	0.1052
	0.1067	0.0945		0.0893	0.0951	0.0944
mean value:	0.0985	0.0924	0.1012	0.0928	0.0985	0.1065

chi-square = 7.33   d.f. = 2   0.02 < P < 0.05   chi-square = 7.99   d.f. = 2   0.01 < P < 0.02

chi-square = 12.46   d.f. = 2   P < 0.01

is found for the mean volume of the erythrocytes but this change
does not increase with age. Therefore only the data for the spleen
cells can be used as an indicator for senescent change. The volume
of the bone marrow cells was not measured as these cell populations
proved to be too heterogeneous in size.

Endogenous Colony Formation

The chance for endogenous colony formation after irradiation
with 800 rad was determined by sacrificing twenty animals 8 and 9
days after irradiation and fixing the spleens in Bouin's solution.
The number of colonies formed was zero, which means that the chance
that an endogenous colony will form during serial transplantation
is low.

Parameters of the Volume Distribution of Spleen Cells in Relation to Transplantation

To investigate whether the cell volume distribution of spleen
cells changes after transplantation in irradiated hosts and whether
eventual changes disappear in time, the parameters of the volume
distributions of erythrocytes and spleen cells were determined at
different times after transplantation of spleen cells (Table V).
The mean volume of the spleen cells appears to be the same before
and after transplantation. No significant difference was found
between the transplanted group and the middle-aged group, while the
difference between the transplanted group and the old age group is
significant with Wilcoxon's test (p $<$ 0.001). For this reason the
mean cell volume is used as an indicator for ageing after serial
transplantations. The mean volume of erythrocytes appears to be
increased in several cases as do the standard deviations of spleen
cells in experiment three. This indicates that even 21 weeks after
transplantation the normal situation has not been established in the
repopulated organs.

Serial Transplantation of Mouse Spleen Cells

We experimented with serial transplantation of spleen cells
with a time interval between transplantations of six months or even
longer. The results obtained so far are shown in Figure 1. The
total number of passages in a three year period is six and although
there are indications of a decline in the survival of the recipients
of the sixth passage quite a few animals are surviving after the
most recent sixth transplantation.

Parameters of the cell volume distributions of erythrocytes
and spleen cells for animals 16-26 weeks after the fifth transplan-

TABLE III.   Means of Cell Volume Distributions (After Transformation) For Erythrocytes from Mice of Three Different Age-Groups

males			females		
8-10 weeks	26-52 weeks	116-131 weeks	8-10 weeks	26-52 weeks	116-131 weeks
1.6128	1.5890	1.5525	1.5953	1.5851	1.6144
1.5886	1.5594	1.5918	1.6129	1.5595	1.5973
1.6231	1.6010	1.6158	1.5977	1.5554	1.5790
1.6077	1.5913	1.6210	1.6128	1.5685	1.5693
1.6014	1.5266	1.5958	1.5898	1.5664	1.5694
1.5963	1.6250	1.6711	1.5959	1.5727	1.5704
1.6314	1.5834	1.6260	1.6415	1.5499	1.6347
1.6113	1.5880	1.5806	1.6596	1.5307	1.5668
1.6153	1.6133	1.6185	1.6200	1.6232	1.6132
1.6115	1.5735	1.6029	1.5822		1.5632

mean value:

| 1.6099 | 1.5850 | 1.6076 | 1.6108 | 1.5679 | 1.5878 |

chi-square = 5.51   d.f. = 2   0.05<P<0.10

chi-square = 10.13   d.f. = 2   P<0.01

chi-square = 12.46   d.f. = 2   P<0.01

TABLE IV. Standard Deviations of Cell Volume Distributions (After Logarithmic Transformation) For Erythrocytes from Mice of Three Different Age-Groups

males			females		
8-10 weeks	26-52 weeks	116-131 weeks	8-10 weeks	26-52 weeks	116-131 weeks
0.0929	0.1080	0.0901	0.0789	0.0849	0.0876
0.0907	0.1025	0.0990	0.0867	0.0891	0.1066
0.0841	0.1031	0.0862	0.0729	0.0875	0.0881
0.0837	0.1017	0.1067	0.0840	0.0920	0.0933
0.0883	0.1003	0.1036	0.0786	0.0891	0.0893
0.0928	0.0980	0.0830	0.0900	0.0880	0.0892
0.0889	0.0920	0.0894	0.0825	0.0843	0.0985
0.0879	0.1065	0.1102	0.0800	0.0979	0.0934
0.0863	0.0990	0.0916	0.0810	0.0863	0.0866
0.0875	0.1155	0.0819	0.0896		0.0910

mean value:

0.0883	0.1027	0.0942	0.0824	0.0888	0.0921

chi-square = 11.78   d.f. = 2   P<0.01   chi-square = 11.23   d.f. = 2 $\bar{P}$<0.01

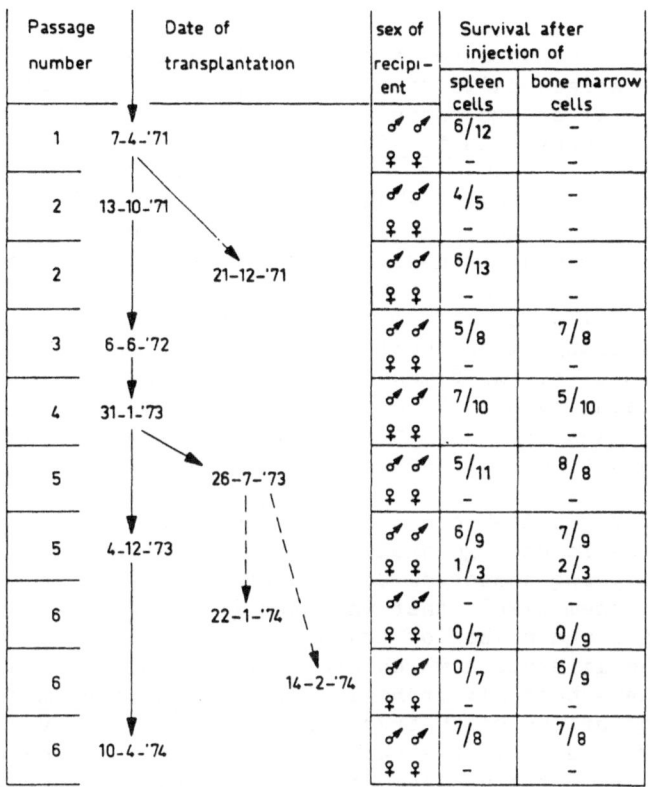

Passage number	Date of transplantation	sex of recipient	Survival after injection of spleen cells	bone marrow cells
1	7-4-'71	♂ ♂	6/12	–
		♀ ♀	–	–
2	13-10-'71	♂ ♂	4/5	–
		♀ ♀	–	–
2	21-12-'71	♂ ♂	6/13	–
		♀ ♀	–	–
3	6-6-'72	♂ ♂	5/8	7/8
		♀ ♀	–	–
4	31-1-'73	♂ ♂	7/10	5/10
		♀ ♀	–	–
5	26-7-'73	♂ ♂	5/11	8/8
		♀ ♀	–	–
5	4-12-'73	♂ ♂	6/9	7/9
		♀ ♀	1/3	2/3
6	22-1-'74	♂ ♂	–	–
		♀ ♀	0/7	0/9
6	14-2-'74	♂ ♂	0/7	6/9
		♀ ♀	–	–
6	10-4-'74	♂ ♂	7/8	7/8
		♀ ♀	–	–

Fig. 1. Survival of irradiated mice after serial transplantation of spleen cells or bone marrow cells.
————————— donor mice received spleen cells
- - - - - - - - donor mice received bone marrow cells

tation are shown in Table VI. Again no difference was found between the transplanted group and the middle-aged group, while a difference is indicated between the transplanted group and the old age group ($p < 0.04$) despite the fact that these cell populations are as old as three years.

DISCUSSION

In the introduction the following problems were stated: 1) is ageing of mouse spleen cells accompanied by an increase in cell volume; 2) do the mouse spleen cells show a limited proliferative capacity when serial transplantation is performed at long time

intervals and is this accompanied by an increase in cell volume.
In our experiments we observed a progressive increase in the mean
volume of mouse spleen cells with advancing age which confirms the
earlier findings of an age-correlated increase in cell volume for
a number of cell types both in vivo and in vitro.

It was also found that the parameters of the cell volume distribution of mouse spleen cells are not the same after the first transplantation. The size distribution of the spleen cells appears more
heterogeneous even 21 weeks after transplantation. As the mean
volume of these cells is the same as in control animals from the
same group this change is probably not due to ageing. Therefore the
conclusion is that the experimental conditions for testing the
proliferative capacity of spleen cells do not completely agree with
the normal conditions for spleen cells in nonirradiated animals,
which may influence the lifespan of spleen cells in this experiment.

Although our results do not provide conclusive evidence as to
whether spleen cells undergo ageing in the course of serial transplantation into young hosts the data point to absence of ageing.
Firstly it does seem likely that when the interval between transplantations is longer the number of transfers is greater than the numbers
published in the literature which are four (22-25) or five (23).
To date there have been six transplantations and since many animals
have survived after the sixth transfer, a successful seventh passage
is not unlikely. Secondly the parameters of the three year old cell
populations at the fifth passage do not show the age correlated
change of normal cells.

An important question is whether the cells we are dealing with
at the sixth transfer are really derived from the original transplant.
The absence of endogenous colony formation after irradiation with 800
rad points in this direction. This is strongly supported by the
chromosomal analysis of bone marrow cells from the fifth transplantation. Eventual surviving cells may show stable structural changes
in their chromosomes due to the heavy irradiation (29). The frequency
for this type of aberration in mouse bone marrow at a dose of 400
rad analyzed 60-100 days after irradiation is known to be 20-27%
(29,30). If linear extrapolation of this figure is allowed the
expectation of the frequency of cells with stable structural changes
in the chromosomes at a dose of 800 rad would be 40%. Of the fifty
metaphases analyzed thirty-nine turned out to be diploid and the
remaining eleven cells did not show structural aberrations but only
numerical changes in the range of thirty-four to forty-six chromosomes. As numerical changes are very rare both in young adult
control and irradiated bone-marrow (30) it appears unlikely that
the original cell population used for the first transplantation
has been replaced by surviving host cells.

TABLE V. Means and Standard Deviations of Cell Volume Distributions (After Logarithmic Transformation) For Spleen Cells and Erytirocytes From Mice At Varying Times After Transplantation

Experiment	Time after transplantation (weeks)	Spleen cells		Erythrocytes	
		mean	stand.dev.	mean	stand.dev.
1	13	2.0328	0.1028	1.6092	0.0826
	18	1.9803	0.0922	1.6537	0.0725
	22	1.9891	0.0896	1.6771	0.0845
	26	1.9999	0.0985	1.6013	0.0930
2	18	2.0285	0.0928	1.6160	0.0894
	22	1.9736	0.0894	1.6063	0.0895
3	6	1.9722	0.1242	1.6953	0.0797
	9	1.9527	0.1213	1.5649	0.1068
	13	1.9858	0.0979	1.6216	0.0851
	17	1.9618	0.1662	1.6228	0.0939
	21	2.0212	0.1132	1.7179 ·	0.0805
	21	1.9938	0.1338	1.7246	0.0719

TABLE VI. Means and Standard Deviations of Cell Volume
Distributions (After Logarithmic Transformation) For Spleen
Cells and Erythrocytes From Mice Twenty-Five Weeks After The
Fifth Serial Transplantation

Spleen cells			Erythrocytes	
mean	stand.dev.		mean	stand.dev.
1.9750	0.0837		1.6190	0.0837
1.9753	0.0639		1.5976	0.0736
1.9781	0.0848		1.5530	0.0768
1.9704	0.0940		1.5542	0.0778
1.9747	0.1211		1.5946	0.0690
2.0715	0.1054		1.5870	0.0727
1.9508	0.1359		1.6001	0.0804
2.0266	0.1145		1.5903	0.0715
2.0498	0.0985		1.6754	0.1079
1.9758	0.0965		1.5787	0.0873
mean value:			mean value:	
1.9943	0.0998		1.5950	0.0801

SUMMARY

The ageing of mouse spleen cells in vivo was investigated by
determining the size distribution of cell volumes. It was found
that ageing is accompanied by an increase in cell volume which is
in agreement with published data on ageing for a number of cell
types both in vivo and in vitro.

After the first transplantation into irradiated hosts the size
distribution of spleen cells proves to be more heterogeneous which
may affect the proliferative capacity. After the fifth transplantation, when the cell population is three years old, the parameters
of the volume distributions of spleen cells agree better with the
data from middle-aged mice than with the data from old-aged mice.
Further it was found that at least six serial transplantations are

possible if the interval between passages is long. Therefore our data do not show an intrinsic limited proliferative capacity of these cells.

ACKNOWLEDGEMENTS

Thanks are due to Dr. C.F. Hollander for providing the old CBA-Rij mice, to the Radiobiological Institute at Rijswijk for providing the irradiated animals, to Dr. P.P.W. van Buul, who kindly carried out the chromosomal analysis and to Prof. Dr. F.H. Sobels for his interest in the work. The technical assistance of Miss I. Doorenberg and Miss S. de Vogel is gratefully acknowledged.

REFERENCES

1. Hayflick, L. and Moorhead, P.S. 1961. The serial cultivation of human diploid cell strains. Exp. Cell Res. 25: 585.

2. Hayflick, L. 1968. Human cells and ageing. Scientific American 218: 32.

3. Martin, G.M., Sprague, C.A. and Epstein, CH.J. 1970. Replicative lifespan of cultivated human cells. Laboratory Invest. 23: 86.

4. Waters, H. and Walford, R.L. 1970. Latent period for outgrowth of human skin explants as a function of age. Journal of Gerontology 25: 381.

5. Goldstein, S. 1969. Lifespan of cultured cells in progeria. Lancet 1: 424.

6. Macieira-Coelho, A., Ponten, J. and Phillipson, L. 1966. The division cycle and RNA synthesis in diploid human cells at different passage levels in vitro. Exp. Cell Res. 42: 673.

7. Pederson, T. and Gelfand, S. 1966. Mitosis in regeneration and ageing in relation to cell division cycle. J. Cell Biol. 81: 84A.

8. Simons, J.W.I.M. 1967. The use of frequency distributions of cell diameters to characterize cell populations in tissue culture. Exp. Cell Res. 45: 336.

9. Cristofalo, V.J. 1970. Metabolic aspects of ageing in
    diploid human cells. Symp. on "Ageing in Cell and Tissue
    Culture." Zinkovy. E. Holeckova and V.J. Cristofalo, eds.,
    Plenum Press.

10. Macieira-Coelho, A. 1970. The decreased growth potential
    in vitro of human fibroblasts of adult origin. Symp. on
    "Ageing in Cell and Tissue Culture." Zincovy. E. Holeckova
    and V.J. Cristofalo, eds., Plenum Press.

11. Simons, J.W.I.M. 1970. A theoretical and experimental
    approach to the relationship between cell variability and
    ageing in vitro. Symp. on "Ageing in Cell and Tissue
    Culture." Zinkovy. E. Holeckova and V.J. Cristofalo, eds.,
    Plenum Press.

12. Lima, L. and Macieira-Coelho, A. 1972. Parameters of ageing
    in chicken embryo fibroblasts cultivated in vitro. Exp.
    Cell Res. 70: 279.

13. Bralow, S.P. and Komarov, S.A. 1962. Parietal cell mass
    and distribution in stomachs of Wistar rats. Am. J. Physiol.
    203: 550.

14. Covell, W.P. 1926. Quantitative cytological studies on the
    renaltubules. The Anat. Record. 34: 61.

15. Illing, G. 1905. Vergleichende histologische Untersuchungen
    uber die Leber der Haussaugetiere. Anat. Anz. 26: 177.

16. Chi Ping. 1921. On the growth of the largest nerve cells
    in the superior cervical sympathetic ganglion of the Norway
    rat. J. Comp. Neur. 33: 281.

17. Chi Ping. 1921. On the growth of the largest nerve cells in
    the superior ganglion of the albino rat from birth to
    maturity. J. Neur. Comp. 33: 313.

18. Simons, J.W.I.M. and Van Den Broek, C. 1970. Comparison of
    ageing in vitro and ageing in vivo by means of cell size
    analysis using a Coulter counter. Gerontologia 16: 340.

19. Orgel, L.E. 1973. Ageing of clones of mammalian cells.
    Nature 243: 441.

20. Krohn, P.L. 1962. Review lectures on Senescence. II.
    Heterochronic transplantation in the study of ageing. Proc.
    Roy. Sco. B. 157: 128.

21. Daniel, C.W. and Young, L.J.T. 1971. Influence of cell division on an ageing process. Lifespan of mouse mammary epithelium during serial propagation in vivo. Exp. Cell Res. 65: 27.

22. Cudkowicz, G., Upton, A.C., Shearer, G.M. and Hughes, W.L. 1964. Lymphocyte content and proliferative capacity of serially transplanted mouse bone marrow. Nature 201: 165.

23. Siminovitch, L., Till, J.E. and McCulloch, E.C. 1964. Decline in colony-formation ability of marrow cells subjected to serial transplantation into irradiated mice. J. Cell Comp. Physiol. 64: 23.

24. Williamson, A.R. and Askonas, B.A. 1972. Senescence of an antibody-forming cell clone. Nature 238: 337.

25. Chen, M.G. 1971. Age related changes in hematopoietic stem cell populations of a long-lived hybrid mouse. J. Cell Physiol. 78: 225.

26. Bekkum, P.W. van and Weyzen, W.W.H. 1961. Repeuplement des tissues irradies. Serial transfer of isologous and homologous hematopoietic cells in irradiated hosts. Path. Biol. 9: 888.

27. Simons, J.W.I.M. 1970. Characterization of somatic cells by determination of their volumes with a Coulter counter. J. Cell Biol. 46: 610.

28. Jonge, H. de. 1963. Inleiding tot de Medische Statistiek. Dell I Verhandeling van het Nederlands Instituut voor Preventieve Geneeskunde XLI, p. 301.

29. Buul, P.P.W. van. 1973. Comparison of frequencies of radiation-induced stable chromosomal aberrations in somatic and germ tissues of the mouse. Mutation Res. 20: 369.

30. Buul, P.P.W. van. Unpublished data.

ON THE AGEING OF INTERMITOTIC CELLS -- INVESTIGATIONS ON ENTEROCYTES AND HEPATOCYTES

G. Leutert, W. Rotzsch and W. Beier

Karl-Marx-Universität

Leipzig

At first, we think, it is necessary to define ageing. Ageing in the sense of Max Burger (1) reflects the physiological, morphological, biochemical and biophysical changes, which occur during the life-span of an individual. These changes are arranged into changes during prenatal development, postnatal development, adolescence and senescence. Ageing takes place at different levels; at the level of the organism, organ and tissue, which includes cells and intercellular substance. The ageing changes are underlying changes at the different levels, because the parts of the organism, its organs, tissues and cells, are aging disproportionately. For that reason the cell represents only one level, but a very important one. Many of the hypotheses now being discussed, put the changes of the cell into the center of consideration. Doubtless there is an ageing process of postmitotic cells concerning lipofuscin, mitochondria, lysosomes, molecular arrangements between DNA and histones, contents of RNA and pattern of enzymes. But it is questionable up to now: Is there an ageing process of intermitotic cells too? At first we have to distinguish between ageing processes starting in single cells and such starting in multicellular arrangements on the level of tissues. The single cell shows changes during its cell cycle. But it is questionable, if there is a difference in the view of rise, maturation and death of intermitotic cells between young and old individuals.

Verzar (2,3) and his co-worker m.Hahn (4) have the opinion that there is no ageing process in intermitotic cells. But Holle (5) says that the intermitotic cells have no eternal youth and many authors (6,7) accent that there is also an ageing process of protozoa. To answer this question we investigated enterocytes of the

small gut of different aged rats and in comparison hepatocytes as
reversible postmitotic cells. Enterocytes of the small gut belong
to a tissue which is characterized by the shortest time of cell
renewal (8-10). The renewal time is 4 days in men and 4.5 days in
the rat. Electron-microscopical investigations of David (11) have
shown, that on the way from the ground of the crypts to the top of
the villi there is a loss of water, followed by concentration of
cytoplasm, condensation of the organelles and the nucleus by which
the cells become smaller. The following cells comprise those cells,
which are desquamated and lose their contact to the basement membrane. During their cell cycle, the enterocytes become more and
more differentiated and in the consequence of this phenomenon they
lose their ability for cell division.

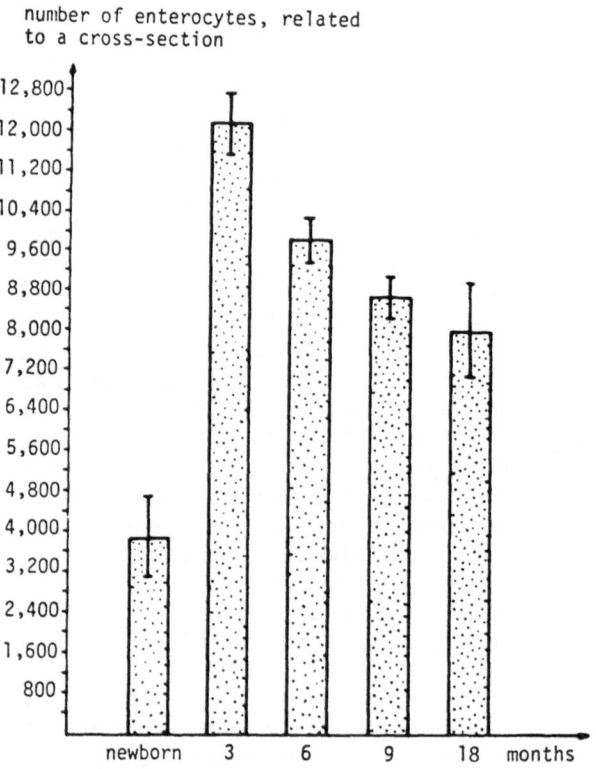

Fig. 1. Number of enterocytes counted in cross-sections of the
small gut of rats in dependence on age.

Our results, concerning the ageing of intermitotic cells as
cell-tissue, include at first some histochemical findings (12). In
the apical part of the cells there exists acid mucopolysaccharides,
proteins and RNA. These substances increase during ageing in rats
up to the 9th month of life. After this point of time the intensity
of the reactions is diminished. The decrease of RNA in old animals
is very clearly shown by the gallocyanin-chromalum-reaction (combined with ribonuclease). Enzyme-histochemical investigations (13)
show that the amount of alkaline phosphatase increases up to the
6th month of life and remains constant after this point of time.
The α-glucosidase decreases after the 9th month of life. Those
enzymes, which hydrolyze compounds of β-glucosides decrease already
during the first month of life. The intracellular oxydo-reductases
(we investigated the succinodehydrogenase, the NADH-diaphorase, the
isocitroacid dehydrogenase and the lactate dehydrogenase) differed:
The amount of enzymes of biological oxidation decreases after
the 6th month of life but the amount of lactate dehydrogenase, which
terminates the anaerobicmetabolism of glucose, increases up to the
6th month of life and becomes constant after this time. These
findings mean that in old age anaerobic glycolysis is preponderant
over biological oxidation.

Further results, with the enterocytes of rats, are: In
old age there is a decrease in RNA, cell water, protein and the
uptake and incorporation of ^{14}C-leucine (14). Number and height are
changing during the life-span too (15). The number of the enterocytes increases from birth to the third month and decreases after
this point of time (Fig. 1), while the height of the enterocytes
becomes larger from birth to the 9th month and smaller in the 18th
month (Fig. 2). This means that at first there is a multiplication
of the enterocytes and in the second line there is the growth of
these cells. These findings agree with those of Mitroiu et al. (16)
and Heim et al. (17). The localisation of the cell-multiplication
is different. In the newborn rats we find in the crypts only few
enterocytes; most of them lie in the villi. But the cell-multiplication, which occurs during the first three months, is starting
especially in the crypts (Fig. 3). The length of the microvilli
becomes longer in the beginning of the life and they retain this
shape in older age (Fig. 2).

Richter and Rotzsch (18) have investigated the adaptation of
enzymes as related to age. Glucose-6-phosphate dehydrogenase
of the rat liver (female BD III-rats) decreases during starvation
in young and old animals in the same manner. Refeeding is accompanied by increase of the enzyme level. On the other hand there are
differences in the time course of recovery after the period of
starvation between young and old rats. The glucose-6-phosphate
dehydrogenase levels increase in young rats significantly after
24 hr, reaching an overshoot 48 hr after refeeding. The ability

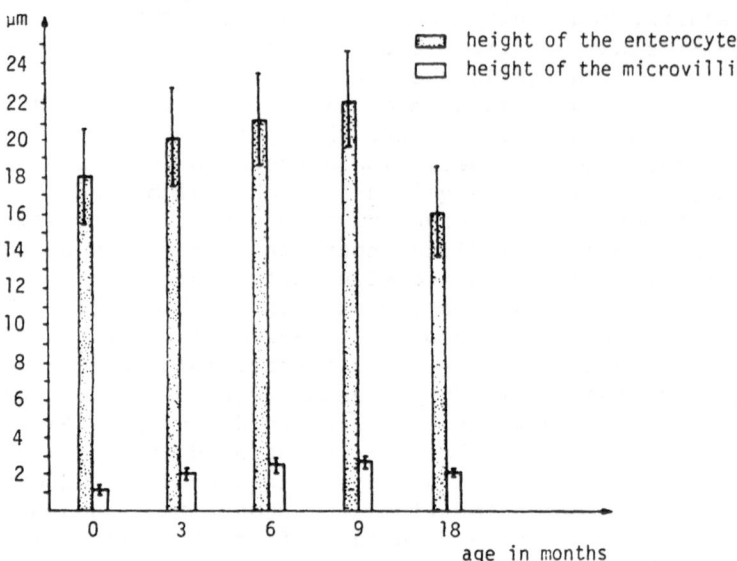

Fig. 2. Height of enterocytes and microvilli of rats in dependence
on age.

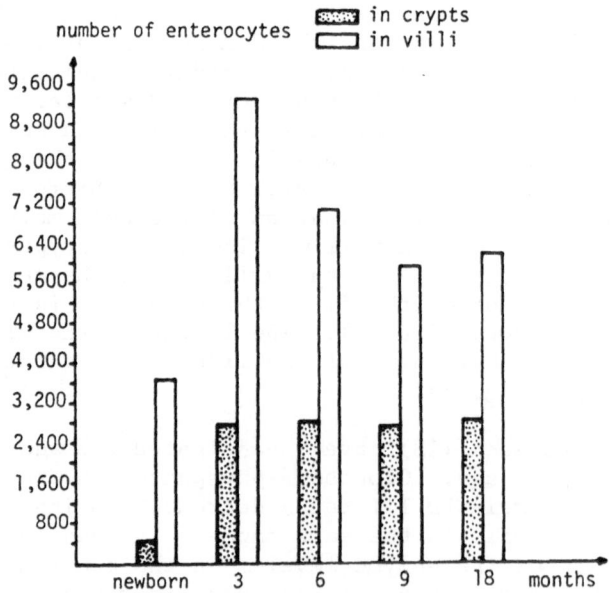

Fig. 3. Number of enterocytes counted in cross-sections of the
small gut of rats in dependence on age related to their localization
in crypts and villi.

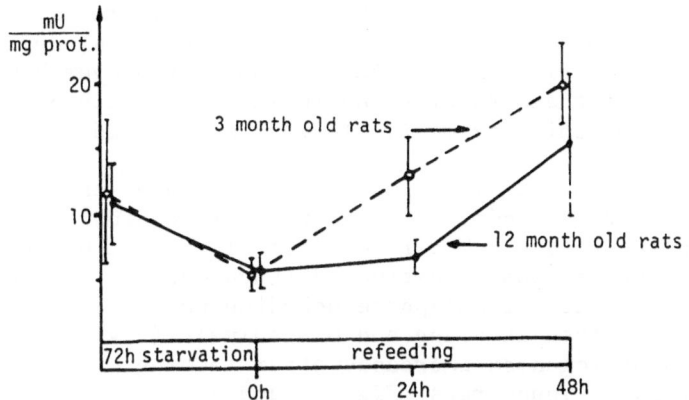

Fig. 4. Age-dependence of the effect of starvation and refeeding
on the level of rat liver glucose-6-phosphate dehydrogenase. Each
point represents the mean ± standard error for at least 5 animals.

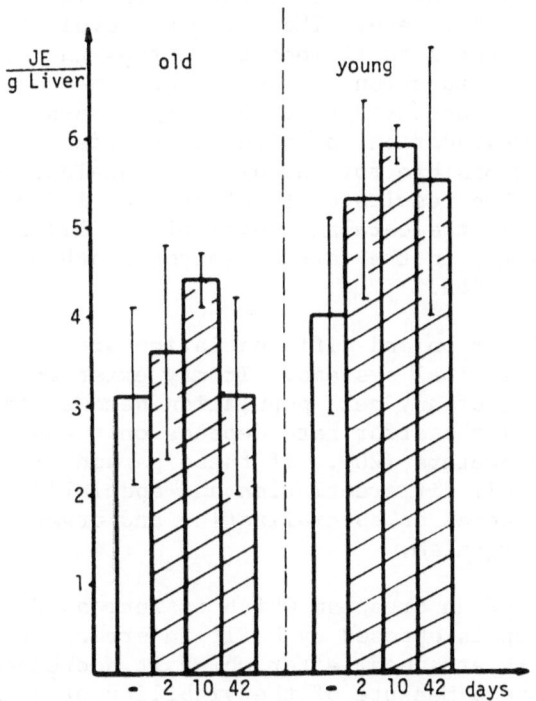

Fig. 5. Activity of glycerine-1-phosphate dehydrogenase of liver
cells after feeding of saccharose in young and old rats.

of old rats to respond similarly is impaired. There is no significant increase in enzyme level after 24 hr refeeding and 48 hr
after refeeding we find an increase but no enzyme overshoot (Fig. 4).
In consequence of these results and in agreement with Adelman (19),
the adaptation of this enzyme in old age is retarded.

Richter and Rotzsch (18) further investigated the adaptation
of some liver enzymes under the condition of feeding of glucose,
fructose and saccharose. Their results: There are no changes of
malate-dehydrogenase and isocitroacid dehydrogenase in young and
old animals. Glycerol-1-phosphate dehydrogenase increases after
feeding of saccharose in young and old animals (Fig. 5) and glucose-
6-phosphate dehydrogenase increases after feeding of fructose and
saccharose only in young rats (Figs. 6 and 7).

During ageing the levels of a number of enzymes in various
organs have been found to change; some increase, others decrease.
The molecular basis of these changes are alterations in either the
rate of enzyme synthesis, or of enzyme degradation or both. Experiments were performed to obtain measures of the relative rates
of synthesis and degradation of rat liver and kidney catalase as a
function of age. The level of catalase in rat liver and kidney is
altered as a function of age. The catalase level of liver increases
from the age of 3 months to 12 months, whereas the kidney catalase
level decreases as a function of age. Since the content represents a steady state achieved by a balance between the rate
of synthesis and degradation, an alteration affecting either or both
rates could be responsible for the altered catalase level as a
function of age. The results of these studies indicate that the age-
dependent changes in the catalase level of rat liver and kidney are
determined by changes in the rate of enzyme synthesis rather than
that of its degradation.

Cell-multiplication and differentiation are expressions of
development in biological systems. It may exist an error rate, μ,
meaning that a part of any cell population becomes faulty per time
unit. In our model the error rate depends on transcription as well
as translation parameters (20). If these parameters are functions
of the degree of cell differentiation and specialization, then the
error rate also depends on specialization and organisation in
subsystems of an organism.

The vitality of an organism which is increased by cell growth
and differentiation is opposed by built-in errors, so that an
optimal vitality is attained, after which it decreases during ageing.
The vitality may be a measure of the viability of biological systems,
which is related to the available number of biological intact cells
and the effectiveness of their interactions. Faulty cells reduce the
functional capacity of a system, and this functional loss is postulated to be the result of the nonsense protein molecules formed in

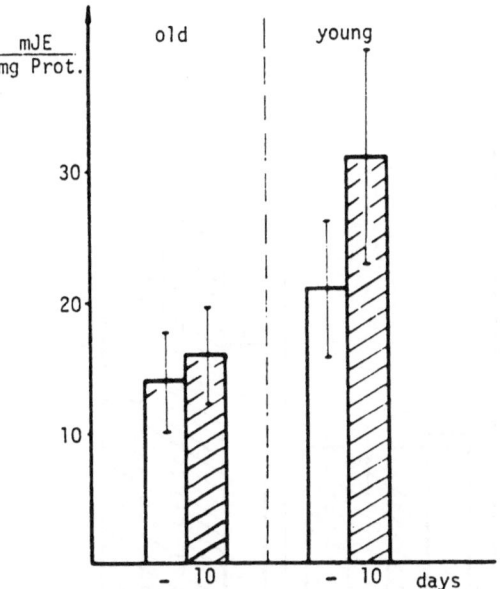

Fig. 6. Activity of glucose-6-phosphate dehydrogenase of liver cells after feeding of fructose in young and old rats.

consequence of incorrect transcription events and production of faulty messengers. It is proposed that whereas growth and differentiation depend on the maintenance of correct gene regulations leading to correct transcription and translation there is a probability of misreading during these processes rather than a precise association of codon and anticodon at any moment. The probability of transcriptional errors is one cause for the occurence of an error rate μ in the system.

We characterize a cell as faulty if this probability exceeds a definite threshold. It is assumed that the effective loss in functional capacity depends both on the quantity of nonsense molecules produced per time unit and the state of differentiation of the cells.

Our model is based largely on the error hypothesis originally proposed by Orgel (21). One indirect cause of the loss of cytoplasmic information related to an accumulation of mis-specified endproducts during translation, as pointed out by Orgel, could be the mis-reading events during transcription discussed in our model. It

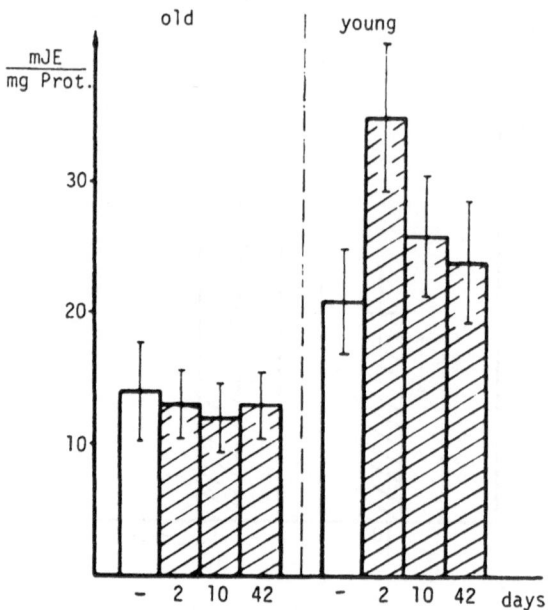

Fig. 7. Activity of glucose-6-phosphate dehydrogenase of liver cells after feeding of saccharose in young and old rats.

is our aim to make an attempt to quantify some fundamental charac-teristics of this hypothesis on the basis of relatively simple assumptions.

In our model we do not discern postmitotic cells from inter-mitotic cells. Our error hypothesis is valid for both cell types. The mathematical expression of our error hypothesis may be given by:

$$F(t) = \mu \int_{o}^{t} E(t) \, dt$$

E(t) represents the development of the biological system which is taken under consideration.

The overall functional activity as defined by Beier et al. (22)

is possible to write:

$$V(t) = E(t) - F(t)$$

If growth and differentiation are fundamental aspects of development, then $E(t)$ is closely allied to the increase in cell number $z(t)$ by cell-multiplication. $F(t)$ then is characterized by the number of faulty cells accumulated in the system from $t = 0$ to the time of observation. Under these assumptions $V(t)$ describes the effective number of available, error-free and functional cells building up the multicellular system.

One can see from our model that a linear decrease in vitality results after a maximum indicating the beginning of senescence has been passed for human beings at about 30 years (23). Our findings are in good agreement with experimental and theoretical findings by Shock (24) and Strehler (6).

From our point of view it seems to be necessary for any progress on the field of gerontology both to continue experimental investigations on all levels of biological organisation and to make further attempts for mathematical modelling of the ageing process.

REFERENCES

1. Burger, M. Altern und Krankheit.
   1. Aufl. VEB Georg Thieme Leipzig, 1947.
   3. Aufl. 1956.

2. Verzár, F. 1971. Physiologie und Gerontologie. Zschr. f.
   Alternsf. 24: 209.

3. Verzár, F. 1972. Regulation of adaptation processes in the
   mechanism of ageing. Vortrag, gehalten auf d.9.Internat.
   Kongreß für Gerontologie in Kiew, 2.-7.7.

4. Hahn, H.P.v. 1969. Genetische Faktoren im cellulären.
   Alternsprozess. Bull.d.Schweiz.Akad.d.Med.Wiss. Bd. 24: 272.

5. Holle, G. 1972. Probleme der gegenwärtigen Alternsforschung.
   In: Handb.d.allg.Path., 6.Bd., 4.Teil, 1, Springer Berlin-
   Heidelberg-New York.

6. Strehler, B.I. 1962. Time, Cells and Aging. Academic
   Press, New York.

7. Comfort, A. 1964. Ageing. The biology of senescence.
   Routledge & Paul, Kegan, London.

8. Bertalanffy, F.D. 1962. Cell renewal in the gastrointestinal tract of man. Gastroenterol. 43: 472.

9. Bertalanffy, F.D. 1963. Aspects of cell formation and exfoliation related to cytodiagnostic. Act. Cytol. 7: 362.

10. Rother, P., Scheller, G., Hellthaler, G. and Dähnert, K. 1971. Uber die Rhythmik, die Dauer und die Häufigkeit von Mitosen im Dünndarmepithel der Ratte. Zeitschr.f.mikr.-anat.Forsch. 83: 399.

11. David, H. 1967. Zum Mechanismus der Zellabstoßung im Bereich der Dünndarmzotten. Virch.Arch.Path.Anat. 342: 19.

12. Tews, K., Jahn, K. and Leutert, G. 1971. Baustein-histochemische Studien am Dünndarmepithel von Ratten verschiedenen Alters. Zschr.mikr.-anat.Forsch. 83: 423.

13. Leutert, G., Jahn, K. and Weise, K. 1973. Altersabhängige Veränderungen der Enzymaktivitäten von Enterozyten - histochemische Untersuchungen. Zeitschr.f.Alternsf. 26: 375.

14. Hellthaler, G., Köhler, H. and Rotzsch, W. 1971. Altern-sabhängige Veränderungen bei Proteinsynthese von Dünndarm-ringen. Zeitschr.f.Alternsf. 24: 243.

15. Leutert, G. and Brandt, E. Morphometrische Untersuchungen Zeitschr.f.Alternsforsch. (im Druck).

16. Mitroiu, P., Lang, W. and Maurer, W. 1968. Autoradiographische Bestimmung des Markierungsindex, der S-Phase und der Gener-ationszeit einiger Zellarten von 2 bis 35 Tage alten Kuken. Z.Zellf. 90: 68.

17. Heim, F., Schwarzlose, W., Truckenbrodt, H. and Mizznegg, P. 1969. Desoxyribonukleinsäure-, Ribonukleinsäure- und Eiweißgehalt, Zellneubildung und Zellgröße in Abhängigkeit von Alter, Geschlecht und Ernährung und deren Beeinflussung durch anabole und antianabole Stoffe. Zeitschr.f.Geront. 2: 25.

18. Richter, V. and Rotzsch, W. 1972/1973. Turnover of rat liver and rat kidney catalase as a function of age. Mech. Age. Dev. 1: 427.

19. Adelman, R.C. 1971. Age dependence effects in enzyme induc-tion, a biochemical expression of aging. Exp. Ger. 6: 75.

20. Wiegel, D., Beier, W. and Brehme, K.-H. 1973. Vitality and
     error rate in biological systems: Some theoretical consi-
     derations. Mech. Age. Dev. 2: 117.

21. Orgel, L.E. 1963. The maintenance of accuracy of protein
     synthesis and its relevance to ageing. Proc. Nat. Acad. Sci.
     USA 49: 517.

22. Beier, W., Rotzsch, W., Leutert, G. and Ries, W. 1970. Zur
     Molekularbiologie des Alterns. 2. Mitteilung: Ein mathe-
     matisches Modell des Alterns eines multizellulären Systems.
     Zeitschr.f.Alternsf. 22: 333.

23. Beier, W., Brehme, K.-H. and Wiegel, D. 1973. Biophysi-
     kalische Aspekte des Alterns multicellulärer Systeme.
     Fortschritte der theoretischen und experimentellen
     Biophysik, Heft 16, VEB Georg Thieme, Leipzig.

24. Shock, N.W. 1961. Physiological aspects of ageing in man.
     Am. Rev. Physiol. 23: 97.

QUESTION TO DR. LEUTERT

Dr. Knook: Since you considered rats of 12 months of age as old
animals I wonder what you observed as the mean lifespan of your
rats? And in relation to this did you use males or females?

Dr. Leutert: At first we have considered rats 18 months of age as
old animals. On the basis of our model the maximum vitality of
the rats investigated by us is at the age of about 6 months.
Therefore older rats were considered as rats showing a decline in
vitality. Both males and females were subjected to investigation
but both groups were kept and evaluated separately.

FAT CELLS IN ONTOGENESIS

Ludmila Kazdová and Pavel Fábry

Research Centre of Metabolism and Nutrition of the
Institute for Clinical and Experimental Medicine,
Prague, Czechoslovakia

In recent years it was revealed that in adipose tissue active
cell proliferation and an increase of cell number takes place; this
process participates in a significant way in the increase of adipose
tissue mass during ontogenetic development (1-4). The growth of
adipose tissue by hyperplasia of adipocytes can be stimulated in
growing animals by some nutritional stimuli, which include refeeding
after previous fasting (5) or hyperphagia in the early postnatal
period (6,7). In recent years we studied intensely this induced
hyperplasia of adipose tissue and found that in the mechanism of
triggering of DNA synthesis insulin may play a role, similar to the
role it plays in some other tissues (8-11). We have demonstrated
that the administration of small doses of insulin for 24-72 hr
leads to an increase of DNA content of adipose tissue. After
isolation of cellular and subcellular fractions from adipose
tissue we found that insulin stimulated the synthesis of nuclear
and not of mitochondrial DNA and caused increased proliferation
of fat and stromavascular cells. In keeping with this was also
the increase in the number of fat and stromavascular cells ascertained by calculation and histological examination of tissue, when
cells in mitotic division were found as well as an increased ratio
of smaller adipocytes (12-14).

In the present paper results of experiments are presented in
which we investigated how, in adipose tissue, the proliferating
activity and sensitivity to factors affecting cell proliferation
changes in relation to the age of the animals. These problems
are interesting from the aspect of the hitherto not elucidated
regulation of growth of adipose tissue and in particular from the
aspect of the pathogenesis of obesity.

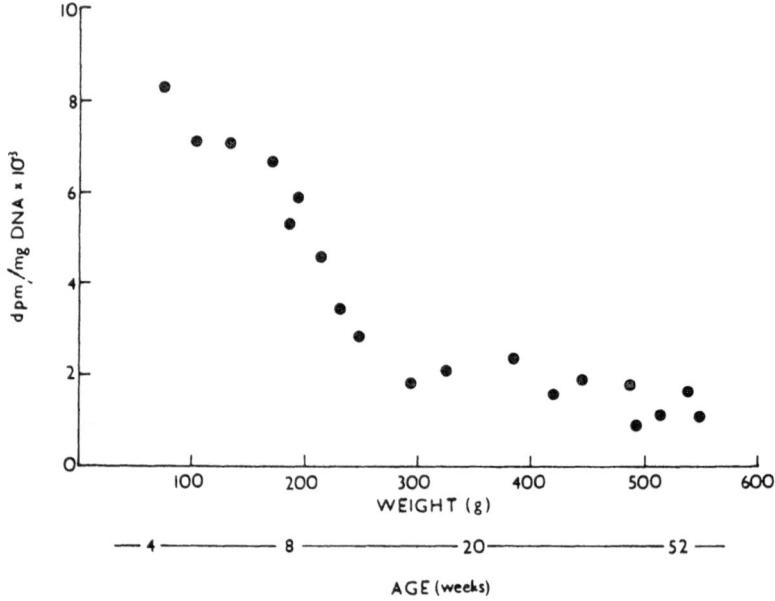

Fig. 1.   Effect of age on incorporation of ^{14}C-2-thymidine into
adipose tissue DNA.   Each point represents average for 6-9 rats.

MATERIALS AND METHODS

For the experiments male Wistar rats were used, 4 weeks to 1
year old.   The proliferative activity was investigated from DNA
synthesis which, as we found, correlates with changes in mitotic
activity (14).   DNA synthesis was measured from the in vitro
incorporation of ^{14}C-2-thymidine (spec. act. 46 mC/mM̄ol, UVVVR,
Czechoslovakia) which is more specific for the investigation of DNA
synthesis than the frequently used ^{3}H-thymidine where the labelled
tritium may be incorporated into lipids and proteins (15,16).   DNA
was extracted from the tissue or individual cellular fractions
after delipidation of tissue (17) using Schneider's method (18).
The number of fat and stromavascular cells was calculated according
to Rakow (19) and Di Girolamo (20).

RESULTS AND DISCUSSION

The results presented in Figure 1 indicate changes in DNA
synthesis in adipose tissue in relation to the animals' age.   It
is apparent that even when the DNA synthesis is reduced from the
period from the 4th to the 10th week by almost half, it remains

fairly high even in one year old animals. As adipose tissue
contains, in addition to adipocytes, a large number of stromavascular cells, we investigated changes in DNA synthesis in these
cellular fractions during aging of the animals. In keeping with
findings of other authors (21,22) we found on histological
examination of different fractions isolated by means of collagenase according to Rodbell's method (23) that perfect separation
of adipocytes from stromasvascular cells is practically impossible.
Therefore during analysis of DNA synthesis, we divided the adipose
tissue after incubation with collagenase not only into layers
of adipocytes and stromavascular fraction but also into an intermediate layer which contains fat and stromavascular cells. From
Figure 2 it is apparent that the reduction of proliferating
activity occurs in all cellular fractions of adipose tissue. The
reduction of DNA synthesis is most marked in the adipocyte fraction
(by 80%) and smallest in the fraction of stromavascular cells

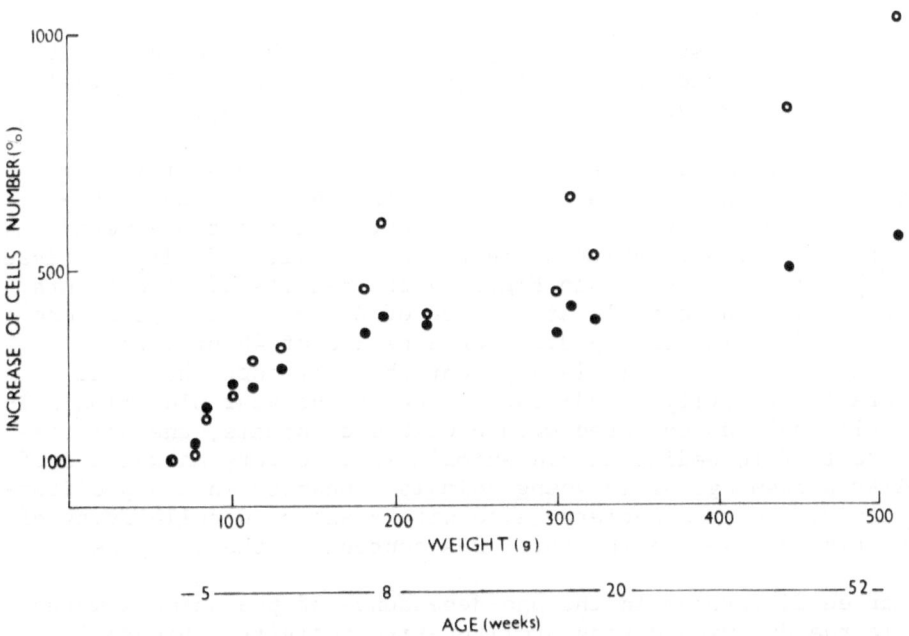

Fig. 2.   The number of fat (●●●) and stromavascular cells (ooo)
in epididymal fat pads of rat of various ages. Each point represents average for 4-6 rats. Relative values: 100% = the number
of cells in rats aged 4 weeks.

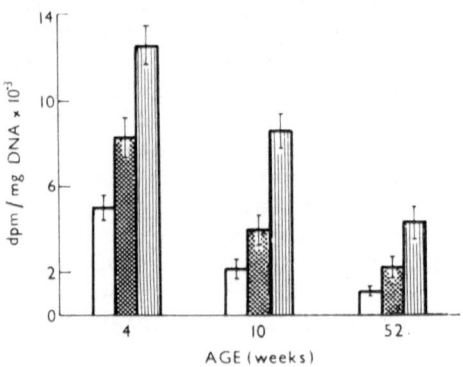

Fig. 3.   Effect of age on incorporation of ^{14}C-2-thymidine into DNA of different fractions of adipose tissue.   Fat cells (☐), intermediary (▨) and stromavascular (▥) fractions were isolated after incubation of adipose tissue with ^{14}C-2-thymidine. Each point represents average for 4-8 rats ± S.E.

(by 60%).   These results are in agreement with the findings that in the course of ontogenesis with the increasing size of fat cells the ratio of stromavascular system increases (Fig. 3).

As we have mentioned before, the proliferative activity of adipose tissue can be stimulated in growing rats by some nutritional or hormonal factors.   In subsequent experiments we were interested in whether during aging the sensitivity of adipose tissue to these stimuli changes.   In Figure 4 the results of experiments are shown where the animals at the age of 8 weeks and 1 year were refed a standard laboratory diet for a period of 48 hr after a previous 3-day fast.   It is apparent that although the basal proliferative activity of adipose tissue in one year old animals is practically half as compared with 8 week old animals, the stimulating effect of refeeding in old animals is also very marked and of a similar percentage as in young animals.   Changes in the proliferative activity were associated also with changes in cellularity of adipose tissue based on the total DNA content in the fat pads.

Marked difference in the age-dependence of the animals were found in the dynamics during proliferative activity.   Figure 5 illustrates the course of DNA synthesis in adipose tissue regenerating after resection of part of the epididymal fat pads.   It is apparent that the maximum DNA synthesis which attained almost the same percentage values in the both age groups is, in one year old animals, shifted by 24 hr as compared with young animals.

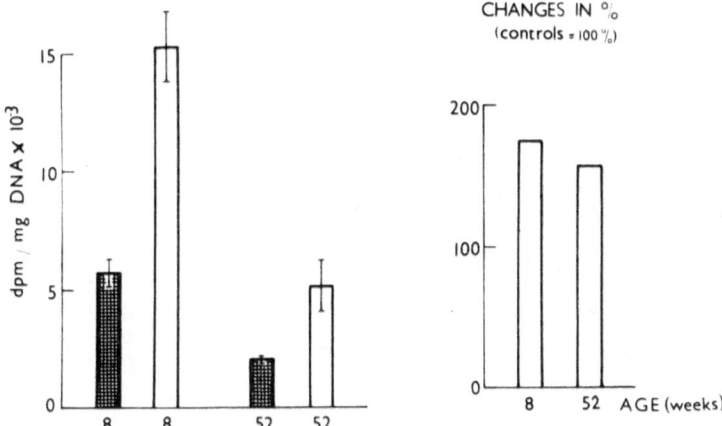

Fig. 4. Effect of age on incorporation of ^{14}C-2-thymidine into DNA of adipose tissue of rats fed ad libitum (▦) and rats refed for 48 hr after a previous 72 hr (☐). Each point represents average of 5-8 animals ± S.E.

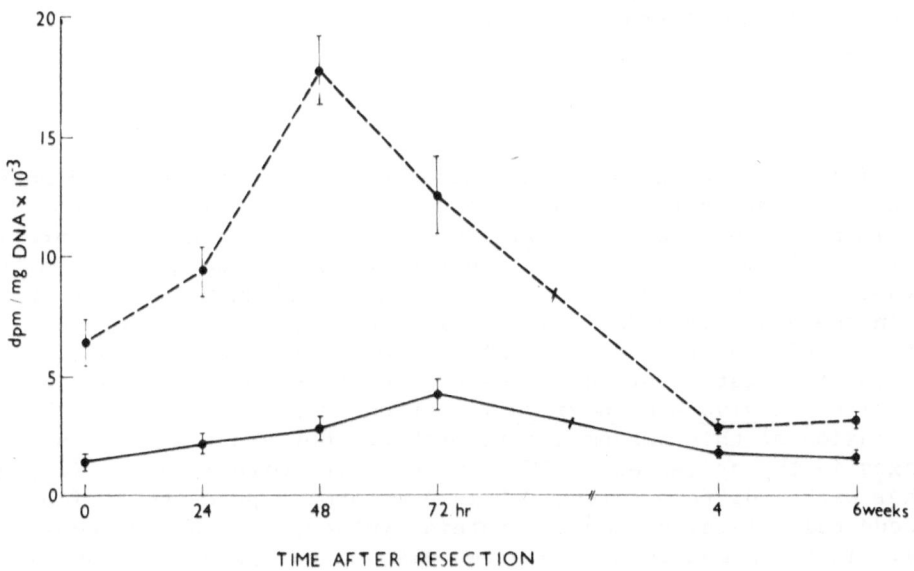

TIME AFTER RESECTION

Fig. 5. Time course of DNA synthesis in regenerating epididymal fat pads of rats aged 6 (●--●) and 52 (●—●) weeks. Each point represents average of 5-8 animals per group ± S.E.

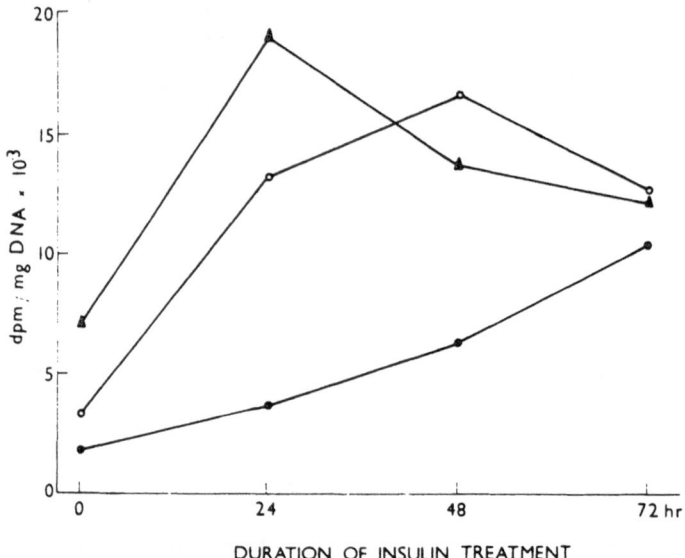

DURATION OF INSULIN TREATMENT

Fig. 6. Effect of insulin treatment on DNA synthesis in adipose tissue of rats aged 5 (Δ—Δ), 8 (o—o) and 52 (●—●) weeks. Each point represents average of 5-10 animals per group. Insulin was injected intraperitoneally, twice daily in a dose of 500 μU/100g body weight.

        Similar differences in the course of DNA synthesis in adipose tissue in relation to the animals' age were found also in the experiment where we used insulin as a stimulus for enhancing proli- feration. From Figure 6 it is apparent that in 5 week old rats the synthesis is highest after 24 hr, in 8 week old animals after 48 hr and in one year old animals after 72 hr or later. It is of interest that the delay of the maximum DNA synthesis or peak of mitotic division to a later period in the course of ageing was found under conditions in vivo and in vitro in various tissues (24-26). The explanation of this phenomenon is probably not associated with changes in the processes of DNA synthesis and mitosis alone (27) but rather with a different level of some metabolic processes which precede cell division such as protein synthesis or RNA synthesis (28). This assumption is supported also by our previous findings that in adipose tissue, protein and RNA synthesis are reduced in old animals and that these processes can be at least partly repaired by the administration of insulin or by refeeding (29).

SUMMARY AND CONCLUSION

It may be concluded that during aging in adipose tissue the
proliferative activity declines. Between different types of cells,
differences exist in the diminution of the proliferating ability.
Even in adipose tissue of one year old animals, there are cells
which are able to respond to some stimuli such as realimentation
after previous fasting, administration of insulin or resection
of part of the fat body by an increased proliferative activity.
In the dynamics of these induced changes of DNA synthesis, there
exist marked differences, depending on the animals' age; with
advancing age the maximum of DNA synthesis shifts to later periods.

REFERENCES

1. Hirsch, J. and Han, P.W. 1969. Cellularity of rat adipose
    tissue: effects of growth, starvation and obesity. J.
    Lipid Res. 10: 77.

2. Hirsch, J. and Knittle, J.L. 1970. Cellularity of obese and
    non-obese human adipose tissue. Feder. Proc. 29: 1516.

3. Lemonnier, D. 1972. Effect of age, sex and site on the cellu-
    larity of adipose tissue in mice and rats rendered obese by
    a high-fat diet. J. Clin. Invest. 51: 2907.

4. Kazdová, L., Fábry, P. and Vrána, A. 1973. Postnatal
    development of adipose tissue in different strains of rats.
    Physiol. bohemoslov. 22: 404.

5. Kazdová, L., Braun, T. and Fábry, P. 1967. Increased DNA
    synthesis in epididymal adipose tissue of rats refed after
    a single fast. Metabolism 16: 1174.

6. Knittle, J.L. and Hirsch, J. 1968. Effect of early nutrition
    on the development of rat epididymal fat pads: Cellularity
    and metabolism. J. Clin. Invest. 47: 2091.

7. Fábry, P., Kazdová, L., Vrána, A. and Koopman, T. 1973.
    Effect of hyperphagia in early postnatal development on the
    cellularity of adipose tissue in adult rats. Physiol.
    bohemoslov. 22: 392.

8. Younger, L.R., King, J. and Steiner, D.F. 1966. Hepatic
    proliferative response to insulin in severe alloxan diabetes.
    Cancer Res. 26: 1408.

9.   De la Haba, G., Cooper, G.W. and Elting, V. 1966. Hormonal
     requirements for myogenesis of stritiated muscle in vitro:
     Insulin and somatotropin. Proc. Soc. Exp. Biol. Med. 56:
     1719.

10. Turkington, R.W. 1968. Hormone-induced synthesis of DNA by
     insulin. Endocrinology 82: 540.

11. Salmon, W.D., Jr., DuVall, M.R. and Thompson, E.Y. 1968.
     Stimulation by insulin in vitro incorporation of ^{35}S-sulphate
     and ^{14}C-leucine into protein polysaccharide complexes, ^{3}H-
     uridine into RNA and ^{3}H-thymidine into DNA of costal cartilage
     from hypophysectomized rats. Endocrinology 82: 493.

12. Kazdová, L. and Vrána, A. 1970. Insulin and adipose tissue
     cellularity. Horm. Metab. Res. 2: 117.

13. Kazdová, L., Fábry, P. and Vrána, A. 1972. Localization of
     insulin stimulated DNA synthesis in rat adipose tissue.
     Physiol. bohemoslov. 21: 404.

14. Kazdová, L., Fábry, P. and Vrána, A. 1974. Effect of small
     doses of insulin in vivo on the proliferation and cellullarity
     of adipose tissue. Diabetologia 10: 77.

15. Bryant, B.J. 1966. The incorporation of tritium from thymidine
     into proteins of the mouse. J. Cell Biol. 29: 29.

16. Kazdová, L., Fábry, P. and Vrána, A. 1972. Comparison of
     the specificity of ^{14}C-2-thymidine and ^{3}H-thymidine as a
     precursor of DNA. Physiol. bohemoslov. 2i: 89.

17. Steele, W.J., Okamura, N. and Busch, H. 1964. Prevention
     of loss of DNA, RNA and protein into lipid solvents. Biochim.
     Biophys. Acta 87: 490.

18. Schneider, W.C. 1957. Determination of nucleic acid in
     tissue by pentose analysis. In: Methods in Enzymology.
     S.P. Colowick and N.O. Kaplan, eds., Academic Press, New York,
     p. 680.

19. Rakow, L., Beneke, G., Mohr, W. and Brauchle, I. 1971.
     Untersuchungen über die Zellvermehrung im weissen Fettgewebe
     der genetisch adipösen Maus (C_{57}BL/6 J ob/ob), Beitr. Path.
     Bd. 143: 301.

20. Di Girolama, M., Skinner, N.S., Hanley, H.G. and Sachs, R.G.
     1971. Relationship of adipose tissue blood flow to fat
     cell size and number. Am. J. Physiol. 220: 932.

21. Mohr, W. and Beneke, G. 1968. Age dependence of nuclear
     content of rat adipose cells. Experientia 24: 1052.

22. Hollenberg, C.H. and Vost, A. 1968. Regulation of DNA
     synthesis in fat cells and stromal elements from rat adipose
     tissue. J. Clin. Invest. 47: 2485.

23. Rodbell, M. 1964. Metabolism of isolated fat cells. I.
     Effects of hormones on glucose metabolism and lipolysis.
     J. Biol. Chem. 239: 375.

24. Marshak, A. and Byron, R.L. Jr. 1945. The use of regenerating
     liver as a method of assay. Proc. Soc. Exp. Biol. Med.
     59: 200.

25. Bucher, N.R.L., Troia, J.F. and Swaffield, M.N. 1961. DNA
     synthesis during hepatic regeneration in rats various ages.
     Feder. Proc. 20: 286.

26. Soukupová, M., Holečková, E. and Hněvkovský, P. 1970. Changes
     of the latent period of explanted tissues during ontogenesis.
     In: Aging in Cell and Tissue Culture. E. Holečková and
     V.J. Cristofalo, eds., Plenum Press, New York, p. 41.

27. Macieira-Coelho, A. 1972. Death and survival at a cellular
     level. Rev. Europ. Études Clin. et Biol. 17: 925.

28. Buetow, D.E. and Gandhi, P.S. 1973. Decreased protein
     synthesis by microsomes isolated from senescent rat liver.
     Exp. Geront. 8: 243.

29. Kazdová, L., Fábry, P. and Vrána, A. 1971. Effect of age
     and insulin on the nucleic acid content and synthesis in
     rat adipose tissue. Physiol. bohemoslov. 20: 68.

QUESTIONS TO DR. KAZDOVA

Dr. Kasten: As a matter of interest for possible biological
models you might wish to consider particular obese mutants in
the mouse: 1) obese is a recessive mutant (obob) which produces
obesity in the homozygous state. 2) Yellow lethal is a dominant
gene (Ay) which produces lethality in homozygous embryos and
obese adults in the heterozygous state.

Dr. Kazdova: Thank you for your comments. Some aspects of growth
of adipose tissue in genetically obese mice (yellow obese and obob)
were studied previously by Dr. Helman et al. and Dr. Rakow et al.

Dr. Rothe: In human beings we meet a surprising fact. If the

body weight exceeds 170% of the normal, fat cells get the capacity
to divide again even in old people. Is there also a dependence of
DNA synthesis on the body weight in rats?

Dr. Kazdova: From our results it was apparent that during normal
ontogenetic development the DNA synthesis in the adipose tissue
decreases with increasing body weight. Since fat cells cannot
continue to expend indefinitely, it is possible that in the extreme
human obesity fat cell enlargement is followed by the formation of
new adipocytes by proliferation and maturation of fat cell precursors. A similar situation has been observed in the genetically
obese Zucker (fafa) rats and in mice made obese by means of a
high fat diet by Dr. Johnson and Dr. Lemonnier. In both cases
there was an increase in fat cell number during later periods of
ontogenesis.

Dr. Leutert: Have you differentiated between brown and white
fat and if you have is there a difference between these kinds of
fat tissue?

Dr. Kazdova: The adipose tissue that we used in our experiments
was only white epididymal tissue. We have no information about
age dependent changes in the fat cell size and proliferative
activity in the brown fat tissue.

AGEING AND THE LOSS OF AUDITORY NEUROEPITHELIUM IN THE GUINEA PIG

Libuše Úlehlová

Otolaryngological Laboratory of Czechoslovak Academy of Sciences, Prague

A general feature of an ageing process in postmitotic cells is a reduction in the number of these cells during the life of an individual. One of the most convenient materials for the study of this decrease in number of cells and its dependence on age is the sensory neuroepithelium of the hearing organ, represented by hair cells.

On the surface of the organ of Corti the hair cells form a regular geometrical pattern. One row is formed by inner hair cells and three rows by the outer hair cells. Hair cells are surrounded by supporting elements. The regular arrangement of acoustic papilla enables an exact evaluation of the number of missing hair cells.

Specimens were prepared by surface specimen technique which is based on the dissection of the tissue "in toto" and which allows after haematoxylin staing to evaluate and to consider the state of every cell and its relations to the neighbouring cells. By means of surface specimen technique the whole spirally curved organ of Corti was dissected out from the surrounding bone. There are four and one-half turns in the guinea pig cochlea.

The full number of hair cells arranged in the regular geo-metrical pattern was taken as a basis for the evaluation and every change in the number of hair cells, i.e., the loss of hair cells, was expressed with respect to this ideal number. The loss of hair cells was calculated in segments of the same length. They were taken from individual turns of the organ of Corti.

Three different age groups of guinea pigs were used for this study. All guinea pigs were bred and kept in a quiet countryside and were not exposed to any experimental conditions, neither drugs nor excess noise.

The youngest group of animals consisted of 53 guinea pigs 6-weeks-old weighing between 300-320 g. The next group consisted of 50 guinea pigs 7 months that is 30 week-old weighing between 440-480 g and the oldest group consisted of 23 guinea pigs 3 yr that is about 156 weeks old weighing between 660-780 g.

By morphological observation of the organ of Corti and by statistical evaluation of the number of missing cells the following results were obtained:

The number of the inner hair cells is almost constant. It decreases with age very slowly. In the group of 3-year-old guinea pigs the loss of inner hair cells was found to be 0.25%. On the other hand the number of outer hair cells is apparently much more age dependent.

In the group of youngest guinea pigs the deviation from the full number was relatively small, the loss of outer hair cells is in average given by 0.73%. In Fig. 1 the distribution of hair cell

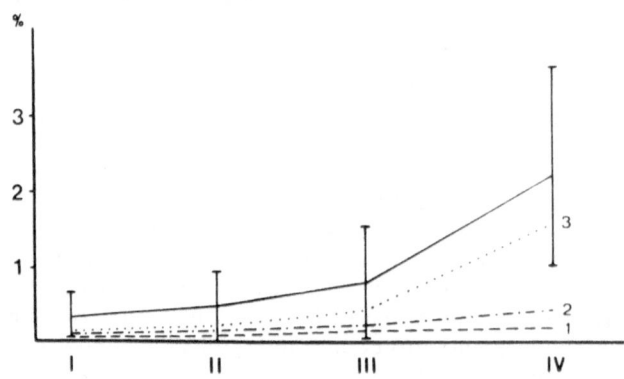

Fig. 1. The distribution of the hair cell loss in percentage in 6 week-old guinea pigs. Roman numerals denote individual turns of cochlea. Full line represents the average loss of outer hair cells. Standard deviations in individual turns for this line are given. By 1, 2, 3, the loss in individual row is given.

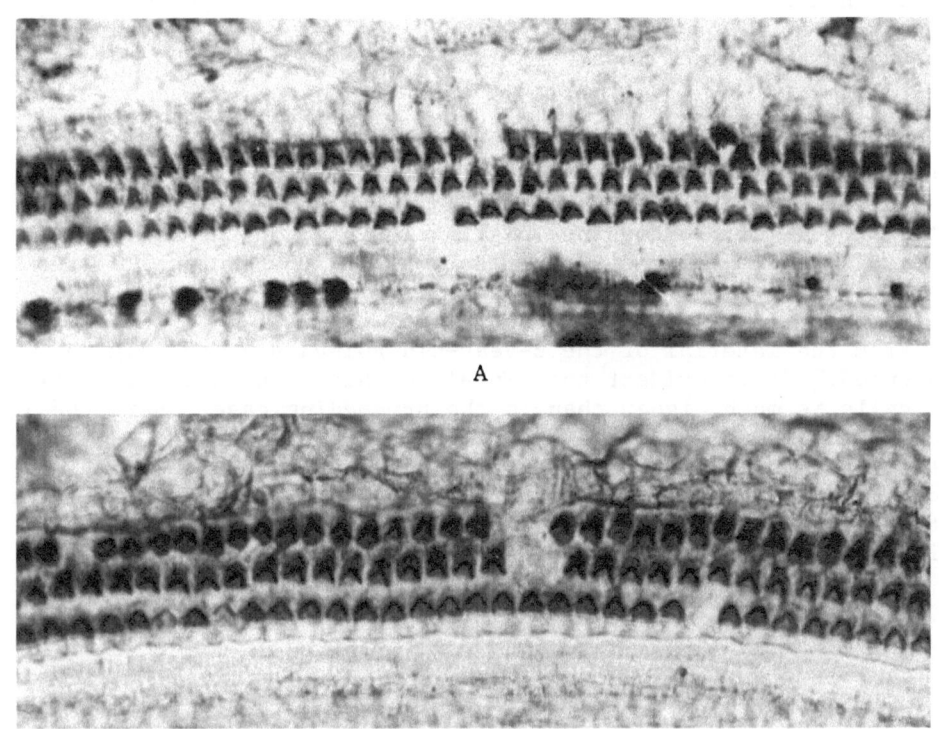

A

B

C

Fig. 2. Cellular pattern of the third turn of the organ of
Corti in guinea pig. Surface specimen, Haematoxylin 500 X.
(A) An animal 6 weeks-old. (B) An animal 7 month-old. (C) An
animal 3 yr-old.

loss is shown.   The smallest loss occurs in the basal turn and the
largest in the fourth,a apical turn.   A typical picture for the
youngest group is presented on Fig. 2a from which one can see that
here and there an individual cell is missing.

     In the group of 7 month-old guinea pigs the loss of cells is
considerably higher as it follows from Fig. 3.   Also in this group
of animals the dependence of the hair cell loss on individual turns
in apparent.   The average loss in this set was 0.85%.   The typical
picture for 7-month-old guinea pig is presented on the Fig. 2b.   Here
no only individual elements but also smaller groups of cells are
missing.

     From the material of the 3-year-old guinea pigs the Fig. 4
is deduced.   It is evident that in all cochlear turns the loss of
hair cells is much higher than in the preceeding case.   Here the loss
is also continuously increasing from the cochlear base to the apex.
In this group similarly to the both discussed above the maximal loss
is observed in the peripheral third row.   A photomicrograph (Fig. 2c)
of the organ of Corti of the 3-year-old guinea pig represents a
typical example for hair cell loss in larger groups.

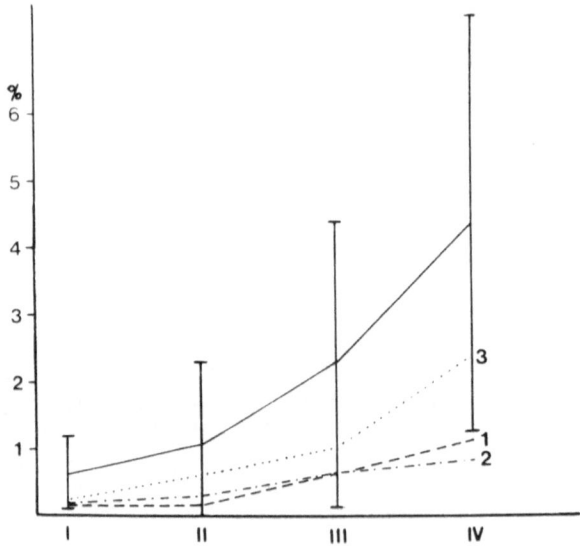

Fig. 3.   The distribution of the hair cell loss in 7 month-old
guinea pigs.   The notation is the same as in Fig. 1.

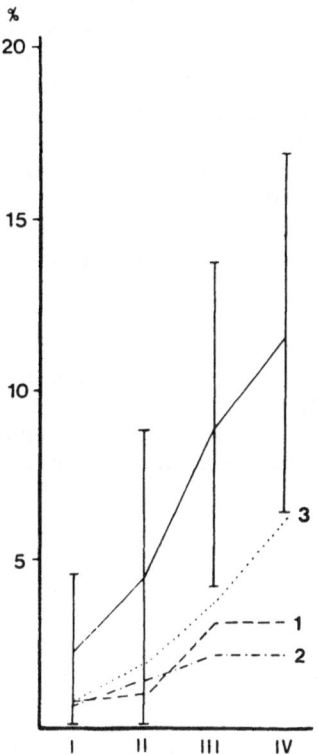

Fig. 4. The distribution of the hair cell loss in 3 yr-old
guinea pigs. The notation is the same as in Fig. 1.

If we compare in the summarizing graph (Fig. 5) in which on the
horizontal the age of guinea pigs is given the hair cell loss in
individual age groups and in individual turns we find that the
hair cell loss increases linearly with the age.

Hence, the hair cell loss is linearly proportional to the age
of animals. However, it is difficult to judge precisely to what
extent this loss of hair cells is caused only by ageing of the organism because the organism is exposed under normal conditions
to many unfavorable influences. The noise of modern civilization
can be one of them.

Bredberg (1) found in material from 74 subjects of different
age that the loss of hair cells in the organ of Corti had been

related to the exposition of these subjects to high intensity noise. Tosen (2) and collaborators examined the noise free population in the Sudan. The isolated primitive tribe of Mabbans whose state of cultural development is the late Stone Age live in a very quiet environment. Rosen compared the hearing of members of this tribe with that of corresponding age group in the United States and he found that people from the silent environment had better hearing in old age than people of the civilized world. There are other studies on this subject and it follows from them, that the exposure to the noise of modern civilization is in direct relation to the atrophy of hearing neuroepithelium.

Although the animals used in this study were taken from a quiet countryside one cannot exclude completely the negative influence of the "normal" noise on the hearing organ. The linearity of the decrease in number of sensory cells supports the idea that the effect observed is induced by the aging of the organism in normal conditions of living.

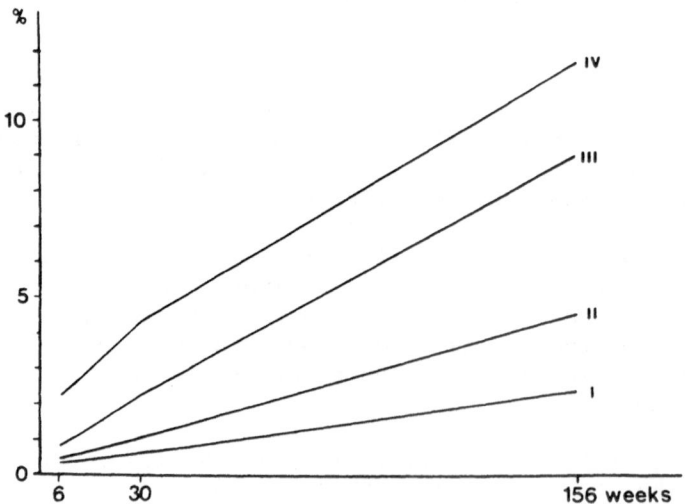

Fig. 5. Dependence of the hair cell loss in percentage on the age of guinea pigs in individual turns (I-IV) of cochlea.

SUMMARY

By microscopic evaluation of the cellular pattern of the organ
of Corti in 3 age groups of guinea pigs (6 weeks, 7 months and 3 yr)
the linear dependence of the loss of sensory cells on the age of
animals was found.   The linearity of the decrease in number of sensory cells supports the idea that the effect observed was induced
by the ageing of the organism in normal conditions of living.

REFERENCES

1. Bredberg, G. 1968. Cellular pattern and nerve supply of the
      human organ of Corti. Acta Oto-laryng. (Stoch.) Suppl. 236.

2. Rosen, S., Bergman, M., Plester, D., El-Mofty, A., Satti, M.H.
      1962. Presbycusis study of a relatively noise-free population
      in the Sudan. Ann. Otol. Rhinol. Laryngol. 71: 727.

QUESTIONS TO DR. ÚLEHLOVÁ

Dr. Franks: Are there structural changes in the hair cells? A
comparison of degenerative changes in animals living in a quiet
environment with those caged in a noisy environment may allow
age changes to be separated from noise-induced changes.

Dr. Úlehlová: There are structural changes in the hair cells but
we were note interested in the intermediate stages.   Such cells
which showed a recognizable degree of degenerative changes and were
not fully functional were considered as dead cells. With respect
to your second question there are a lot of papers dealing with
the action of noise of different intensities and different
characteristics on the sensory neuroepithelium of the organ of
Corti. The aim of this study was to assess the basic data for the
control uninfluenced group of animals.

Dr. Martin: I wonder if you have an explanation for the striking
"clustering" distribution of the cell loss? On the basis of the
Orgel hypothesis, for example, and assuming the extent of metabolic
cooperation (intercellular communication) between these cells, one
might expect a spread of "lethal" cytoplasm to contiguous neighbouring cells. Your model system is a very favourable one because
the geometry allows one to ask a variety of interesting questions
concerning the pattern of cell loss.

Dr. Úlehlová: In my opinion the explanation may run as follows:

The upper surface of the organ of Corti is formed by the reticular membrane which is the important border between endolymph and perilymph. According to several authors the hair cells kept in the endolymph undergo degeneration and die very quickly. It may be that when the hair cell die, the hole in the reticular membrane is not closed fast enough by the phalangeal processes of the supporting Deiters' cells and the endolymph attacks the neighbouring hair cells.

THE PHAGOLYSOSOMAL SYSTEM OF THE RETINAL PIGMENT EPITHELIUM IN

AGEING RATS

P.M. Leuenberger

University Eye Clinic

Geneva, Switzerland

The retinal pigment epithelium is responsible for the removal of the continuously shedded terminal disks of rod outer segments. As such, in mammalian retinas, each pigment epithelial cell engulfs and destroys thousands of disks every day (1). The pigment epithelium possesses a highly developed phagolysosomal system, which has to digest these considerable amounts of exogenous material for a lifetime. In contrast to phagocytic cells of the reticuloendothelial system, the pigment epithelium has a very low rate of mitosis and is likely to accumulate with time a great deal of non-digestible material, which in turn might impair cellular function and be at the origin of some of the retinopathies observed in the ageing eye (2).

In the course of a study on age-correlated vascular changes in the retina (3) we observed conspicuous differences in the morphology of the phagolysosomal system in the pigment epithelium between young and old animals. Increased numbers of lipofuscin granules and dense bodies were encountered in old animals and the lipofuscin granules appeared to be generally of larger size.

Microperoxisomes (4), catalase-rich organelles closely related to smooth endoplasmic reticulum (ER) and GERL (Golgi Endoplasmic Reticulum Lysosome) (5), a region at the inner aspect of Golgi apparatus rich in acid phosphatase (acPase), have recently been demonstrated in the pigment epithelium of mammalian retinas (6,7). Microperoxisomes are involved in aspects of lipid metabolism, GERL is considered as being a site of packaging of lysosomal enzymes. Close functional relationships between lipofuscin granules, microperoxisomes, lysosomes and smooth ER have been reported (8). The purpose of this study is the demonstration of age-correlated

265

ultrastructural and cytochemical changes in the phagolysosomal
system of retinal pigment epithelium.

MATERIALS AND METHODS

Eyes of Wistar-derived albino-rats of different ages (3 days,
2 weeks, 3, 10 and 26-31 months old) were enucleated under ether
anesthesia and immediately fixed for 90 min in either cold 3%
glutaraldehyde-0.1 M cacodylate or a mixture of 2.5% glutaraldehyde
and 2% formaldehyde (9). The eyes were opened at the ora serrata
in the cold fixative and the posterior half was further processed
in the following manner.

Electron Microscopy

Small blocks of tissue of less than 0.2 mm thickness were cut
out with razor blades under a dissecting microscope, briefly rinsed
in buffer, postfixed in cold 1% OsO_4 (phosphate buffered), en bloc
stained with 0.5% uranyl-acetate (10), dehydrated in a graded
alcohol series and embedded in Epon 812 (11). Thin sections were
cut with diamond knives on A porter Blum MT-2 microtome and stained
with lead citrate (12).

The sections were observed and photographed in a Philips EM300
at either 80 or 100 kV, using 30 μm objective apertures.

Cytochemistry

1. Light microscopy: 15 μm frozen sections of prefixed tissue
(3% glutaraldehyde 0.1 M cacodylate) were incubated at 37°C in a
shaking incubator for nucleosidediphosphatase activity (using thiaminepyrophosphate [Sigma] as substrate) (13) and acid phosphatase
activity (using cytidinemonophosphate [Sigma] as substrate) (14).
The reaction product was visualized with ammonium sulfate and the
sections were mounted with glycerogel on glass slides. Controls
were performed by omitting the substrate in the medium or by
heating the sections prior to incubation to 60°C.

2. Electron microscopy: 30 μm tissue slices were cut on a
Smith-Farquhar tissue sectioner (15) and incubated for either acid
phosphatase (14) or nucleosidediphosphatase (13) activity or in an
alkaline 3.3'-diaminobenzidine medium (DAB, Sigma) (16) in a
shaking incubator at 37°C. Incubation time was determined by
monitoring of sections with light microscopy, after ammonium sulfide
treatment. After a twofold wash in 7.5% sucrose the tissues were
osmicated and further processed as in the Electron Microscopy
section.

RESULTS AND DISCUSSION

The first 4 micrographs (Figs. 1a,b; 2a,b) demonstrate that under our conditions of assessment the intensity of staining for acPase in the pigment epithelium seems to decrease with age. Even though _in situ_ cytochemistry does not accurately allow quantitative determination of enzymatic activity, the differences are obvious and strongly suggest a lower level of acPase in the pigment epithelium of old animals. A comparable phenomenon may be observed in ganglionic cells (Figs. 5 and 6). Furthermore, differences with respect to shape and size of the acPase-positive organelles are discernible. In young animals, acPase-rich sites often appear ellipsoid or elongated rather than circular, as it may be observed in the old (cf. Figs. 1a, 2a, and 5).

AcPase-cytochemistry at the ultrastructural level reveals in young animals the presence of reaction product over dense bodies, over ER-cisternae on the inner aspect of the Golgi-apparatus (corresponding to GERL) and over a multitude of small vesicles (Figs. 10 and 11). These acPase-rich areas and their interrelations are readily demonstrable in "thick" sections (0.5 to 1 μm) (17) of incubated material. In old animals these areas as well as the major part of dense bodies showed, with the method used, hardly any reaction product (compare Fig. 12 to Figs. 10 and 11).

In the absence of any correlative biochemical data on the activity of different acid hydrolases in subcellular fractions of ageing retinal pigment epithelium, the interpretation of our rather unexpected findings must remain purely speculative. In most cell types assessed, an increase in both cytosomal and lysosomal acPase-activity has been reported in senescent cells (18). Concomitantly, Golgi cisternae and vacuoles have been reported to be more prominent in old cells (19), a phenomenon that we did not observe in pigment epithelial cells (Figs. 7, 8 and 9).

Lipofuscin granules are present in greater number and are of larger size in the old animals (Figs. 3 and 4). In both young and old animals these organelles showed invariably acPase-activity (20). The reaction product was localized over the electron-dense granular component of lipofuscin granules (Figs. 11, 12, and 16a). The globular, vacuole-like, relative electron-dense area of lipofuscin granules (Figs. 15 and 16b), containing presumably the lipids characteristic of "lipofuscin" are free of reaction product. In old animals the lipid content of lipofuscin granules appeared to be increased and frequently partially or completely segregated by membranous elements (Figs. 13 and 14). A comparable observation has been reported in ageing rat myocardial cells (21) and was considered as morphological support for the possibility that the composition of lipid changes with age.

No acPase-activity was found over the third component of lipofuscin granules (Fig. 15c), consisting of electron lucent areas of
irregular granular appearance.

Identical areas of lipofuscin granules that showed acPase-
activity were also stained when incubated in an alkaline DAB-medium
(Figs. 15 and 16). It is not yet known whether this DAB-reactivity
is enzymatic (peroxidase or catalase, which reacts peroxidatically)
or involves a non-enzymatic oxidation of DAB, as it has been suggested
in human lipofuscin granules (22). If lipid degradation within
lysosomes should occur by partial autoxidation (23) with consecutive
accumulation of peroxides, this peroxidative activity might serve to
relieve the cell from the damaging effects of peroxides. Catalase
might also be involved in lipid metabolism by acting as a lipid-
peroxidation catalyst (24).

To conclude we ought to explain the occurrence of diminished
acPase activity as revealed by the cytochemical methods used, as
well as the accumulation of large secondary lysosomes in the pigment
epithelium of our ageing rats. The apparent "overengorgement" of
secondary lysosomes that we observed might be caused by a (relative)
lack of acid hydrolases within the lysosomes. But the defect would
not reside in leaky membranes that allow enzymes to escape (25),
but rather in a decreased availability of normal enzymes due to a
decreased rate of synthesis of normal enzymes or the synthesis of
defective enzymes, as it has been demonstrated in I-cell disease
(26).

We think that on the basis of our preliminary findings a more
detailed biochemical and cytochemical assessment of age-induced
changes in the phagolysosomal system of the retinal pigment epithelium seems promising and might lead to a better understanding of
certain retinopathies in the ageing eye.

                          ACKNOWLEDGEMENTS

We wish to express our gratitude to Professor F. Verzar
(Institute of Experimental Gerontology, Basle) for the generous
gift of eyes from ageing animals and to Professor Jean Babel for
his support and continuous encouragement. Thanks goes also to
Miss M.-L. Beauchemin for assistance in part of this work. This
work was supported by SNSF grant No. 3.1150.73.

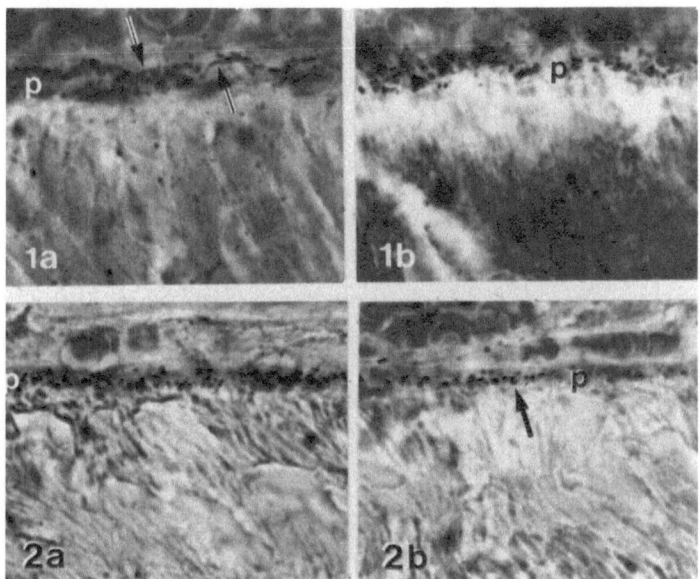

Figs. 1 and 2. 15 μm prefixed frozen sections of retinas, incubated
for acPase activity for 40 min. (1200 X).
1a) 36 hr   1b) 15 days   2a) 10 months   2b) 29 months old
There is an apparent decrease in the number of acPase-positive
organelles and a change in their shape from elongated-ellipsoid in
the young (Fig. 1a, 1b) to circular in the older animals.
p = pigment epithelium

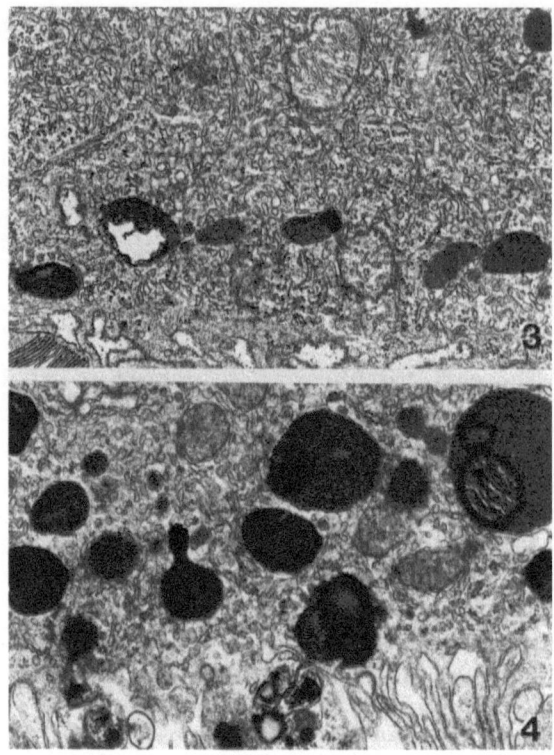

Figs. 3 and 4. Pigment epithelial cells of a 10 month (Fig. 3)
and a 31 month old animal. Note the increased number and size of
lipofuscin granules and other secondary lysosomes in the old
animal. Rough surfaced ER seems, on the contrary, more apparent
in the young animal (14,500 X).

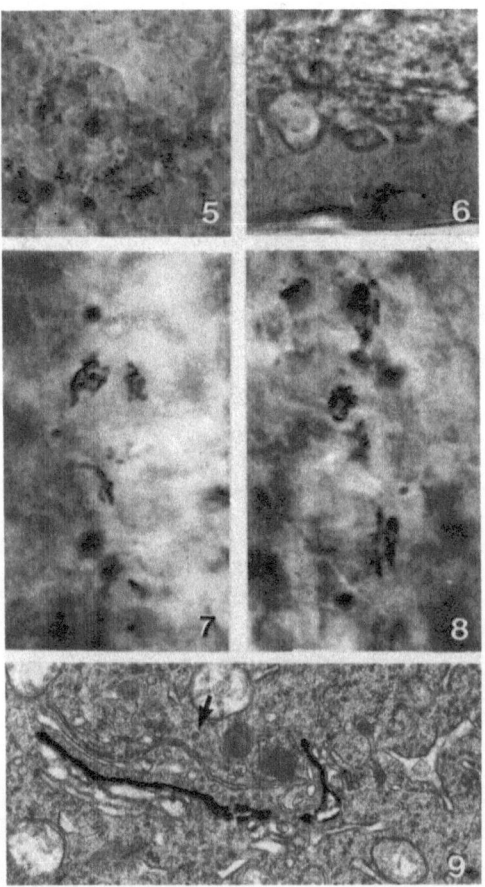

Figs. 5 and 6. Acid phosphatase activity in retinal ganglion cells.
Fig. 5 - 15 days old; Fig. 6 - 29 months old. There are fewer, but
larger acPase-positive particles visualized in the old animal.
15 μm prefixed frozen section (900 X).

Figs. 7 and 8. Golgi apparatus of retinal ganglion cells in young
(Fig. 7) and old (Fig. 8) animals, visualized by incubation in
a TPP-ase medium. There is no noticeable difference in either size,
shape or distribution of this organelle between young and old
animals. 15 μm prefixed frozen section (1700 X).

Fig. 9. TPP-ase preparation of ganglion cell demonstrating the
diphosphatase-rich elements of the Golgi-apparatus. No reaction
product is seen over GERL (G), which lies at the inner aspect of
Golgi-apparatus (in the picture above). Arrow indicates coated
vesicle, surrounded by several dense bodies (15,600 X).

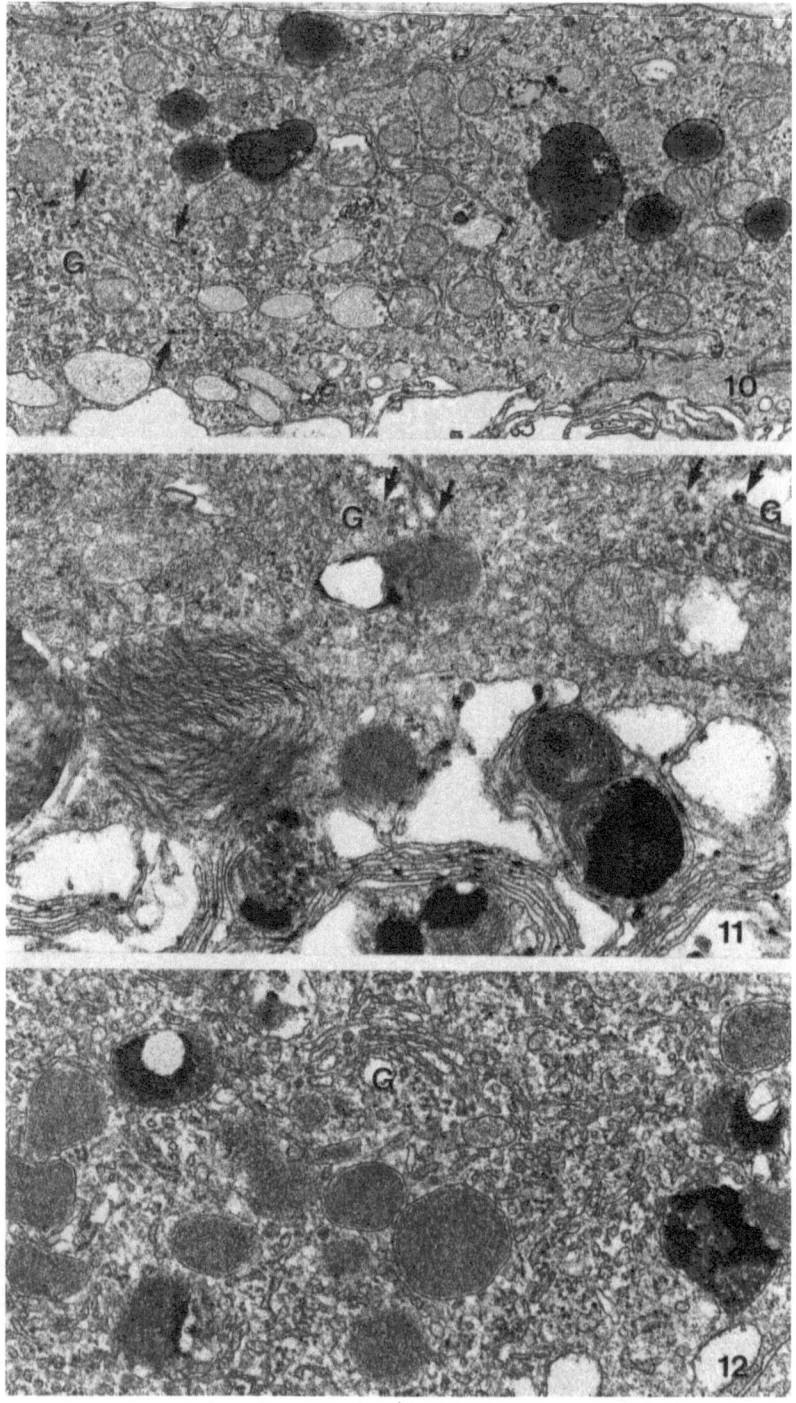

Figs. 10, 11 and 12. Pigment epithelia of 36 hr old (Fig. 10),
10 month old (Fig. 11) and 29 month old animals (Fig. 12).
Incubation for acPase activity for 20 min. The number of dense
bodies displaying no acPase activity is much higher in the old
animal (Fig. 12). Note the absence of reaction product over
GERL (G) in the old animal (Fig. 12) and the presence of numerous
acPase-positive vesicles in the younger animals (Figs. 10 and 11,
arrows). (Fig. 10, 20,200 X; Fig. 11, 29,200 X; Fig. 12, 31,200 X).

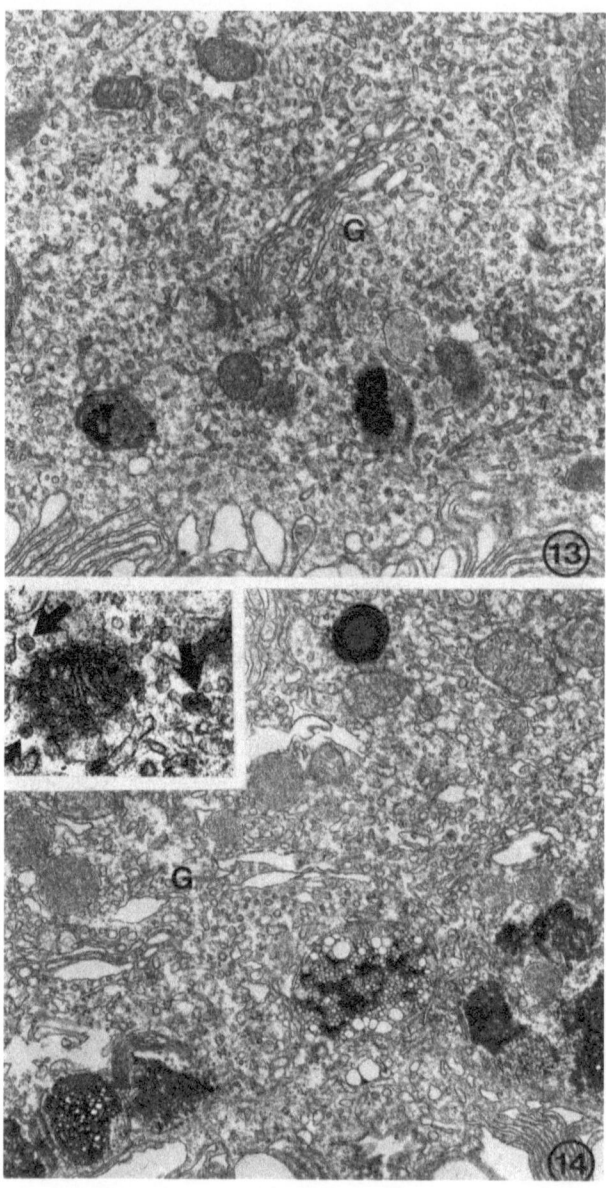

Figs. 13 and 14. Pigment epithelia of a 10 month (Fig. 13) and
29 month (Fig. 14) old animal. Incubation for 90 min in an alkaline
DAB-medium. There is a conspicuous increase in the size, staining
intensity and presumable lipid-content of the lipofuscin granules in
the older animal (Fig. 14). In the same figure numerous dense bodies
surrounding GERL (G). Inset shows staining of microperoxisomes
(arrows). Figs. 13 and 14, 15,600 X, inset 31,500 X.

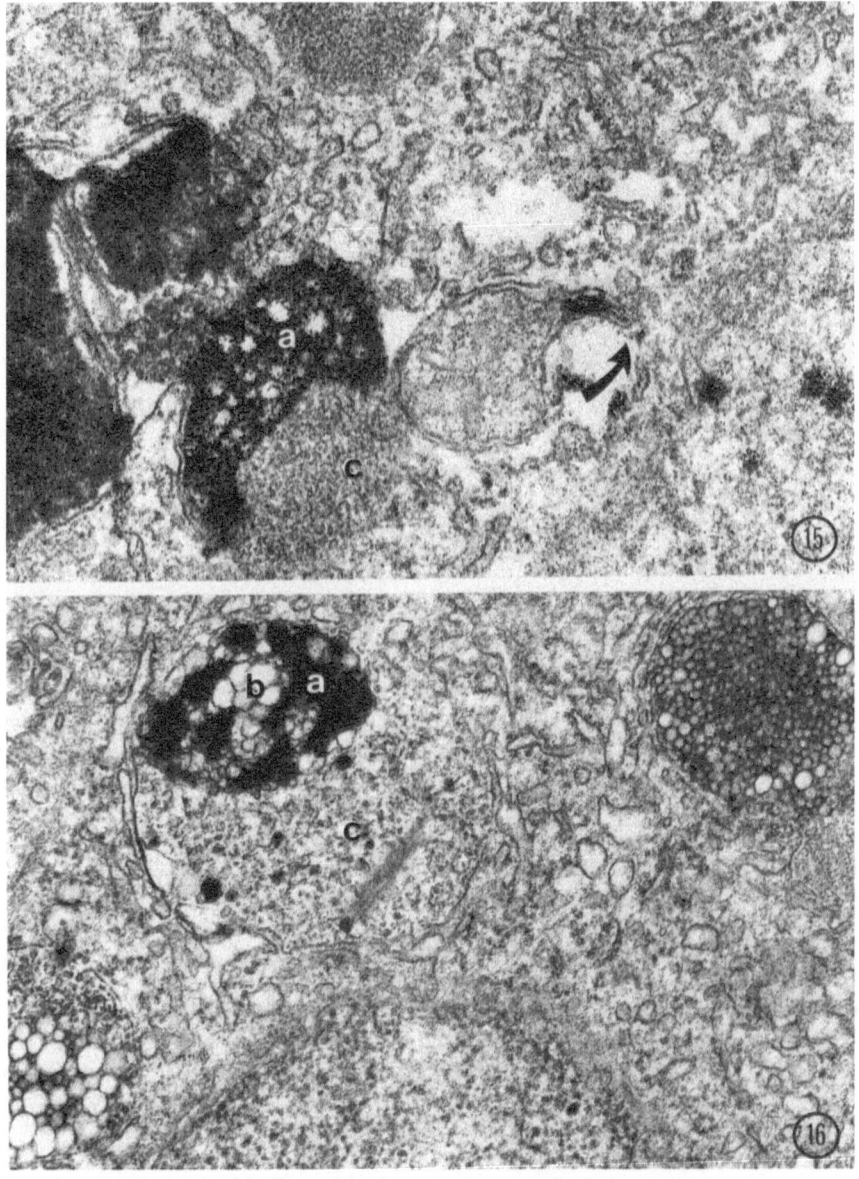

Figs. 15 and 16. Lipofuscin granules in old (29 month) animals.
Incubation for acPase (Fig. 15) and in alkaline DAB (Fig. 16).
Reaction product is localized over identical components of lipofuscin granules ("a") with both methods. Arrow indicates possibly
an autophagic vacuole. Fig. 15, 64,500 X; Fig. 16, 46,500 X.

REFERENCES

1. Young, R.W. 1971. The renewal of rod and cone outer segments
    in the rhesus monkey. J. Cell Biol. 49: 303.

2. Feeney, L. 1973. The phagolysosomal system of the pigment
    epithelium. A key to retinal disease. Invest. Ophthal.
    12: 635.

3. Leuenberger, P.M. 1973. Ultrastructure of the ageing retinal
    vascular system with special reference to quantitative and
    qualitative changes of capillary basement membranes.
    Gerontologia 19: 1.

4. Novikoff, P.M. and Novikoff, A.B. 1972. Peroxisomes in
    absorptive cells of mammalian small intestine. J. Cell. Biol.
    53: 432.

5. Novikoff, A.B. 1964. GERL, its form and functions in neurons
    of rat spinal ganglia. Biol. Bull. 127: 358A.

6. Leuenberger, P.M. and Novikoff, A.B. 1973. Microperoxisomes
    in the retina of pigmented and albino rodents. (abstract)
    Exp. Eye Res. 17: 399.

7. Leuenberger, P.M. and Novikoff, A.B. 1973. Anomalous
    lysosomes in the retina of the beige mouse, a homologue of
    the Chediak-Higashi-syndrome. ARVO abstract, manuscript in
    preparation.

8. Novikoff, A.B., Novikoff, P.M., Quintana, N. and Davis, C.
    1973. Studies on microperoxisomes. IV. Interpretations
    of microperoxisomes, endoplasmic reticulum and lipofuscin
    granules. J. Histochem. Cytochem. 21: 1010.

9. Miller, F. and Herzog, V. 1969. Die Lokalisation von
    Peroxidase und sauer Phosphatase in eosinophilen Leukocyten
    während der Reifung. Elektronenmikroskopisch-cytochemische
    Untersuchungen am Knochenmark von Ratte und Kaninchen.
    Z. Zellforsch. Mikrosk. Anat. 97: 84.

10. Farquhar, M.G. and Palade, G.E. 1965. Cell junctions in
    amphibian skin. J. Cell. Biol. 26: 263.

11. Luft, J.H. 1961. Improvements in epoxy resin embedding
    methods. J. Biophys. Biochem. Cytol. 9: 409.

12. Reynolds, E.S. 1963. The use of lead citrate at high pH
    as an electron-opaque stain in electron microscopy. J. Cell.
    Biol. 17: 208.

13. Novikoff, A.B. and Goldfischer, S. 1961. Nucleosidediphos-
     phatase activity in the Golgi apparatus and its usefulness
     for cytological studies. Proc. Nat. Acad. Sci. USA 47: 802.

14. Novikoff, A.B. 1963. Lysosomes in the physiology and pathology
     of cells: crntributions of staining methods. Ciba Found.
     Symp. Lysosomes, p. 36.

15. Smith, R.E. and Farquhar, M.G. 1965. Preparation of non-frozen
     sections for electron microscope cytochemistry. RCA Scti.
     Instr. News 10: 13.

16. Novikoff, A.B., Novikoff, P.M., Davis, C. and Quintana, N.
     1972. Studies on microperoxisomes. II. A cytochemical
     method for light and electron microscopy. J. Histochem.
     Cytochem. 20: 1006.

17. Rambourg, A. and Chretien, M. 1970. L'appareil de Golgi:
     examen en microscopie electronique de coupes epaisses
     (0.5-1 μ) apres impregnation des tissus par le tetroxyde
     d'osmium. C. R. Acad. Sci. (Paris) 270: 981.

18. Cristofalo, V.J. 1970. Metabolic aspects of aging in diploid
     human cells. In: Aging in Cell and Tissue Culture. E.
     Holeckova and V.J. Cristofalo, Eds., Plenum Press, New York,
     p. 83.

19. Lipetz, J. and Cristofalo, V.J. 1972. Ultrastructural
     changes accompanying the aging of human diploid cells in
     culture. J. Ultrastruc. Res. 39: 43.

20. Essner, E. and Novikoff, A.B. 1961. Localization of acid
     phosphatase activity in hepatic lysosomes by means of electron
     microscopy. J. Biophys. Biochem. Cytol. 9: 773.

21. Travis, D.F. and Travis, A. 1972. Ultrastructural changes
     in the left ventricular rat myocardial cells with age. J.
     Ultrastruc. Res. 39: 124.

22. Goldfischer, S., Villaverde, H. and Forschirm, R. 1966.
     The demonstration of acid hydrolase, thermostable reduced
     diphosphopyridine nucleotide tetrazolium reductase and
     peroxidase activities in human lipofuscin pigment granules.
     J. Histochem. Cytochem. 14: 641.

23. Hochschild, R. 1971. Lysosomes, membranes and aging. Exp.
     Gerontol. 6: 153.

24. Black, V.H. and Bogart, B.I. 1973. Peroxisomes in inner
     adrenocortical cells of fetal and adult guinea pigs. J.
     Cell Biol. 57: 345.

25. Frank, A.L. and Christensen, A.K. 1968. Localization of
     acid phosphatase in lipofuscin granules and possible
     autophagic vacuoles in interstitial cells of the guinea pig
     testis. J. Cell Biol. 36: 1.

26. Hickman, S. and Neufeld, E.F. 1972. A hypothesis for I-cell
     disease: Defective hydrolases that do not enter lysosomes.
     Biochem. Biophys. Res. Comm. 49: 992.

QUESTIONS TO DR. LEUENBERGER

Dr. Wilson: In some of our aged cultures we have found changes in
surface enzymes. One was a loss in alkaline phosphatase and
another was the occurence of acid phosphatase on the surface after
a short incubation period. Have you ever seen acid phosphatase on
the surface of old retina pigment epithelial cells?

Dr. Leuenberger: After prolonged incubation we observed some acid
phosphatase activity on cytoplasmic membranes mostly on the apical
cell prolongations. However, there was no obvious difference in
cytoplasmic membrane between young and old animals.

Dr. Wilson: Have you tried any other lysosomal enzymes or the use
of different substrates for the demonstration of acid phosphatase
as Dr. Maggi has shown a differential distribution of acid phosphatases in mouse cells.

Dr. Leuenberger: We tried, up to now, the demonstration of β-
glucuronidase and glucosaminidase activities; the former remains
unchanged with age, the latter showed a decrease. We are going to
look for acylsulphatase at the EM-level and to do biochemical
determinations of lysosomal, cytosomal, and nuclear acid phosphatase activity. The use of other substrates poses, at the EM-level,
problems of diffusion artifacts.

Dr. Martin: Have you considered two additional possible interpre-
tations of your acid phosphatase data: 1) an accumulation of an
inhibitor in the aged cells and 2) an increased rate of degradation
of essentially normal phosphatase molecules in the aged cells.

Dr. Leuenberger: There are certainly numerous possibilities for
the interpretation of our findings of decreased acid phosphatase
activity. We merely wanted to point out that a decrease in lysosomal
acid phosphatase activity must not necessarily be due to a leakage

of such enzymes. Still another interesting explanation in
addition to the ones mentioned by Dr. Martin and myself is the
non-recognition of enzymes at the receptor organelle.

THE ROLE OF RETINOL IN, AND THE ACTION OF ANTI-INFLAMMATORY DRUGS

ON, HEREDITARY RETINAL DEGENERATION

A. J. Dewar and H. W. Reading

MRC Brain Metabolism Unit, Department of Pharmacology

University of Edinburgh

Inherited retinal dystrophy in the rat is transmitted by an autosomal recessive gene and blindness occurs after birth as a result of degeneration of the photoreceptor cells (1,2). This resembles the situation in certain forms of inherited retinitis pigmentosa in man. Degeneration of the retina in the strain of albino dystrophic rats known as "Campbells" (1) is associated with an increase in the level of "free" lytic enzymes which appear to originate from lysosomes in the adjacent pigment epithelium (3). It has been suggested (4) that the degeneration of the visual cells is produced by breakdown of lysosomal membranes in pigment epithelium and retina. This breakdown is thought to be due to an abnormal build up of vitamin A alcohol (retinol) in the pigment epithelium which arises from the action of light on an unusually labile type of visual pigment (5).

We have attempted to provide further biochemical evidence for the above hypothesis and, to devise methods of retarding the degeneration by:

1. Studying the effect of light deprivation on the progress of retinal degeneration.
2. Studying the effect of retinol on retinal lysosomal stability in vitro.

There is evidence that retinol damages liver lysosomes (6) and bovine retinal and pigment epithelial lysosomes (7).

3. Studying the effect of anti-inflammatory drug acetylsalicylic acid on lysosomal stability in vitro and investigating its effect on retinal degeneration in vivo.

The Effect of Light Deprivation on the Progress of
Retinal Degeneration

The effect of eye pigmentation. In the albino dystrophic rats
visual cell degeneration can be retarded by raising the animals in
the dark (2). This may have been due to a reduction in the amount
of retinol released by the action of light on the visual pigment
and on this basis it seemed likely that a similar delay in the onset
of blindness could be obtained by the presence of melanin in the
dystrophic eye.

A strain of pigmented dystrophic rats was bred in our laboratory
by crossing the white-coated, tan hooded, pink eyed Campbell rats,
homozygous for retinal degeneration, with sighted pigmented Piebald
Virol Glaxo (PVG) rats. This new true breeding strain (known as
"Hunter") has been established for over two years and the viability
of the rats does not differ significantly from the PVG, albino Wistar
or Campbell strains. The full details of the breeding programme have
been published elsewhere (8). The four different rat strains are
illustrated in Figure 1.

A full scale comparative study of the histology and biochemistry
of the two dystrophic strains and the sighted PVG and albino Wistar

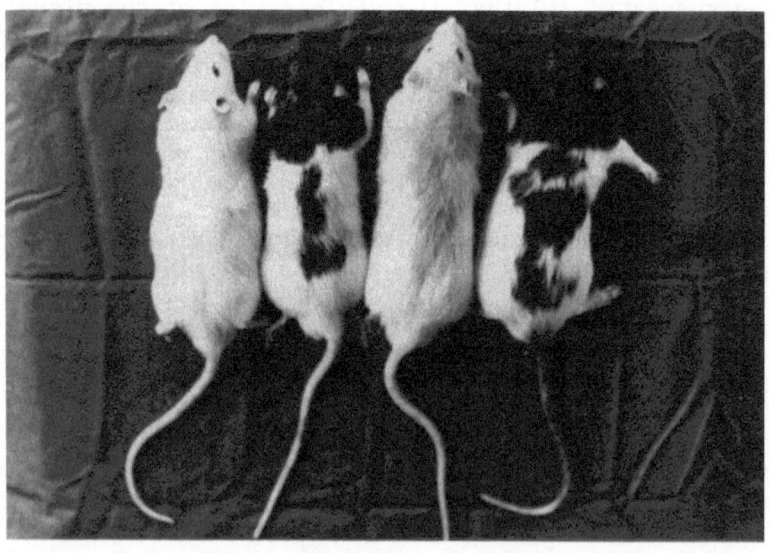

Fig. 1. Left to right: Albino Wistar, PVG, Campbell and Hunter
rats, all male and aged 12 weeks.

strain has been made (8). It was found that the retinal RNA and
DNA levels mirrored the progress of the degeneration assessed
histologically. RNA and DNA levels provide useful indices of cytoplasmic volume and cell number respectively (9). The retinal RNA
and DNA levels of the two dystrophic strains and the two sighted
strains are shown in Figures 2 and 3. It was apparent that the onset
and progression of the lesion in the pigmented dystrophic rats were
delayed by about one week compared with the albino dystrophic rats.
The fall in RNA accompanies the degeneration of the photoreceptor
cell nuclei and the reduction in retinal DNA accompanies the loss
of the photoreceptor cell nuclei (the outer nuclei). After degeneration the retinal RNA/DNA level was considerably higher (Fig. 4)
thus suggesting that the photoreceptor cells have a lower RNA/DNA
ratio and thus a smaller cytoplasmic volume than the retinal cells
not affected by the dystrophy.

The RNA and DNA contents of the posterior cup of the eyes did
not differ significantly in the four strains nor did the RNA and DNA
contents of whole brains differ. However we have demonstrated
marked biochemical changes in the visual cortex of adult dystrophic
rats (10-12).

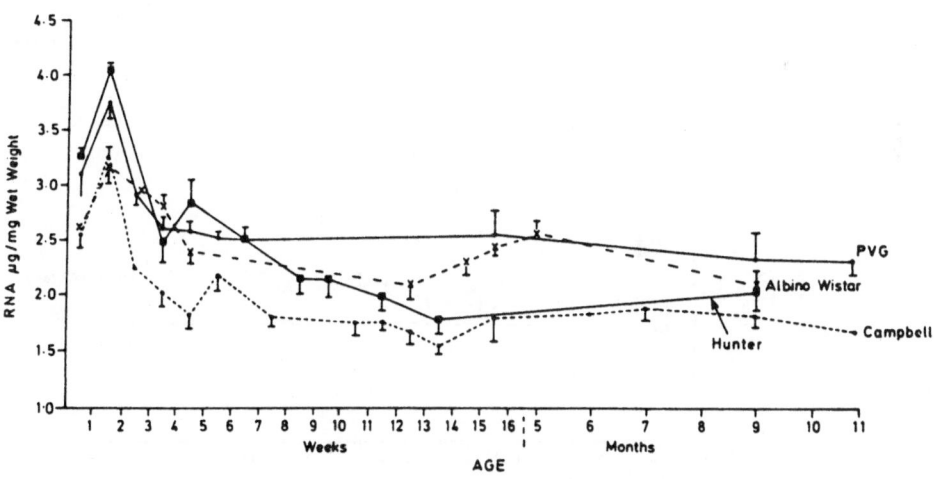

Fig. 2. The retinal RNA content of PVG, Albino Wistar, Hunter and
Campbell rats. Each point represents the mean ± S.E.M. (where
applicable) of 4-26 determinations. Where no S.E.M. is shown, the
point represents the mean of duplicates.

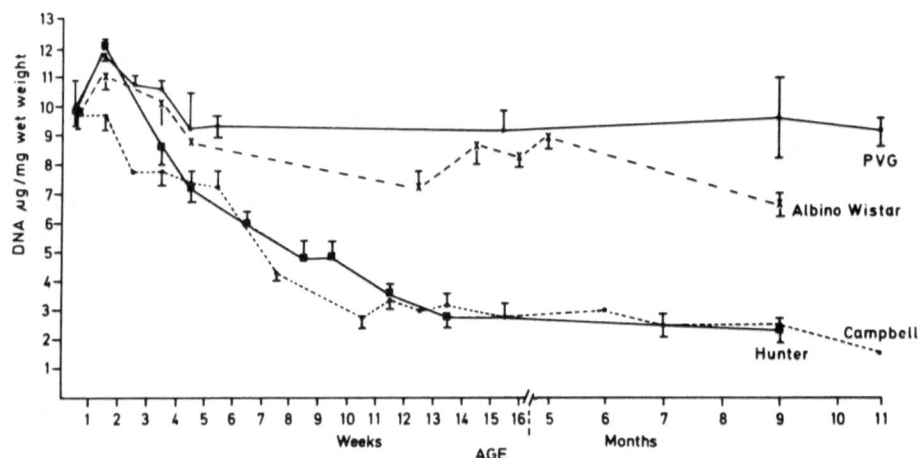

Fig. 3.   The retinal DNA content of PVG, Albino Wistar, Hunter and
Campbell rats.   Each point represents the mean ± S.E.M. (where
applicable).   For n, see Fig. 2.

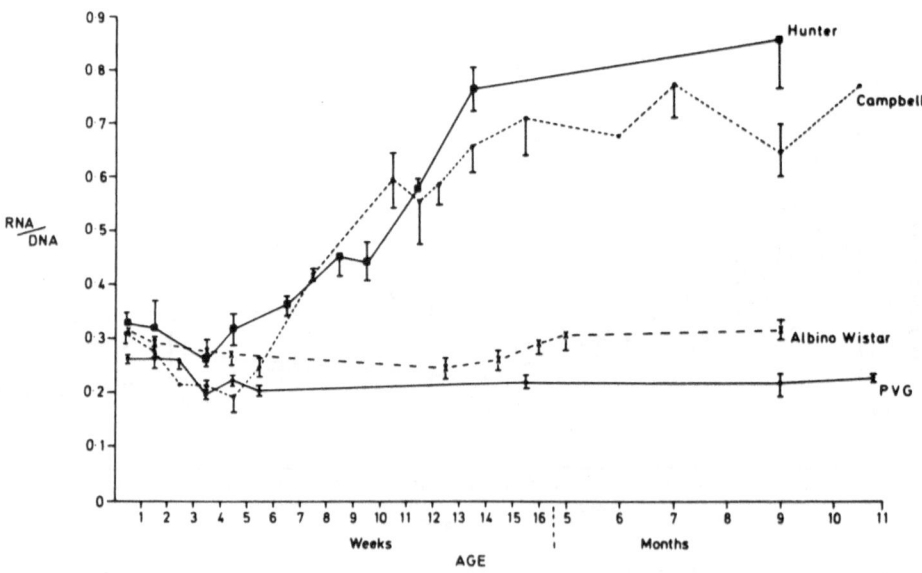

Fig. 4.   The retinal RNA/DNA ratio of PVG, Albino Wistar, Hunter
and Campbell rats.   Each point represents the mean ± S.E.M. (where
applicable).   For n, see Fig. 2.

The effect of dark rearing. After the demonstration that the
presence of pigment in the eye retards the degenerative process we
examined the effects of dark rearing on retinal degeneration in
both dystophic strains.

Litters of Hunter and Campbell rats were maintained in a well
ventilated light-proof box from birth until 8 weeks of age, the
dams being removed after weaning at 21 days. All animal maintenance
procedures were performed with the aid of a photographic safety
lamp. The retinal RNA and DNA were extracted and measured as previously described (8), and compared with the corresponding values
obtained from Campbell, Hunter and PVG rats kept under standard
light/dark cycles for 8 weeks. The results are shown in Table I.
Dark rearing significantly reduced the decrease in the albino dystrophic strain but dark rearing appeared to have no significant effect
on the DNA and RNA/DNA in the pigmented dystrophic strain. These
results confirmed our earlier finding that the loss of retinal DNA
in dystrophic rats exposed to normal light/dark cycles proceeds
more slowly in the pigmented strain. However the retinal DNA level
in both the pigmented dystrophic and light-deprived albino dystrophic
strain is still considerably lower than the retinal DNA of a normal
sighted rat of 8 weeks old. Thus light deprivation can retard
degeneration but not abolish it altogether.

The Effect of Retinol on Retinal Lysosomal Stability In Vitro

Methods. The effect of retinol on the stability of lysosomes
from normal retina was investigated. The retinae were obtained from
albino Wistar or PVG rats of either sex, and of an age range of 1-3
months. Similar results were obtained from all the retinae.

After killing the rats by decapitation the retinae were removed
and homogenated in 12 ml ice-cold 0.25 M sucrose/g wet weight for
0.5 min at 1,000 rev/min. The supernatant was then split into
0.5 ml aliquots (each containing the equivalent of 46 mg of original
tissue) and these were added to tubes containing 2.3 ml 0.25 M sucrose
plys either 0.1% v/v Triton X-100 or 0-2.5 µg/mg original tissue
retinol. The retimol was dissolved in a minimum volume of ethanol.
This volume of ethanol was also added to the tubes not containing
retinol. All tubes were incubated at 37°C for 45 min under a continuous flow of nitrogen. After incubation each tube was cooled to
0°C and then centrifuged at 15,000 g for 20 min to pellet the intact
lysosomes. To assess the damage to the incubated lysosomes the
activity of lysosomal enzymes in the supernatant was measured. In
initial experiments the lysosomal enzyme measured with β-glucuronidase but in later experiments acid phosphatase, β-galactosidase
and hexosaminidase were also investigated. All activities were
expressed as a percentage of the total releasable activity, i.e., as

TABLE I. The Retinal Nucleic Acid Content of 8 Week Old Normal, Dystrophic and Light-Deprived Dystrophic Rats

	RNA μg/mg wet weight	DNA μg/mg wet weight	$\dfrac{RNA}{DNA}$
PVG (normal sighted)	2.23 ± 0.07 (24)	9.33 ± 0.32 (24)	0.24 ± 0.008 (24)
Campbell (albino dystrophic)	2.14 ± 0.10 (18)	2.67 ± 0.14 (18)	0.80 ± 0.04 (18)
Campbell - Light deproved	2.15 ± 0.12 (20)	4.81 ± 0.23 (20)	0.45 ± 0.02 (20)
	N.S.	$t = 7.42$ $P < 0.0001$	$t = 7.21$ $P < 0.0001$
Hunter (pigmented dystrophic)	1.95 ± 0.16 (14)	4.33 ± 0.29 (14)	0.46 ± 0.03 (14)
Hunter - Light deprived	2.28 ± 0.16 (19)	4.55 ± 0.29 (14)	0.50 ± 0.03 (19)
	N.S.	N.S.	N.S.

Values represent mean $\pm$ S.E.M.   n shown in brackets.

a percentage of the activity in the supernatant derived from the
tubes treated with 0.1% Triton X-100.

β-Glucuronidase activity was measured by incubating 1 ml of
each supernatant with 0.2 ml 0.1 M phenophthalein mono-β-glucuronate
in 1.6 ml 0.1 M sodium acetate buffer, pH 4.5, for 19 hr at 37°C.
Protein was then precipitated with 1 ml 6% trichloracetic acid and
2 ml 0.4 M glycine buffer (pH 10.45) added to 2 ml of the protein-
free supernatant. After 30 min the extinction at 540 nm was determined. This method was derived from that of Vento and Cacioppo (7).

Acid phosphatase was assayed by adding 1 ml aliquots of diluted
lysosome-free supernatant to 1 ml citrate buffer, pH 5.5, containing
0.1 M para nitrophenyl phosphate preincubated for 10 min at 37°C.
The reaction was stopped after 15, 30 and 45 min incubation at 37°C
by addition of 1 ml 0.5 M NaOH. The p-nitrophenol was measured
spectrophotometrically at 400 nm.

β-Galactosidase was measured by adding 0.5 ml of lysosome-
free supernatant to 1 ml of 0.1 M glycine-HCl buffer pH 3 containing
1×15^3 M 4-methylumbelliferyl-galactoside (MUG) plus 0.5 ml water
which had been preincubated at 37°C for 10 min. After incubation
at 37°C for 60 min, 2 ml 0.1 M Na_2CO_3 was added and the fluorescence
read at an excitation wavelength of 365 nm and a fluorescence wavelength of 450 nm on a Perkin-Elmer spectrophotofluorimeter. This
method was based on that of Yates (13).

Hexosaminidase was measured by adding 0.2 ml of the supernatant
to 0.8 ml of 2.5 mM 4-methylumbelliferyl-2-acetamide-2-dioxy-β-D-
glucopyranoside in a buffer composed of 50 parts 0.2 M Na_2HPO_4 and
50 parts 0.1 M Citric Acid. After 30 min incubation at 37°C the
reaction was stopped by addition of 3 ml 0.1 M Na_2CO_3. The fluorescence was read at an excitation wavelength of 370 nm and an emission
wavelength of 450 nm.

Results. Figure 5 shows the effect of retinol on retinal lysosome enzyme release. Retinol significantly increased the release of
β-glucuronidase, β-galactosidase and hexosaminidase, the maximum
release being elicited by 1.5 μg/mg wet weight. However, 0.5 μg/mg
caused an increase only in the release of β-glucuronidase.

The effect of retinol was not as convincingly demonstrated
using acid phosphatase as a lysosomal enzyme marker. Even in the
absence of retinol the bulk of the lysosomal acid phosphatase
entered the supernatant fraction. There is evidence, however, that
acid phosphatase can leak through intact lysosomes (14).

It was possible that the apparent increase in release of lysosomal enzymes was not a release phenomenon at all but due to retinol

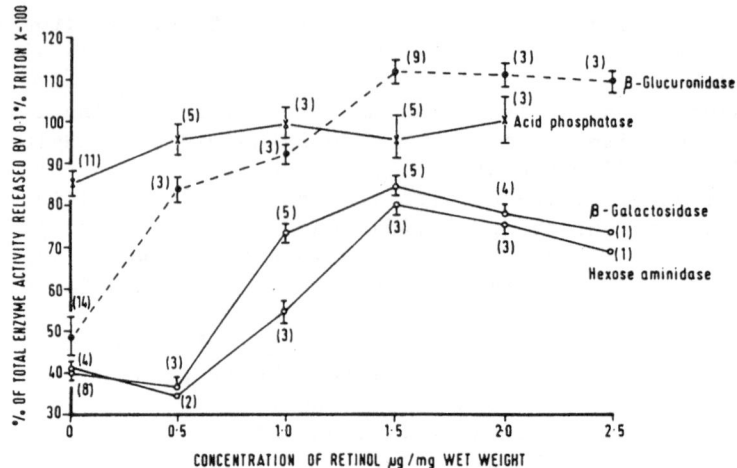

Fig. 5. The effect of retinol on the release of enzymes from retinal
lysosomes during a 45 min incubation at 37°C. The levels of significance (t test) of the increased release produced by 1.5 µg/mg
were: β-glucuronidase t = 7.66, P < 0.001; hexosaminidase t = 4.35,
P < 0.01; β-galactosidase t = 6.42, P < 0.001. The increase in acid
phosphatase release was not statistically significant.

activating the free lysosomal enzymes. This possibility was excluded
by performing a series of incubations in the presence of 0.1% Triton
X-100 and estimating the total released lysosomal enzyme activities
in the presence of 0-2.5 µg/mg retinol. It was found that no concentration of retinol altered the activities of the released enzymes,
thus demonstrating that retinol increased the release of lysosomal
enzymes rather than activating them after release.

The Effect of Acetylsalicylic Acid on Retinal
Lysosome Stability In Vitro

The stability of lysosomes from normal retinae was examined in
the presence of 0-1 mM acetylsalicylic acid and in the presence of
0-1 mM acetylsalicylic acid plus 1.5 µg/mg retinal tissue retinol.
The methods of preparation, incubation and enzyme assay were identical to those described above.

Results. Acetylsalicylic acid at concentrations of 0.25-0.50
mM significantly reduced the rate of release of all the lysosomal
enzymes tested (Tables II and III). At concentrations less than

TABLE II.   The Effect of Acetyl Salicylic Acid on β-Glucuronidase Release From
Retinal Lysosomes During a 45 Min Incubation at 37°C

Concentration of ASA, mM	% Release in absence of retinol	% Release with 1.5μg/mg Retinol in incubation medium
0 (Control)	49.5 ± 22 (14)	112.1 ± 9.8 (9)
0.05	54.1 ± 2.6 (3)	104.2 ± 33.2 (3)
0.10	65.0 ± 1.4 (3)	119.3 ± 14.7 (4)
0.25	17.1 ± 5.4 (5) +	54.3 ± 9.2 (4) ++
0.50	21.5 ± 11.5 (6) *	26.6 ± 15.9 (6) ++
1.00	34.9 ± 13.9 (3)	88.5 ± 8.8 (4) **
2.50	24.1 ± 6.1 (3) ∅	72.6 (2)
5.00	71.3 (2)	70.0 (2)
7.50	79.6 (2)	70.0 (2)
10.00	123.0 ± 4.9 (3)	121.6 ± 1.08 (4)

∅ $P < 0.05$ ** $P < 0.005$
* $P < 0.025$ ++ $P < 0.0001$
+ $P < 0.01$

Values represent mean ± S.D.   n shown in brackets.

TABLE III. The Effect of Acetyl Salicylic Acid on Acid Phosphatase, β-Galactosidase, and Hexosaminidase Release From Retinal Lysosomes During a 45 Min Incubation at 37°C

Conc. ASA mM	% Release of acid phosphatase		% Release of β-Galactosidase		% Release of Hexosaminidase	
	In absence of Retinol	+ 1.5 μg/mg Retinol	In absence of Retinol	+ 1.5 μg/mg Retinol	In absence of Retinol	+ 1.5 μg/mg Retinol
0	86.3 ± 15.1 (11)	96.2 ± 16.1 (5)	41.3 ± 10.8 (8)	78.4 ± 5.6 (5)	41.6 ± 5.4 (4)	71.6 ± 5.3 (3)
0.05	87.3 ± 10.3 (4)	77.5 (2)	46.3 ± 5.7 (3)	53.0 ± 14.1 (3) a	34.3 ± 4.7 (3)	68.0 ± 1.4 (3)
0.10	82.6 ± 27.7 (5)	63.5 (2)	41.7 ± 7.1 (4)	44.0 ± 15.5 (3) b	30.3 ± 1.9 (3) a	73.6 ± 4.4 (3) b
0.25	47.9 ± 14.0 (11) e	43.1 ± 6.6 (5) d	11.8 ± 8.9 (5) e	13.1 ± 6.4 (4) e	0.37 ± 0.44 (3) e	19.0 ± 12.7 (3) b
0.50	61.0 ± 16 (5)	45.7 ± 4.9 (5) d	22.4 ± 13.8 (5) a	23.2 ± 2.2 (4) e	9.2 ± 0.94 (3) e	19.0 ± 2.5 (3) d
1.0	93.5 ± 15.5 (4)	53.0 ± 21.1 (3)	50.6 ± 7.5 (3)	45.3 ± 1.8 (3) e	15.1 ± 0.47 (3) e	39.3 ± 2.6 (3) b

Values represent mean ± S.D. where appropriate, n shown in brackets

a $p < 0.05$ c $p < 0.010$
b $p < 0.025$ d $p < 0.005$
          e $p < 0.001$

0-10 mM it had no effect except on hexosaminidase release. The
release of β-glycuronidase was reduced by acetylsalicylic acid
concentrations as high as 2.5 mM but at higher concentrations the
acetylsalicylic acid appeared to have a labilizing effect on the
lysosomes. Acetylsalicylic acid had a very marked effect in
stabilizing the lysosomes against the labilizing action of 1.5 µg/mg
retinol. In the absence of acetylsalicylic acid, retinol produced
an enzyme release of between 78-112% but in the presence of the
optimum concentration of acetylsalicylic acid this was reduced
to between 13.1-43.1%. For β-glucuronidase 0.25-7.50 mM acetylsalicylic had a stabilizing effect which was maximal at 0.50 mM.
Above 7.50 mM there was no stabilizing effect. In the case of the
other lysosomal enzymes, 0.25 mM acetylsalicylic acid appeared to be
the most effective concentration.

The Effect of Acetylsalicylic Acid on the Progress of
Retinal Degeneration In Vivo

On the basis of the in vitro effect of acetylsalicylic acid,
one would expect, if the lysosome hypothesis is correct, that
treatment with acetylsalicylic acid may retard the progress of
retinal degeneration in vivo. To attempt to demonstrate this,
litters of Campbell rats were injected for 5 days each week i.p.
with acetylsalicylic acid in saline. The doses used within each
litter ranged from 0.01-100 mg/kg/day. Controls in the litter
received saline. The aspirin was solubilized by an addition of an
equimolar amount of N-methylglucamine. The resulting N-methylglucamine salt had the desired solubility in saline. The injections
were commenced 10 days after the birth of the litter (i.e. 2-3 days
before eye opening) and the rats were killed when they were 8 weeks
old. The retinal RNA and DNA were extracted as before. The results
of preliminary experiments on 5 litters are shown in Table IV. The
aspirin treated animals in this experiment had a significantly
higher retinal DNA than the controls and a significantly lower RNA/
DNA ratio. These results resemble those obtained by maintaining
the rats in complete darkness. However, rather surprisingly, the
effects produced did not appear to be related to the dose of aspirin
administered. Preliminary results obtained using one litter of 11
Hunter rats indicated a significant decrease ($p < 0.001$) in the
retinal RNA/DNA ratio in the aspirin-treated animals but the size
of the decrease was considerably smaller than that seen in the
Campbells.

DISCUSSION

The demonstration that retinol alters the stability of retinal
lysosomes, thus causing release of lysosomal enzymes, provides
further evidence for the hypothesis outlined earlier. In addition,

TABLE IV. The Retinal Nucleic Acid Content of 8 Week Old Albino Dystrophic Rats Treated With Acetyl Salicylate

Dose mg/kg/day	RNA µg/mg	DNA µg/mg	$\frac{RNA}{DNA}$
0 Control	1.75 ± 0.19 (10)	2.46 ± 0.81 (10)	0.71 ± 0.15 (10)
0.01	1.81 ± 0.12 (5)	3.85 ± 1.19 (5) [+]	0.47 ± 0.004 (5) [*]
0.1	1.79 ± 0.11 (5)	3.56 ± 0.42 (5) [*]	0.50 ± 0.07 (5) [*]
1.0	1.50 ± 0.19 (5) [+]	3.00 ± 0.80 (5)	0.50 ± 0.11 (5) [*]
10	1.57 ± 0.21 (4)	3.21 ± 0.42 (4)	0.49 ± 0.03 (4) [*]
100	1.43 ± 0.18 (4) [*]	3.76 ± 0.99 (4) [+]	0.38 ± 0.07 (4) [ø]
Mean of all treated animals	1.62 ± 0.35 (23)	3.48 ± 1.01 (23) [**]	0.46 ± 0.08 (23) [++]

[+] $P < 0.05$
[*] $P < 0.025$
[**] $P < 0.010$
[ø] $P < 0.005$
[++] $P < 0.001$

Values represent mean $\pm$ S.D.
n shown in brackets

the fact that the rate of degeneration can be retarded by measures
designed to reduce the amount of light impinging on the retina provides corroborative evidence for the important role of light in the
degenerative process. However, the finding that dark rearing only
retards but does not halt the degeneration in albino dystrophic rats
and has no effect on the pigmented dystrophic rats suggests that
there may well be a secondary mechanism which is light-independent.
Against this view one could argue that the exposure of the dark-
reared rats to the minimal stimulation from the photographic safety
lamp (during the maintenance procedures lasting for approximately
15 min every 2-3 days) is sufficient to trigger off the degenerative
process. As yet we are unable to distinguish between these two
possibilities.

There is evidence that anti-inflammatory drugs have a stabilizing effect on lysosomal membranes and can retard the labilizing
effect of histamine in inflammation (15). We have demonstrated
that at concentrations of 0.25-0.50 mM acetylsalicylic acid significantly stabilizes retinal lysosomes and dramatically reduces the
labilizing effects of retinol. Aspirin has been reported to stabi-
lize liver lysosomes in vitro (16), and it is interesting that the
optimum concentration was found to be 0.5 mM—comparable to our
present findings in the retina. However, other investigators have
been unable to demonstrate any effect of aspirin on liver lysosome
stability in vitro (17).

Our preliminary in vivo results have shown that prolonged
administration of acetylsalicylate appears to retard the loss of
retinal DNA in dystrophic rats. However, it is premature to draw
any firm conclusion from this observation since its significance
from the histological and functional point of view remains to be
assessed. In 1964 a clinical study revealed that diabetic patients
suffering from rheumatoid arthritis and requiring continuing large
doses of salicylates had a significantly reduced incidence of
retinopathy when compared with the general diabetic population (18).
This observation taken in conjunction with our present in vitro and
in vivo findings suggest that anti-inflammatory drugs merit further
investigation as possible agents for combating retinal degeneration.

ACKNOWLEDGEMENT

The authors wish to thank Mrs. Gillian Barron for excellent
technical assistance.

SUMMARY

Light deprivation retarded retinal degeneration in albino
dystrophic rats. In pigmented dystrophic rats the presence of

pigment in the eye retarded the degenerative process. Retinol
labilized rat retinal lysosomes in vitro. Acetylsalicylic acid
stabilized retinal lysosomes even in the presence of the concentration of retinol which produced the maximum labilization. The
effect of acetylsalicyclic acid was concentration dependent, maximum
stabilization being produced by 0.25-0.50 mM. The results provide
further evidence for the hypothesis that hereditary retinal degeneration in rats is mediated by an increased amount of retinol
(produced by the action of light on an unusually labile type of
visual pigment) causing a premature release of lysosomal enzymes.

REFERENCES

1. Bourne, M.C., Campbell, D.A. and Tansley, K. 1938. Hereditary
    degeneration of the rat retina. Brit. J. Ophthal. 22: 613.

2. Dowling, J.E. and Sidman, R.L. 1962. Inherited retinal
    dystrophy in the rat. J. Cell Biol. 14: 73.

3. Burden, E.M., Yates, C.M., Reading, H.W., Bitensky, L. and
    Chayen, J. 1971. Investigation into the structural integrity
    of lysosomes in the normal and dystrophic rat retina. Exp.
    Eye Res. 12: 159.

4. Reading, H.W. 1970. Biochemistry of retinal dystrophy. J.
    Med. Genetics 7: 227.

5. Reading, H.W. 1966. Retinal and retinol metabolism in heredi-
    tary degeneration of the retina. Biochem. J. 100: 34P.

6. Fell, H.B., Dingle, J.T. and Webb, M. 1962. Studies on the
    mode of action of excess of Vitamin A. 4: The specificity
    of the effect on embryonic chick-limb cartilage in culture
    and on isolated rat liver lysosomes. Biochem. J. 83: 63.

7. Vento, R. and Cacioppo, F. 1973. The effect of retinol on
    the lysosomal enzymes of bovine retina and pigment epithelium.
    Exp. Eye Res. 15: 43.

8. Yates, C.M., Dewar, A.J., Wilson, H., Winterburn, A.K. and
    Reading, H.W. 1974. Histological and biochemical studies on
    the retina of a new strain of dystrophic rat. Exp. Eye Res.
    18: 119.

9. Hess, H.H. and Thalheimer, C. 1965. Microassay of biochemical
    structure in nervous tissue I: Extraction and partition of
    lipids and assay of nucleic acids. J. Neurochem. 12: 193.

10. Dewar, A.J. and Reading, H.W. 1973. A comparison of RNA
    metabolism in the visual cortex of sighted rats and rats
    with retinal degeneration. Exp. Neurol. 40: 216.

11. Dewar, A.J., Winterburn, A.K. and Reading, H.W. 1973. RNA
    metabolism in subcellular fractions from rat cerebral cortex
    of rats with retinal degeneration. Exp. Neurol. 41: 133.

12. Dewar, A.J. and Winterburn, A.K. 1973. Metabolism of nuclear
    and cytoplasmic proteins in the visual cortex of sighted rats
    and rats with retinal degeneration. Exp. Neurol. 41: 584.

13. Yates, C.M. 1974. Personal communication.

14. Brunk, U.T. and Ericsson, J.L.E. 1972. Cytochemical evidence
    for the leakage of acid phosphatase through ultrastructurally
    intact lysosomal membranes. Histochemical J. 4: 479.

15. Chayen, J., Bitensky, L. and Ubhi, G.S. 1972. The experi-
    mental modification of lysosomal dysfunction by anti-
    inflammatory drugs acting in vitro. Beitr. Path. Bd. 147: 6.

16. Tanaka, K. and Lizuka, Y. 1968. Suppression of enzyme release
    from isolated rat liver lysosomes by non-steroidal anti-
    inflammatory drugs. Biochem. Pharmacol. 17: 2023.

17. Robinson, D. and Willcox, P. 1969. Interaction of salicylates
    with rat liver lysosomes. Biochem. J. 115: 54P.

18. Powell, E.D.U. and Field, R.A. 1964. Diabetic retinopathy
    and rheumatoid arthritis. Lancet 2: 17.

EFFECT OF AGE ON KIDNEY HYPERPLASIA IN THE RAT AFTER UNILATERAL

NEPHRECTOMY

Milena Soukupová, Přemysl Hněvkovský and Jiří Najbrt
Department of General Biology, Medical Faculty, Charles
University, and Institute of Clinical and Experimental
Medicine, Prague

Adaptive growth of the mammalian kidney after nephrectomy of
the contralateral organ is a firmly established phenomenon. Both
increase in cell size and cell number take part in the compensatory
growth and the greatest changes appear in kidney cortex. It is not
clear, however, whether some ontogenetic differences in renal growth
after unilateral nephrectomy exist, as there is evidence pro as well
as contra the relation between the age of the individual and the
degree and character of kidney growth and restoration of normal
function (1-4). Therefore we studied the main characteristics
of kidney growth after unilateral nephrectomy in very young, adult
and ageing, but not yet senile, rats comparing kidney weight and
number of cells obtained from the kidney by trypsinization.

MATERIALS AND METHODS

Inbred rats, males and females of the Lewis strain aged 3 weeks,
3 months and 12 months were used, 6 animals in each age group. The
right kidney was removed aseptically under light ether anesthesia,
weighed and prepared for trypsinization. After 5 days, the nephrectomized animals were killed and the remaining kidney weighed and
used for trypsinization in the same way as the right, "control"
kidney.

Trypsinization was done with kidneys freed from connective
tissue structures, minced with fine scissors and subjected to 0.2%
trypsin in PBS, pH 2.2-7.4, on an electromagnetic stirrer at 26-
28°C. We used 20 ml of trypsin per 1 g kidney tissue. The first
fractions containing tissue debris and red cells were discarded
after 5 min of trypsinization in the 3 week old rats and 10 min in
the 3 and 12 month old rats. Further fractions were collected
after 20 min and the action of trypsin was stopped by an addition

of 0.5 ml calf serum to 100 ml of cell suspension. The kidneys
were trypsinized till no visible tissue fragments remained, and
the cell suspensions were stored during the time of trypsinization
at 4°C. After filtration through a fine cloth, the cells were
slowly centrifuged, the supernatant removed and the cells resuspended in Michl's medium (5) for counting in a hemocytometer.
Blood cells, if present, were not counted.

    The weight of the kidneys before and after nephrectomy, absolute
and per 100g body weight, and the number of cells per whole kidney
were calculated. Further, the percentage of weight increase of
the kidney and the percentage of the increase in the cell number
per kidney were calculated. The results were subjected to statistical evaluation either by Student's t test or by Wilcoxon's test
(matched-pairs signed rank test).

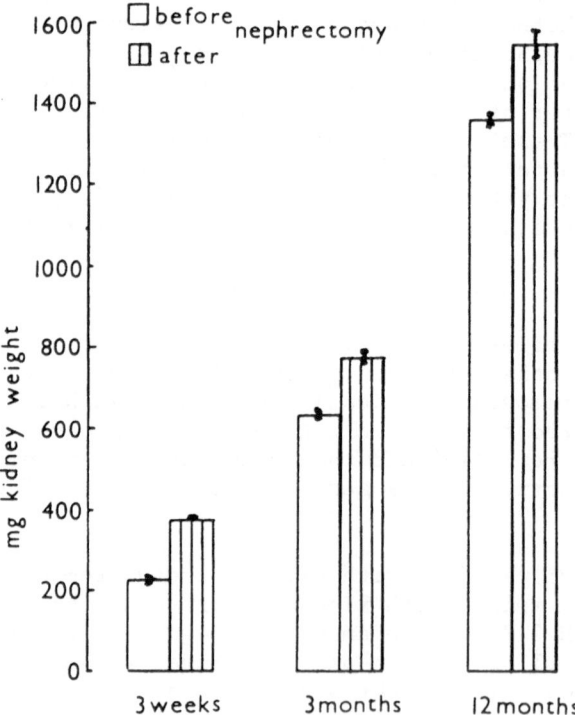

Fig. 1. Kidney weight in mg before and after nephrectomy at 3
weeks, 3 months and 12 months rats.

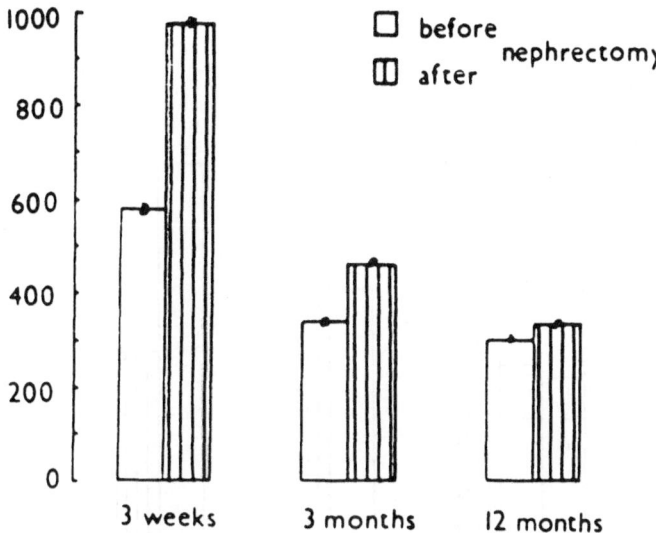

Fig. 2. Mg kidney weight per 100 mg body weight before and after
nephrectomy at 3 weeks, 3 months and 12 months rats.

The DNA was measured spectrophotometrically after Schmidt and
Thannhauser (6) and the DNA values calculated per whole kidney.

RESULTS

Figure 1 shows the increase in kidney weight before and after
unilateral nephrectomy in the three age groups of rats. The values
increased significantly during ageing, and the difference between
the kidney weight before and after nephrectomy was significant at
the 5% level in all three age groups.

Figure 2 demonstrates mg kidney weight per 100g body weight.
This value decreases significantly from the 3 week old rats to the
3 month old animals, but later, the difference disappears. After
nephrectomy, the increase in the relative kidney weight is very
high in the youngest rats (significant at the 1% level), and smaller
but still significant (1% level) in the 3 month old rats. In the
oldest rats, the difference disappears.

Figure 3 shows the changes in the number of cells freed by
trypsinization from the kidneys before and after nephrectomy. The
number of cells increases during ontogenesis, but after nephrectomy

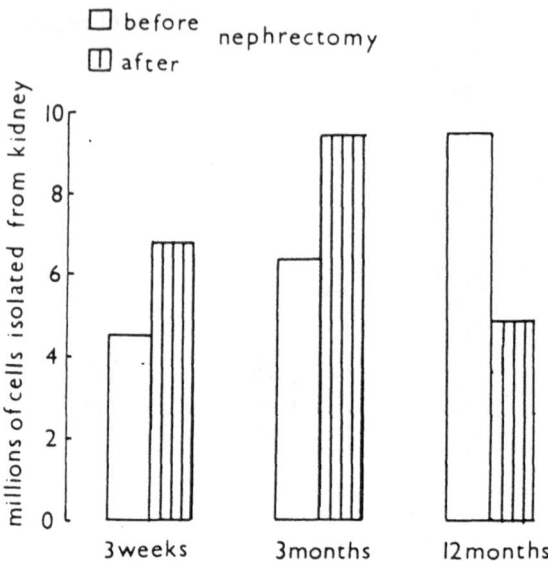

Fig. 3. The number of the cells expressed in millions freed by trypsinization from the kidneys before and after nephrectomy at 3 weeks, 3 months and 12 months rats.

an increase occurs only in the 3 week old and the 3 month old animals. In the rats aged 12 months, less cells were obtained than in the rats of the two younger groups, although the kidney was very large and contained the most cells before nephrectomy.

Figure 4 summarizes these results expressed as percentage of mean values before nephrectomy. The growth of the kidney decreases significantly from the] week old to the 3 month old rats and from the 3 month old to the 12 month old rats. The number of cells remains the same in the two younger age groups, and significantly falls in the oldest age group.

Figure 5 shows our preliminary results with DNA measurements in the kidneys of the 3 week old and 12 month old rats. There is an increase in the DNA values per kidney after nephrectomy in the 3 week old rats larger (increase of 131%) than in the 12 month old ones (increase of 12%), practically no change in the older rats. All our results are summarized in the Table.

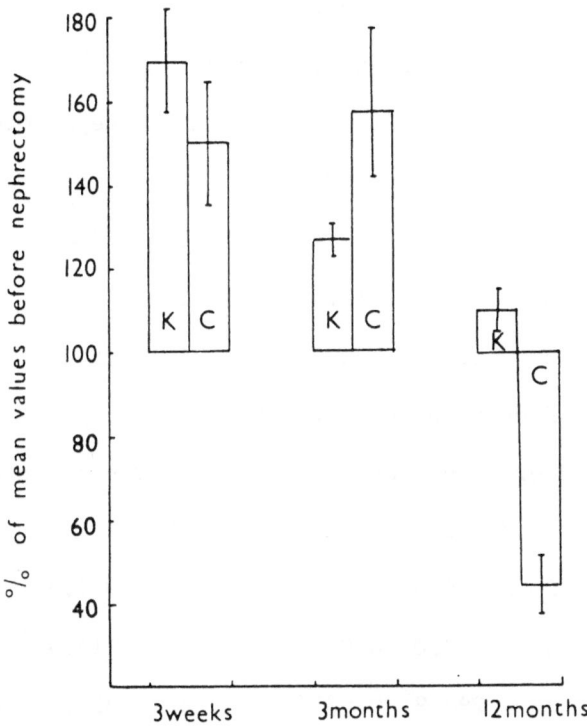

Fig. 4. The change of the kidney weight and the number of the
cells after nephrectomy at 3 weeks, 3 months and 12 months rats
expressed by percentage of mean values before nephrectomy. K =
kidney weight, C = number of cells.

DISCUSSION

     Our results are in agreement with the results of the authors
who found a greater compensatory renal hypertrophy in younger
than in older animals, and who also found a continuous decrease in
kidney weight per unit body weight from young to adult rats (2).
Our rats in the oldest age group are by no means too old or senile
animals and would be considered as adults. Their kidney, however,
compensates the nephrectomy with greater difficulties than that
of the young animals. The age changes are evident in the unexpec-
tedly low yield of cells from the kidneys of 12 month old rats
after nephrectomy, where the numbers of cells freed by trypsinization are significantly lower than before nephrectomy. This finding
is still more interesting when we consider our, though preliminary,
DNA measurements. In the kidneys of the 12 month old rats, the DNA

M. SOUKUPOVA ET AL.

TABLE I

Kidney Weight in mg

Age of Rats	Before nephrectomy			After nephrectomy		
	N	Y	SE	N	Y	SE
3 weeks	6	221	17.4	6	370	23.3
3 months	6	629	12.4	6	772	32.3
12 months	6	1347	32.3	6	1506	70.1

Mg Kidney Weight per 100mg Body Weight

Age of Rats	Before nephrectomy			After nephrectomy		
	N	Y	SE	N	Y	SE
3 weeks	5	579.3	0.249	5	971.3	0.713
3 months	5	342.0	0.056	5	414.4	0.0898
12 months	6	298.9	0.0816	6	333.5	0.134

The Number of Cells in Millions

Age of Rats	Before nephrectomy			After nephrectomy		
	N	Y	SE	N	Y	SE
3 weeks	6	4.577	0.329	6	6.844	0.753
3 months	6	6.463	0.525	6	9.406	1.003
12 months	6	12.547	3.075	6	4.512	1.081

% of Mean Values Before Nephrectomy

Age of Rats	Kidney Weight			Number of Cells		
	N	Y	SE	N	Y	SE
3 weeks	6	71.16	12.71	6	51.42	15.67
3 months	6	22.6	3.58	6	51.98	19.86
12 months	6	11.75	4.60	6	-55.70	11.44

N = the number of animals; Y = arit. average; SE = standard error

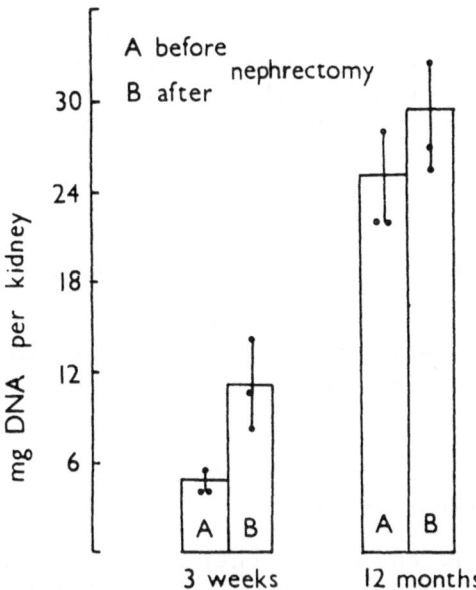

Fig. 5.   DNA mg in the whole kidney before and after nephrectomy at 3 weeks and 3 months rats.

content did not decrease after nephrectomy, it remained practically the same. This means that before trypsinization the number of cells in the kidney after nephrectomy was the same as before the operation but that the cells were probably more vulnerable and were destroyed by the procedures not dangerous for the cells of younger individuals. Another possibility is a firmer bounding of the kidney cells in collagenous structures of the older rat's kidneys, but the greater vulnerability seems to us to be more probable because the trypsinization of the kidneys before and after nephrectomy took the same time till complete dissolution, of course longer than in the younger rats.

SUMMARY

1.   The absolute weight of the rat's kidney increases from 3 weeks to 3 months and from 3 months to 12 months. The relative weight decreases. The number of cells freed from the kidney by trypsinization increases.

2.   After nephrectomy, the kidney weight increases in all age groups. The increase is mery high in the youngest rats, lower but still significant in the 3 month old rats and still lower in the

12 month old rats. The number of cells freed from the kidney after
nephrectomy is again very high in the young and 3 month old rats,
but very low in the 12 month old rats. In these animals, the cell
number is lower after nephrectomy than before, and a greater vulnerability of the cells is suggested as the cause of this phenomenon.

REFERENCES

1. Dicker, S.E. and Shirley, D.G. 1971. Mechanism of compensatory
    renal hypertrophy. J. Physiol. 219: 507.

2. Dicker, S.E. and Shirley, D.G. 1973. Compensatory renal growth
    after unilateral nephrectomy in the new-born rat. J. Physiol.
    228: 193.

3. Hollander, C.F. 1970. Functional and cellular aspects of organ
    ageing. Exp. Geront. 5: 313.

4. Arndt, J., Boigt, R. and Unverricht, A. 1973. Zum information-
    swert des isotopennephrogramms bei kompensatorischer hyperplasie
    der restniere nach nephrektomie. Fortschr. Rontgenstr. 118: 197.

5. Michl, J. 1961. Metabolism of cells in tissues culture in vitro.
    Exp. Cell Res. 23: 324.

6. Schmidt, G. and Thannhauser, S.J. 1945. A method for the
    determination of desoxyribonucleic acid, ribonucleic acid and
    phosphoproteins in animal tissues. J. Biol. Chem. 161: 83.

QUESTIONS TO DR. SOUKUPOVA

Dr. Holeckova: I would rather like to make a comment than to ask
a question. We measured the vitality of trypsinized kidney cells
two days after nephrectomy and we also found lower numbers of cells
capable of surviving explantation and able to multiply in vitro
after as compared with before nephrectomy in the oldest animals
we had.

Dr. Martin: Have you considered the possibility that the decreased
yield of cells from post-nephrectomized older animals may be
attributable to alterations in the stromal connective tissue which
might interfer with trypsinization?

Dr. Deyl: In the past we have measured the proportion of collagen
in kidneys and we always found an increase in both aged animals
and in animals which were subjected to nephrectomy. This, however,

was 6 weeks to 2 months after operation. The striking fact was
that in spite of taking all precautions with respect to the
specificity of basement membrane collagen we had very low amounts
solubilized and what was solubilized was always material with high
molecular weight according to Sepharose chromatography. I also
would like to stress the problems arising from calculating the
proportion of collagen in kidneys from hydroxyproline estimations
since in the case of high proportion of basement membrane collagen
due to the lower level of hydroxylation of this particular collagenous species the results may be highly erroneous.

EFFECT OF AGE ON KIDNEY HYPERPLASIA IN THE RAT DURING COLD

ACCLIMATION

Emma Holečková and Marie Baudysova

Institute of Physiology, Czechoslovak Academy of
Sciences, Prague

The decreased adaptability of old organisms to different
stressing agents is a well known phenomenon, usually considered as
one of the main indicators of ageing. Regeneration proceeds at a
slower rate, even if the final size of the regenerate may sometimes be the same as in younger individuals. Adaptive growth of
the kidney compensating unilateral nephrectomy in rats is inhibited
in older individuals (1).

Increase in kidney weight caused by cell multiplication can
be induced in intact rats exposed to low temperature. Holeckova
et al. (2) have shown that the kidneys of adult laboratory rats
acclimated to 5°C are significantly larger than the kidneys of the
controls living at 24°C, and that the main weight increase occurs
during the first week of acclimation. The increase in the weight
of the kidneys is closely paralleled by an increase of the DNA
content, an increase of the cell number per kidney and by an increased ability of kidney cells to survive explantation and divide
in culture (3).

In the present work, we studied the effects of cold exposure
and acclimation on the weight of the kidneys and the reaction of
kidney cells to explantation in adult and old laboratory rats. We
also studied the effect of unilateral nephrectomy on the kidney of
control and cold-treated animals.

MATERIALS AND METHODS

Three month-old Wistar and LEW.AVN females were kept in single

307

cages on metal wire with food and water ad libitum at 5°C for 5,
90 and 150 days. The experiments were started with 6-9 cold treated
and 6-9 control rats in one group. The control animals lived for
further 5, 90 and 150 days at 24°C. For the old age group, 18
month old males only were available.

Nephrectomy and cell cultures: the right kidney was removed
aseptically under light ether narcosis immediately after cold exposure. The corresponding controls were nephrectomized simultaneously. The kidney was weighed, freed from connective tissue
structures, minced with fine scissors and trypsinized in 0.2%
trypsin in PBS (20 ml per 1 g kidney tissue) on an electromagnetic
stirrer. Five 30 min fractions were made, the cell suspension
collected, centrifuged, the supernatant removed and the cells resuspended in Michl's medium with 5% calf serum and antibiotics (3).
The cells were counted in a hemocytometer and set up for culture in
Muller flasks. At least 4 flasks containing 5 ml of medium with
8×10^5 cells were used for one kidney. After 8 days at 36°C, the
cells were released from the glass with 0.2% trypsin in PBS and
counted. Two counts were made for every sample.

After two days at 24°C, the nephrectomized rats were killed,
the remaining kidney removed and prepared for culture in the same
way.

All results were analyzed statistically with Student's t test.

RESULTS

All rats exposed to the cold for 5 and 90 days survived well.
Of the 6 animals exposed for 150 days, 3 died of intercurrent
respiratory infection. All rats exposed to the cold at the age of
18 months died during the first day of exposure. All adult and old
rats survived the unilateral nephrectomy.

Figure 1 shows the relation between the length of the cold
exposure and the weight of the kidneys as well as the behavior
of the kidney cells in vitro. The results are expressed as percentage of controls, whose mean values are taken as 100%. The control and experimental animals aged during the experiment and their
kidneys increased, but the experimental animals had larger kidneys,
especially after long-term acclimation. Cells isolated from 3
and 6 month old rats could withstand explantation better than cells
of control rats of corresponding age, but in the oldest age group,
the enlarged kidneys yielded cells whose reaction to explantation
was worse than that of the control animals' cells.

Two days after nephrectomy, no significant changes in the

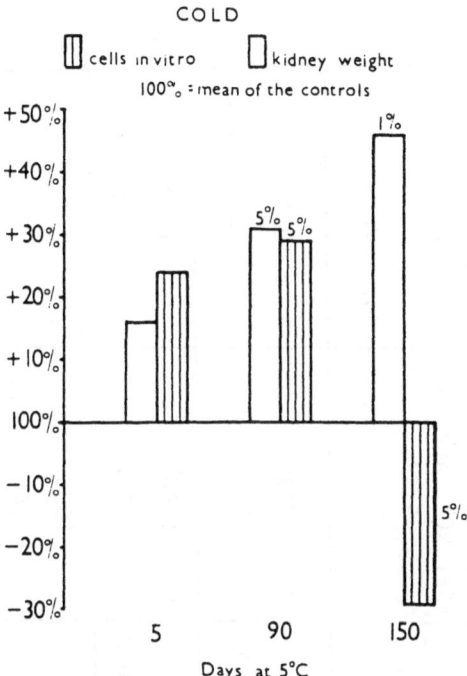

Fig. 1. The influence of 5, 90, and 150 days at 5°C on the weight of the kidneys and on the number of kidney cells surviving 8 days in vitro. Mean values of the control rats living at 24°C = 100%. Differences between the cold-treated and the control animals significant on 5% and 1% level as indicated.

weight of the kidney appeared, as was expected, but the cells of the old rats were significantly less vital in vitro after nephrectomy than before the operation (Fig. 2).

Combined effect of cold exposure followed by nephrectomy was studied in rats aged 3 and 6 months. Two stressors applied to the organisms increased the reaction in a positive way in the two groups of younger animals. Their kidney after nephrectomy was larger than that of the cold-exposed or nephrectomized animals, and also the vitality of their cells in culture was the best. This comparison could not have been done in the group of the old rats,

Table I

Mean Kidney Weight in Milligrams

Age of rats in months	Control			Nephrectomy			Cold (5[a], 90[b], and 150[c] days)			Cold + nephrectomy		
	No.	Wt.	SE	No.	Wt.	SE	No.	Wt.	SE	No.	Wt.	SE
3 Wistar	9	755 100%	$\pm$5.6	6	840 111%	$\pm$9.4	9	877[a] 116%	$\pm$6.7	6	1011 134%	$\pm$10.2
6 Wistar	4	990 100%	$\pm$9.0	4	925 93%	$\pm$10.2	5	1296[b] 131%	$\pm$7.7		1140 123%	$\pm$12.1
11 LEW.AVN	3	700 100%	$\pm$0.6		--		3	1020[c] .46%	$\pm$6.5		--	
18 Wistar	6	1698 100%	$\pm$2.2	6	1850 109%	$\pm$4.8	6	dead			--	

Mean Number of Cells In Vitro X 10^3

	No.							
3	6	322	6	346	9	399^a	6	429
		100% +14.8		113% +4.4		124% +10.2		133% +8.0
6	4	150	4	169	5	194^b	6	272
		100% +2.4		113% +1.6		129% +10.5		160% +3.8
11	3	194		—	3	138^c		—
		100% +26.3				71% +19.0		
18	6	134	6	109	6	dead		—
		100% +15.7		+ 81% +3.3				

No. = Number.

Wt. = Weight.

SE = Standard Error.

NEPHRECTOMY

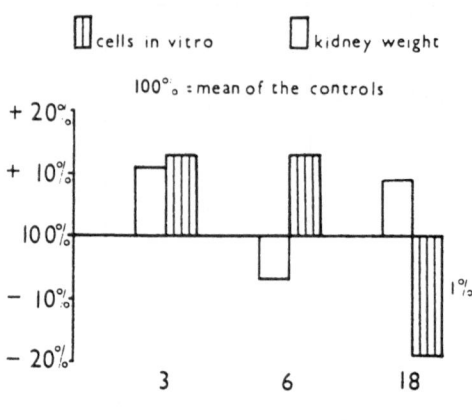

Age of rats in months

Fig. 2. The changes in kidney weight and in the number of kidney cells surviving 8 days in vitro 48 hr after nephrectomy in 3, 6, and 18 month-old rats. Mean values found before nephrectomy = 100%. Statistically significant difference on 1% level as indicated.

because they did not survive more than 18 hr in the cold (Fig. 3, Table I).

DISCUSSION

The decreased adaptability of old organisms to stressing agents was demonstrated both with unilateral nephrectomy and with cold exposure and acclimation. The most striking event was the early death of rats aged 18 months and exposed to 5°C. Nephrectomy appeared to be a milder stress to our aged animals, which survived the operation without difficulties.

With increasing age, the weight of the kidney increases and the number of cells capable to survive explantation and live in vitro decreases. The weight of the kidney did not change after unilateral nephrectomy, although in rats aged 3 and 6 months, explanted cells were slightly stimulated. This is in aggreement with the increased DNA synthesis soon after nephrectomy. Cells of the old rats, however, survived after nephrectomy in lesser numbers than before. We do not have any explanation of this loss of vitality of the explanted kidney cells.

Exposure to cold is also a powerful stimulus to kidney growth.

The weight of the kidney increased with the length of time spent
at low temperature, and the explanted cells of 3 month old rats
exposed for 5 days as well as those of 6 month old rats exposed
for 90 days were more vital in vitro than the cells of control
rats of corresponding age. Some flasks prepared from the cells
of these animals were cultured for 2 months and were evidently
able to form in vitro strains. In animals aged 11 months, living
in the cold for 150 days, the vitality of the cells indicated again
a serious damage of unknown character.

In rats aged 3 and 6 months, nephrectomized after 5 days in
the cold, both stimulators of hyperplasia combined so that the
increase in the number of cells capable to survive in vitro was
larger than in rats subjected to cold or nephrectomy only. The
weight of the kidneys was increased only in the 3 month old
animals. In the 6 month old rats, cold and nephrectomy combined
lead to larger kidney than nephrectomy only, but not than cold
only.

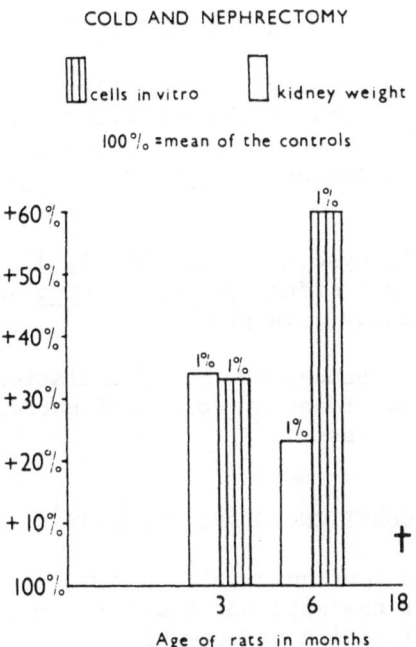

COLD AND NEPHRECTOMY

cells in vitro   kidney weight

100°/₀ =mean of the controls

Age of rats in months

Fig. 3. Combined effect of cold exposure and nephrectomy on kidney
weight and number of kidney cells surviving 8 days in vitro in
3 and 6 month-old rats. Exposure to 5°C for 5 days, nephrectomy,
explantation 48 hr after nephrectomy. All differences from mean
values in control rats of the same age statistically significant
on 1% level.

Our material shows that combined stressors may lead in younger organisms to a more favorable response than isolated stressors, at least when adaptive kidney growth after nephrectomy and cold is being considered. In older organisms, some isolated stressors, not dangerous for younger organisms, may be lethal.

SUMMARY

In adult rats acclimated to 5°C, the kidney weight increases with the length of the exposure. Kidney cells of younger rats survive explantation better than the cells of control rats of the same age. Kidney cells of older rats after cold acclimation and after unilateral nephrectomy have a lower vitality in vitro than corresponding control cells. Unilateral nephrectomy performed in rats aged 3 and 6 months, exposed previously to low temperature, leads to a better survival of explanted kidney cells than nephrectomy or cold only. Old rats do not survive exposure to 5°C for more than 18 hr.

REFERENCES

1. Soukupova, M., Hnevkovsky, P. and Najbrt, J. Effect of age on kidney hyperplasia in the rat after unilateral nephrectomy. Proceedings of the ECBO Symposium, Impairment of cellular functions during ageing in vivo and in vitro, Plenum Press, in press.

2. Holeckova, E., Baudysova, M. and Michl, J. Effect of cold acclimation on rat kidney growth in vivo and in vitro. Physiol. bohemoslov., in press.

3. Holeckova, E. and Baudysova, M. Stimulation of DNA synthesis in rat liver and kidney by cold acclimation. Physiol. bohemoslov., in press.

QUESTION TO DR. HOLEČKOVÁ

Dr. Franks: Have you continued the experiments for a longer period to see whether the apparent reduction in growth capacity is due to a delay in initiation of growth?

Dr. Holečková: It is evident that cells of the older rats cultured in vitro need more time to recover from the shock of trypsinization and they also were not able to form monolayers in a time period that was sufficient for the cells of the younger adult rats. We did not measure the time course of the changes in the kidneys directly.

AGE-RELATED CHANGES OF CELLS INVOLVED IN IMMUNE RESPONSES

G. M. Butenko and A. F. Andrianova

Institute of Gerontology AMS USSR, Kiev, USSR

SUMMARY

The interaction between T- and B-lymphocytes in the immune response to sheep reythrocytes was revealed by estimating the direct plaque-forming cells in mice CMA of various age. Investigations with a transfer of cells to irradiated and intact recipients were performed. We found a 23-fold diminution of the immune response in mice aged 26-28 months, when compared with that in mice aged 8 months. The first signs of a disturbed activity of the immune system were found in animals whose age did not exceed the half of their average lifespan. The injection of B-lymphocytes from young animals does not give rise to the immune response of old non-irradiated recipients. The injection of young B-lymphocytes to old irradiated recipients, being primed by antigen using Playfair's technique before irradiation, results in the highest response than it is in other combinations. The effect of disturbances in the interaction between T- and B-cells on the mechanism of age-related changes of the immune response is discussed.

It is common knowledge now that the immune system undergoes a series of changes with age. In general, this is a decrease of humoral and cellular responses to antigen, and an increased number of cases of autoimmune and allergic reactions (1-6). It should be emphasized that many of these events occur in a relatively early age when no noticeable changes in the structure of such organs as lymph nodes or spleen can be found. At the same time, it is known that immunological events are dependent on the activity of cells of the lymphoid system and other cellular elements of the connective

315

tissue. It is obvious, therefore, that one should seek the causes
of these changes in the state of cells to be involved in the immune
response.

They consider now that immune reactions are provided by the
function and interaction of the three main types of cells: macrophages (A-cells), thymus-derived lymphocytes (T-cells), and bone
marrow-derived/or bursa-dependent lymphocytes (B-cells) (7).
Moreover, the stromal elements of lymphoid organs and of a number
of other cells, e.g. polymorphonuclears, are, possibly, involved in
the effector link of immune reactions. But the main link of the
immun response is connected with the interaction of the first
three elements, the involvement of which is differently manifest
in the reactions of various types and in various antigens.

As far as macrophages are concerned, most of the investigators
indicate relatively minor changes in their involvement in the
induction of the immune response in old animals as compred with
young one (8,9). Having in mind the fact that the cell environment also makes but a small part of a summary effect of aging (1),
we may conclude that changes in the lymphoid cells play a main
role in a decrease of the immunological activity with age.

The purpose of this paper is a study of the effect of aging
on the function of separate cell components in the humoral immune
response. Investigations with a transfer of cells to the intact and
irradiated recipients of various age were performed. Such well-
known facts were taken into consideration as an essential decrease
of the number of colony-forming units (CFU) – actual precursors of
antibody-producing cells (APC) and, mainly, the unknown position of
T-lymphocytes in old subjects (11,12).

The preliminary series of experiments showed a 23-fold decrease
of the number of APC in mice CVA aged 26-28 months (old) in comparison
with that in 8-month old (adult) in the response to a single intravenous injection of sheep red blood cells (SRBC) in the dose 1.10^8
(Table I).

TABLE I

The number of direct plaque-forming cells (PFC) in a spleen of
adult and old mice CBA on the 4th day after the primary
immunization

Groups	PFC/5.10^6 nuclear cells	PFC/spleen
Adult	1234	51100
Old	52	2184
p*	0.01	0.01

* Here and in other cases, the statistical significance was estimated
by means of nonparametric criteria. For illustration, medians are
given.

Using another age combination -- young (4 months) and elderly
(12 months) animals, we failed to obtain any distinct differences
in the number of PFC per one dose of SRBC 5.10^8. While comparing,
however, the ratio of the level of hemolytic antibodies to the number of PFC in spleen, we observed a significant decrease of this ratio
in adult animals on the 4th day after immunization followed by its
subsequent correction at a later period. This fact suggests some
delay in the antibody synthesis of APC in elderly mice as compared
with that of young animals. It was also interesting that a 10-fold
increase of antigen dosage leads to a rise of the number of APC
in young mice, whereas it produces their obvious decrease in elderly
animals (Table II).

TABLE II

The number of direct PFC in spleen of mice aged 4 and 12 months
following intravenous injection of various SRBC doses

Antigen dose	5.10^8 SRBC	5.10^9 SRBC	P
4-month mice	32400	55655	0.01
12-month mice	44737	29950	0.01
P	0.05	0.01	

The above data suggest that changes in the immune response can be
found already before animals reach a half of their average lifespan.

The establishment of these facts served the basis for the

performance of the main experiment in order to elucidate the question
of the state of T-dependent component of the immune system in old
age. Heterogeneity of T-cells population and difficulties in a
separation of the homogenous function group of cells enabled us to
carry out the experiment with T-cells being kept in the organism of
irradiated or intact animals. We used the idea of Playfair (13) on
a prior antigen priming of the animals followed by the lethal irradiation 3 days after antigen administration. In this case, antigen
activated T-lymphocytes pass through the wave of mitosis and become
radioresistent. Concomitantly, B-cells, which have the maximal mitotic activity on the 4th day after antigen priming, must be killed
for irradiation. Followed after the irradiation, the restorative
bone marrow cells injection allows to reveal not only the functional
activity of T-cells, but also their capacity for the interaction with
newly-administered B-cells. The data obtained are presented in Table
III.

As is seen from this table, the smallest number of antibody-
forming cells are in the combination of T-cells from old animals
with B-cells of the same age. The response is decreased in the
combination of T-cells from adult mice with old B-cells, too. Somewhat unexpected and difficult for explanation were the results in the
combination of adult B-cells with old T-cells. They were the highest.
If we consider that together with adult bone marrow cells enter a
great number of B-cells (according to our data obtained with the help
of the method by Till and McCulloch (14), the number of CFU in old
animals was about 7 times as less as in the adults: 4 and 29 per
100,000 nuclear bone marrow cells respectively), then we should be

TABLE III

The number of PFC per 5.10^6 karyocytes in mice spleen in various
combinations of T- and B-cells (6 animals are in each group)

Source of T-cells and environment	Source of B-cells	
	Adult	Old
Adult	77	47
Old	123	27

aware of a large pool of T-cells in old animals which are capable
of interacting with the injected B-cells.

Taking into account the fact that the B-cells remainders in
the old organism are few, and that such side effects as irradiation
and varying radioresistence of cells of different age are difficult
to be controlled, the same experiment was performed once more on
normal intact animals, as recipients. The old mice were first
administered antigen, and then 3 days later, they were given B-cells
from young animals, followed by antigen again. The control group
was made of young and old mice, with a single or 2-fold antigen
administration (Table IV).

TABLE IV

The number of direct PFC in spleen of old mice after infection
of young bone marrow cells

I. SRBC 1.10^8	Young adult BMC $1.5.10^7$	II. SRBC 1.10^8	PFC/5.10^6
+	+	+	288
+	-	+	320
-	+	+	74
-	-	+	90

In such an experiment as described above, it is impossible to
demonstrate any essential effect of adult B-cells upon the old
recipient. This is, probably, accounted for by the fact that the
host T-cells are engaged with interacting with their own B-cells.

Thus, in the cell ensembel which provides the immune response, the most pronounced changes were found in lymphocytes of T- and B-systems. These changes are, primarily, reflected in their number. If we can say for sure that the number of B-cells in uniformly diminishing judging by the diminished CFU in the bone marrow, but as for T-cells, the question remains contradictory enough. On the one hand, there is a significant atrophy of thymus -- the site of production of these cells, as well a diminution of the bone marrow precursors of thymocytes. There are some evidences on the reduction of the thymus-dependent areas in spleen and lymph nodes, and a decrease of O-bearing lymphocytes in the circulation (15,16). On the other hand, there are many events which allow to suggest a rise in their number, at least, at a definite age (5,17). A series of facts shows that the main functions in these cells such as velocity multiplication rate and capacity of protein synthesis are also changed in the side of a delay and dimuntion. It is clear that the impairment of mechanisms providing the vital activity of cell underlies all these events. But there is one more aspect of cell viability in the multicellular organism:-- the capacity to exert influence, by means of certain signals, on other cells and then to answer the regulatory signals of this kind. It is found that these are likely to be new functions in the evolutionary aspect. Here we can speak not only about general regulatory influences, like hormones or mediators, but about those influences which enable the interaction of cells in more specialized cell populations, for example, the immune system. Many facts show that the earliest signs of aging of the immune system are the disturbances in the intrasystemic regulation connected with such a disturbed interaction between T- and B-cells. And though it is little known now about the nature of such connections, their sources, signal substances, receiving devices, as well as about pathways of realization of these signals in a cells, their indisputable existence shows that they may be exclusively important in the solution of some problems on the possible age-related changes in an organism.

REFERENCES

1. Ram, J.S. 1967. Aging and immunological phenomena. J. Geron- 22: 92.

2. Walford, R.L. 1967. The general immunology of aging. IN: Advances in Gerontological Research (Strehler, B.L., editor), Academic Press, N.Y.-London, p. 159.

3. Dobrowolski, L.A. 1970. Allergy in the aged. Giorn. Geront. 18: 709.

4. Makinodan, T. 1972. The immune system -- a model for the
    study of aging of precursor cells in a differentiating tissue.
    Repts. of 9th Intern. Congress of Gerontol., Kiev 1: 41.

5. Heidrick, M.L., and Makinodan, T. 1972. Nature of cellular
    dificiencies in age-related decline of the immune system.
    Gerontologia 18: 305.

6. Ricken, D. 1972. Immunopathologie des hoheren Lebensalters2.
    Actuelle Gerontologie 2: 29.

7. Talmage, D.W., Radovich, J., and Hemminscen, H. 1970. Cell
    interaction in antibody synthesis. IN: Advances in Immunology,
    Academic press, New York-London 12: 271.

8. Jaroslow, B., and Larnick, J.W. 1973. Clearance of foreing
    red cells from the blood of aging mice. Mech. Ageing Develop.
    2: 23.

9. Heidrick, M.L., and Makinodan, T. 1973. Presence of impairment
    of humoral immunity in nonadherent spleen cells of old mice.
    J. Immunol. 111: 1502.

10. Makinodan, T., Perkins, E.H., and Chen, M.G. 1971. Immunologic
    activity of the aged. IN: Advances in Gerontological Research,
    Academic Press, New York-London 3: 171.

11. Claman, H.N., Chaperon, E.A., and Triplett, R.F. 1966. Thymus-
    marrow cell combinations. Synergism in antibody production.
    Proc. Soc. Exp. Biol. Med. 122: 4.

12. Halsall, M.H., Heidrick, M.L., Deitchman, J.W., and Makinodan, R.
    1973. Role of suppressor cells in age-related decline in the
    proliferative capacity of spleen cells. Gerontologist 13: 46.

13. Playfair, J.H.L. 1972. Response of mouse T- and B-lymphocytes
    to sheep erythrocytes. Nature New Biol. 235: 115.

14. Till, J.E., and McCulloch, E.A. 1961. A direct measurement of
    the radiation sensitivity of normal mouse bone marrow cells.
    Radiation Res. 14: 213.

15. Gelfand, M.C., and Steinberg, A.D. 1973. Mechanism of allograft
    rejection in New Zealand mice. I. Cell synergy and its age-
    dependent loss. J. Immunol. 110: 1652.

16. Olsson, L., and Claësson, M.H. 1973. Studies on subpopulations
    of theta-bearing lymphoid cells. Nature New Biol. 244: 50.

17. Price, G.B. 1972. Immunological senescence. Gerontologist
     12: 29.

QUESTIONS TO DR. BUTENKO

Dr. Balazs: If the number of CFU cells in old animals is smaller
by a factor of 7 than in the young animals, the cells in the old
animals must be carrying out haemopoiesis more actively because
they produced the same number of blood cells as CFU cells of the
young animals. Do you agree?

Dr. Butenko: Yes, this may be so. However, I believe that the
main factor in these events is a diminution of the influence of
the feedback mechanism which maintains the cell population at its
constant level. In this case it may be a sign of weakness of control mechanism in cell-to-cell interactions.

Dr. Gutmann: Is there sexual differentiation of the immune response
in your experiments in which you crossed the T and B cells?

Dr. Butenke: Our experiments were performed on females only. As
far as males are concerned, it is known that their immune response
is less than that of females.

OBSERVATIONS OF AGE AND ENVIRONMENTAL INFLUENCES ON THE THYMUS KEPT

IN TISSUE CULTURE

O. Török, S.U. Nagy and G. Csaba

Department of Biology, Semmelweis University of Med.

Budapest IX, Hungary

In our earlier examinations we have demonstrated the presence of Gomori-positive cells in the tissue cultures of thymus of Wistar CB rats of various ages. Besides Gomori-positivity, these cells gave positive PAS-reaction, too, and exhibited histamine and serotonine fluorescence. Tissue cultures of the thymi of older rats contained these cells in greater number. The influence of TSH increased the number of these cells (4).

On the basis of our earlier studies (1) we have supposed that for the iodine uptake of the thymus these cells alone could be responsible. For that very reason, we examined, in the present experiments, whether the Gomori-positive cells in the thymus cultures of Wistar CB rats of various ages are capable of iodine uptake either spontaneously, or after TSH stimulation.

MATERIALS AND METHODS

Pieces of thymi of newborn, 15 day old and 100 g rats were explanted to cover glass stripes coated with coagulated hen plasma containing chicken embryo extract. The stripes with the explants were then placed into H-tubes. As cultivating medium the following mixture was used: synthetic medium of No. 199 (Parker), calf serum, lactalbuminhydrolysat (0.5% in Hanks) and chicken embryo extract in the ratio of 75% + 10% + 10% + 5%. This medium contained also penicillin (100 IU/ml). After a week of cultivation the cultures were incubated for 10 min in the above medium containing ^{125}I (potassium iodide) 2 μCi/ml of TSH (Ambinon, Organon) 1 IU/ml in the following experimental arrangements:

1. cultures incubated with ^{125}I for 10 min,
2. cultures incubated with TSH for 10 min and then with ^{125}I
    for 10 min,
3. cultures incubated with ^{125}I for 10 min and then with TSH
    for 10 min.

After incubation the cultures were fixed in Carnoy's fixative. Thereafter, the glass stripes carrying the cultures were stuck
to slides with Canada balsam and coated with Ilford G 5 emulsion. The
emulsion was diluted with double-distilled water (1:1) and then
kept in a water bath regulated to 42°C for 30 min. The slides coated
with the emulsion were dried at angles of 45°C and then exposed for
one week and two in the dark, at 4°C. After exposure the preparations were developed in ORWO A 49 developer for 5 min and then
fixed in acidic fixative. After this, PAS-reaction and hematoxylin
staining were performed on the preparations.

RESULTS

Above the region of the explanted pieces a large number of
grains was observable. In the migration zone only the PAS-positive
cells accumulated the iodine (Figs. 1 and 2). The number of the
grains was greater above the rounded PAS-positive cells than above
the flattened and processed ones. At the same time the other cells
remained unmarked. The number of the labelled PAS-positive cells
was greater in the cultures prepared from the thymus of newborn
rats in comparison with the cultures made from 15 day old animals

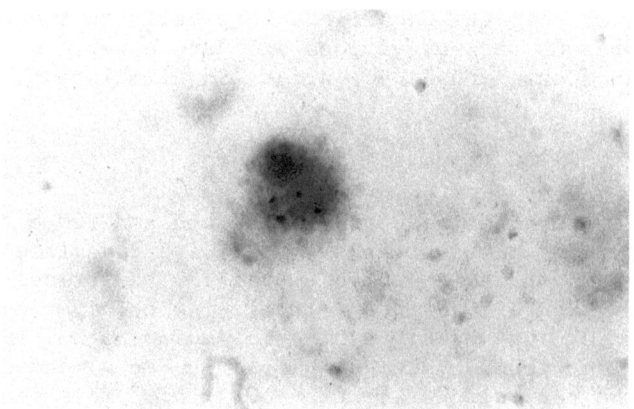

Fig. 1. Tissue culture of the thymus of the adult rat. PAS-positive
cell filled with secretion. Grains are above the cytoplasm. PAS –
H, X 320.

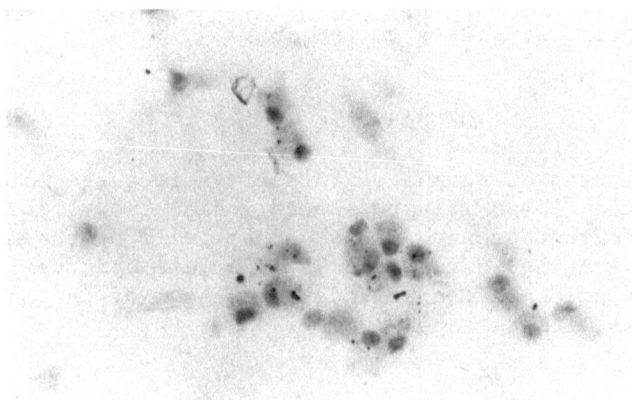

Fig. 2. Tissue culture of the newborn rat treated with TSH. The
grains are only over the PAS-positive cells. PAS - H, X 160.

and in those from animals weighing 100 g. As the quantity of iodine
taken up by the PAS-positive cells, the TSH treatment had only a
slight effect which was rather a negative one.

DISCUSSION

The capacity of iodine accumulation characterizes not only some
specialized glands of entodermal origin, but it represents a general
property of the entoderm. Thus, the thymus (1), the trachea (2),
the lungs (3), the salivary glands and the gastric glands (5) all
are capable of accumulating the iodine. From birth, with the pro-
gressing development of the organism, this capacity becomes more
and more restricted to the thyroid gland. The thymus and the
thyroid gland of the newborn rat show, however, only a slight
difference in the quantity of the incorporated iodine. The iodine
uptake by the newborn thymus is 50 times that of the adult thymus.
The decrease of the iodine accumulating capacity of the thymus can
be observed not only during ontogenesis, but under in vitro
circumstances, too. In the migration zone, namely, the incorporation of iodine was performed only by the PAS-positive cells. Since
the explanted piece contains these cells in a greater number than
the migration zone, the greater number of grains above it can be
explained. All this points to that the PAS-positive cells are exclusively responsible for the iodine incorporating capacity of the
newborn thymus. In our present experiments no appreciable TSH effect

could be demonstrated. To explain this fact one may consider either
that the timing of the experiments was not adequate, or the thymus
cells had no receptors sensitive to the TSH.

SUMMARY

In tissue cultures of thymus of rats of various ages only the
cells of Gomori- and PAS-positivity showed iodine incorporation as
demonstrated by radioautography. Above other cells of the migration
zone no grains indicating iodine accumulation were observable. The
capacity of iodine accumulation depends on the age of the animal,
namely it decreases with the ageing.

REFERENCES

1. Csaba, G., Kiss, J. and Nagy, S.U. 1973a. Comparative studies
    on the ^{125}I uptake of the thyroid and thymus. Experientia 29:
    357.

2. Csaba, G., Nagy, S.U., Bombera, G. and Mándics, R. 1973. Germ
    layer specificity of iodine accumulation. Endocrin Experiment,
    in press.

3. Nagy, S.U. and Csaba, G. 1973. Adatok az elöbél entoderma
    általános jódfelvevö képességéhez. Biol. Közl., in press.

4. Csaba, G. and Török, O. 1974. Examinations of the Gomori-
    positive cells of the thymus in tissue culture. Endocrin
    Experiment 8: 3.

5. Ingbar, S.H. and Woeber, K.A. 1968. The thyroid gland. In:
    Textbook of Endocrinology. R.H. Williams, ed., Saunders,
    Philadelphia.

QUESTIONS TO DR. TÖRÖK

Dr. Franks: What function has the iodine?

Dr. Török: The PAS-positive cells in the thymus are capable of
accumulating the iodine. However, this has no specific function
in the thymus but merely reflects the general involutive process.

Dr. Cristofalo: What medium do you use? Have you studied the
effect of hormones on proliferation? I am particularly interested
in the role of hydrocortisone. Whitefield and his coworkers have
shown a stimulation of cell proliferation by low concentration of

hydrocortisone and this effect was modulated, in part, through cyclic AMP. Do you have any further information on this point?

Dr. Török: I used a semisynthetic medium, consisting of WO 199 TC, lactalbuminhydrolysate, calf serum and chicken embryo extract (in the ratio described in the paper). We have not examined the effect of hydrocortisone on the proliferation of the Gömöri and PAS-positive cells. TSH treatment stimulated the number of these cells (G. Csaba and O. Török: Examinations of the Gömöri-positive cells of the thymus in tissue culture. Endocrin. Experiment. 8: 3-12, 1974).

Dr. Martin: I may have missed this point in your presentation. But what is known about the kinetics of the iodine concentrating ability of thymus cells postnatally?

Dr. Török: Both in vivo and in vitro the capacity for iodine accumulation decreased.

Dr. Kasten: Did you observe any organization in vitro of Hasall's corpuscles?

Dr. Török: Formation of Hasall's corpuscles was not observed but the cysts were formed by these PAS-positive cells.

Dr. Stoklosowa: In our laboratory Dr. Stadnicka has cultured the thymus gland as an organ culture. After approximately one week in culture the epithelial parts of the glands degenerated and then there was renovation of this tissue with simultaneous formation of typical Hasall's bodies.

Dr. Török: In tissue cultures of the rat thymus we observed only the formation of cysts. In other unpublished work on mice we have demonstrated the formation of Hasall's corpuscles after treatment with hydrocortisone.

FIDELITY IN THE COLLAGEN SYNTHESIZED AND MODIFIED BY AGING FIBROBLASTS IN CULTURE

Paul M. Gallop and Mercedes A. Paz
Dept. of Biological Chemistry and Orthopaedic Surgery,
Harvard Medical School and Harvard School of Dental
Medicine and The Children's Hospital Medical Center
Boston, Massachusetts 02005

It has been shown by Hayflick and Moorehead (1) and confirmed in other laboratories that diploid cell strains have a finite lifespan in culture. Lifespan is measured in the average number of population doublings before the so-called phase III or "senescent" state of the culture is reached. It is reasonable to propose that certain types of error continually or increasingly accumulated in the synthetic apparatus and reflected in each generation by increasingly defective proteins, may be causative of the "aging" of the culture. Such a concept has been proposed by Orgel (2) and preliminary evidence from Holliday and Tarrant (3) appears to lend support, although some controversy exists in the literature. In our estimation, the Hayflick phenomenon or the so-called "programmed death" of diploid cell strains allowed the planning of unique experiments in which error accumulation in proteins can be evaluated. The errors may be random, reflected in a generalized and increasing degree of protein imperfection, broadened or narrowed enzymatic specificity and increased thermal lability. Furthermore, one might expect significant changes in the responses to various metabolic controls of the type usually seen with larger enzymes composed of precisely tuned interacting subunits. The errors may not be entirely random in that certain sites may be particularly susceptible to spontaneous changes, perhaps by a selection mechanism, and that aging could be carefully orchestrated in movements in the sense that with increasing passage levels, defined off-on enzyme systems come into play. Perhaps both random and preordained changes occur simultaneously so that biological variation in the senescent phenomenon in regard to the average extent and average time course results.

We have chosen as our model cells the defined and extensively studied diploid human cell fibroblast strain, WI 38, which can

329

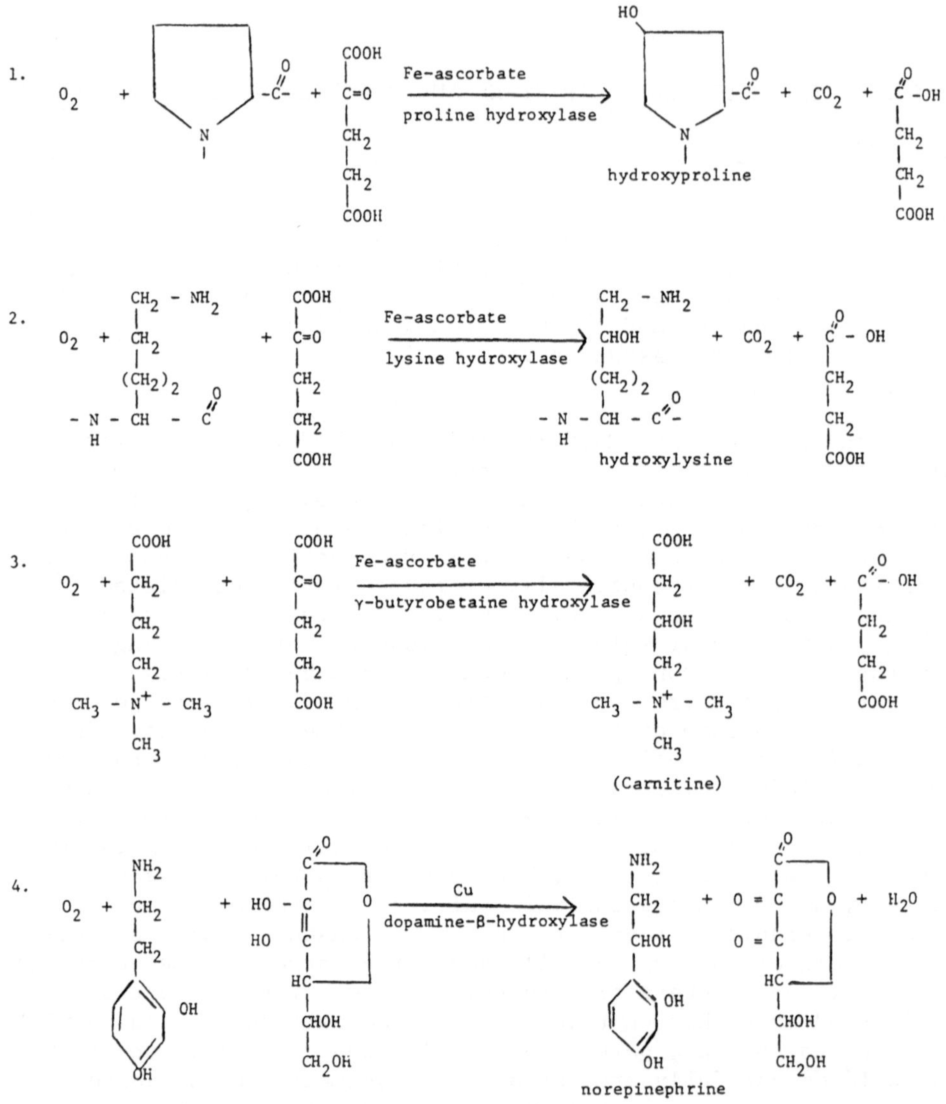

1. O_2 + [pyrrolidine ring with $-C-$ (C=O)] + $C=O$ (COOH, $C=O$, CH_2, CH_2, COOH) → Fe–ascorbate / proline hydroxylase → [HO–pyrrolidine ring] $-C-$ + CO_2 + $C-OH$ (CH_2, CH_2, COOH)   hydroxyproline

2. O_2 + (CH_2-NH_2, CH_2, $(CH_2)_2$, $-N-CH-C$, H) + (COOH, $C=O$, CH_2, CH_2, COOH) → Fe–ascorbate / lysine hydroxylase → (CH_2-NH_2, CHOH, $(CH_2)_2$, $-N-CH-C$, H) + CO_2 + $C-OH$ (CH_2, CH_2, COOH)   hydroxylysine

3. O_2 + (COOH, CH_2, CH_2, CH_2, $CH_3-N^+-CH_3$, CH_3) + (COOH, $C=O$, CH_2, CH_2, COOH) → Fe–ascorbate / γ-butyrobetaine hydroxylase → (COOH, CH_2, CHOH, CH_2, $CH_3-N^+-CH_3$, CH_3) + CO_2 + $C-OH$ (CH_2, CH_2, COOH)   (Carnitine)

4. O_2 + (NH_2, CH_2, CH_2, [benzene ring with OH, OH]) + (HO–C, HO–C, HC, CHOH, CH_2OH) → Cu / dopamine-β-hydroxylase → (NH_2, CH_2, CHOH, [benzene ring with OH]) + (O=C, O=C, HC, CHOH, CH_2OH) + H_2O norepinephrine

Fig. 1.   Some mixed function oxidase (MFO) having preferential requirements for ascorbate.   MFO, αKG cosubstrate (1–3); ascorbate mediator; Fe.   MFO, ascorbate cosubstrate (4); Cu.

undergo 50 $\pm$ 10 population doublings before reaching phase III, the stage prior to the death of the culture. The life of these cultures can be extended significantly by the continual addition of hydro-cortisone to the cultures as shown by Cristofalo (4), adding another useful experimental parameter to this system.

We are looking at the production and enzymatic modifications of collagen produced in large amounts by these cells. Collagen is a long fibrous protein made of three polypeptide chains and extruded by the cells into the media. Before extrusion and after extrusion, unique post-translational enzymatic modifications of the collagen occurs at several points. Accordingly, if one is able to examine the detailed structure and properties of the collagen produced, focusing on the relevant sites, it might be possible to determine if certain enzymes produced by the aging cells have become less perfect in structure and function and otherwise modified with increasing passage level.

Initially for examination, we have chosen for study the collagen hydroxylating enzymes. These are mixed function oxidases (MFO) of at least two types; namely prolyl hydroxylase and lysyl hydroxylase. These hydroxylases require ferrous iron as cooxidant with molecular oxygen as oxidant and with α-ketoglutarate as an obligatory cosub-strate and reductant (5,6). Ascorbate is an effective mediator but its mode of action is not completely understood and, in addition to a possible role in oxidation reduction mediation, it may be an activator of an inactive precursor of prolyl hydroxylase (7). These reactions which lead to hydroxyprolyl and hydroxylysyl residues are depicted in Figure 1 with two other unusual MFO added for compar-ison. Carnitine synthetase (γ-aminobytyric acid, betaine β-hydrox-ylase) is included in the figure since, at present, it is the only other MFO known to employ α-ketoglutarate as a cosubstrate (8). Dopamine β-hydroxylase is included since it is an unusual example of an MFO in which ascorbate is the putative cosubstrate and core-ductant (9).

In the collagen system, selected prolyl and lysyl residues are hydroxylated so that a typical collagen may have, as an example, 120 prolyl and 90 hydroxyprolyl residues and 30 lysyl and 6 hydroxylysyl residues per α-chain (92,000). We expect to determine the number and distribution of hydroxylysyl residues in the collagen produced by WI 38 cells with increasing passage level by examining collagen α-chain fingerprints with methods specifically designed to detect the sites of hydroxylysyl residues. The purpose of this experiment is to check the lysyl hydroxylase specificity. This program has just begun.

Our major effort for the past two years has been examination of the extent of proline hydroxylation, and the role of ascorbic

acid in the collagen hydroxylation process in the aging WI 38 cultures.
A word on methodology is added here to facilitate understanding of our
experimental approach. The WI 38 cells were cultured in Eagle's basal
medium supplemented with 10% fetal calf serum. Collagen is isolated
from the media and cell layers after one week during which time the
various cultures have been kept at confluency. Ascorbic acid (50
µg/ml) was or was not present, the ascorbic acid being added every
other day. Twenty-four hr before harvest, labeled proline is added
so that both the prolyl and hydroxyproline positions in the collagen
will be labeled. The collagen from the cells and the media (separated
from the cell layers) is treated with bacterial collagenase and the
characteristic peptide digests examined by ion-exchange chromatography. Collagen which would be normally hydroxylated would contain
a high level of the tripeptide gly-pro-hyp and little gly-pro-pro.
On the other hand, poorly hydroxylated collagen contains higher
levels of gly-pro-pro and lower levels of the gly-pro-hyp. The
most common tripeptide in both hydroxylated and underhydroxylated
collagen is gly-pro-ala. Without going into further detail, it is
possible with this methodology by employing the ratio of gly-pro-
hyp to the sum of gly-pro-ala and gly-pro-pro to accurately measure
small levels of underhydroxylation.

RESULTS AND DISCUSSION

    The data shown in Tables I and II relate to three groups of
cells divided according to the number of population doublings or
passages. Passages 26 and 41 originated from the same batch of
young cells and the cultures were carried until they reached phase
III at about passage 51. The passage classified as passage 41
"old" came from another initial batch, and the cells at this
passage level already showed evidence of "senescence". With
"aging" by increasing passage levels (Table I) a decrease in the
amount of DNA and total protein of the cells is observed corresponding to a decrease in the total number of cells. The decrease
of proliferative capacity in "aging" cells is probably associated
with an increased number of non-dividing cells. Much less protein
and DNA were found in the passage 41 "old" cells, but the level
of collagen associated with the cells expressed as mg collagen/mg
DNA did not differ significantly from that associated with the
younger cells. Twice as much collagen was associated with the
young and old cells when ascorbic acid was present. This collagen
was found to be very insoluble and is highly crosslinked.

    In the younger cells, ascorbic acid has an overall inhibitory
effect on total collagen protein synthesis as is shown by the 60%
reduction in the amount of collagen protein extruded into the
medium by passage 26 cells in the presence of ascorbic acid (Table
II). The collagen produced by young confluent cells in the

TABLE I. WI 38 Cells -- Collagen Content in the Cells and Cell Layer†

Passage Levels	26	26	41	41	41(old)	41(old)
Ascorbic Acid	+	-	+	-	+	-
Total Protein (mg)	114	123	77	85	30	56
Total DNA (mg)	2.6	2.8	2.0	2.2	1	1.9
Total Collagen (mg)	2	0.9	1.9	1.1	0.7	0.8
mg Collagen/mg DNA	0.8	0.3	0.9	0.5	0.7	0.4
Degree of hydroxylation (%)	94	50	96	70	82	76
Collagen (cells) (%) / Total Collagen Produced	23	5	25	16	18	29

†Cultures kept one week at confluency 20T75 flasks were used in each group

TABLE II. WI 38 Cells -- Collagen Content in the Medium[+]

Passage Levels	26	26	41	41	41(old)	41(old)
Ascorbic Acid	+	-	+	-	+	-
Total Collagen (mg)	6.8	16.8	5.6	5.9	3.2	2
mg Collagen/mg DNA	2.6	6.0	2.8	2.7	3.2	1
Degree of hydroxylation (%)	100	48	99	76	77	77
Collagen (medium)/Total Collagen Produced (%)	77	95	75	84	82	71

[+]Cultures kept one week at confluency 20T75 flasks were used in each group

presence of ascorbic acid is 100% hydroxylated in contrast to that
found in the absence of the vitamin which is only 50% hydroxylated.
This underhydroxylated collagen is very soluble and is probably
deficient in some crosslinkages, only 5% of the total collagen
produced by passage 26 cells in the absence of ascorbate is associated
with the cells, the rest (95%) is present in soluble form in the
medium. Ascorbic acid is required by young cells to achieve complete
collagen hydroxylation. The collagen synthesis and proline hydroxylation systems of the older cells are less ascorbate dependent,
about 15% less collagen was synthesized by P41 cells in the presence
of ascorbic acid. Collagen which is 70% hydroxylated can be produced by passage 41 cells in the absence of the vitamin. The cells
which were close to "senescence" showed no dependence on ascorbate
for hydroxylation and the collagen produced by these cells in the
presence or absence of the vitamin was equally hydroxylated to
about 80% of the normal degree of hydroxylation. In the presence
of ascorbate, passage 41 "old" cells secreted as much collagen per
mg of DNA into the medium as the younger cells. In the absence of
ascorbic acid, the "old" cells extruded significantly less collagen
into the medium. There is considerable evidence for variations in
the hydroxylation of lysine and proline in bone collagen (10) and
skin collagen (11,12) with age. Collagen produced by WI 38 fibroblasts at different passage levels also showed an age related
variation in the hydroxylation of proline. The hydroxylation of
proline was incomplete in collagen produced by senescent cells
whether or not ascorbate was present. There was also a variation in
the dependence for ascorbic acid of the hydroxylating enzymes in
young versus old cells. Young cells require ascorbic acid to
achieve complete hydroxylation, in the absence of the vitamin the
collagen produced by young cells is only 50% hydroxylated. Similar
results were obtained with 3T6 fibroblasts (13) and L929 cells (14)
which synthesize an underhydroxylated form of collagen in the absence
of ascorbic acid. Complete hydroxylation is not a requirement for
extrusion, totally hydroxylated collagen as well as underhydroxylated
collagen were present in the medium. A certain level of hydroxylation
may be necessary to control the synthesis of new collagen by a feedback type of mechanism. This is suggested by the fact that much
more collagen of 50% underhydroxylation is secreted by the young
cells in the absence of ascorbate acid. Older cells show less
dependence on ascorbate for hydroxylation and can achieve 70% of
the optimum hydroxylation. More collagen is secreted into the
medium in the absence of ascorbate, but the difference is not as
dramatic as with younger cells.

Accordingly, the prolyl hydroxylase system and its various
controls is clearly a cell "age" dependent system. The loss of
ascorbic acid dependency with "age" may result from defined or
programmed changes in the enzyme(s). On the other hand, it may
result from a gradual increase in age dependent random imperfections

leading ultimately to a hydroxylase system which at cell senescence is clearly unable to bring about full hydroxylation even in the presence of ascorbic acid. Clearly, the enzymology and protein chemistry of the hydroxylation system isolated from different "aged" cell cultures is warranted.

REFERENCES

1. Hayflick, L. and Moorehead, P.S. 1961. The serial cultivation of human diploid cell strains. Exp. Cell Res. 25: 585.

2. Orgel, L.E. 1963. The maintenance of the accuracy of protein synthesis and its relevance to ageing. Proc. U.S. Nat. Acad. Sci. 49: 517.

3. Holliday, R. and Tarrant, G.M. 1972. Altered enzymes in ageing human fibroblasts. Nature 238: 26.

4. Cristofalo, V.J. 1973. Factors modulating cell proliferation in vitro. Symposium on the Molecular and Cellular Mechanisms of Aging. December, Paris (France).

5. Cardinale, G.J., Rhoads, R.E., Udenfriend, S. 1971. Simultaneous incorporation of ^{18}O into succinate and hydroxyproline catalyzed by collagen proline hydroxylase. Biochem. Biophys. Res. Commun. 43: 537.

6. Kivirikko, K.I., Shudo, K., Sakakibara, S. and Prockop, D.J. 1972. Studies on protocollagen lysine hydroxylase. Hydroxylation of synthetic peptides and the stoichiometric decarboxylation of α-ketoglutarate. Biochemistry 11: 122.

7. Stassen, F.L.H., Cardinale, G.J. and Udenfriend, S. 1973. Activation of prolyl hydroxylase in L929 fibroblasts by ascorbic acid. Proc. U.S. Nat. Acad. Sci. 70: 1090.

8. Lindstedt, G. and Lindstedt, S. 1970. Cofactor requirements of γ-Buryrobetaine hydroxylase from rat liver. J. Biol. Chem. 245: 4178.

9. Friedman, S. and Kaufman, S. 1966. An electron paramagnetic resonance study of 3,4-dihydroxyphenylethylamine β-hydroxylase. J. Biol. Chem. 241: 2256.

10. Miller, E.J., Martin, G.R., Piez, K.A. and Powers, M.J. 1967. Characterization of chick bone collagen and compositional changes associated with maturation. J. Biol. Chem. 242: 5481.

11. Barnes, M.J., Constable, B.J., Morton, L.F. and Kodicek, E.
     1971. Hydroxylysine in the N-terminal telopeptides of skin
     collagen from chick embryo and newborn rat. Biochem. J. 125:
     925.

12. Barnes, M.J., Constable, B.J., Morton, L.F. and Royce, P.M.
     1974. Age-related mariations in hydroxylation of lysine and
     proline in collagen. Biochem. J. 139: 461.

13. Bates, C.J., Prynne, C.J. and Levene, C.I. 1972. The synthesis
     of underhydroxylated collagen by 3T6 mouse fibroblasts in
     culture. Biochim. Biophys. Acta. 263: 397.

14. Peterkofsky, B. 1972. The effect of ascorbic acid on collagen
     polypeptide synthesis and proline hydroxylation during the
     growth of cultured fibroblasts. Arch. Biochem. Biophys. 152:
     318.

QUESTIONS TO DR. GALLOP

Dr. Hay: Have any comparative studies been made on the rates of
collagen synthesis by fibroblasts in vitro versus the similar activity in connective tissue present in a young, growing animal?

Dr. Gallop: Nothing that I can accurately recall. However my
feeling is that certain organ cultures can make more collagen per
cell although one would really not know how many fibroblasts are
present in an organ culture.

Dr. Courtois: Did you try to use your fingerprint method to
study collagen synthesis and its degree of hydroxylation at
different points of the cell cycle?

Dr. Gallop: In a very preliminary manner. Young cells with
ascorbate present seem to make hydroxylated collagen during log
phase, but most of the collagen is made after confluency.

Dr. Franks: Do you add α-keotglutarate as well as ascorbate to
your medium? A colleague of mine, R.G. Hodge, finds that in organ
cultures this gives better preservation of the stroma.

Dr. Gallop: α-ketoglutarate is not added. It is present in the
cells as an intermediary metabolite. Its level depends on the
extent of the Krebs cycle, glutamic dehydrogenase and glutamic
α-ketoacid trasaminase.

Dr. Deyl: What are the details of your fingerprinting technique?
Could you specify that a bit?

Dr. Gallop: Collagen in the culture is labelled with tritiated proline 24 hr prior to harvest and fractionated from the cell layers and the medium with hot TCA.   Both hot TCA soluble and insoluble material is treated with bacterial collagenase and then fingerprinted on the long column of an automatic amino acid analyzer equipped with a split stream arrangement so that part of the outflow goes to the flow cell of a scintilation counter.

Dr. Franks: What is the nature of the cell associated collagen?

Dr. Gallop: This is both collagen within cells (perhaps newly synthesized) and collagen deposited around the cell.   Most of the collagen has been released by the cells into the medium.

BIOSYNTHESIS OF COLLAGEN DURING THE LIFE CYCLE OF HUMAN DIPLOID

CELL LINES

J. Hurych, M. Macek, K. Smetana, F. Beniač and
D. Řezáčová
Centre of Industrial Hygiene and Occupational Diseases
Charles University, Czechoslovak Academy of Sciences
and Institute of Sera and Vaccines, Prague

The senescence of human diploid fibroblasts and their limited lifespan in vitro, shown by Hayflick and Moorhead (1), suggested a possible relationship between these events and ageing in vivo. In this paper we report our observations concerning the problems of collagen synthesis of human diploid cell strains during their long term in vitro cultivation.

MATERIALS AND METHODS

We examined 78 fibroblast strains from 58 donors. We analyzed strains initiated from 18 embryos, 13 foetuses, 27 children and 3 adults of the age not higher than 30 years. The investigated strains were derived from 23 different tissues: umbilical cord, skin, gingiva, muscle, diaphragm, rib, bone marrow, pericardium, peritoneum, pleura, amnion, heart, liver, spleen, thymus, kidney, adrenals, amniotic fluid cells, foetal blood, brain, eye, spinal cord, solid tumors.

The tissue cultures were started from very small tissue fragments (2), and cultivated in medium EPL (3), supplemented by 5-10% of human or calf serum. The embryonal cells were grown in this medium without serum. The methods of cultivation of foetal peripheral blood and amniotic fluid cell cultures are described in detail in our previous papers (4,5).

The collagen synthesis was ascertained by the determination of hydroxyproline (Hyp) and by the ratio of soluble to insoluble collagen. Nitrogen (N) was determined after combustion by Nesslerization (4-8). For the characterization of collagen formed, the amniotic fluid cells were cultivated the last 3 days in the presence of β-aminopropionitrile (50 μg/ml). Sixteen hr before finishing the cultivation, ^{14}C-proline (0.52 μCi/ml) was added (9). Collagen from the medium and cells were extracted into 1 M NaCl pH 7.4. The extracts were centrifuged at 100,000 g for 1 hr. Soluble radioactive proteins were mixed with carrier collagen (20 mg). The collagen and radioactive proteins were precipitated by dialysis against 0.02 M disodium phosphate. Proteins from the medium were denatured at 60°/10 min and purified by chromatography on Sepharose 6 B (110 X 2.2 cm). The elution was carried out at 60°C with 0.05 M Tris-HCl buffer pH 7.5 and 1 M $CaCl_2$.

Proteins extracted from the cells were, after addition of carrier collagen, also precipitated. Aliquot was hydrolyzed and ^{14}C-Hyp and ^{14}C-Pro were separated by ionex chromatography (Dowex 50x8, 30x1 cm).

Radioactivity was assayed by liquid scintillation spectrometry (Packard-Tricarb, Model 3365).

The specimens fixed in glutaraldehyde (10) and osmium tetroxide (11) were dehydrated in ethanol containing uranyl acetate (12) and embedded in Durcupan-Epon mixture (13). Ultrathin sections were cut with a Reichert OMU2 ultramicrotome and were stained with uranyl acetate followed by lead citrate (14) and observed with a Philips 3000 electron microscope.

RESULTS

The study of collagen synthesis during 14 months of cultivation of fibroblast strains derived from the skin, gingiva and umbilical cord showed that it decreased after 8-9 months of cultivation (Fig. 1). This decrease correlates with their transition to the third degenerative phase of growth. During the active phase of growth there was no difference between them in the rate of collagen synthesis, expressed by the ratio μg Hyp/mg N and μg Hyp/10^7 cells, as calculated from the regression equations.

We have demonstrated a permanent increase of protein concentration with the length of in vitro cultivation as expressed by the ratio of cell nitrogen to 10^7 cells (Fig. 1).

The investigation of the ratio of soluble to insoluble collagen in different strains examined between the 7-38th passage brought evidence that the average value is 1.1 ± 0.33. This ratio was significantly increased in fibroblast cultures from mother and her child affected by Marfan syndrome (Fig. 2). In these cultures, the increase in the proportion of insoluble collagen was demonstrated after 18 passages of cultivation.

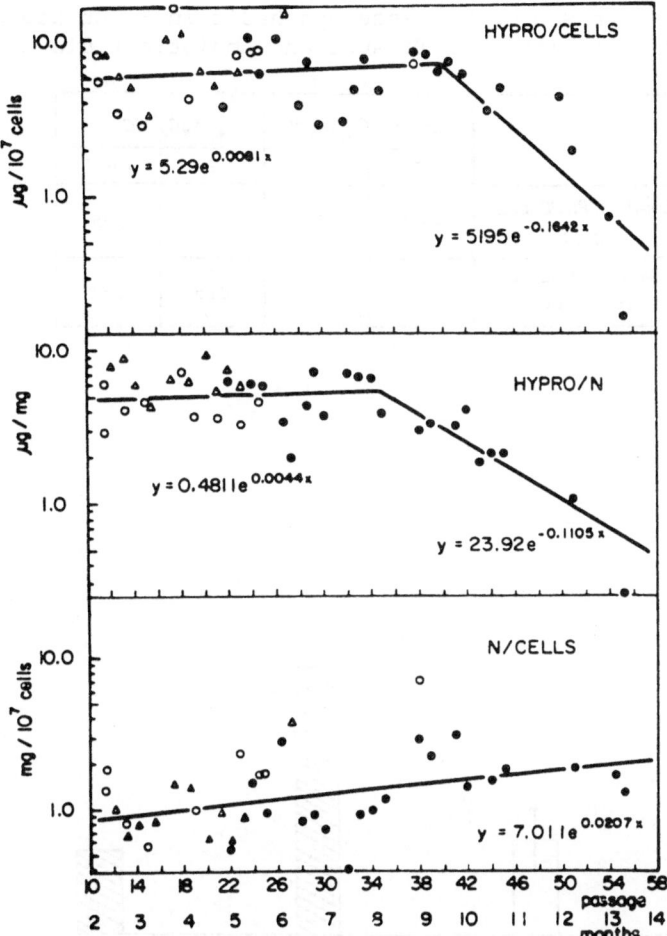

Fig. 1. Changes in hydroxyproline and nitrogen in fibroblasts
during cultivation. o skin fibroblasts (passage 10–25);
● gingiva fibroblasts (passage 10–55); Δ umbilical cord fibroblasts (passage 12–23). All data are pooled and regression
equation was calculated.

The comparison of the rate of collagen synthesis in fibroblast
cultures, derived from different tissues of embryos, foetuses,
children and adults during the active phase of growth did not show
significant differences (Fig. 3).

TABLE I.   Comparison of Collagen Synthesis in Human Diploid
Cell Lines From Different Tissues and Amniotic Fluid Cells

	μg HYPRO /mg N		NSC/ISC	
	$\bar{x}$	± ts $\bar{x}$	$\bar{x}$	± ts $\bar{x}$
HUMAN DIPLOID CELL LINES	4.995	0.4	1.24	0.35
AMNIOTIC FLUID CELL CULTURES	3.21	1.40	0.88	0.53

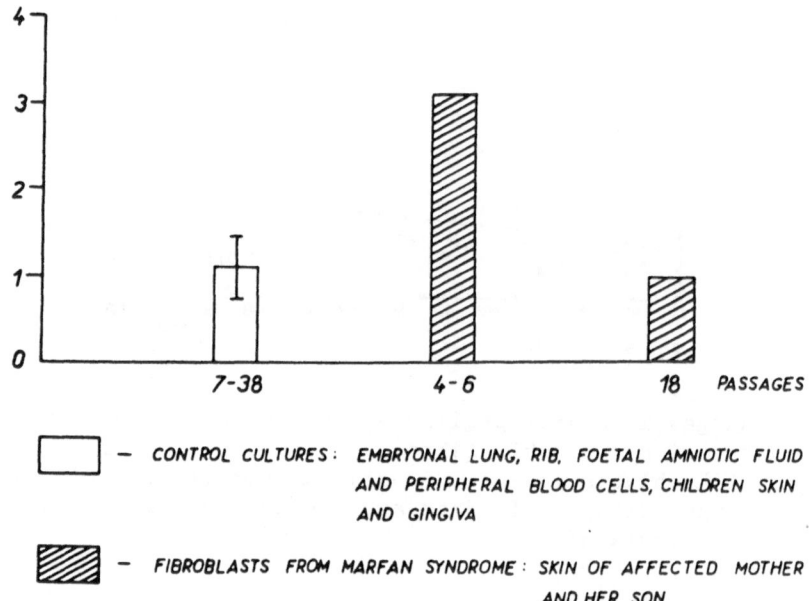

- CONTROL CULTURES: EMBRYONAL LUNG, RIB, FOETAL AMNIOTIC FLUID
  AND PERIPHERAL BLOOD CELLS, CHILDREN SKIN
  AND GINGIVA

- FIBROBLASTS FROM MARFAN SYNDROME: SKIN OF AFFECTED MOTHER
  AND HER SON

Fig. 2.   Ratio of soluble to insoluble collagen in control
fibroblast cultures and cultures from Marfan syndrome. $ts_{\bar{x}}$ =
95% upper and lower confidence limit of arithmetic mean.

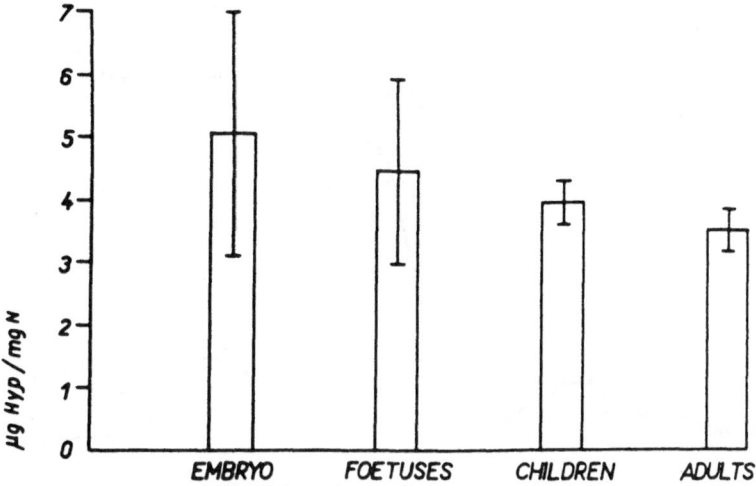

Fig. 3. Synthesis of collagen in human diploid cultures from different periods of development.

The investigation of collagen synthesis in long term amniotic fluid cell cultures was carried out. We proved that these fibroblast cells produce collagen in vitro at the same rate as the cells from fibroblast cultures, initiated from other tissue in the praenatal or postnatal phase of development (Table I). The same is true for the ratio of soluble to insoluble collagen.

The elution curve of proteins secreted to the cultivation medium by amniotic fluid cell culture resembles that of collagen α-chains (Fig. 4). The detection of labeled Hyp confirmed the synthesis of collagen in cultivated amniotic fluid cells (Fig. 5). Further characterization of this collagen is in progress.

The investigated ultrathin sections revealed the cells characterized by the presence of fine filaments of the diameter 90–100 Å

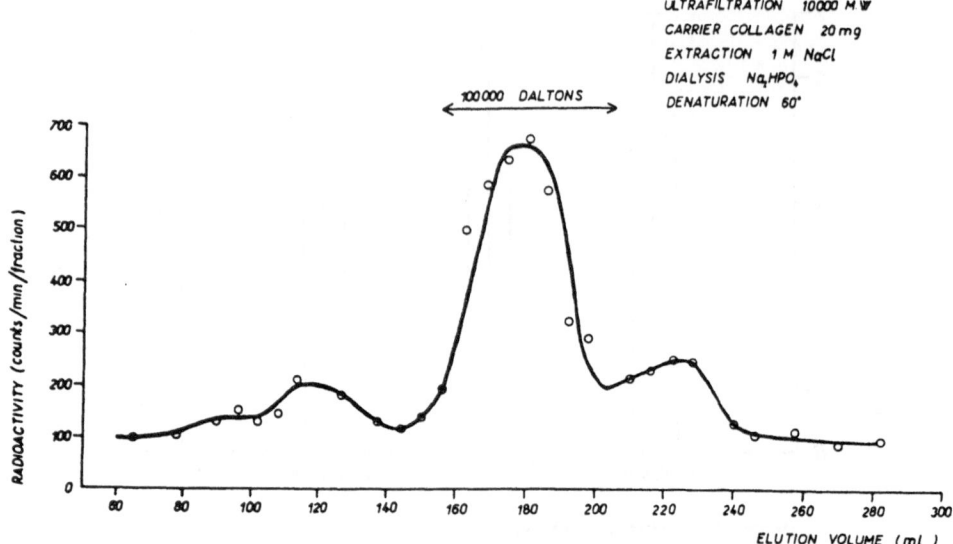

Fig. 4. Amniotic fluid cell culture. Fractionation of cultivation medium on Sepharose 6B. The elution was carried out at 60° with 0.05 M Tris-HCl pH 7.5 and 1 M CaCl₂.

in cytoplasm (Fig. 6). These filaments were also proved in inter-cellular space. Their diameter was 300-500 Å (Fig. 7). In large cells rough endoplasmic reticulum was observed (Fig. 8).

DISCUSSION

The decrease in collagen synthesis reaches a maximum during the transition into the degenerative phase III. We consider it important that this trend is not dependent on the absolute length of _in vitro_ cultivation, but on the onset of the symptoms of cell senescence _in vitrr_, which appears at different times in particular lines (8) cultivated under the same conditions.

Our observation of the increase of the cell nitrogen and thus cell protein with the length of _in vitro_ cultivation agrees with the recent finding of Lewis and Holliday (15) obtained with certain mutant strains of Neurospora and supports Orgel's theory (16) of ageing as a result of accumulation of errors in protein syn-thesis.

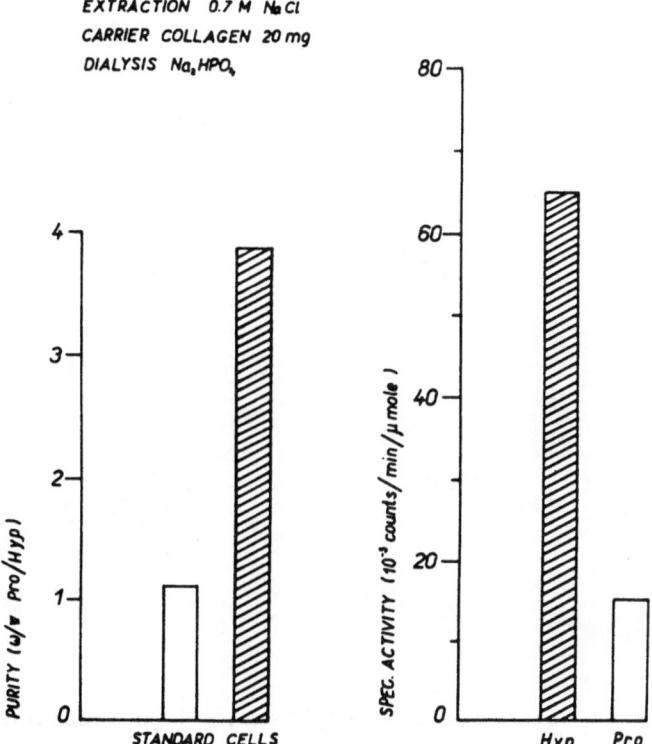

Fig. 5. Amniotic fluid cell culture. Characteristics of the cell collagen. Separation of ^{14}C-hydroxyproline was performed on a Dowex 50 X 8 column (30 X 1 cm).

The average value of 1.1 for the ratio of soluble to insoluble collagen (4-6,8) in cultivated fibroblasts from different tissues during the active phase of growth indicates relative stability of the collagen maturation in vitro. The increased proportion of soluble to insoluble collagen in cultures originating from Marfan syndrome (6) was recently confirmed by Priest et al. (17). The increase of insoluble collagen in long term cultivated fibroblasts from this syndrome suggests that ageing in vitro might be accompanied also by the increase of structural stability of collagen as observed in vivo (18).

The comparison of collagen from tissue cultures of embryos, foetuses, children and young adults (7,8) did not prove significant differences in the rate of collagen synthesis in our sample under investigation. This however does not preclude detection of a gradual

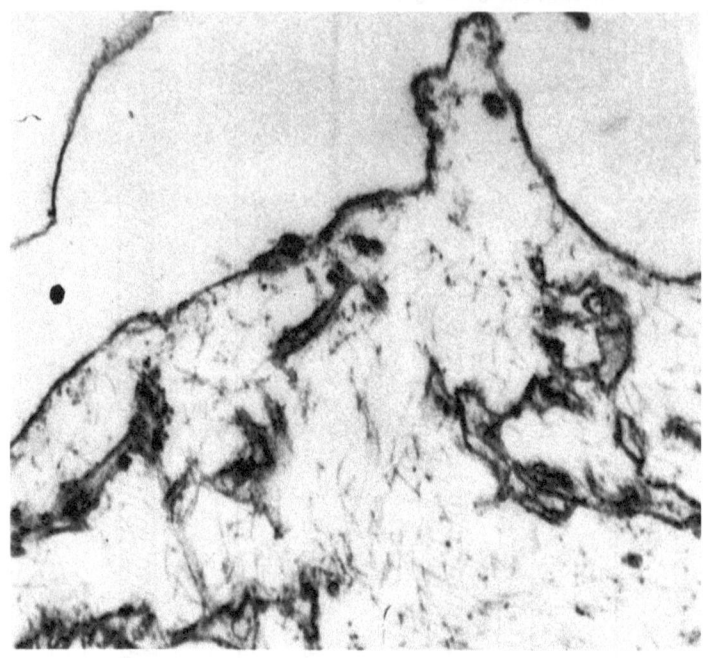

Fig. 6.   Fine filaments in cytoplasma.   Magnification 33,700.

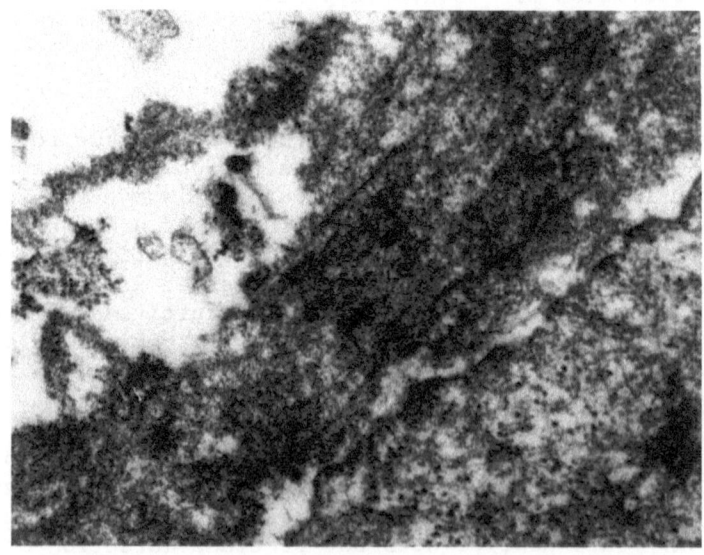

Fig. 7.   Extracellular filaments.   Magnification 30,000.

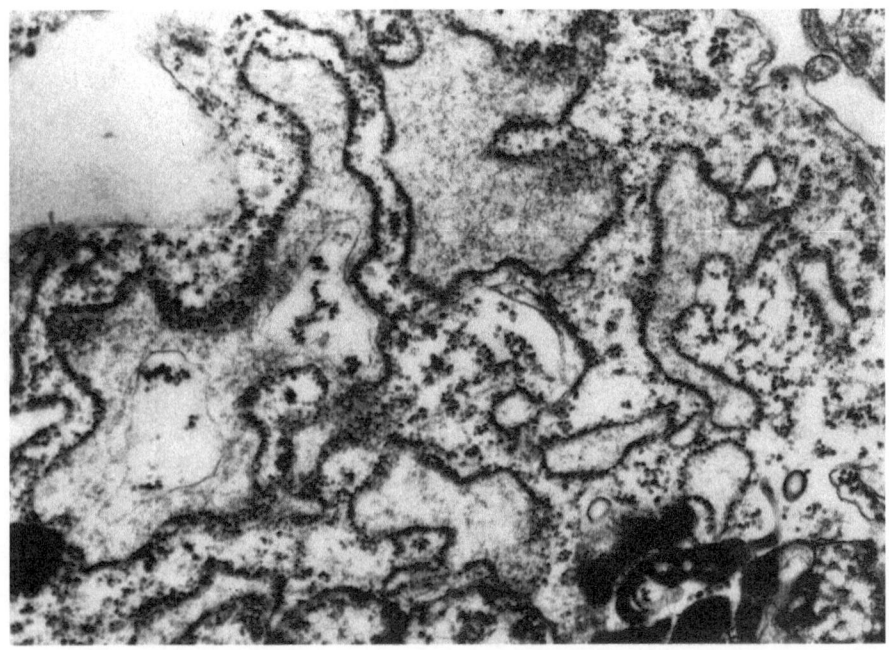

Fig. 8.   Rough endoplasmic reticulum.   Magnification 30,000.

decrease of collagen synthesis with the age of donor if more sensitive methods are used and when tissue cultures of older individuals
are studied.

Nevertheless, we consider that our observation demonstrates
the possibility to investigate the errors of collagen metabolism
in tissue cultures initiated in embryonal or foetal phase of development.   Therefore, we attempted to find out whether the amniotic
fluid cells cultivated in long term culture and characterized by
fibroblastoid morphology produce collagen.   We have been able to
prove it (5) and demonstrate that its rate and proportion of soluble
and insoluble collagen does not differ from other human diploid
cell strains.   The separation and partial characterization of radioactive proteins secreted by amniotic fluid cells into the culture
medium and the analysis of collagen, associated with cell layer
confirmed the synthesis of collagen.   The findings of thin intracellular fibrills and their extracellular growth corresponds to the
biochemical analysis.

These results bring evidence that there is a real possibility
for prenatal study of collagen metabolism and inborn errors of

connective tissue on the cellular level by means of the present
methods of amniocentesis and cultivation of amniotic fluid cells.
Further study of collagen in vitro might provide new information
concerning the problems of ageing in vitro and in vivo.

SUMMARY

        In the course of the active phase of growth collagen synthesis
appears relatively stable and decreases when the culture passes
into the degenerative phase III. The increase of cell nitrogen
indicates the increasing accumulation of cell proteins with
their aging in vitro. The ratio of soluble to insoluble collagen
is relatively constant in fibroblast cultures from different
tissues, periods of development during the active phase of
growth. The increase of insoluble collagen in long term cultivated
Marfan syndrome fibroblasts indicates the possible relationship
between increasing stability of collagen and ageing of fibroblasts
in vitro. Further studies confirmed the significant differences
between control fibroblast and fibroblast derived from tissues
of patients affected by Marfan syndrome. No differences were found
in collagen synthesis and in the ratio of soluble to insoluble collagen
in fibroblast strains from embryos, foetuses, children and young
adults. The collagen synthesis and the ratio of soluble to insoluble collagen does not differ in fibroblast cultures derived from
amniotic fluid cells and other human diploid strains from the prae-
or post-natal period of life. The collagen synthesis in amniotic
fluid cells was confirmed by analysis of cell proteins and proteins
produced into the cultivation medium. The ultrastructural analysis
of these cells reveals formation of intracellular fibrills and
their growth in the extracellular space. The possibility of prenatal
investigation of inborn errors of collagen metabolism is pointed
out.

ACKNOWLEDGEMENTS

        The excellent technical assistance of H. Zemanová, H. Duchková
and J. Mašková is highly appreciated. The statistical analysis
was performed by L. Pellar.

REFERENCES

1. Hayflick, L. and Moorhead, P.G. 1961. The serial cultivation
        of human diploid cell strains. Exp. Cell Res. 25: 585.

2. Macek, M. and Michl, J. 1964. Contribution to the cultivation
        of human diploid cells. Acta Univ. Carol. Med. (Prague) 10:
        518.

3. Michl, J. 1961. Metabolism of cells in tissue culture <u>in</u>
    <u>vitro</u>. Exp. Cell Res. <u>23</u>: 324.

4. Macek, M., Hurych, J. and Smetana, K. 1973. The collagen
    protein synthesis in long-term cultures of human foetal
    peripheral blood. <u>In</u>: Biology of Fibroblast. E. Kulonen
    and J. Pikkarainen, eds., Academic Press, London and New York,
    p. 127.

5. Macek, M., Hurych, J. and Řezáčová, D. 1973. Collagen syn-
    thesis in long-term amniotic fluid cell cultures. Nature
    (Lond) <u>243</u>: 289.

6. Macek, M., Hurych, J., Chvapil, M. and Kadlecová, V. 1966.
    Study on fibroblasts in Marfan's syndrome. Humangenetik
    <u>3</u>: 87.

7. Macek, M., Hurych, J. and Chvapil, M. 1967. Total hydroxy-
    proline in post-mortem tissue cultures of human fibroblasts
    strains. Cytologia (Tokyo) <u>32</u>: 308.

8. Macek, M., Hurych, J. and Chvapil, M. 1967. The collagen
    protein formation in tissue cultures of human diploid strains.
    Cytologia <u>32</u>: 426.

9. Trelstad, R.L., Kang, A.H., Cohen, A.M. and Hay, E.D. 1973.
    Collagen synthesis <u>in vitro</u> by embryonic spinal cord epithe-
    lium. Science (Wash.) <u>179</u>: 295.

10. Sabatini, D.D., Bensch, K. and Barrnett, R.J. 1963. Cytochem-
    istry and electron microscopy. The preservation of cellular
    ultrastructure and enzymatic activity by aldehyde fixation.
    J. Cell Biol. <u>17</u>: 19.

11. Palade, J.E. 1952. A study of fixation for electron micro-
    scopy. J. Exp. Med. <u>95</u>: 285.

12. Smetana, K. 1970. Electron microscopy of lymphocytes.
    Methods Canc. Res. <u>5</u>: 455.

13. Mollenhauer, H.H. 1964. Plastic embedding mixtures for use
    in electron microscopy. Stain Technol. <u>39</u>: 111.

14. Venable, J.H. and Coggeshall, R. 1965. A simplified lead
    citrate stain for use in electronmicroscopy. J. Cell Biol.
    <u>25</u>: 407.

15. Lewis, C.M. and Holliday, R. 1970. Mistranslation and
    ageing in neurospora. Nature (Lond), <u>228</u>: 877.

16. Orgel, L.E. 1963. The maintenance of the accuracy of protein
     synthesis and its relevance to aging. Proc. Nat. Acad. Sci.
     USA 49: 517.

17. Priest, R.E., Moinuddin, J.F. and Priest, J.H. 1973. Collagen
     of Marfan syndrome is abnormally soluble. Nature (Lond) 245:
     264.

18. Verzar, F. 1964. Aging of the collagen fibre. Intern. Rev.
     Conn. Tissue Res. 2: 243.

19. Hurych, J., Macek, M., Smetana, K., Beniač, F. and Řezáčová,
     D. In preparation.

QUESTION TO DR. HURYCH

Dr. Martin: Do you have any information as to the types of collagen
chains synthesized by these various fibroblast lines?

Dr. Hurych: At this time we have no information of the type of
collagen chains in various fibroblast lines. We began with the
characterization of collagen type in amniotic fluid cell cultures
and these experiments are in progress.

AGEING PROCESSES IN COLLAGENS FROM DIFFERENT TISSUES OF RATS

M. Juřicová and Z. Deyl

Czech. Acad. Sci.

Prague, Czechoslovakia

At the beginning of my talk I would like to go back to 1973 and to recall the closing speech of Dr. Gardell at the Varberg Symposium on connective tissue ageing. At this opportunity, Dr. Gardell stressed the fact that among people who are involved in connective tissue ageing most of them are limited to the main protein component e.g., collagen and most of these people do not recognize the existence of cells in connective tissue. Although I am not going to speak very much about cells, I still recognize their existence and you perhaps would excuse my talking about extracellular proteins as these seem to play a very specific role in ageing processes. The nature of the connective tissue research has rapidly changed since the elucidation of the primary structure of the α_1 polypeptide chain of collagen and since the elucidation of the nature of the major cross-links which for long have been believed to be to the main element in ageing of the extracellular protein components. I would like to stress that our picture of ageing connective tissue is that which has been formulated in 1972 by F.M. Sinex (1). In this view ageing of the connective tissue starts with the fact that the fibroblasts gradually lose their capability for metabolic activity and mitosis with age. Given the stimulation however, fibroblasts will proliferate in older animals, although not always in the most orderly or constructive way. As older tissue becomes more dormant the cross-linkage is completed and glycosaminoglycans perhaps will decrease or will themselves become aggregated. The fibrous protein which remains may either deteriorate or become tougher depending on its environment. It is believed, though it is certainly not proved, that these changes may play an important role in transport processes and in the ability of cells to survive. This view is based on the fact that these changes in intercellular matrix may cause substantial changes in the microenvironment in which every cell has to live and

to carry out the basic processes which allow this cell to survive.
In addition to a general statement about an increase in cross-links
with age, proposed some decades ago by Verzár (2) and followed by
a number of investigators in the past years, we know nowadays that
the formation of cross-links in connective tissue proteins is a
very specific enzyme controlled process. We also know that the
so-called unusual cross-links which are of other nature than those
lysine derived are perhaps very rare if not absent at all.

        In the view of connective tissue ageing drawn by Sinex (1) to
which we have contributed in our previous paper (3) there are two
important things which should be distinguished. The first is the
level of cross-linking of connective tissue proteins in a particular
organ, and the other is the overall proportion of connective tissue
compared to the parenchyma. In a previous paper we came to the conclusion that it is very likely that the overall proportion of connective tissue in organs is much more important from the point of age
than the level of cross-linking. Since the knowledge about connective
tissue proteins has developed very much lately and since it seems that
collagen represents rather a category of compounds than a single
protein, it seemed quite plausible to investigate to what extent
individual collagen preparation, obtained from different tissues
of a single animal, were different. In other words whether there are
collagens within a single living body which, judged according to
their level of cross-linking, are of different biological or chemical
age.

                        MATERIALS AND METHODS

        Rats, 4 and 26 months old, were subjected to investigation. All
our animals were kept on pelleted Larsen diets. Collagen concentration in different organs was determined by Stegemann's procedure
(4). Collagens from corresponding organs, i.e. from liver, lung,
kidney, heart and skin, were prepared by a procedure that has been
published before (3).

                    Neutral Salt-Soluble-Collagen

        This was prepared from pooled samples by stirring for 24 hr
with about ten volumes of a 0.14 M NaCl-phosphate buffer (pH 7.4),
containing 8.24 g NaCl, 1.96 g $Na_2HPO_4 \cdot 12 H_2O$ and 0.16 g KH_2PO_4
per liter. Afterwards extraction was performed with a 0.45 M NaCl
phosphate buffer (pH 7.4, ionic strength 0.5).

                        Acid-Soluble-Collagen

        This was prepared from the residue after salt extraction which

was first washed by dialysis against tap water and then extracted by
stirring with about ten volumes of citrate buffer, pH 3.7 (10.9 g
citric acid monohydrate, 100 ml 1.0 M NaOH and 50 ml HCl per liter).

All extraction procedures were repeated four times.

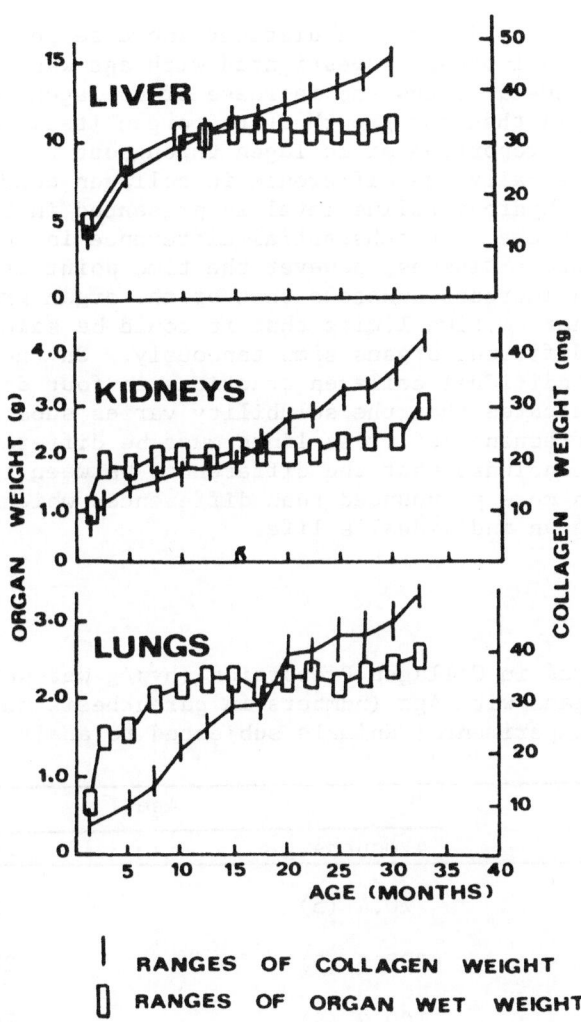

Fig. 1.   Comparison of collagen concentration with organ wet
weight.

Urea-Soluble-Collagen

After washing of the residue by dialysis against tap water, extraction was carried out with 6 M urea. The residue after urea extraction was washed with tap water. Operations were conducted below 5°C with the exception of urea extraction which was done at room temperature.

RESULTS AND DISCUSSION

As indicated in Figure 1, a distinct increase in the collagen concentration in all organs investigated with age was observed. At the age between 10-15 months the increase of collagen concentration becomes more rapid than the growth of the organ itself which means that the general proportion of collagen throughout the organ is also increased. Numerically the difference in collagen concentration as reflected by the hydroxyproline level is presented in Table I. It is obvious that there is a substantial difference in collagen concentration in different tissues, however the time point at which the rate of collagen increase enhances that of the organ growth is observed in so narrow time limits that it could be said that it happens in all different organs simultaneously. On the other hand, the balance of individual collagen fractions in four different collagen preparations indicates that the solubility varies substantially and hence also the frequency of cross-links must be different (Table II). It can be also concluded that the differences between individual tissues are much more pronounced than differences which accumulate step wise during an individual's life.

TABLE I. Changes in Collagen Proportion (hyp/g wet weight) In Different Organs with Age (numbers in parenthesis indicate number of experimental animals subjected to analysis)

Organ	Age	
	4 months	26 months
Liver	70.4 (5)	85.7 (2)
Lung	202.6 (5)	270.- (3)
Kidney	100.2 (5)	143.- (3)
Heart	77.3 (5)	170.- (3)

TABLE II. Solubility of Collagen in Different Tissues in Relation to Age

	No. of animals	Percentage of Collagen				
		Neutral salt soluble collagen in 0.15M NaCl	Neutral salt soluble collagen in 0.45M NaCl	Acid soluble collagen	Insoluble collagen — 6M urea extractable collagen	Insoluble collagen — Residue
Skin 4 months	5	0.36	4.50	2.95	34.7	53.2
26 months	5	traces	2.00	2.52	21.6	68.3
Kidney 4 months	5	traces	traces	0.8	15.2	84.0
26 months	5	traces	traces	0.9	16.3	82.8
Heart 4 months	5	traces	traces	0.5	18.2	81.3
26 months	5	traces	traces	0.5	18.4	81.1
Liver 4 months	5	0.20	4.30	3.05	34.6	57.8
26 months	5	0.10	1.80	2.50	26.8	31.2

It is therefore conceivable to assume that the final degree of collagen cross-linking is not very important from the view point of tissue ageing, as has been proposed by Deyl et al. (3). The present results support the original working hypothesis that the important role of connective tissue in ageing is based on its proportion to parenchyma rather than on the frequency of cross-linking interactions. The frequencies of cross-links within a particular preparation of collagen obtained from a particular tissue reflect both the tissue specificity of collagen and the intensity of processes which govern cross-link formation in the tissue under investigation; the activity of lysyl oxidase could be taken as a good example of this until there is some additional evidence available it is suggested not to use the degree of cross-linking of collagen as a measure of the biological age of the animal or of a particular tissue.

We should also refer to the previous data of Adam et al. (5) who were capable of introducing a considerable increase in the number of cross-links in collagen by treating the experimental animals with sodium gold thiosulphate without a distinct effect upon their survival.

SUMMARY

It has been demonstrated by solubility data, that the differences in cross-linking and degree of polymerisation in connective tissue from different organs of a single animal are much more pronounced than the increase which occurs with advancing age. Judged from the cross-linking analysis one can therefore conclude that in a single animal there are connective tissues which differ in their degree of maturity. It is therefore conceivable to suggest that the level of cross-linking should not be taken as the measure of the biological age of the animal.

REFERENCES

1. Sinex, F.M. Ninth International Congr. Geront., July 1972, Kiev, U.S.S.R.

2. Verzár, F. 1964. Aging of the collagen fiber. In: Inter. Rev. of Connective Tissue Res. D. Hall, ed., Academic Press, N.Y.

3. Deyl, Z., Juřicová, M., Rosmus, J. and Adam, M. 1971. The effect of food deprivation on collagen accumulation. Exp. Geront. 6: 383.

4. Stegemann, H. 1958. Mikrobestimmung von Hydroxyprolin mit
    Chloramin-T und p-Dimethylaminobenzenaldehyd. Hoppe-Seylers
    Zeitshrift für Physiologische Chemie 311: 41.

5. Adam, M., Deyl, Z. and Rosmus, J. 1966. The intravital
    influence of cross-linkages formation of collagen. Med.
    Pharmacol. 14: 129.

QUESTION TO DR. JUŘICOVÁ

Dr. Leutert: Is there an increase of collagen in heart muscle?

Dr. Juřicová: The collagen content in both heart and striated
muscle changes quite dramatically during life. In contrast to
other organs, in muscle the collagen content is high in foetuses
and in young animals. It drops down as the animal ages but in
old age it starts to increase again. Also the way in which this
collagen is cross-linked is very specific as can be revealed from
the absence of the beta fraction in polyacrylamide gel electrophoresis.

THE EFFECT OF NUTRITIONAL REGIMES UPON COLLAGEN CONCENTRATION AND

SURVIVAL OF RATS

Z. Deyl, M. Juřicová and E. Stuchlíková

Czech. Acad. Sci. and Fourth Med. Dept. Charles Univ.

Prague, Czechoslovakia

It is generally believed that food restriction of experimental animals in the adult age does not alter the expected lifespan of the animals. Already the early experiments of McCay and coworkers (1) showed that no further increase in the lifespan could be brought about by restricting the food intake in adult animals, namely in rats beyond one year of age. Some other data on the same topics were published by Barrows and Roeder (2) who reported that rats that were subjected to limited food intake in the adult age did not survive longer than their ad libitum fed counterparts. Similar effects were observed also in insects. So for instance Kopeć (3) was unable to prove the increase in the lifespan of the imago of Drosophila if the imagos were subjected to reduced dietary intake. In connection with this latter experiment Northrop (4) proved that the overall increase in lifespan of Drosophila can be accounted for by the changes which occur during the larval stage. In a series of more recent experiments Barrows (5) studied the overall lifespan of some rotifers using different types of restricted diet. This author observed that in some cases animals which were food restricted throughout their whole lifespan lived shorter than those which were kept on a food restricted regime after cessation of egg production. These latter animals increased their lifespan by approximately one quarter. Barrows (5) concludes that under certain conditions the lifespan of animals can be increased even when dietary restriction is imposed later in their life. In order to explain this latter statement, Barrows (5) proposes that our present knowledge due to the paucity of data regarding the effect of the degree of dietary restriction and the age at which the dietary restriction is imposed to alter the lifespan can hardly offer a univocal statement about life prolongation. Another problem which is connected with the increase of the lifespan of animals deals with the mechanism and proposals that

359

could be offered to explain the increased lifespan of food restricted
animals. It is possible that the programmed genetic events which
occur throughout the life of an organism are influenced by sequential synthesis of specific regulatory RNA and proteins and that the
regulation of the genetic information transfer therefore may possibly
be controlled by the synthesis of specific RNA and proteins at various
times throughout the lifespan of an organism (5). From this presump-
tion he proposes that a possible way to delay senescence especially
if it results from the genetic expression would be to decrease the
synthesis of specific RNA or protein during the life of an organism.
From this point of view one could expect that restricting of the
dietary intake would delay the occurence of these specific events
during the early life and consequently result in life prolongation.
It should be said, of course, that this is a conceivable speculation
but nothing more than speculation since there are no direct experimental data that would prove this statement.

Among the different specific events which occur during the
individual lifespan and can be considered as markers of aging is the
rate of accumulation of fibrous proteins of the intercellular matrix.
In a previous paper we have been able to show (6) that there is a
distinct increase in the collagen concentration in lung, liver and
kidney in rats around 15 months of age. In organs investigated in
that paper no distinct organ specificity was observed. In undernourished animals by which we mean fifty per cent food-restricted
individuals the same phenomenon was delayed until the age of 25
months. The starting point of collagen accumulation is followed by
an evident increase in the mortality rate. The shift of the starting
point of collagen accumulation between control animals and food
restricted animals is exactly the same as shift of the average lifespan of undernourished and fully fed animals.

MATERIALS AND METHODS

Male rats of the Wistar strain, golden hamsters and mice were
used throughout this investigation. All animals were kept in separate
cages populated with a single animal only. Undernourished animals
were given exactly one half of the amount of food consumed by controls. The experiment was divided into four parts. Control animals
were fed ad libitum throughout their whole life (Group No. 1). In
the group No. 2 animals were food restricted throughout the whole
life. In the group No. 3 animals were kept on full diet when young
and food restricted when adult and in group No. 4 the regime was
reversed. Animals were food restricted when young and fully fed
when adult and old. Each experiment started with 100 animals except
for rats which were 25 at the beginning. Survivals and in the case
of rats weights were registered throughout the whole life and
results are summarized in the Results.

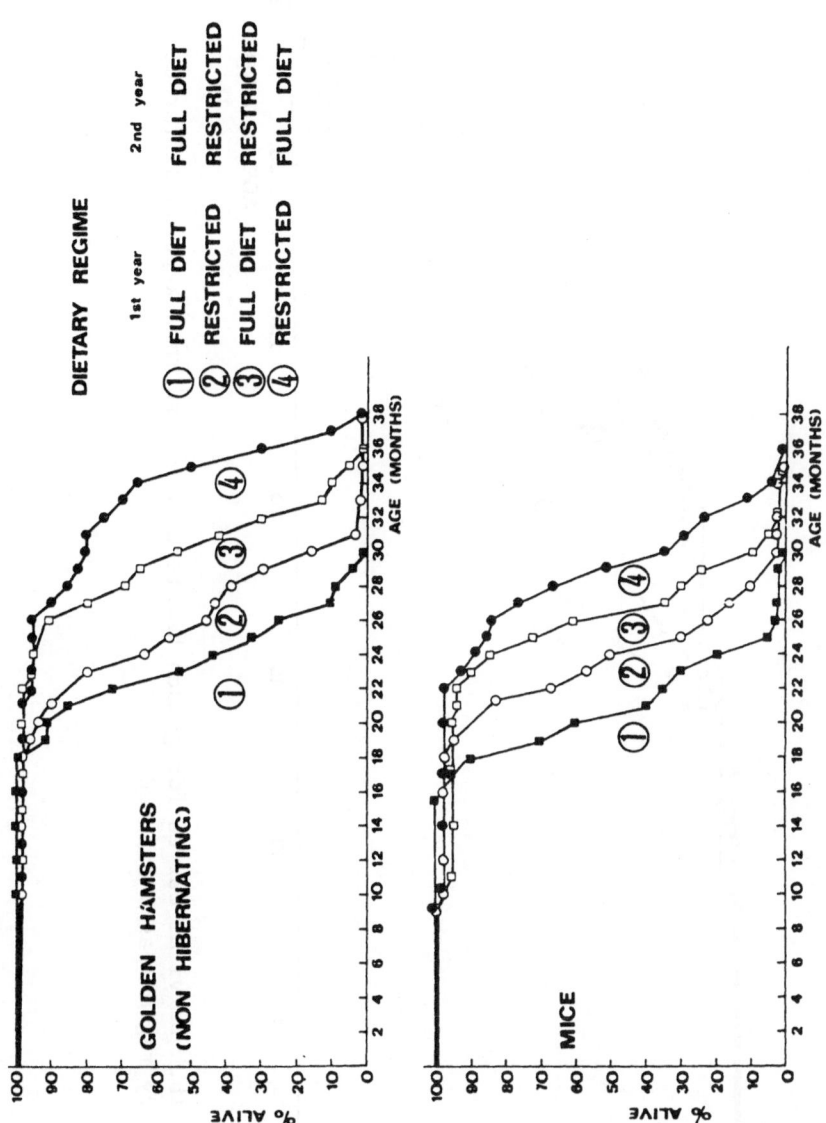

Fig. 1a. Survival curves for golden hamsters, mice and rats kept on different dietary regimes as indicated.

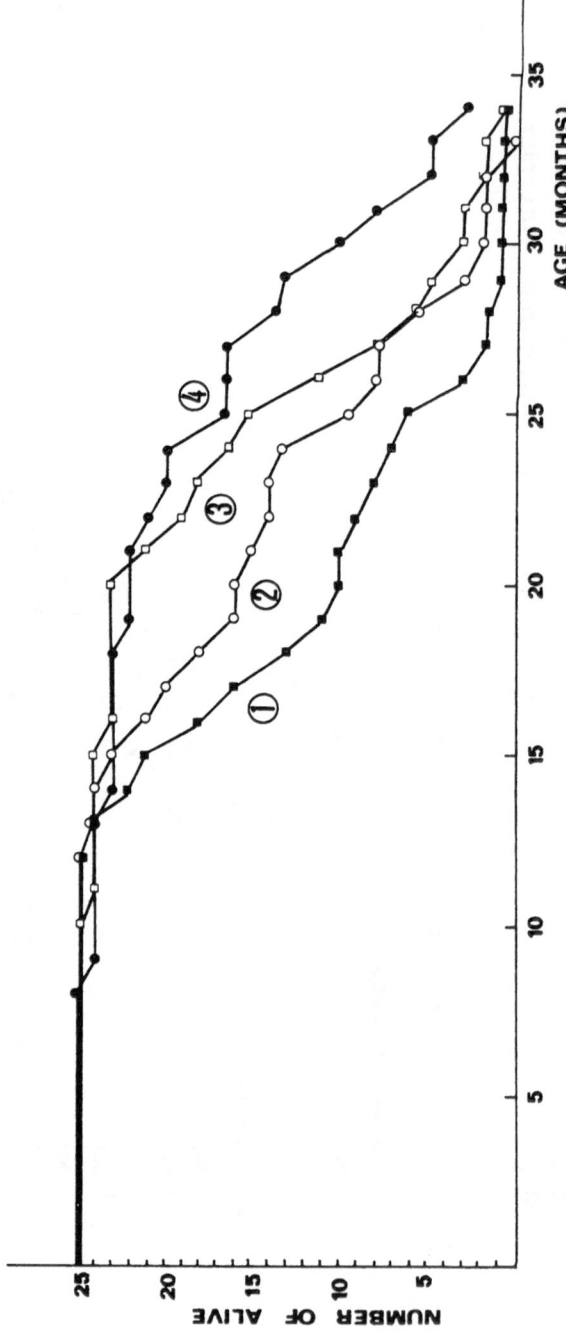

Fig. 1b. Survival curves for golden hamsters, mice and rats kept on different dietary regimes as indicated.

Another series of rats divided into groups (100 animals each) one fed ad libitum and the other food restricted during the first year and fully fed thenafter was used for the estimation of collagen concentration in kidneys and liver. Animals were killed at definite intervals (always three randomly selected individuals) and the collagen level in the respective organs and their weights were estimated. For collagen determination the procedure published by Stegeman (7) was used.

RESULTS AND DISCUSSION

As it is evident from Figure 1, food restricted animals had always a better survival than animals kept on a full diet. In agreement with previously published data no substantial increase in the maximum survival was observed in any of the three studied species. However, this is due to the fact that some few of the ad-libitum fed animals survived much longer than the vast majority of the population.

Fig. 2. A typical portrait of a rat food restricted during the first year of life and fully fed thereafter. The individual was 28 months old when the picture was taken.

TABLE I. 50% Survivals in Golden Hamsters, Mice and Rats Kept on Different Dietary Regimes. Numbers in Parenthesis Indicate the Increase in the Age of 50% Survivals in Months.

Regime 1st year 2nd year	Age of 50% survival (months)				Relative increase to fully fed (%)		
	Full diet Full diet	Restricted Restricted	Full diet Restricted	Restricted Full diet	Restricted Restricted	Full diet Restricted	Restricted Full diet
Golden hamsters	23.3	25.5 (2.2)	30.3 (7.0)	35 (11.7)	9.44	30.1	50.1
Mice	20.5	24 (3.5)	26.4 (5.9)	28.4 (7.9)	26.8	28.8	38.5
Rats	18.3	24.2 (5.9)	26.1 (7.8)	29.4 (11.1)	32.2	42.6	60.6

The average lifespan and 50% survival in undernourished animals was
increased quite distinctly.   Though the data of the relative increase
of the 50% survival are somewhat different when comparing rats, mice
and hamsters as indicated in Table I, the general order of individual
dietary regimes in all species investigated is the same.   By far,
the best results with respect to survival were observed in animals
which were food restricted during the first year of life.   In
all cases the shortest survival was in animals which were fully fed
throughout the whole life.   The increase in the average lifespan of
those animals which were food-restricted when adult and old is about
thirty per cent.   Those animals which were food restricted when young
and fully fed when old, developed a certain degree of hyperphagia when
they were allowed unlimited amounts of food.   These individuals grew
up very fat as indicated in the attached figure (Fig. 2).   Their
average weight after one month of realimentation was over 700 g and
in this experimental group there were animals (rats) which approached
the total body weight of one kilogram (Fig. 3).   This extreme body
weight caused of course a considerable restriction in their ability
to move, but in spite of this, these animals did very well from the
viewpoint of survival.   The increase in 50% survivals was always
over 35%.   In a preliminary experiment which followed these long
term studies just reported, we have used two different types of
reducing agents in order to elucidate the question whether the food
restriction is somewhat related to the oxidative free radical attack
on long chain macromolecules.   Both half per cent (0.5% w/w) addition
of ethoxyquin to the standard diet and reduced diet and similar
addition of tert-butyl hydroxytoluene resulted in a considerable
increase not only of the average lifespan but also of the maximum
lifespan.   In control animals which were offered free access to the
antioxidant enriched diet the increase in maximum lifespan was almost
two hundred days and was somewhat lower in 50% survival.   In animals
which were subjected to some of the food restricted regimes as indicated in Figure 4 the effect of reducing agents was quite similar.
To our opinion several conclusions can be drawn out from the above
reported experiments.   It can be concluded that from the environmental
factors nutrition is quite important and it is clearly demonstrated
that the onset of food restriction is always very efficient.   It is
also demonstrated that the effect of undernutrition is not identical
with that of reducing agents since combination of both induced an
additive increase in lifespan.

When comparing the collagen concentrations in kidneys and liver
of fully fed animals and those which were fully fed during the first
year and food restricted thereafter the onset of the increase in
collagen concentration has been postponed accordingly so that it
appeared near the age of 30 months while in controls it could be
observed between 10 and 15 months of age.   In accordance with our
previous experiments (6) this onset appears roughly ten months
before 90% of the population dies out.

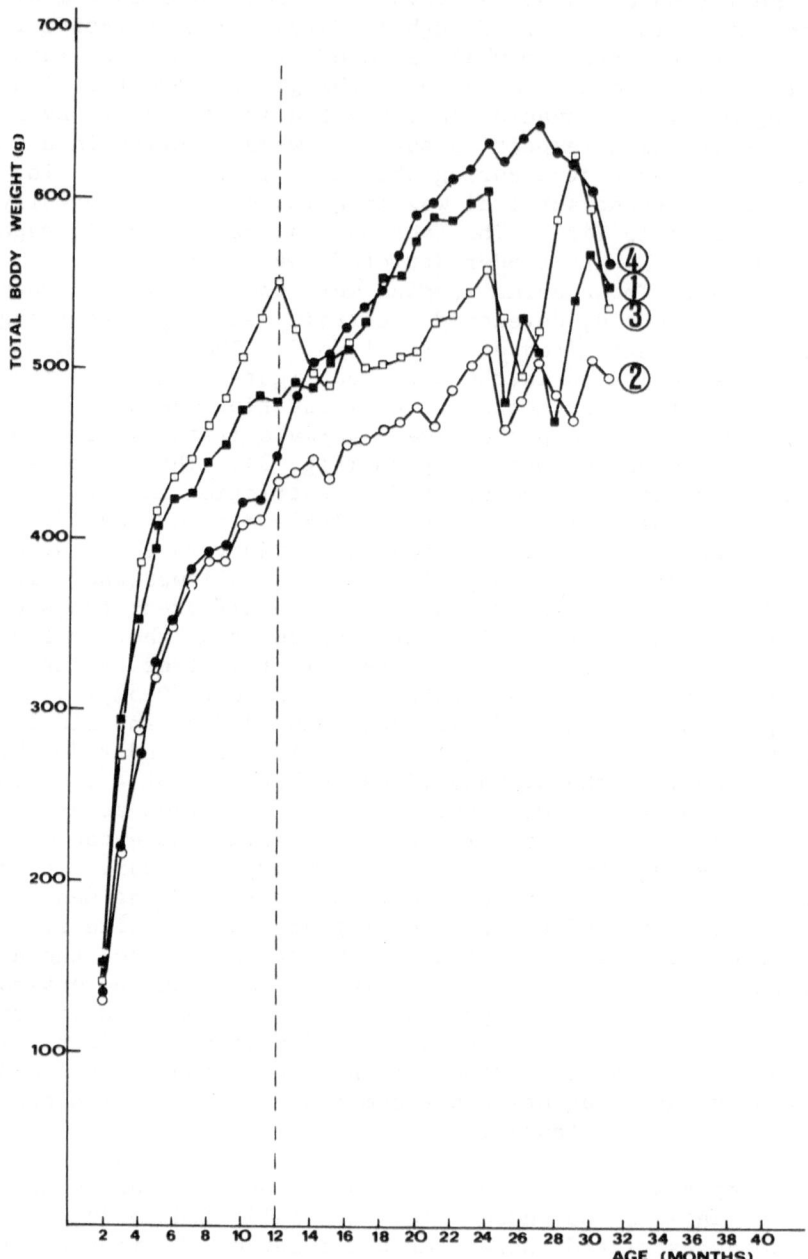

Fig. 3.   Average body weights of rats kept on different dietary
regimes.

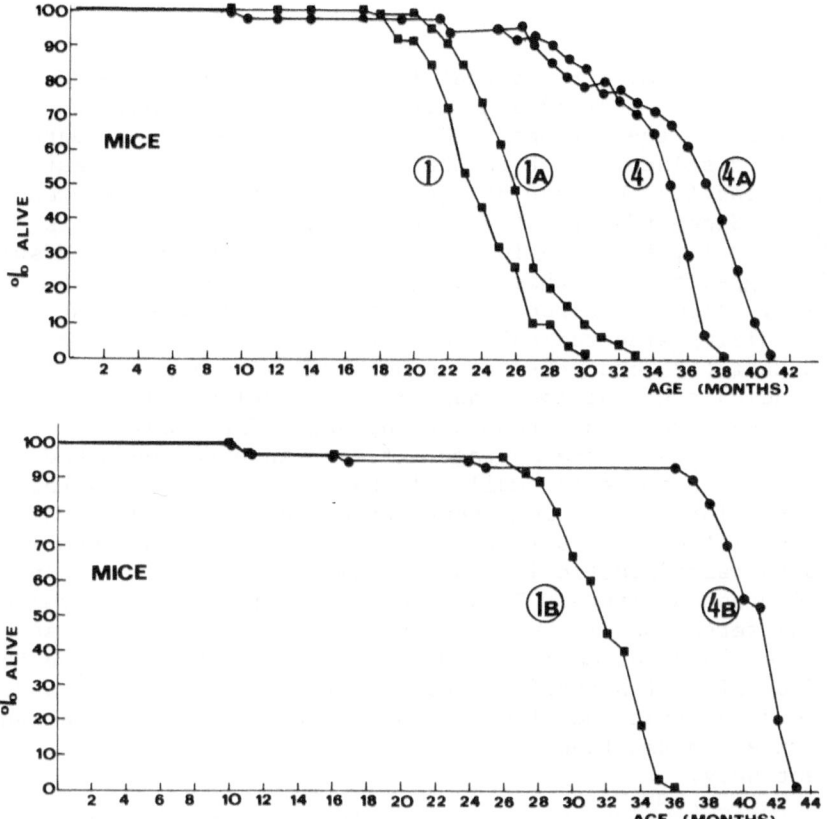

Fig. 4.   Survival curves of mice kept on full access to food (1) and restricted during the first year of life (4) in comparison with animals kept on the same regimes to which 0.5 w/w per cent ethoxyquin (indicated by A) and the same amount of tert–butyl toluene (indicated by B) was added.

The data accumulated throughout this work are indicative of the fact that a slowdown in metabolic processes, deprivation of redundant energy and possibly more effective use of energy supplies including more efficiently used transcriptions at the DNA–RNA level are the most easily accessible ways of life prolongation in experimental animals. However, in order to be fair with the published data, no data have been accumulated which would allow a rational explanation of the observed phenomena.

SUMMARY

It has been demonstrated that food restriction put upon animals at any stage of the individual's life, if chronic, produces a distinct increase in the lifespan. This can be effected in youth and still have a distinct effect in old age even if the food restriction is cut down and for the rest of the life the animal is allowed food ad libitum. Since the effect upon the aging of the animal is delayed on the time scale, it is obviously an effect which is stored somewhere and it is suggested that this storage occurred somewhere along the DNA-RNA pathway. Also the effect of undernutrition is not identical with that of oxidizing free radical blocking agents and therefore it is concluded that the food deprivation does not minimize the attack of the free radicals on the long chain macromolecules and as a matter of fact it seems that the proportion by which free radicals contribute to the changes in the average lifespan in undernourished and fully nourished animals is small. It has been also demonstrated that the addition of reducing agents to normal diet and to the diet of food restricted animals increases the average and maximum lifespan in both cases practically to the same extent, which supports the idea expressed before. This feeding effect has been observed in three different species of rodents and no extrapolation has been done to other types of mammals. Due to the data published on this topic and dealing with rotifers and some insects (2) it is conceivable to conclude that the effect of undernutrition is general and is not limited to the food restriction in the early stages of development only.

Collagen starts to accumulate in the kidneys and liver of experimental animals roughly ten months before 90% of the population dies out. Thus an increase in collagen concentration can be indicative of involutional changes in the organ (and perhaps organism). These data are i- good agreement with previously published results on the relation between collagen accumulation and chronic food deprivation in rats (6). It can be also concluded that food deprivation induced in the adult, though not so effective in life prolongation as the food restriction during early development, still can increase survival in experimental animals to a high degree.

REFERENCES

1. McCay, C.M., Crowell, M.F. and Maynard, L.A. 1935. Effect of retarded growth upon length of life-span and upon ultimate body size. J. Nutr. 10: 63.

2. Barrows, C.H. Jr. and Roeder, L. 1965. The effect of reduced dietary intake on enzymatic activities and lifespan of rats. J. Gerontol. 20: 69.

3. Kopeč, S. 1928. Influence of intermittent starvation on
    longevity of imaginal stage of Drosophila melanogaster. Brit.
    J. Exp. Biol. 5: 204.

4. Northrop, J. 1917. The effect of prolongation of the period of
    growth on the total duration of life. J. Biol. Chem. 32: 123.

5. Barrows, C.H. Jr. 1972. Nutrition, aging and genetic program.
    Amer. J. Clin. Nutr. 25: 829.

6. Deyl, Z., Juricova, M., Rosmus, J. and Adam, M. 1971. The
    effect of food deprivation on collagen accumulation. Exp.
    Geront. 6: 383.

7. Stegeman, H. 1958. Mikrobestimmung von Hydroxyprolin mit
    Chloramin-T und p-Dimethylaminobenzenaldehyd. Hoppe-Seylers
    Zeitschrift für Physiologische Chemie 311: 41.

QUESTIONS TO DR. DEYL

Dr. Martin: If I read the slides correctly, there was an increase
in mean survival for the rat with your various regimes. There
did not appear to be an increase in the maximum lifespan in
contrast to the findings with mice and hamsters.

Dr. Deyl: I would say that the only effect is the increase in
the average lifespan. The difference in the slides I showed is
based on the fact that in a population of animals you always find
a single one which is long lived. When drawing survival curves
you always come to a stage when you are waiting for this last
animal to die and it keeps living. You can not circumvent these
effects and they spoil the appearance of survival curves a little
bit.

Dr. Cristofalo: What do the animals die from? Do they die from
the same reason? Are you familiar with the work of Ross in which
he could modify, by dietary manipulation, the incidence of diseases
with age? Does B.H.T. have an anorexic effect on your animals?

Dr. Deyl: I have no idea what the animals die from. As far as
B.H.T. is concerned our animals ate everything that they were
offered.

BONE AGING

U. J. Schmidt, Irmgard Kalbe, and F. Sielaff

First Medical Clinic of the Medical Division
(Charite Hospital) of Humboldt University, Berlin

From about the age of 60 onwards, the balance existing between osteoblastic and osteoclastic activities in younger life is
consdierably disturbed at the expense of the former. Consequently,
the balance between the formation and the resporption of the organic
substance of bone also ceases to exist, and so does the so-called
remodeling.

In old age, the number of incompletely absorbed osteones,
which remain as interstitial lamellae, increases. The extending
mosaic pattern of the bone resulting from this process is characteristic of elderly people. The advancing process of resorption leads
to an enlargement of the marrow spaces and of the whole marrow
cavity with concomitant thinning of the corticalis. The ocrres-
ponding Haversian canals are enlarged and contain a high proportion
of yellow bone marrow. The process as a whole is called osteo-
porosis in old age and is known to be more pronounced and to manifest itself earlier in women and all female mammals.

SOME COMMENTS ON "BONE COLLAGEN"

Previous morphological studies on age-specific changes in the
collagenous fibres of the interstitial substance have been unsatis-
factory. Biochemically, there is with advancing age only a slight
decrease in the total amount of soluble collagen fractions in the
bone as well as a reduced metabolism, whereas the half-life may
increase proportionally. The biological half-life of the insoluble
collagen as well as its percentage in the total bone-collagen content increases. The metabolism of this fraction decreases. The

371

shift in the relation occurring in favour of the insoluble fraction
together with a strong interlacing due to cross-linking corresponds to that of other connective tissues. There is in different
species a difference in the biological half-live of the bone
collagen. Even in very old age the bone collagen is not inert.
The metabolic rates are prolonged and exceed those of cartilage.
Despite the disturbed balance between osteoblastic and osteoclastic
activities, the overall decrease in the metabolism prevents a
noticable increase in the elimination of free and bonded hydroxyproline. To relate these elimination values to the bone-collagen
metabolism alone is incorrect.

The increase in the biological half-life of the insoluble
bone collagen to more than 400 days is partly due to increased
cross-linking and hardening. Thus, the insoluble collagen is
withdrawn from resorption, and consequently from metabolism. At
the end of the period of maturation the elimination of hydroxyproline, and in particular of free hydroxyproline, is at its highest.
In maturity it is reduced by half. This reduction in hydroxyproline elimination is related to a reduction of the soluble collagen fraction. It is not certain whether there is an age-related
decrease in the serum content of apolarly attacking collagen pep-
tidases.

THE GROUND SUBSTANCE

Proteoglycane synthesis in the bone is reduced, and there is
a trend towards a general reduction of protein synthesis in the
bone. The S-sulphate incroporation rates decrease rapidly.

The biological half-life of the glycosamine-glycane of the bone
(chondriotine-4-sulphate) as produced by the osteoblasts, has been
insufficiently studied with regard to its aging-relatedness as have
the hybridizations of chondroitinesulphate and kerata sulphate.

In the corticalis of bone, the total content of glycosamine-
glycane (which is mainly known through determinations of the total
content of hexosamine) amounts ot about 0.2-0.4% of the dry weight.

The hydroxyproline or collagen proportion of the dry weight
is about 15%. In the process of aging, the hexosamine and uronic
acid content and the hexosamine-hydroxyproline ratio decrease,
while the hydroxyproline content remains constant or increases
slightly. The metabolism of the proteoglycanes of the bone is
reduced. Changes occurring in the proteoglycane destruction of
the interstitial substance have not been analyzed sufficiently as
far as the biochemical parameters are concerned. Histochemical
glycoridasis determinations do not permit any quantitative

statements to be made.

THE BONE CELLS

In advanced age the metabolic rates of both osteoblasts and osteoclasts are reduced. However, the question whether an increased deposition in the presence of a normal resorption is typical, or whether resorption is increased in the presence of a normal deposition in old age, remains to be answered.

It is quite understandable if a decrease in the osteoblastic activity and an increase in the osteoclastic activity are recorded. The transformation of compact bone into spongy bone leads to the above-mentioned changes and, at the same time, to a reduction of the specific bone weight. As a results total content of interstitial substance changes and so does the calcium-salt bond and, hence, the mineral content of bone. In this connection it should be noted that mineralization depends on the content of interstitial substance, and in particular on the content of proteoglycane and the proteoglycane pattern. Even in advanced age there is a certain proliferation and differentiation of bone cells together with an increased synthesis of proteoglycans and collagen. It is, however, uncertain whether or not aging differences are inherent in this process.

REFERENCES

1. Chatterji, S., Wall, J.C., and Jeffery, J.W. 1972. Changes in the degree of orientation of bone materials with age in the human femur. Experientia 28: 156.

2. Dequeker, J., Remans, J., Franssen, R., and Waes, J. 1971. Age patterns of trabecular and cortical bone and their relationship. Calc. Tiss. Res. 7: 23.

3. Fazekas, I. Gy., and Szendrenyi, J. 1973. Corticosteroid-fraktionen in der menschlichen Knochenhaut mit besonderer Rücksicht des Lebensalters und saisonaler Veranderungen Endorinologie 61: 276.

4. Franke, H. 1973. Aktuelle probleme der gerontologie bzw. Geriatrie Klin. Wschr. 51: 151.

5. Gebhardt, M., and Münzenberg, K.J. 1970. Kristallinitatsgrad und mineralarten des knochens in Abhängigkeit von alter und lokalisation. Z. Orthop. u. Grenzgeb. 108: 104.

6. Gilbert, B.M., and McKern, Th.W. 1973. A method for aging
    the female os pubis. Amer. J. Phys. Anthrop. 38: 31.

7. Helela, T. 1970. Combined cortical thickness and cortical
    index in various age groups. Ann. Clin. Res. 2: 240.

8. Hernborg, J., and Nilsson, B.E. 1973. Age, and sex incidence
    of osteophytes in the knee joint. Acta Orthop. Scand. 44:
    66.

9. Hernborg, J., and Nilsson, B.E. 1973. The relationship
    between osteophytes in the knee joint, osteoarthitis and
    aging. Acta Orthop. Scand. 44: 69.

10. Kormao, M. 1971. Radiographic appearance of the pubic
    symphysis in old age and rheumatoid arthritis. Acta Rheum.
    Scand. 17: 286.

11. Krieg, H. 1972. Quantitav erfassbare Veranderungen am Knochen
    und Knochenmark in verschiedenen Altersklassen. Actuelle
    Geront. 2: 531.

12. Papworth, D.G., and Vennart, J. 1973. Retention of 90Sr in
    human bone at different ages and the resulting radiation
    doses. Phys. Med. Biol. 18: 169.

13. Patnaik, B.K., and Padhi, S.C. 1972. Age related changes in
    the solubility of collagen in the bone and skin of the garden
    lizard, Calotes versicolor. Exper. Geront. 7: 143.

14. Puccinelli, A.D., Laurant, G., and Crehalet, Y.L. 1973. De-
    termination simplifiée de l'âge osseuc á partir de la sur-
    face du grand os. J. Radiol. Electrol. 54: 57.

15. Schmidt, Y.J., and Kalbe, I. 1973. Aging changes of calcium
    and mucopolysaccharides of the bones and arterial vessel
    wall. Varberg, IVth EBS-Symposium of IAG.

16. Stather, J.W. 1974. Distribution of 32P, 45Ca, 85Sr and 133Ba
    as a function of age in the mouse skeletion. Health Physics
    26: 74.

17. Tonna, E.A., and Weiss, R. 1972. The cell proliferation
    activity of parodontal tissues in aging mice. Arch. Oral.
    Biol.17: 969.

18. Vejlens, L. 1971. Glycosaminoglycans of human bone tissue.
    I. Pattern of compact bone in relation to age. Calc. Tiss.
    Res. 7: 175.

RELATIONS BETWEEN DEVELOPMENT OF THE CAPILLARY WALL AND MYO-

ARCHITECTURE OF THE RAT HEART

B. Oščadal, T.H. Schiebler and Z. Rychter

Institute of Physiology Czechoslovak Academy of Sciences,
Institute of Anatomy, University of Wörzburg (FRG) and
Institue of Histology, Medical Faculty, Charles University,
Prague, Czechoslovakia

The oxgen supply of the heart depends on the proportion between
the mount of oxygen consumed by the cardiac cell and that offered
to the cardiac cell. Since, under normal conditions, the cornary
arteriovenous difference is considerably high, the heart is apparently
not able to increase its oxygen supply to any appreciable extent
by raising the oxygen extraction from the blood (1). Thus, the
dominant factor affecting the supply of oxygen to cardiac sturctures
is governed by the type and overall capacity of the cardiac terminal
vascular bed.

In this report we tried to present a short survey of the results
of some of our experiments dealing with the evolution of the terminal
vascular bed and its relationship to the development of cardiac
myoarchitecture.

COMPARATIVE ASPECTS OF THE MYOCARDIAL STRUCTURE/BLOOD
SUPPLY RELATIONSHIPS

Changes in the inner structure of the myocardium during
phylogenetic development are accompanied by corresponding changes
in its blood supply. Whereas the heart of adult homoiotherms consists entirely of compact musculature with coronary blood supply,
in poikilotherm animals the cardiac musculature consists either
entirely of the spongious type or its spongious musculature is
covered by an outer compact layer (e.g., 2-5). The spongy-like
musculature is supplied predominantly by diffusion from the intertrabecular spaces. They belong to the chamber cavity and have a ·

375

continuous endothelial lining. Nevertheless, in some species of
fish (elasmobranchs: Torpedo, Scyllium) and reptiles (turtle,
Testudo Hermani) capillaries can also be found in some trabecles
of the spongy-like musculature. There is no structural difference
between the capillaries in compact and spongious musculature.
The capillary wall is formed by relatively thin endothelial cells;
it is always closed and has a continuous basal membrane. Relatively
frequently pericytes occur; they were found lying between the two
sheaths of the basal membrane (6). No ultrastructural differences
have been observed between the copillaries in the heart of cold
blooded and warm blooded animals (7,8).

It can be summarized that in the animal kingdom generally
there exist four types of myocardial blood supply (Fig. 1):

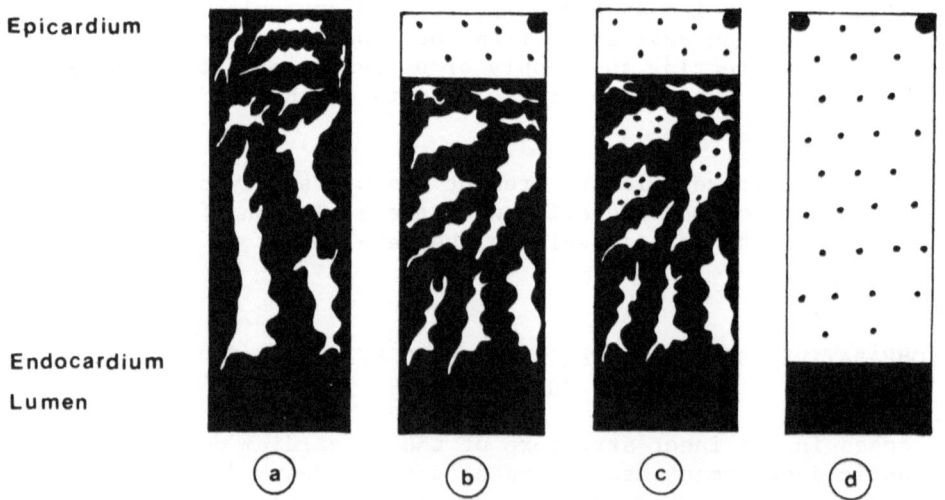

Fig. 1. Different types of the myocardial blood supply: (a)
spongious musculature supplied from the ventricular lumen; (b) the
inner spongious layer is covered by an outer compact musculature with
vascular supply; (c) as (b), but capillaries are present also
in some trabecels of spongy-like musculature; (d) compact musculature supplied from coronary vessels. (Reproduced, with permission
from Ostadal et al., ref. 38.)

a) spongious musculature only, which is supplied from the ventricular cavity;

b) the inner spongious layer is covered by an outer compact musculature with a vascular supply;

c) as b), but capillaries are also present in some trabecles of spongy-like musculature;

d) compact musculature only, which is supplied from coronary vessels.

It has been shown (7,9) that the total mass of the compact layer in different species of cold blooded animals is related more to the heart weight than to the phylogenetic position of the animal; the size of the compact layer evidently increases with increasing heart weight. The same trend also exists within the individual species (carp, turtle). The increase of compact musculature is thus obviously necessary for the maintenance of balanced blood pressure conditions in the larger hearts (10).

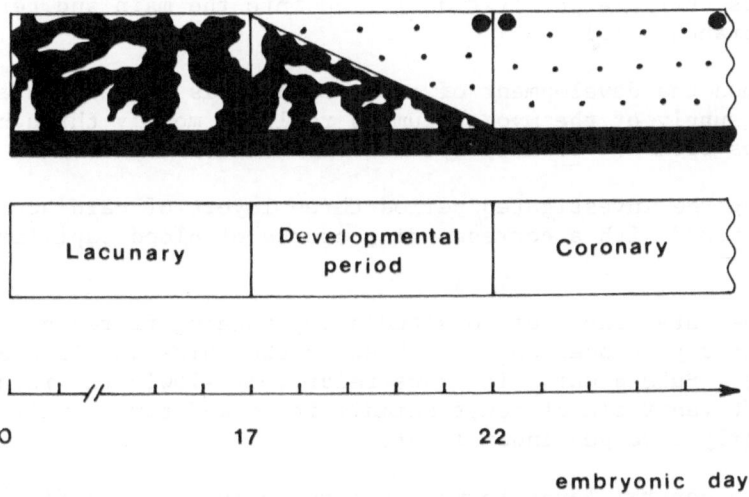

Fig. 2. Schema of the development of the terminal vascular bed in the rat heart. Developmental relationships between the 2 cardiac nutritive systems. (Data from Ošťadal et al, ref. 18).

ONTOGENETIC DEVELOPMENT OF THE TERMINAL VASCULAR
BED IN THE CARDIAC MUSCLE

The changes in heart size in the course of ontogenetic
development of homoiotherms are accompanied by the formation of
interventricular septum together with the gradual transformation of
the vascularless spongious musculature into a compact myocardium,
supplied through coronary arteries (e.g., 11-17). The development
of the myocardial blood supply is closely associated with that of
muscle fibres. It can be divided into three periods (18,19; Fig. 2):

1) Up to the development of coronary arteries (chick embryo
on the 7th embryonic day - ed -, rat embryo on the 17th ed), the
myocardium is entirely spongious and supplied by means of a system
of intertrabecular and sinusoidal spaces which communicate with
the ventricular cavity.

2) From the beginning of the coronary arterial stem development to the time when the definite cornoary ramification is formed,
the myocardium may be supplied both from intertrabecular and sinusoidal spaces communicating with the ventricular cavities, as well as
from the developing capillary bed. The type of the myocardial blood
supply depends in this period on the topography of the development
of coronary vascularization. The development of the cardiac venous
system is of great importance for this period (20). It develops
before the cardiac arterial system appears. Coronary artery pri-
mordia join the preformed primitive terminal vascular bed of the
venous nets that are partly transformed into the main and terminal
coronary branches (21).

3) When the development of coronary vessels is terminated,
the blood supply of the myocardium is realized mostly through
capillaries from the cornary arteries.

During the investigated period three layers of cardiac musculature are fromed with a corresponding course of blood capillaries
(18,22).

a) The outer layer of longitudinally running fibres and capillaries may already be seen on the 9th ed in the chick and 19th ed
in the rat. Subsequently it grows relatively slowly in width and
in the left ventricle of adult animals it is 6-7 times smaller than
the similarly arranged inner layer.

b) The central layer formed by circularly oriented fibres;

c) the inner layer of longitudinal fibres begins to be formed
on the 13th ed in the chick and on the 22nd ed in the rat embryonic
heart respectively. While its relative size in the right ventricle

remains practically the same during further development, it reaches half the thickness of the cardiac wall in the left ventricle. The final arrangement of muscle cells and blood capillaries is terminated on the 15th ed in the chick heart and during the second postnatal week in the rat.

In spite of the fact that the main role in myocardial oxygen supply is played by the terminal vascular bed, some questions dealting with the ultrastructural development of capillaries in the rat cardiac muscle should also be mentioned (23). As has already been shown the myocardium is entirely spongious and supplied from the ventricular lumen up to the development of coronary arteries (in the rat up to the 17th ed). Two types of primitive mascular bed are characteristic for this period (15th–16th ed):

a) intertrabecular spaces, which penetrate deeply into the ventricular wall as direct continuation of the endocardium (24), and b) intramyocardial clefts without endothelial lining; blood cells are thus in direct contact with the myocardial cell (Fig. 4). During further development of the terminal vascular bed, the out-growth of endothelial cells into the myocardial clefts is important (Fig. 3). However, the vascular wall is not yet closed and has only a very thin and discontinuous basal membrane. The cytoplasm of endothelial cells is light and has a relatively large number of ribosomes. The first capillaries, characterized by a closed but irregular and relatively thick endothelial wall can be observed on the 18th day of embryonic life (Fig. 5). At this time varous developmental stages ot he terminal blood bed can be observed simultaneously. Within the following period (20th–21st ed) the thick capillary walls become narrow and pericytes appear. The process of differentiation spreads in both ventricles and the septum from the base to the cardiac apex and is practically finished by the 14th day of postnatal life. At a later stage, only adult types of myocardial capillaries can be observed, i.e. capillaries with a narrow endothelial wall, continuous basal membrane and peri-cytes (25–27).

In our experiments we also paid attention to the development of permeability of the capillary wall. We used ferritine (as an example of a substance with a large molecule, cca 110 A (28) and peroxidase (small molecule, cca 24–30 A (29). It was found, that in contrast to a virtually uniform distribution of ferritin within the endothelial cells in the prenatal period, an accumulation of ferritin in pinocytic vesicles can be observed between 5 and 14 days of postnatal life. The main route for the transport of per-oxidase through the capillary wall, i.e., the intercellular clefts, is already present in foetal capillaries, similarly as in adult animals (Fig. 6).

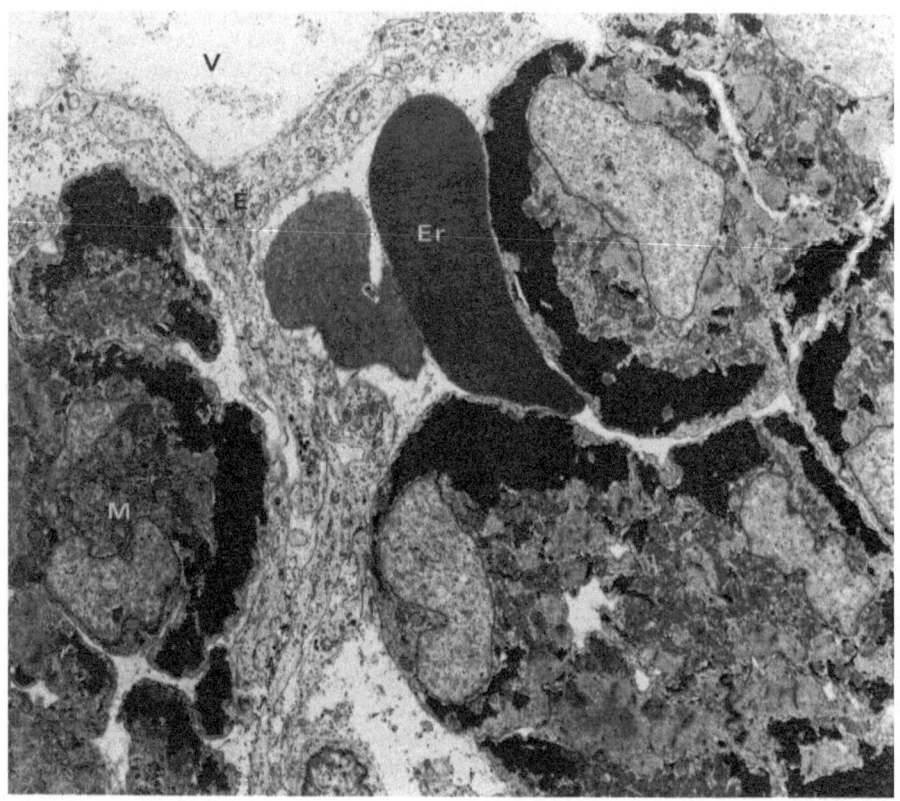

Fig. 3. Rat heart, 16th day of embryonic life. The outgrowth of
endothelial cells (E) into the intramyocardial clefts; Er -
erythrocyte, V - ventricular cavity, M - muscle cell. Magnification
4,000 X. (Data from Oštadal and Schiebler, ref. 23).

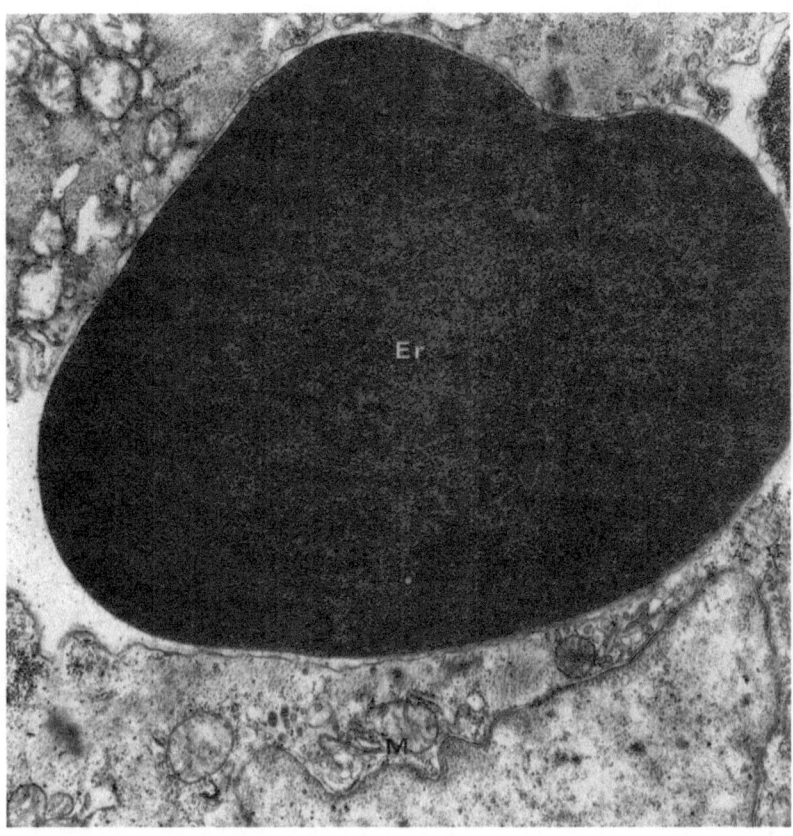

Fig. 4. Rat heart, 17th day of embryonic life. Intramyocardial cleft
without endothelia llining; erythrocyte (E) is in direct contact
with the myocardial cell (M). Magnification 15,800 X. (Data from
Ošťadal and Schiebler, ref. 23.)

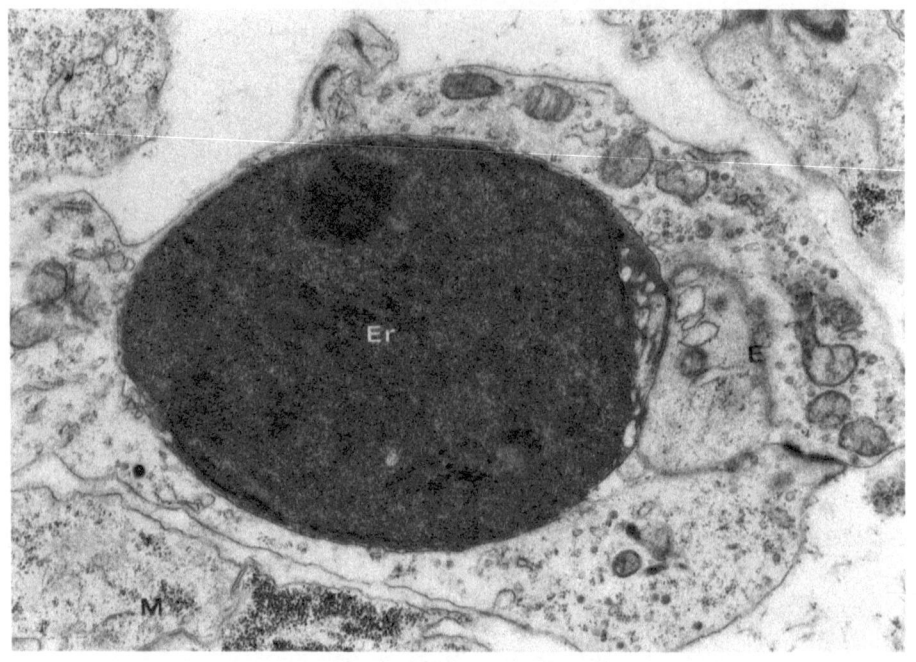

Fig. 5. Rat heart, 18th dayoof embryonic life. Fetal capillary
with closed but irregular and relatively thick endothelial wall
(E); Er - erythrocyte, M - muscle cell. Magnification 15,800 X.
(Data from Oṥtadal and Schiebler, ref. 23.)

Our observations dealing with the ultrastructural development of cardiac musculature are in good agreement with the results of Schiebler and Wolff (30), Forsmann et al. (31) and Melax and Leeson (32). On the 15th day of embryonic life irregular orientation of myofibres can still be observed. The basal membrane appears on the 16th, i.e., a little earlier in comparison with the basal membrane of endothelial cells. The longitudinal arrangement of myofibres starts between the 20th and 22nd day of embryonic life; intercalar discs may be observed at this period. The ultrastructural differentiation of myofibres together with the development of sarcoplasmic reticulum is terminated during the 4th postnatal week.

On the basis of the above mentioned observations it may be concluded, that during development the type of myocardial blood supply appears to be closely related to structural arrangement of cardiac musculature. This view is supported by the findings of postnatal persistence of spongious myocardium with embryonic blood supply in human specimen (e.g., 33–36). Ostadal and Rychter (37) have demonstrated, that intraamnial administration of iso-proterenol may cause a variety of malformations in the chick em-bryonic heart; they were related to the selected period of cardiac development. The blockade of coronary artery development connected with the persistence of spongious musculature with lacunary blood supply was characteristic for one such period.

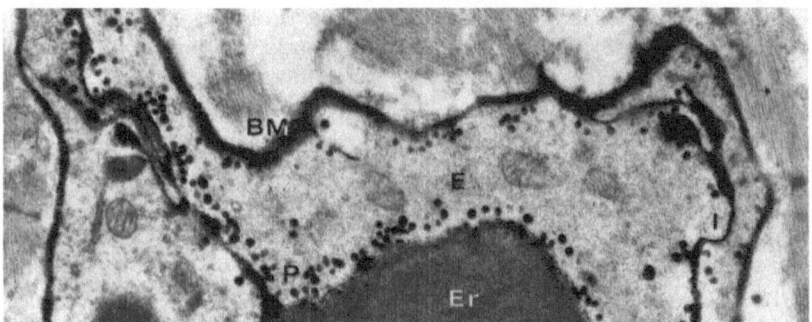

Fig. 6. Rat heart, 21st day of embryonic life. Transport of peroxidase across the capillary wall. The substance can be observed in pinocytotic vesicles (P), in the basal membrane (BM), in the intercellular junctions (I); E – endothelial cell, Er – erythrocyte in the capillary lumen. Magnification 16,800 X. (Data from Ostadal and Schiebler, ref. 23).

SUMMARY

The type of the blood supply to the myocardium appears to be
closely related to its structural arrangement. The heart of adult
poikilotherm animals is either entirely spongious, supplied from the
ventricular cavity or its spongious musculature is covered by an outer
compact layer with vascular supply. The size of the compact layer
increases with increasing heart weight.

The changes in the heart size during the ontogenetic development of homoiotherms are accompanied by the gradual transformation
of the vascularless spongious musculature into a compact myocardium
supplied thrugh coronary vessels. Up to the development of coronary
arteries (in the rat up to the 17th day of embryonic life - ed)
the myocardium is entirely spongious and supplied from mentricular
cavity. Two types of primitive vascular bed are characteristic for
this period: a) intertrabecular spaces, which penetrate deep into
the ventricular wall as direct continuation of the endocardium, and
b) intramyocardial clefts without endothelial lining. During further
development of the terminal mascular bed, the outgrowth of endothelial cells into the myocardial clefts is important. The first
capillaries with closed endothelial wall can be observed on the 18th
ed. At this time various developmental stages o the terminal blood
bed can be observed simultaneously. Within the following period
(20-21 ed) the thick capillary walls become narrow and pericytes
occur. The process of differentiation spreads in both ventricles
and in septum from the base to the cardiac apex and is practically
finished by the 14th day of postnatal life.

The longitudinal orientations of myofibres starts between the
20th and 22nd ed. The final arrangement of muscle cells and capillaries into three layers (outer and inner longitudinal, central
circular) is terminated during the second postnatal week.

REFERENCES

1. Rushmer, R.F. 1961. Cardiovascular Dynamics. W.B. Saunders,
       Philadelphia.

2. Grant, R.T., and Regnier, M. 1926. The comparative anatomy
       of the cardiac coronary vessels. Heart 13: 283.

3. Clark, A.J. 1927. Comparative Physiology of the Heart.
       Cambridge University Press.

4. Benninghoff, A. 1933. Herze. IN: Hanb. vergleich. Anat.
       Wirbeltiere (Bolk, L., Göppert, E., Kallius, E., Lubosch, W.,

editors). VI. Band, Berlin, Wien.

5. Oštadal, B., Rychter, Z., and Poupa, O. 1970. Comparative aspects of the development of the terminal vascular bed in the myocardium. Physiol. Bohemoslov. 19: 1.

6. Poirier, J., and Nguyen, H. Anh. 1967. L'ultrastructure des capillaires sanguins. La Presse Med. 75: 1469.

7. Oštadal, B., and Schiebler, T.H. 1971. Über die terminale Strombahn in Fischherzen. Z. Anat. Entwickl.-Gesch. 134: 101.

8. Oštadal, B., and Schiebler, T.H. 1971. Die terminale Strombahn im Herzen der Schildkrote (Testudo Hermanni). Z. Anat. Entwickl.-Gesch. 134: 111.

9. Bass, A., Oštadal, B., Pelouch, V., and Vítek, V. 1973. Differences in wieght parameters, myosin ATPase activity and the enzyme pattern of energy supplying metabolism between the compact and spongious cardiac musculature of carp (Cyprinus carpio) and turtle (Testudo Horsfieldi), Pflügers Arch. 343: 65.

10. Bauerseisen, E. 1953. Vergleichende physiologie des Sauerstofftransportes im Wirbeltierherzen. Naturwissenschfaten 40: 352.

11. Martin, H. 1894. Recherches anatomiques et embryologiques sur les arteres coronaires au coeur chez les vertebres. (Steinheil, editor), Paris.

12. Grant, R.T. 1926. Development of the cardiac coronary vessels in the rabbit. Heart 13: 261.

13. Bennet, H.S. 1936. The development of the blood supply to the heart in the embryo pig. Amer. J. Anat. 60: 27.

14. Licata, R.H. 1954. Human embryonic heart in ninth week. Amer. J. Anat. 94: 73.

15. Dbaly, J. 1964. Development of the branching of arteriae coronariae cordis in chick embryo. Čs. Morfol. 13: 401.

16. Dbaly, J., Oštadal, B., and Rychter, Z. 1968. Development of the coronary arteries in rat embryos. Acta Anat. 71: 209.

17. Rychter, Z., Jelínek, R., and Marhan, O. 1971. Progress

of vascularization of the ventricular myocardium in the
rat embryo. Physiol. Bohemoslov. 20: 257.

18. Ostadal, B., Rychter, Z., and Poupa, O. 1968. Qualitative
    development of the terminal coronary bed in the preinatal
    period of the rat. Folia Morphol. 16: 116.

19. Rychter, Z., and Ostadal, B. 1968. Periodixation of the
    blood supply development of the myocardium in chick embryo.
    Physiol. Bohemoslov. 17: 485.

20. Voboril, Z., and Schiebler, T.H. 1969. Über die Entwicklung
    der Gefässversorgung des Rattenherzens. Z. Anat. Entwickl.-
    Gesch. 129: 24.

21. Rychter, Z., and Ostadal, B. 1971. Mechanism of the develop-
    ment of coronary arteries in chick embryo. Folia Morphol.
    19: 113.

22. Ostadal, B., Rychter, Z., and Rychterová, V. Comparison of
    the different sensitivity of the chick and rat embryonic
    heart to isoproterenol during the prenatal and postnatal
    development. Acta Univ. Carol., in press.

23. Ostadal, B., and Schiebler, T.H. 1971. Die Kapillarentwicklung
    im Rattenherzen. Elektronenmikroskopische Untersuchungen. Z.
    Anat. Entwickl.-Gesch. 133: 288.

24. Henningsen, B., and Schiebler, T.H. 1970. Zur Frühentwick-
    lung der herzeigenen Strombahn. Elektronenmikroskopische
    Untersuchung an der Tatte. Z. Anat. Entwickl.-Gesch. 130:
    101.

25. Moore, D.H., and Ruska, H. 1957. Fine structure of capillaries
    and small arteries. J. Biophys. Biochem. Cytol. 3: 457.

26. Palade, G.E. 1961. Blood capillaries of the heart and other
    organs. Circulation 24: 368.

27. Bruns, R.R., and Pallade, G.E. 1968. Studies on blood
    capillaries. I. General organization of blood capillaries
    in muscle. J. Cell Biol. 37: 244.

28. Florey, H. 1967. The uptake of particulate matter by endo-
    thelial cells. Proc. Roy. Soc. (London), Ser. B. 166: 375.

29. Karnovsky, M.J. 1967. The ultrastructural basis of capillary
    permeability studied with peroxidase as a tracer. J. Cell
    Biol. 35: 213.

30. Schiebler, T.H., and Wolff, H.H.   1966.   Elektronenmikro-
    skopische Untersuchungen am Herzmuskel der Ratte wahrend
    der Entwicklung.   Z. Zellforsch. 69: 22.

31. Forssmann, W.G., Siegrist, G., and Girardier, L.   1966.
    ULtrastrukturelle Befunde an embryonalen Rattenherzen.
    Verh. Anat. Ges. 120: 71.

32. Melax, H., and Leeson, T.S.   1969.   Fine structure of de-
    veloping and adult intercalated discs in rat heart.   Cardio-
    vasc. Res. 3: 261.

33. Grant, R.T.   1926.   An unusual anomaly of the coronary vessels
    in the malformed heart of a child.   Heart 13: 273.

34. Feldt, R.H., Rahimtoola, S.H., and Davis, G.D.   1969.
    Anomalous ventricular myocardial patterns in a child with
    complex congenital heart disease.   Amer. J. Cardiol. 23:
    732.

35. Singer, H., Bayer, W., Reither, M., and Hinuber, G.v.   1973.
    Koronargefässanomalien und persistierende Myokardsinusoide bei
    Pulmanalatresie mit intaktem Ventrikelseptum.   Basic Res.
    Cardiol. 68: 153.

36. Dušek, H., Ošťadal, B., and Dušková, M.   Postnatal persistence
    of spongy myocardium with embryonic blood supply.   Arch.
    Path., in press.

37. Ošťadal, B., and Rychter, Z.   1972.   The effect of prenatal
    administration of isoproterenol on the chick and rat embryonic
    heart.   IN: Les Surcharges Cardiaques (Heart Overloading),
    (Hatt, P.Y., editor), Paris.

38. Ošťadal, B., Rychterová, V., and Procházka, J.   1973.   The
    development of the myocardial blood supply and the origin
    of the acute experimental cardiac necrosis (in Czech).
    Čs. Fysiol. 22: 65.

QUESTIONS TO DR. OSTADAL

Dr. Morasco: In your last slide you ahowed us a very short flash
on the effects of isoproterenol on the development of coronary
vessels. Can you give us more detailed information on this particular
topic.

Dr. Ošťadal: The effect of isoproterenol on the chick heart was studied from the embryonic day to the time of hatching. Intraamnial administration of this substance (2 X 80 mg/kg b.w./48 hr) causes different types of changes in different developmental periods (e.g., total body edema, haemorrhagic edema of the occipital region, cardio-megaly, transposition of big vessels, etc). The retardation of coronary vascularization was regularly observed between the 7th-14th day of embryonic life, i.e., during the period of the coronary blood supply development. (Ostadal and Rychter, ref. 37.)

Dr. Kasten: From your experience would you tell us something about the relative availability of fibroblasts and endothelial cells for trypsinization from newborn rat and early chick embryo hearts.

Dr. Ostadal: Unfortunately, we have no experience with trypsiniza-tion of the rat and chick embryonic hearts. Our results are based on the in vivo observations.

FUNCTIONAL CAPACITY OF NEONATAL MAMMALIAN MYOCARDIAL CELLS DURING AGING IN TISSUE CULTURE

Frederick H. Kasten

Department of Anatomy, Louisiana State University
Medical Center, New Orleans, Louisiana   70119

I.   INTRODUCTION

Human aging is accompanied by a logarithmic increase in death rate from disease, especially from cardiovascular disorders and decreased efficiency in homeostatic mechanisms. Comfort (1) speculates that is throughout life man had the same resistance to stress and disease as at age 20, one-half of us could expect to live to the age of 700. Heart diseases are the leading causes of death in people over the age of 55. Since the best way to prolong life in humans is to remove the pathologic causes of death, it would be the most effective course of action. The heart is responsive to a multiplicity of external influences impinging on it through the circulation and by the nervous system. Within the heart, self-regulation occurs by means of pacemaker tissue and the Purkinje fiber system (2). Because of these factors which complicate studies of the cellular changes of myocardial cells during aging, we have chosen to work with ventricular newborn rat heart cells in culture as an experimental model. The initial studies on this material were made by Harary and Farley (3) and later by Mark and Strasser (4). We find that the individual myocardial cells are self-contracting units which rapidly link up in vitro to form beating networks. In long-term primary cultures which are kept as long as 100 days, these networks aggregate during this period and produce "mini-hearts" and fibers which are visible to the eye. Cultured cells can be sucessfully stored at liquid nitrogen temperature and recultured several times. These and other results were presented at the meeting, partly in the form of two 16 mm movie films --- "Mitosis and Differentiated Properties of Mammalian Myocardial Cells in Culture", and "Contractile Behavior of Myocardial

Cells <u>In Vitro</u>". For the purpose of documentation, still photomicrographs, which are based on the same biological material as
that employe- in the films, are used in the manuscript.

II. MATERIALS AND METHODS

The usual source of heart tissue is 3- to 4-day newborn rats.
Occasionally, older animals are employed, but these yield few viable
cells with the ordinary trypsinization procedures employed. Right
and left mentricles are dissected free, washed repeatedly in Gey's
Balanced Salt Solution (BSS), and cut into small pieces. These are
trypsinized with 0.125 % trypsin at 37°C for 15 to 20 min. The
first trypsinized digest is discarded. At least 4 to 5 subsequent
trypsinization are done sequentially and all cells isolated are combined and washed. To remove most of the endothelial cells present
in the cell mixture, the cells are put into a 250 ml Falcon flask
and permitted to settle for 90 min. Endothelial cells are selectively
attached to the surface, leaving a suspension of almost pure myocardial cells. This "Differential Attachment Technique" has been
described (5,6) and can be employed using other variation. For
example, the mixed cell population is injected directly into Rose
chambers (7). At the end of 90 min, each chamber is inverted so
that the myocardial cells fall to the opposite glass surface. A
"karate-chop" on the lab table with the edge of the hand while
holding the chamber in the palm, helps remove myocardial cells
which may be attached to the initial surface. Since the "Differential Attachment Technique" has general significance in tissue
culture, the details are shown in Figures 1 and 2. Cells are grown
in a modified Eagle's MEM. Further details of the cell isolation,
tissue culture, and photomicrographic techniques are given elsewhere (5,6,8).

III.   RESULTS AND DISCUSSION

A. Cell Isolation

Trypsin treatment of neonatal rat ventricle produces two effects
(9). First, there is a progressive detachment of cells from each
other and second, a penetration of the enzyme mixture into myocardial but not endothelial cells. Figure 3 is an electron micrograph of a section of intact neonatal ventricle before trypsinization.
Myofibrils are seen in longitudinal and transverse profiles. Other
structures in the field include mitochondria, nuclei, and a variety
of membranes. A group of cells fixed in osmium tetroxide immediately
following trypsinization is shown in Figure 4. In addition to causing
the isolated cells to assume a spherical shape, the isolation
procedure induces other alterations. Nuclei of myocardial cells

are centrifuged to the cell edges and the myofibrils are no longer
present as intact organelles. Instead, only their constituent myofilaments are seen (Fig. 5). Some remnants of Z lines are present.
In other micrographs taken from this material, abnormal granules
(trypsin?) are observed directly under the plasma membrane. Trypsin
has been reported to attack and degrade myosin (10). Endothelial
cells in the same cell isolate, derived from coronary vessels
(11), fail to reveal any damage due to the enzyme treatment or the
centrifugation (1000 rev/min).

B. Early Cultures

     When the damaged myocardial cells are put into culture medium,
they do not begin to attach for about 15 hr. However, even before
this time the rounded myocardial elements often pulsate weakly in
an rratic fashion. A group of three such cells is shown in a
living state 6 hr after isolation (Fig. 6). The basis for such
contractions seems to lie in the fact that the myofilaments rapidly
reorganize and regenerate new myofibrils. A micrograph taken from
a 24 hr culture (Fig. 7) shows the longitudinal assembly of myofilaments containing dense zones. These dense areas resemble emerging
Z lines. The pattern of development here bears a remarkable resemblance to what is seen in development of skeletal muscle during
embryogenesis (12). Legato (13) also believes that Z substance is
essential for the production of new sarcomeres. A cultured cell in
a more advanced stage of myofibril development is shown in Figure
8. Typical sarcomeres are present as well as intercalated discs
between two interconnecting cells. Contractile activity begins in
the rat embryo at about the 9th day (14). The neonatal rat exhibits
fully differentiated cardiac tissus in terms of contractile function.
The organelles responsible become "artificially dedifferentiated"
due to the action of trypsin. With the assembly of new myofibrils
in vitro, a rapid "redifferentiation" ensues. As will be discussed
below, these spontaneously contracting cells maintain their specialized function for the remainder of their lifetime in culture.

C. Morphology and Behavior of Cell Cultures During Aging

     The spherical myocardial cells rapidly produce new membranes and
attach to the surface. There is no active migration, in contrast
with endothelial cells which may be present. The myocardial cell
enlarges and presents a broad or elongated shpae in the first 24 to
48 hr. The cells interact with each other upon contact through growth
and form firm connections, as is seen in Figure 9. Practically all
the cells in the field are myocardial. Note the dense mitochondria
or sarcosomes which contrast with the thin endothelial cells in the
corner. Further cytologic details are seen at high magnification

(Fig. 10) where an outer membrane is seen to encircle the nuclear
envelope. The outer membrane is a specialized Golig apparatus, which
is termed NAGA (nuclearassociated Golgi apparatus) and found exclusively in myocardial cells (15). Another special cytologic
feature of these cells is the presence of numberous myofibrils which
course through the cell along the long axis (Fig. 11). All major
components of the sarcomere are seen. Film records taken at natural
or high speed demonstrate peculiar oscillatory movements of myofibrils in non-contracting cells (16). In other cases, active
contractions of sarcomeres occur to produce rhythmic and arrhythmic
cell contractions (17).

     With aging in culture, a dense mat of cells fills the chamber.
The myocardial cells can be distinguished from non-myocardial
cells by the cytologic criteria already mentioned. Also, myocardial
elements have smaller nuclei and are binucleated (Fig. 12). A
drawing of a cardiac cell in vitro is shown in a three-dimensional
view in Figure 13. Ultrastructural studies of cultured cells have
been reported (18.19). A critical period in the life of the culture
occurs at about one week. There is a steady increase in the number
of non-myocardial cells with a tendency to overgrow the myocardial
cells. If this occurs, the number of viable myocardial cells decreases and they may disappear altogether. A high initial seeding
of myocardial cells helps to prevent such overgrowth and leads to
viable and more complex contracting networks. The cells produce
long myopodial processes and develop thin spider-like shpaes
(Fig. 14). The cells produce complicated networks or colonies which
attach to each other through myopodial extensions. Figure 15 is
a montage of such a system which extended throughout the culture
chamber, or 2.5 cm. The culture was photographed after 28 days
in vitro. Such networks frequently contract synchromously. In
older cultures, the networks coalesce and form unusually large masses
of contractile tissue or "mini-hearts" (Fig. 16) which are large
enough to be seen with the unaided eye (Fig. 17). The masses shown
in Fig. 17 were grown in Falcon flasks and appeared viable up to 100
days, although they had stopped contracting earlier. Each mass
measured about 2 to 4 mm in size and was connected by non-viable
tissue. In the Rose chamber "mini-hearts" and thick strands are
found surviving at the edge of the chamber, close to the silicone
gasket. Apparently, there is an anoxic condition in the center of
the chamber in very old cultures. This problem should be easily
overcome by perfusing oxygen into the chamber. In Falcon flasks
kept in CO_2 incubators, there is no oxygen deficiency and heart
masses grow readily. Studies of the optimal oxygen concentration
for maintaining heart cells were reported (20). Harris (21) points
out that the aged human heart has a decreased ability to utilize
oxygen. Synthetic strands of cardiac muscle have been induced in
very young cultures from chick embryos by depositing them in thin
grooves (22).

Contractile rates of individual cells in early cultures display a spectrum ranging from 15 to more than 100 beats per min. Mark and Strasser (4) showed that when a fast-beating cell contacts a slow beater, the final beat is that of the fast one. The same is true when a non-myocardial cell connects two myocardial cells. These results were confirmed by Goshima and Tonomura (23). The situation apparently is not this simple since one other report (24) indicates that the synchronized rate connot be predicted from a knowledge of the rates of the two cells before contact is made. Our results indicate that networks usually contract at 50 to 60 beats per min, which is in the intermediate range. If fast beaters always pace the slow ones, one would expect older networks to exhibit fast beating and this is not the case. As cultures age and networks become more complex with larger numbers of cells in each mass, the beat rates tend to slow down to less than 50 per min and finally stop in the largest cardiac masses. The present results are difficult to interpret since they may represent aging changes to some extent but could be explained as well by postulating that diffusion of oxygen, metabolites and toxic products in and out of networks is reduced as the tissue mass in vitro increases with age. When a more efficient system of maintaining long-term cultures is developed, the interpretation of contraction rates with aging should be less ambiguous.

D. Mitosis and Differentiation

The question as to whether differentiated cells are capable of division is an old one. Often, the argument revolves around the definition of what is meant by a differentiated cell. A reasonable definition is that a differentiated cell performs a specialized function and is no longer capable of changing to another cell type. Until recently, the myocardial cell was regarded as an example of a differentiated cell which could no longer divide. Indeed, the increase in heart size associated with postnatal growth, aging, and response to stress has been regarded as due largely to hypertrophy of muscle cells, although some increase in cell number of interstitial tissue is considered to occur (25).

Rumery and Rieke (26) reported that DNA synthesis in cultured chick heart cells is due to myoblasts, the undifferentiated myocardial precursors. Later (4,27) it was reported that differentiated ray myocardial cells in vitro are capable of cell division. This has been confirmed and new observations made that contractions cease in mid-mitosis; later the daughter cells resume contractions (5). Frequently, there is a failure of cytokinesis, leading to the development of binucleated cells. It was also shown that myocardial cells undergo DNA synthesis. Apparently, Z lines disintegrate during mitosis and reorganize or regenerate when mitosis is complete. This

is an example of modulation of differentiated function. Figure 18
shows successive stages of division of a myocardial cell from a rat
culture. Numerous reports indicate that mitosis and DNA synthesis
occur in differentiated cardiac cells in vivo (28-31). The presence
of myofibrils in a dividing cell is the kind of evidence presented
(Fig. 19). During postnatal growth of rat, there is a rapid increase in the number of cardiac cells up to one month of age. This
is followed by a gradual increase up to two years (32). It is
suggested by Claycomb (33) that the cessation of DNA synthesis during
the postnatal period is due to a loss or inactivation of DNA polymerase. During aging of rat and human hearts, a definite increase
in polyploidization occurs (34,35). It is not certain that hypertrophy is the only explanation for increases in heart mass as a
result of aging and stress. When adult cardiac tissue can be routinely cultured, it should be possible to investigate these phenomena
further.

IV.   SUMMARY

Results are presented which demonstrate the feasibility of using
neonatal rat heart cells and networks in culture as a model system
for aging. Results presented here demonstrate the following:

1.   An improved technique for isolating and cultivating
mammalian myocardial cells.

2.   Reversible damage induced by trypsinization and centrifugation, used to isolate these cells.

3.   Morphology, growth, and contractile behavior of differentiated cells in primary cultures.

4.   Development of synchronized networks and giant cardiac
masses or "mini-hearts" in long-term cultures maintained for up to
100 days.

5.   Differentiated myocardial cells are capable of division
produce differentiated daughter cells or a single binucleated
cells. Contractions cease in mid-mitosis as organized myofibrils
disappear. Contractions resume in daughter cells.

ACKNOWLEDGEMENTS

This work was supported by the United States Public Health
Service Grants NS-09524 from the National Institute of Neurological
Diseases and Stroke, CA-12067 from the National Cancer Institute,
HL-15103 (Specialized Center of Research) from the National Heart

and Lung Institute, and United States Public Health Service Training
Grant 5-T01-DE-0024 from the National Institute of Dental Research.

The author would like to acknowledge the technical assistance
and pleasant cooperation of Phil Constantin, Mary Vaughn, and
Dominic Yip. The manuscript was typed by Susan Leitz and Jo Ann
Cambre.

REFERENCES

1. Comfort, A. 1964. The Process of Ageing. Signet Science
    Library (New American Library), New York.

2. DeHaan, R.L. 1965. Development of pacemaker tissue in the
    embryonic heart. Annals N.Y. Acad. Sci. 127: 7.

3. Harary, I., and Farley, B. 1960. In vitro organization of
    single beating rat heart cells into beating fibers. Science
    132: 1899.

4. Mark, G., and Strasser, F.F. 1966. Pacemaker activity and
    mitosis in cultures of newborn rat heart ventricle cells.
    Exp. Cell Res. 44: 217.

5. Kasten, F.H. 1972. Rat myocardial cells in vitro: Mitosis
    and differentiated properties. IN: Symposium on Functional
    Differentiated Culture Systems (Kasten, F.H., editor),
    In Vitro 8: 128.

6. Kasten, F.H. 1973. Mammalian myocardial cells. IN: Tissue
    Culture Methods and Applications (Kruse, P.F. Jr, and
    Patterson, M.K. Jr., editors), p 72, Academic Press, New York.

7. Rose, G. 1954. A separable and multiplurpose tissue culture
    chamber. Tex. Rep. Biol. Med. 12: 1074.

8. Kasten, F.H. 1971. Cytology and cytochemistry of mammalian
    myocardial cells in culture. Acta Histochem., Suppl. 9:
    785.

9. Kasten, F.H. 1966. Electron microscope studies of the combined
    effects of trypsinization and centrifugation on rat heart cells,
    with observations of early cultures. J. Cell Biol. 31: 131A.

10. Mihalyi, E., and Szent-Gyorgyi, A.G. 1953. Trypsin digestion
    of muscle proteins. J. Biol. Chem. 201: 189.

11. Oštádal, B., Rychter, Z., and Poupa, O. 1968. Qualitative

development of the terminal coronary bed in the perinatal period of the rat. Folia Morph. 16: 116.

12. Allen, E.R. 1973. Immunoghemical and ultrastructural studies of myogenesis. IN: The Striated Muscle (Pearson, C.M., and Mostofi, F.K., editors), p. 40, Williams & Wilkins Co., Baltimore.

13. Legato, M. 1970. Sarcomerogenesis in human myocardium. J. Molec. Cell Cardiol. 1: 425.

14. Goss, C.M. 1938. The first contractions of the heart in rat embryos. Anat. Rec. 70: 505.

15. Kasten, F.H., and Cerda-Olmedo, N. 1971. Solution to the mystery of the apparent "double nuclear membrane" observed in living myocardial cells in culture. Anat. Rec. 169: 353.

16. Kasten, F.H. 1968. Film recording of oscillatory myofibrils and rhythmic and arrhythmic contractile patterns of cultured mammalian myocardial cells and a method for quantitative analysis. J. Cell Biol. 39: 148A.

17. Kasten, F.H. 1969. High resolution filming of rhythmic and arrhythmic contractile behavior of cultured myocardial cells with a method for quantitative analysis. In Vitro 4: 150.

18. Kasten, F.H. 1967. Dynamic cytology of beating heart cells in cultures. J. Ultrastruct. Res. 21: 163.

19. Legato, M.J. 1972. Ultrastructural characteristics of the rat ventricular cell grown in tissue culture, with special reference to sarcomerogenesis. J. Molec. Cell. Cardiol. 4: 299.

20. Karsten, U., Kössler, A., Halle, W., Janiszewski, E., and Schulze, W. 1972. Kultiveierung spontan schlagender Herzzellen unter definiertem Sauerstoff-Partialdruck. Acta Biol. Med. Germ. 1041.

21. Harris, R. 1970. The Management of Geriatric Cardiovascular Disease. J.B. Lippincott Co., Philadelphia.

22. Purdy, J.E., Lieberman, M., Roggeveen, A.E., and Kirk, R.G. 1972. Synthetic strands of cardiac muscle. J. Cell Biol. 55: 563.

23. Goshima, K., and Tonamura, Y. 1969. Synchronized beating of embryonic mouse myocardial cells mediated by FL cells in monolayer culture. Exp. Cell Res. 65: 387.

24. DeHaan, R.L., and Hirakow, R. 1972. Synchronization of pulse
     rates in isolated cardiac myocytes. Exp. Cell Res. 70: 214.

25. McMillan, J.B., and Lev, M. 1962. The aging heart: Myo
     cardium and epicardium. IN: Biological Aspects of Aging
     (Shock, N.W., editor), p. 163, Columbia University Press,
     New York.

26. Rumery, R.E., and Rieke, W.O. 1967. DNA synthesis by cultured
     myocardial cells. Anat. Rec. 158: 501.

27. Kasten, F.H., Bovis, R., and Mark, G. 1965. Phase-contrast
     observations and electron microscopy of cultured newborn rat
     heart cells. J. Cell Biol. 27: 122A.

28. Manasek, F.J. 1968. Mitosis in developing cardiac muscle.
     J. Cell Biol. 37: 191.

29. Weinstein, R.B., and Hay, E.D. 1970. Deoxyribonucleic acid
     synthesis and mitosis in differentiated cardiac muscle cells
     of chick embryos. J. Cell Biol. 47: 310.

30. Oberpriller, J., and Oberpriller, J.C. 1971. Mitosis in adult
     newt ventricle. J. Cell. Biol. 49: 560.

31. Rumyantsev, P.P. 1973. Post-injury DNA synthesis, mitosis and
     ultrastructural reorganization of adult frog cardiac myocytes.
     Zts. Zellforsch. 139: 431.

32. Sasaki, R., Watanabe, Y., Morishita, T., and Yamagata, S. 1968.
     Estimation of the cell number of heart muscles in normal
     rats. Tohoku J. Exp. Med. 95: 177.

33. Claycomb, W.C. 1973. DNA synthesis and DNA polymerase activity
     in differentiating cardiac muscle. Biochem. Biophys. Res.
     Commun. 54: 715.

34. Klinge, O. 1970. Karyokinese und Kernmuster im Herzmuskel
     wachsender Ratten. Virch. Arch. Abt. B. Zellpath. 6: 208.

35. Fischer, B., Schluter, G., Adler, C.P., and Sandritter, W. 1970.
     Zytophotometrische DNS-, Histon- un Nicht-Histonprotein-
     Bestimmungen an Zellkernen von menschliche Herzen. Beitr.
     Path. 141: 238.

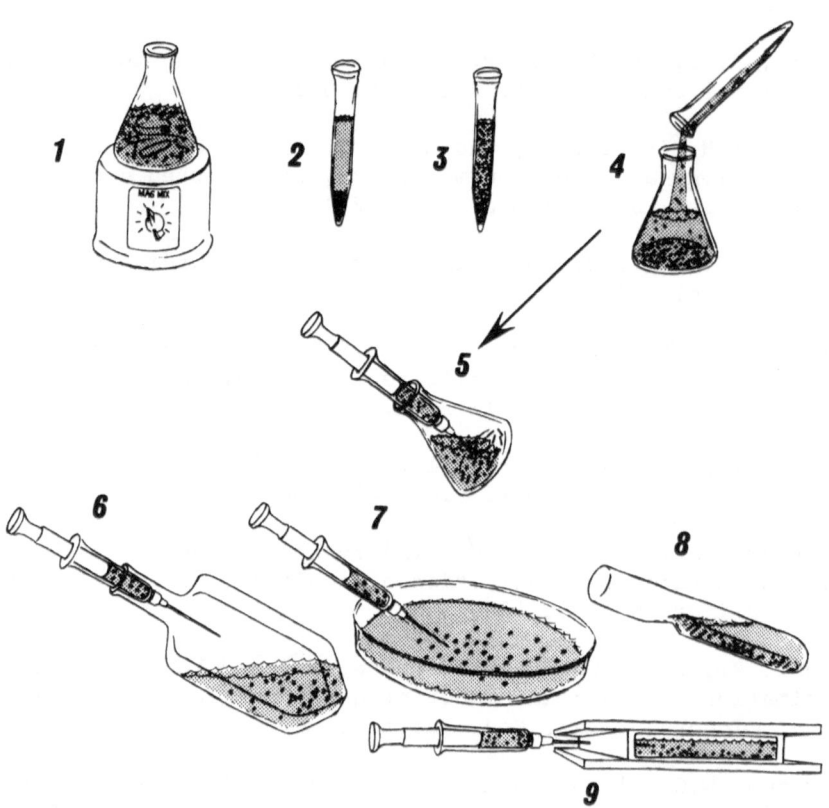

Fig. 1. Illustration of "differential attachment technique" used to separate myocardial from endothelial cells prior to cultivation. Principle depends on the rapid and preferential adhesion of endothelial cells to the substrate. Various methods are shown of culturing the isolated-myocardial cells. See text and refs. 17, 18 for description.

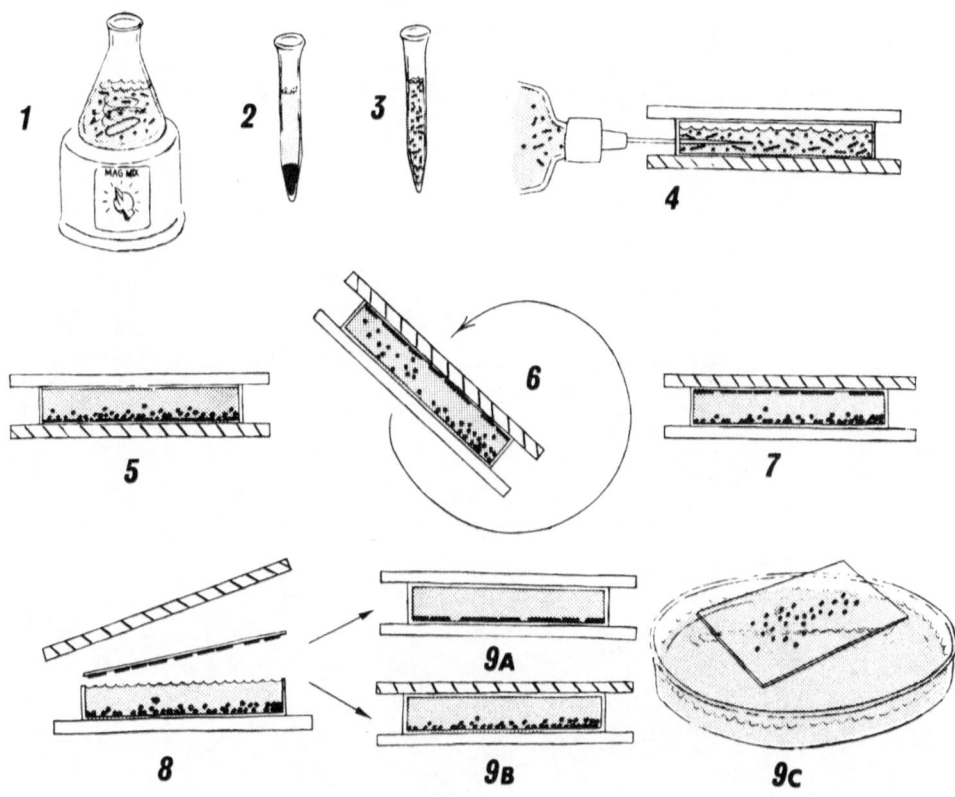

Fig. 2. "Differential attachment technique" applied to Rose chamber for separation of myocardial from endothelial cells. A "flip-flop" procedure is used to obtain more myocardial cells. Either myocardial or endothelial-rich chambers can be obtained. See text and refs. 17, 18 for description.

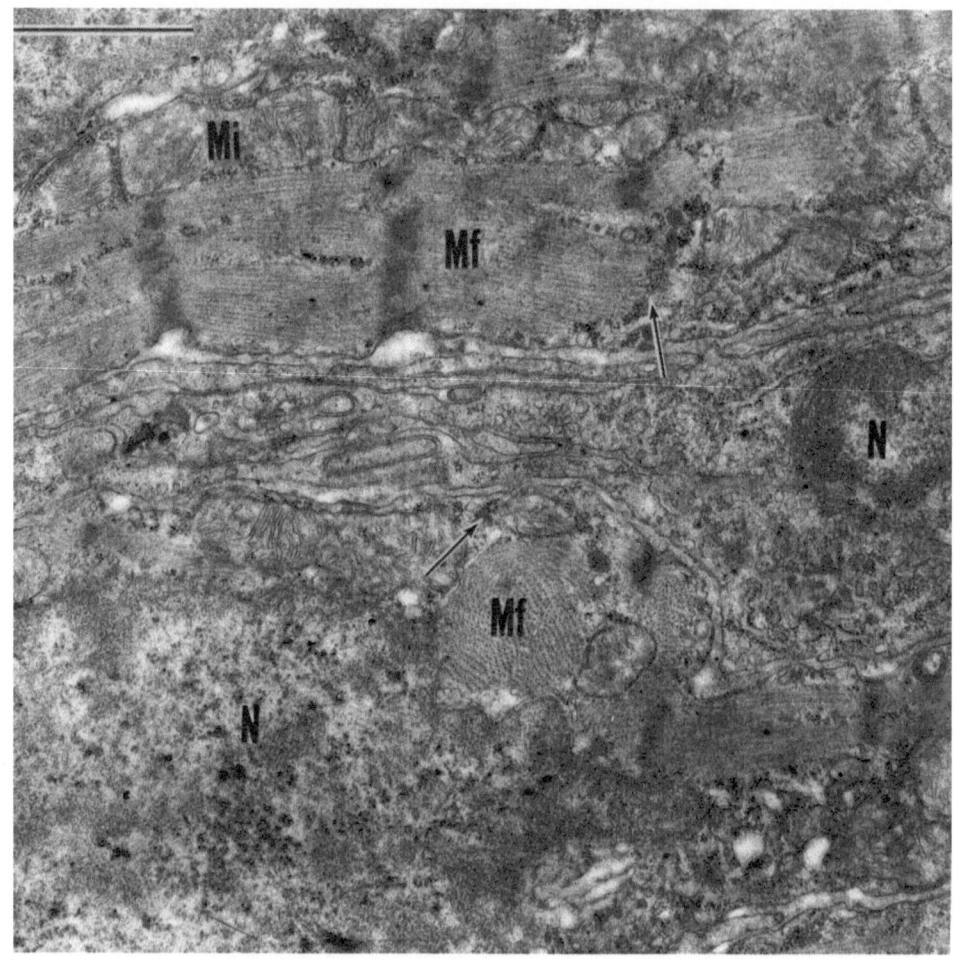

Fig. 3. Electron micrograph of thin section of newborn rat heart ventricle. The tissue exhibits considerable organization and is typical of material used for tissue culture. Well differentiated myofibrils (Mf) are seen surrounded by columns of tightly packed mitochondria (Mi). Other organelles include nuclei (N) and ribosomes (arrows). OsO$_4$ fixation. Uranyl acetate and lead citrate staining. X22,800. Marker = 1μ (ref. 16).

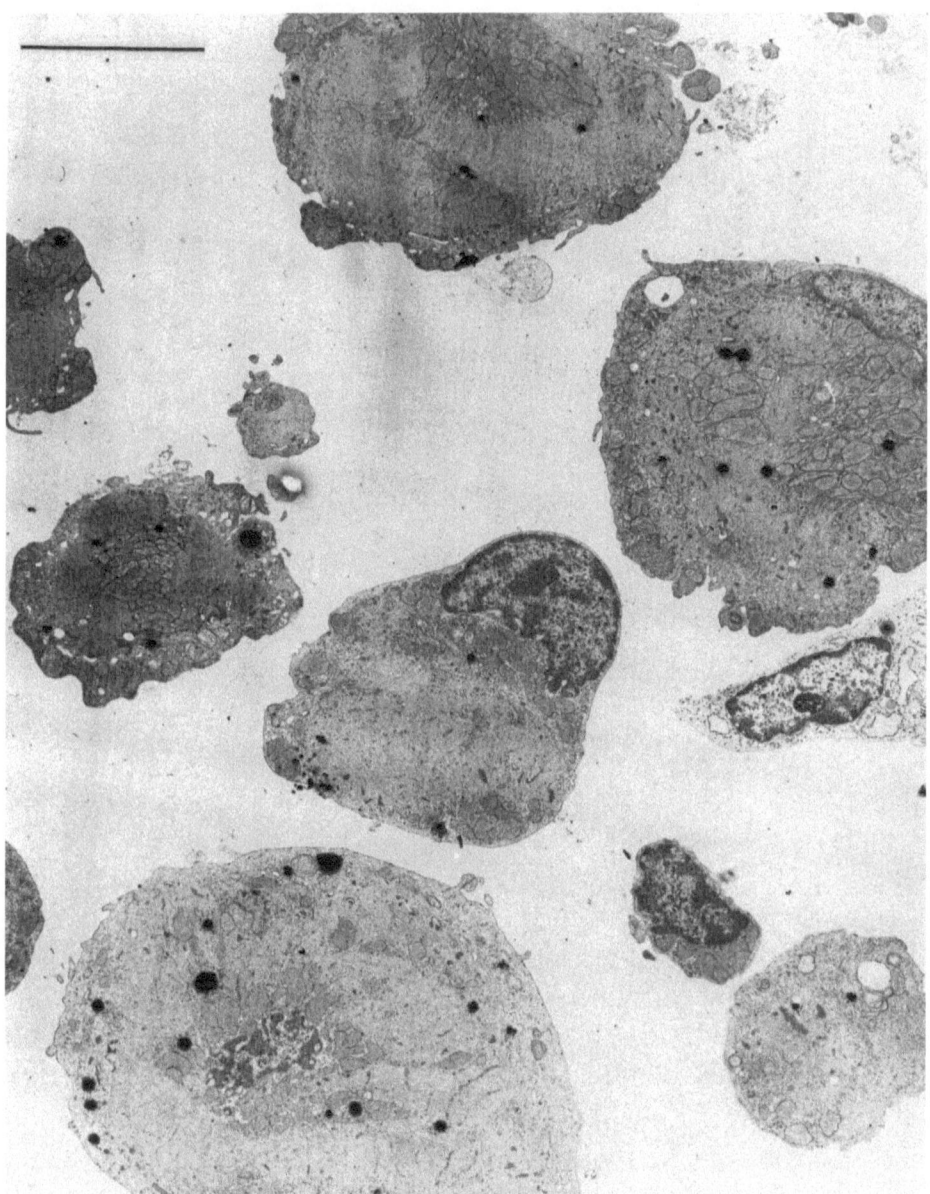

Fig. 4. Low-power electron micrograph of isolated heart cells
directly after multiple trypsinization-centrifugation procedure.
Myocardial cells are rounded and show internal damage to myofibrils,
packing of mitochondria, and displacement of nuclei. Same fixation
and staining as in Figure 3. X4,800. Marker = 5μ (ref. 16).

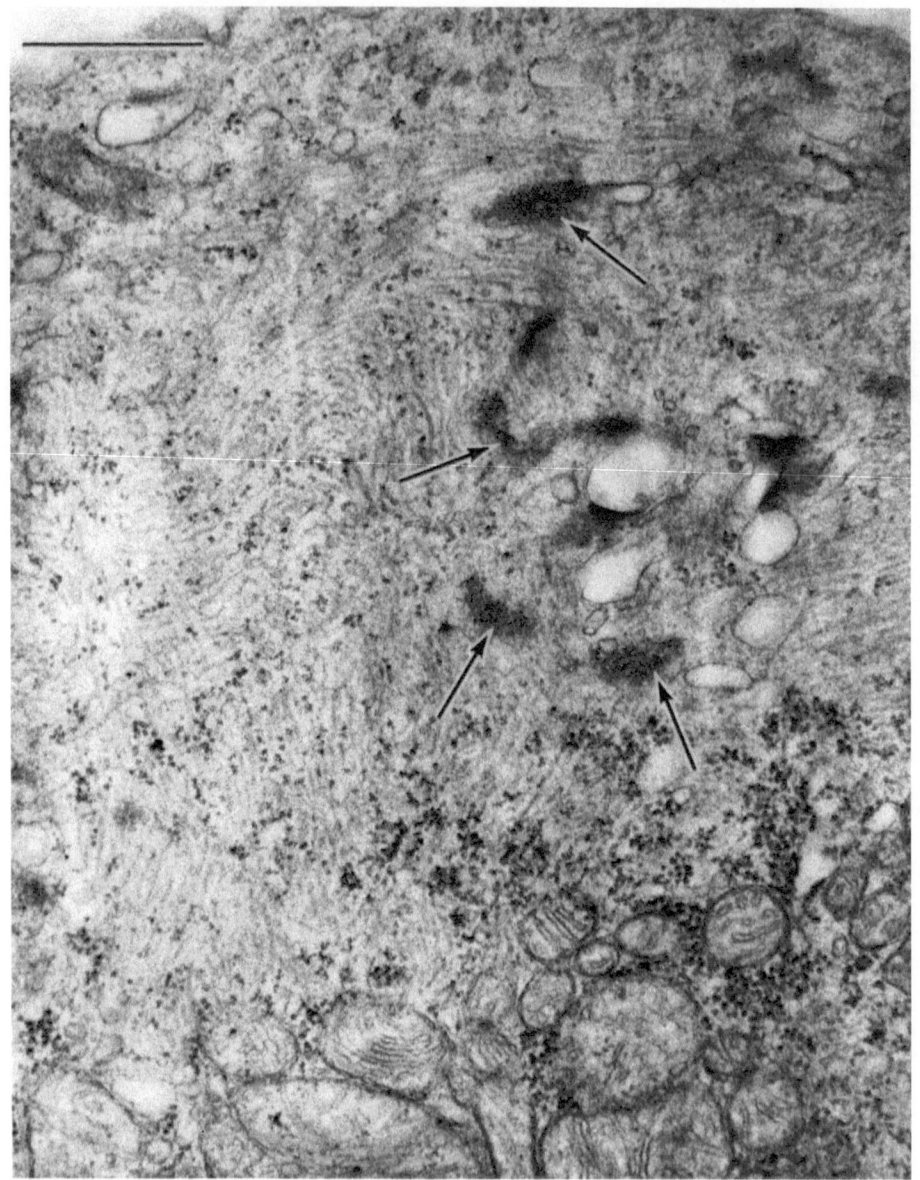

Fig. 5. Trypsinization used to dissociate myocardial cells also causes preferential digestion of myofibrils, as seen in this electron micrograph. Several dense areas (arrows) represent residual Z lines enveloped by a mass of myofilaments. Some vesicles, ribosomes, and intact mitochondria are seen. Same fixation and staining as in Figure 3. X48,000. Marker = 0.5μ (ref. 16).

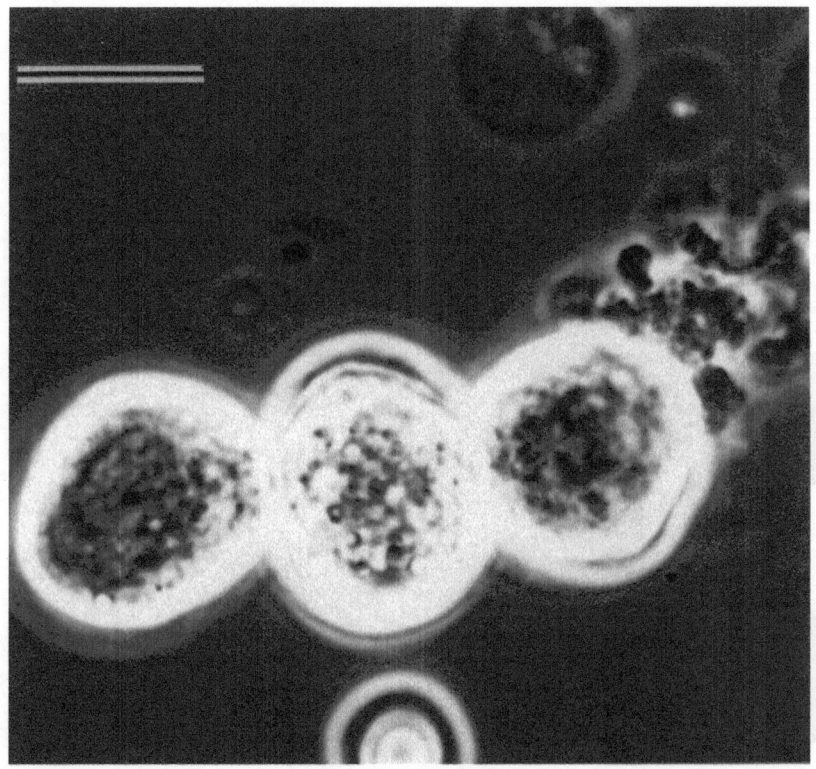

Fig. 6.   Six hr culture of trypsinized rat heart cells.   Three
rounded myocardial cells are observed.   They are already pulsating
erratically at this time.   Zeiss Neuofluor phase optics.   X100
XI.25 Optovar.   (X480) Marker = 50μ.

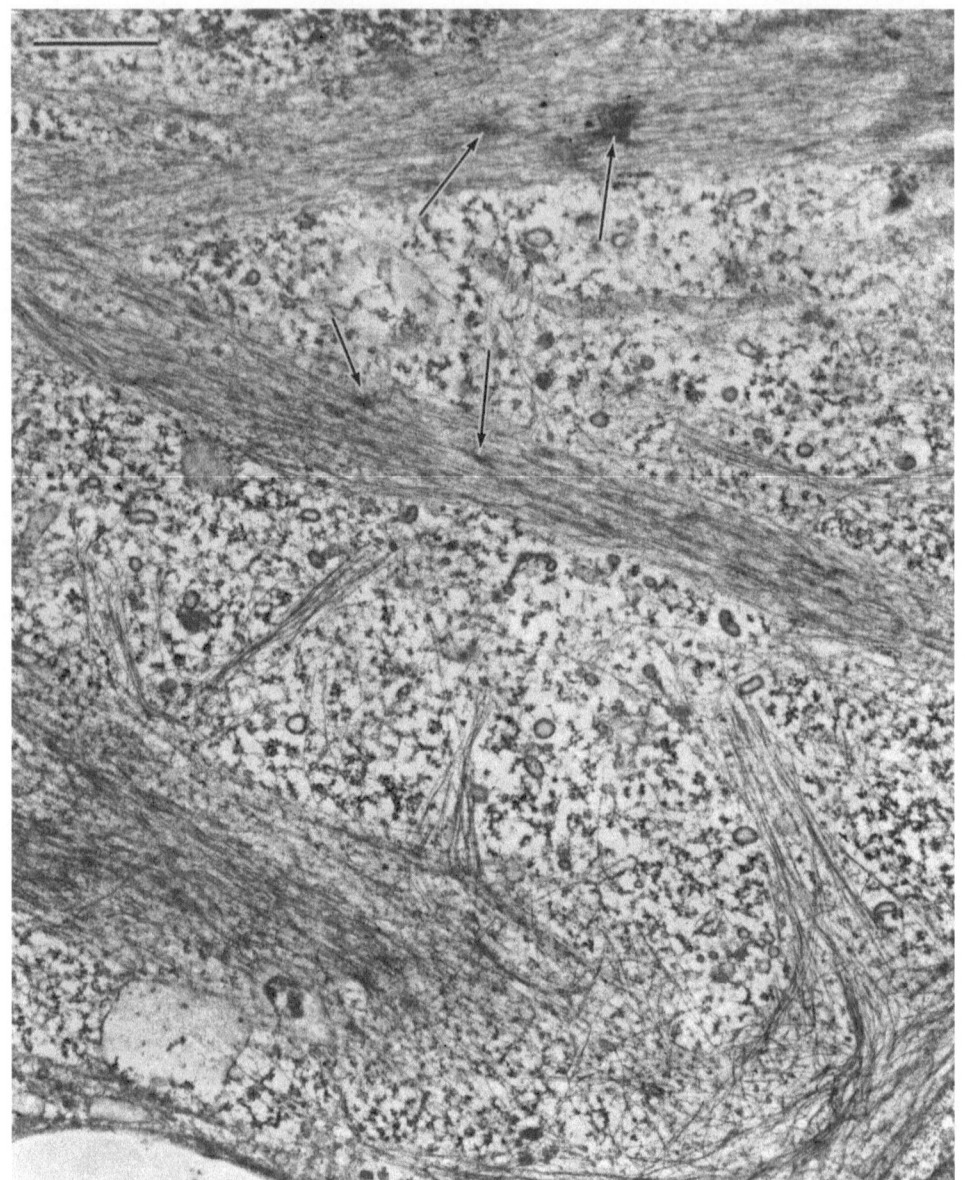

Fig. 7. Electron micrograph of a myocardial cell from a one-
day heart culture. Material fixed and flat-embedded directly on glass.
Sectioning done parallel to plane of growth to preserve normal relationships. Field shows integration of most myofilaments into
differentiating myofibrils. Within each mass of parallel-
aligned myofilaments, dense regions are seen (arrows) which probably
are regenerative centers for protein synthesis and Z line formation.
Same fixation and staining as in Figure 3. X16,000. Marker = 1μ (16)

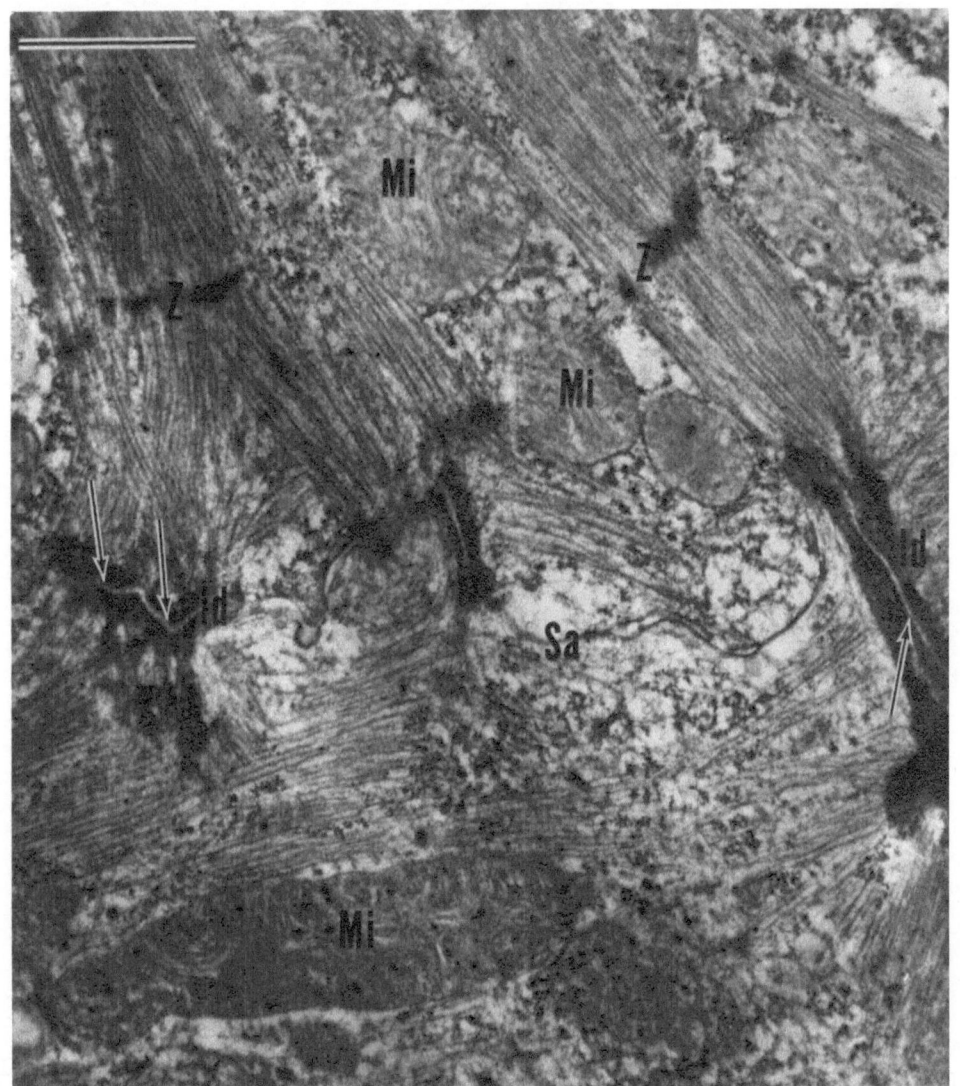

Fig. 8. Micrograph from a one-day culture as described for Figure 7.
The upper cell exhibits a more advanced stage of redifferentiation
than in the preceding figure. Myofibrils in the upper cell show
well-defined Z lines (Z). NOte reorganization of mitochondria (Mi)
into a column as in the original heart tissue (see Fig. 3). Upper
cell is attached tightly to a second myocardial cell at the sarcolemma (Sa) and includes three intercalated discs (Id) at arrows,
lower, cell shows groups of myofilaments in a more primitive stage
of redifferentiation. Note several groups of myofilaments at lower-
right which connect to dense plaque at intercalated disc. Same
fixation and staining as in Figure 3. X24,000. Marker = 1 μ (ref.
16).

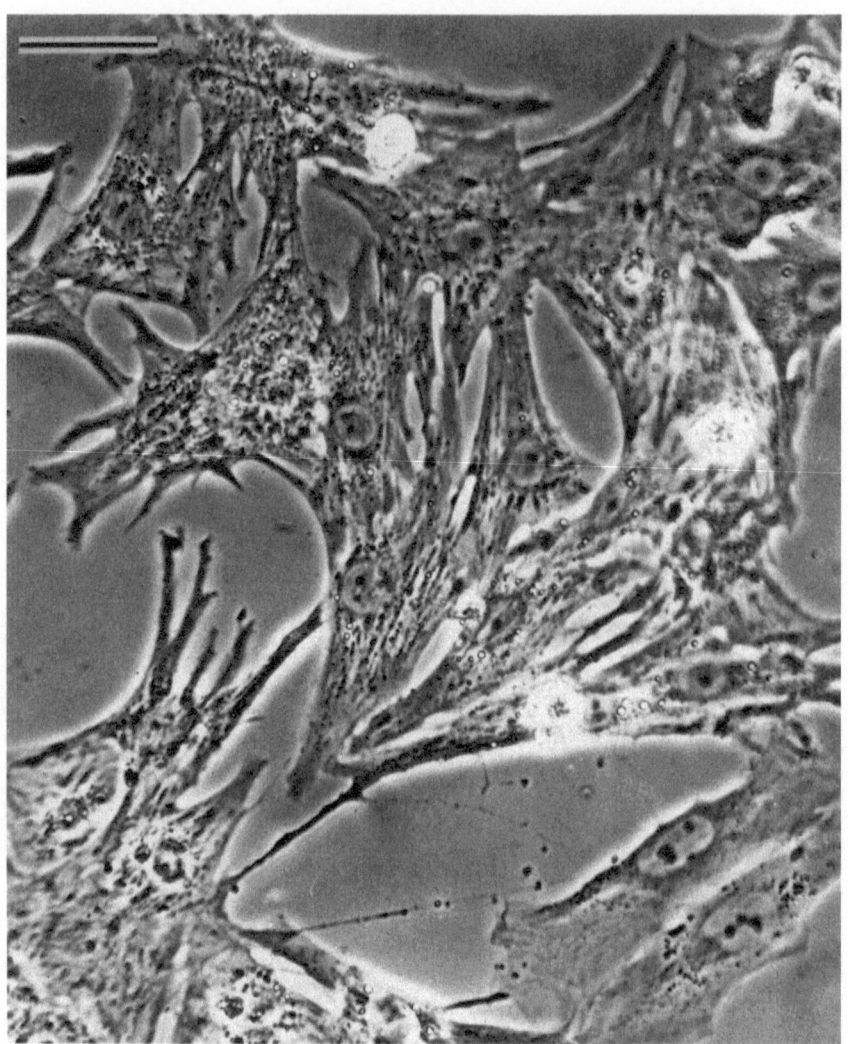

Fig. 9.   Four day culture of myocardial cells following removal
of 90-95 percent endothelial cells by "differential attachment
technique".   Myocardial cells contain numerous, dense mitochondria.
Two endothelial cells (in corner of photograph) contain thin, filmy
cytoplasm.   Zeiss Neofluor phase optics.   X25 X1.25 Optovar.
(X185) Marker = 100μ.

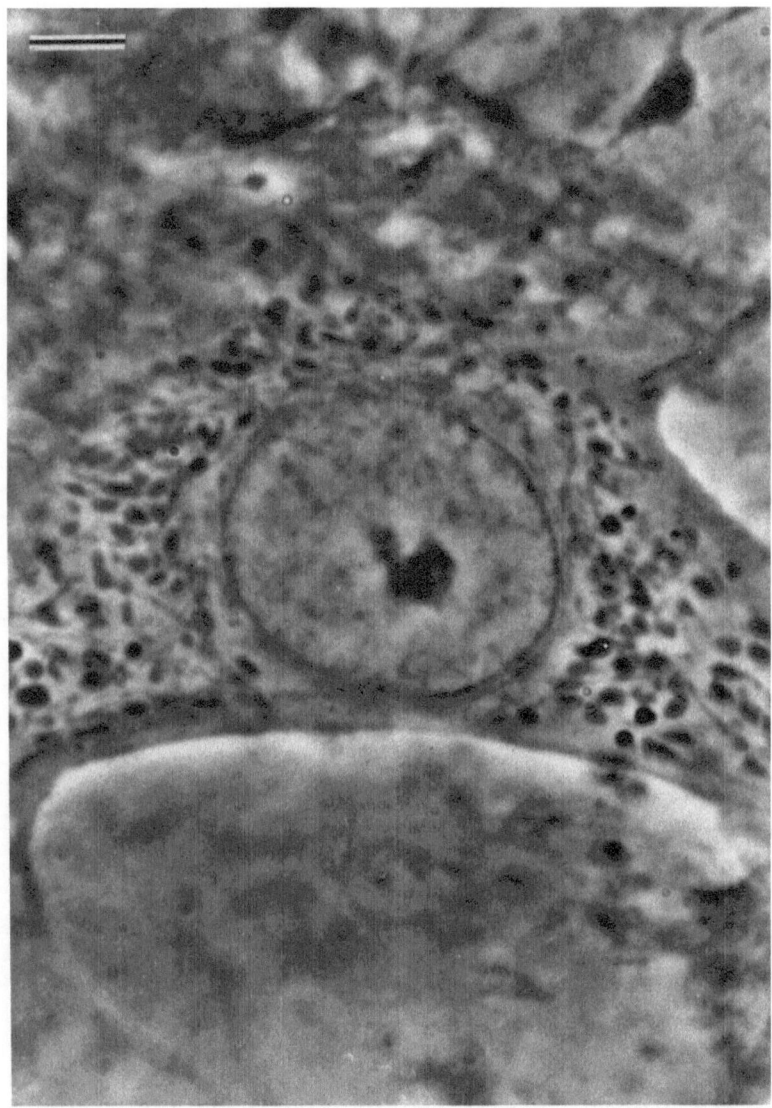

Fig. 10. High magnification view of myocardial cell from a 12-
day culture. Spherical nucleus has a single nucleolus and is
surrounded by a thin membrane, the nuclear-associated Golgi
apparatus (NAGA). Dense bodies in cytoplasm are mitochondria.
Cell was contracting 25 beats per min at time picture was taken.
Zeiss Neofluor phase optics. X100 X20 Optovar. (X1,300) Marker =
10 μ.

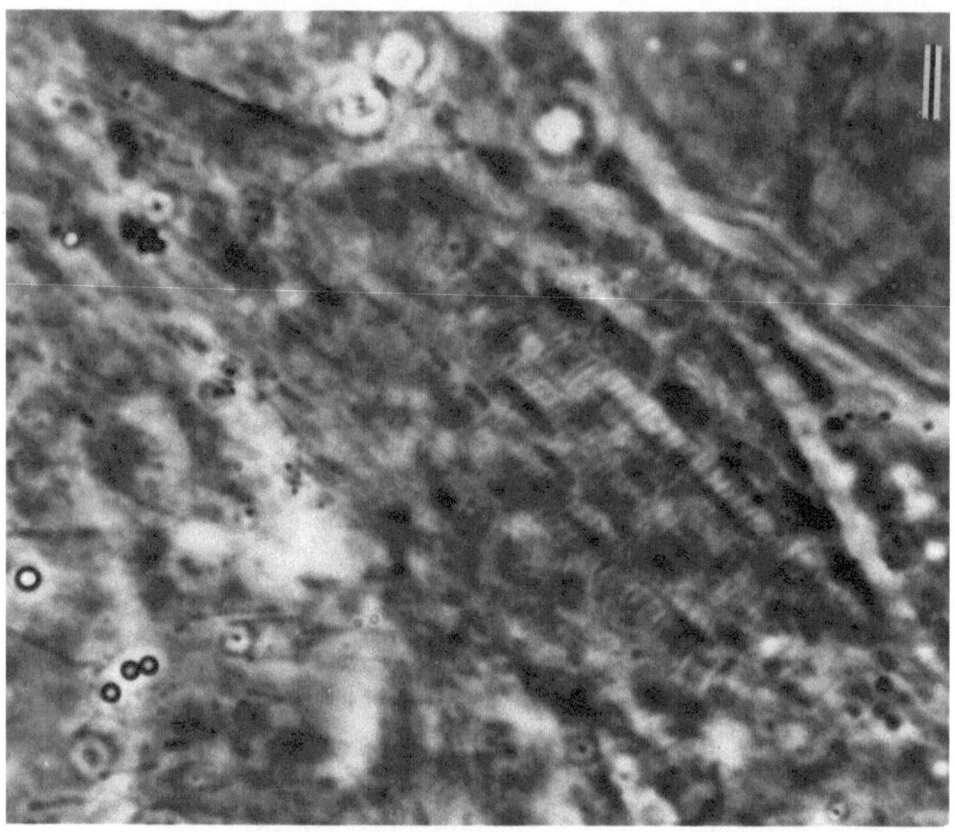

Fig. 11. Bands of myofibrils are seen in a well-differentiated myocardial cell from a 14-day culture. Typical banding patterns of
sarcomeres are present. Zeiss Neofluor phase optics. X100 X1.6
Optovar. (X960) Marker = 10μ.

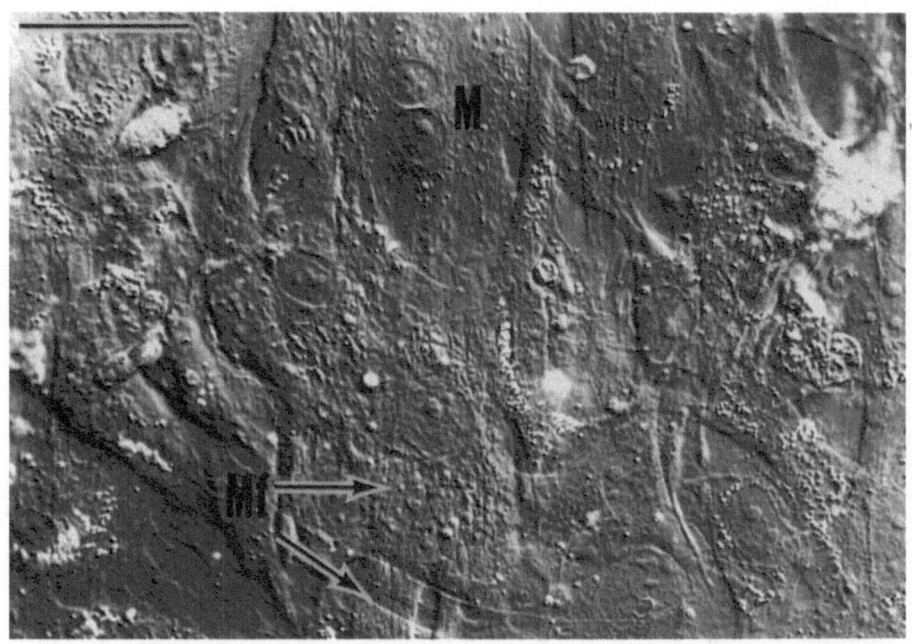

Fig. 12. Seven-day culture reveals complex multilayered relationship of myocardial and endothelial cells. Myofibrillar networks
(mf) are seen. Note binucleated myocardial cell (M). Phase-
interference optics (Nomarski). X40 X1.25 Optovar. (X290) Marker =
100μ .

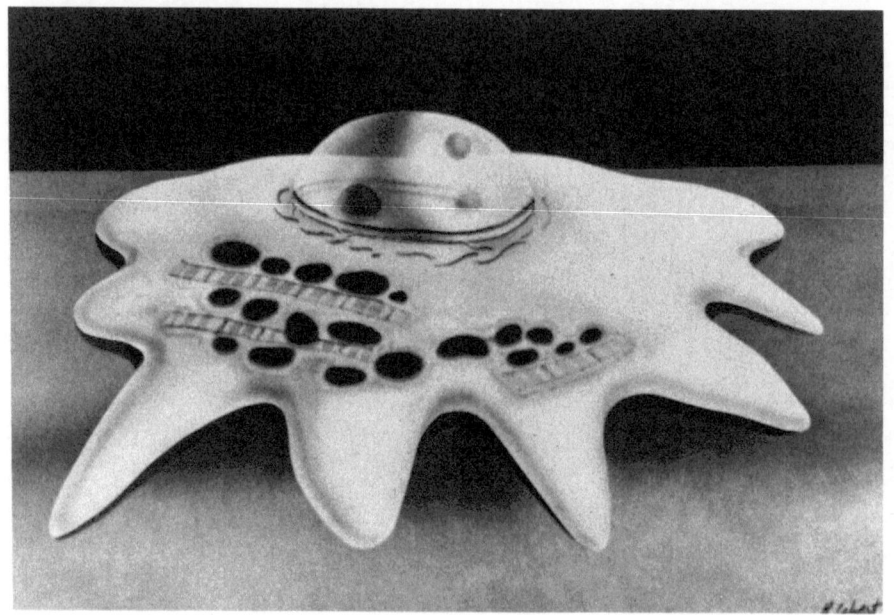

Fig. 13. Three-dimensional drawing of typical cardiac cell in culture. Typical cell is approximately 100 in length. See text for description.

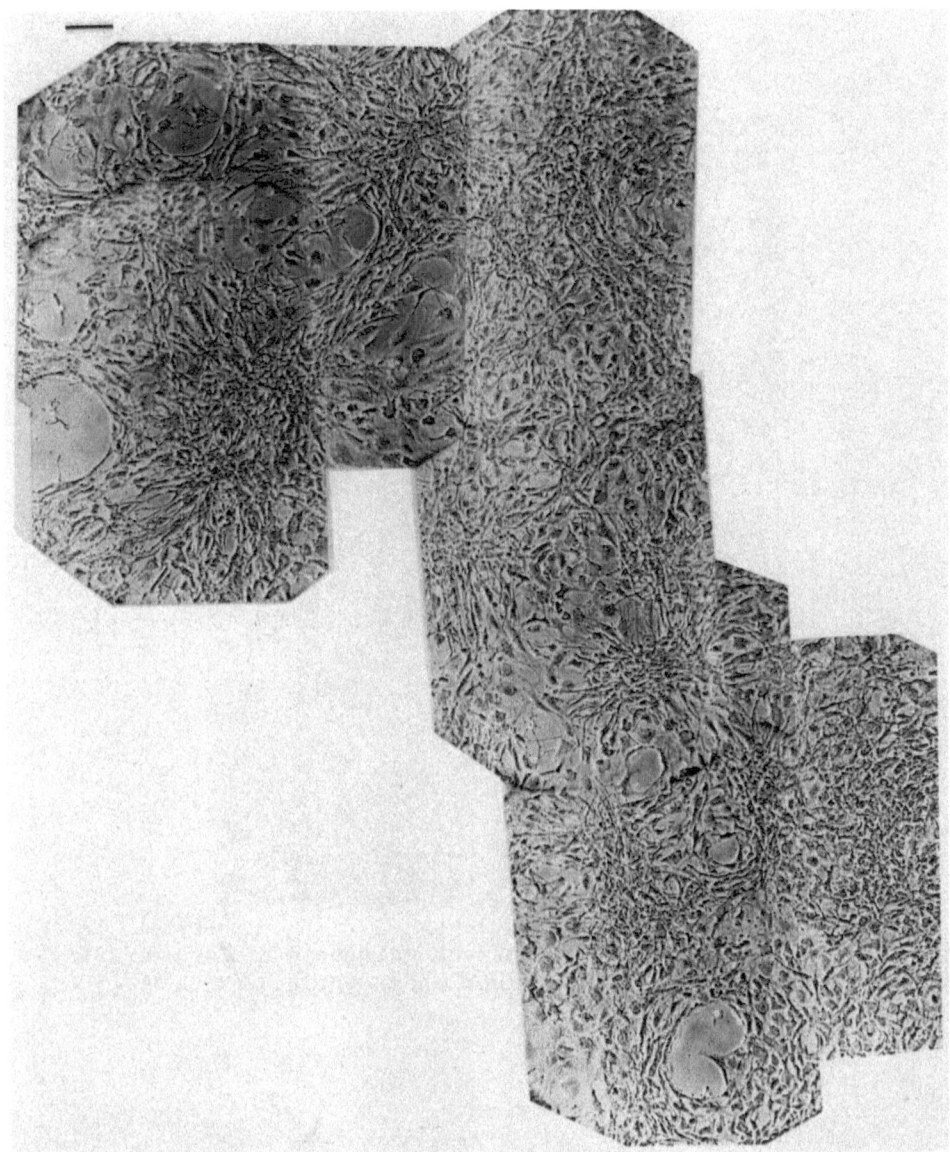

Fig. 14.   Twenty-eight-day culture consisting of multiple foci of
interconnecting beating networks.   Montage made with Nikon phase
objective.   X10 X6.   (X145) Marker = 100μ.

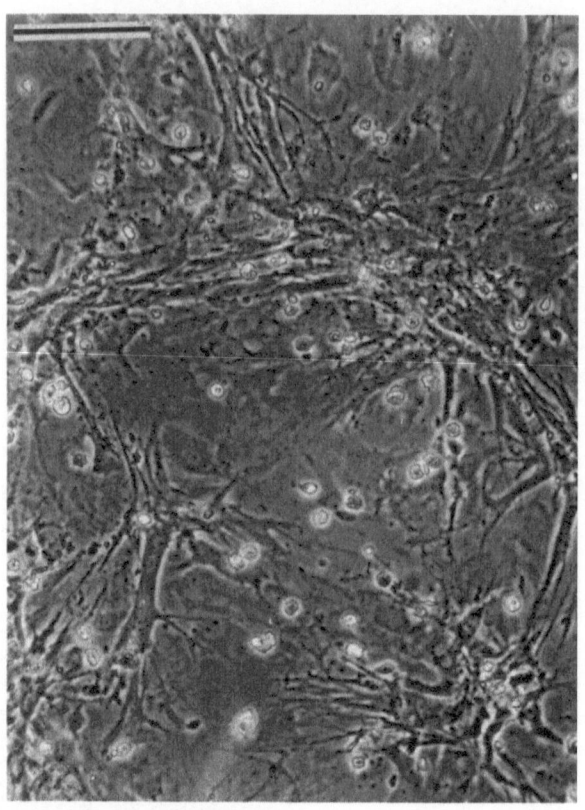

Fig. 15.   Ring of contracting networks connected with extensions.
Eleven-day culture.   Zeiss Neofluor phase optics.   X16 X1.25
Optovar (X90) Marker = 200μ.

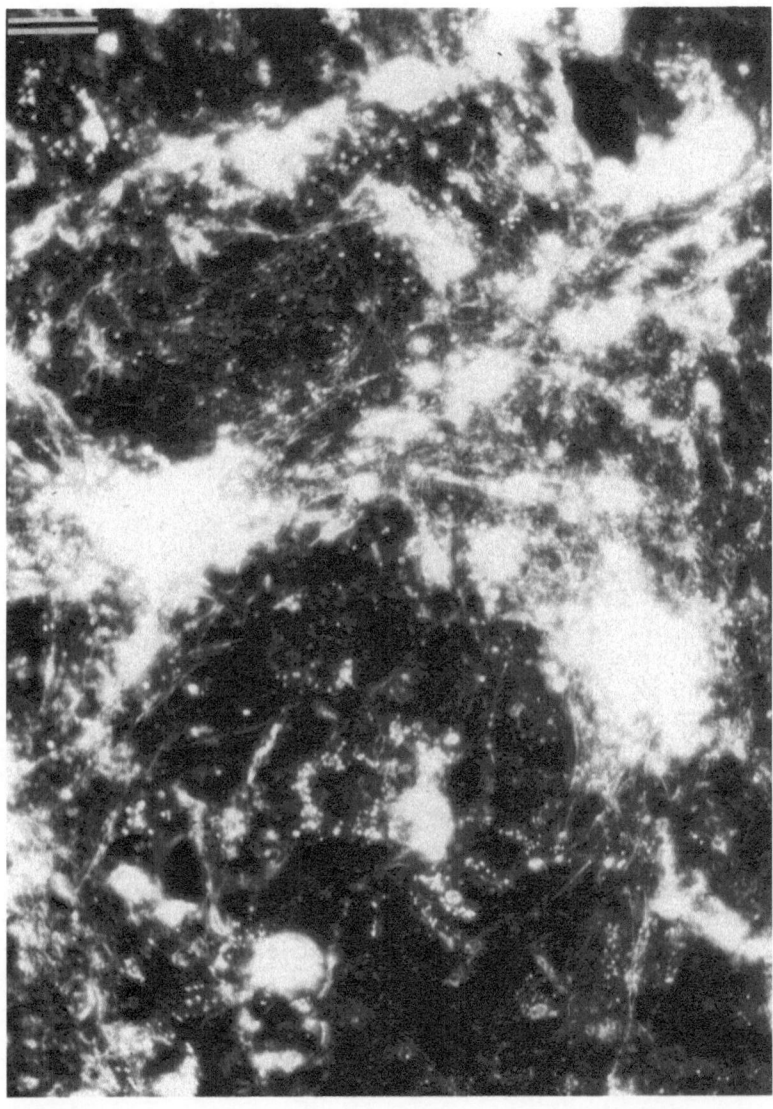

Fig. 16. Concentration of myocardial cells in networks is shown in this darkfield micrograph from a 23-day culture. Zeiss Neofluor objective. X16 X1.25. (X115) Marker = 100μ.

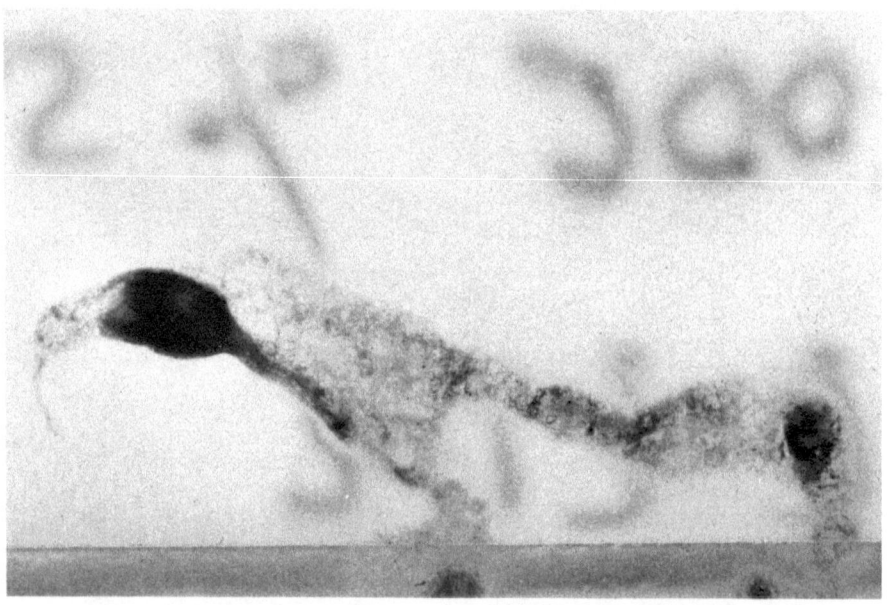

Fig. 17. Two masses of cardiac tissue or "mini hearts" following aggregation during 100-day culture in Falcon flask. The heart masses are 2 to 4 mm long and are connected by acellular fibrous strands.

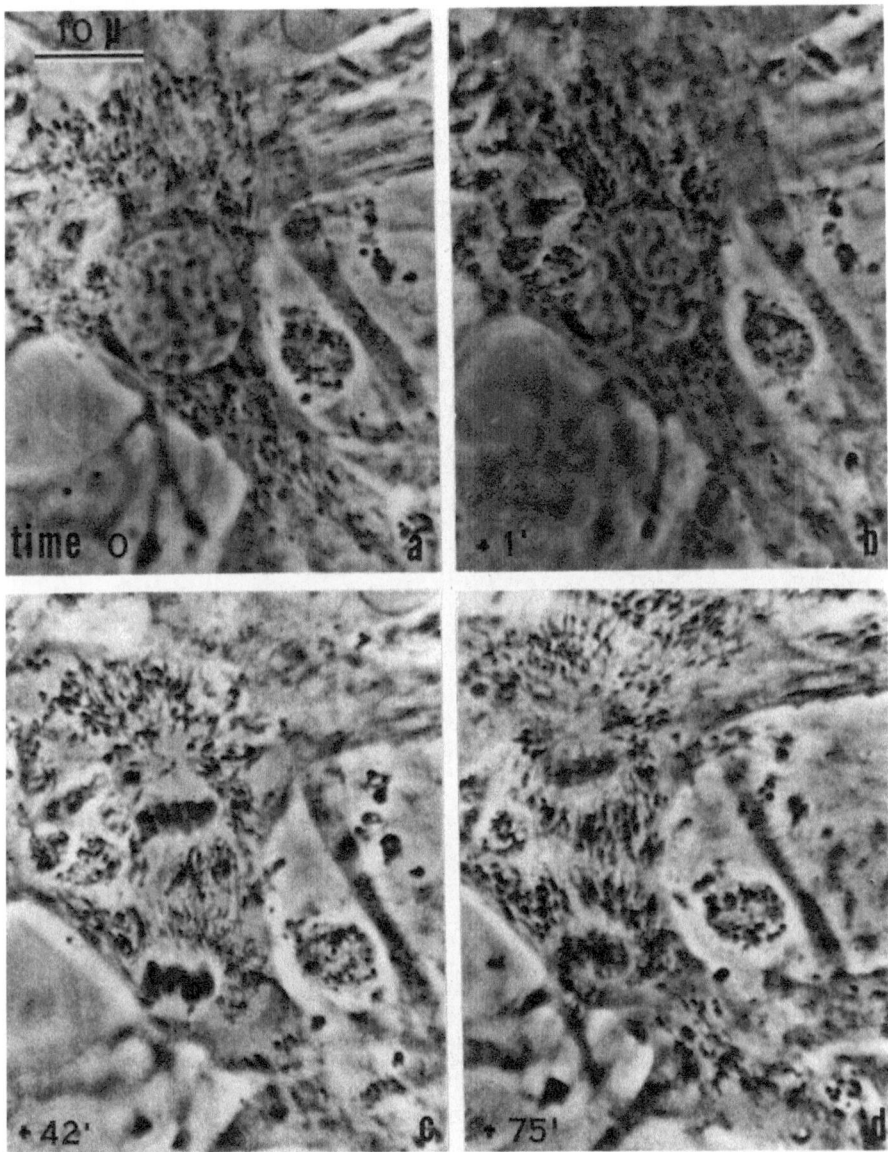

Fig. 18. Dividing myocardial cell photographed from prophase to telophase during 75-min period. Cell remains flattened and attached to other cells during mitosis. Zeiss Neofluor phase objective. X100 X1.25. Abstracts from 16 mm film originally photographed at 16 frames per second.

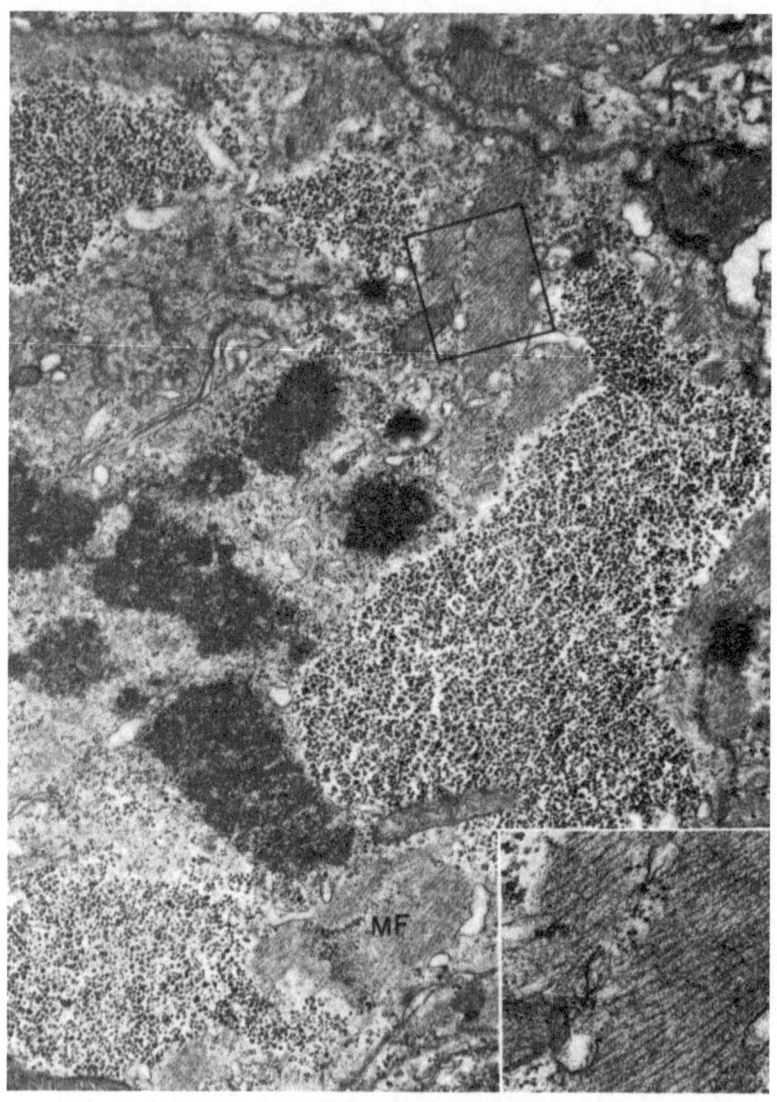

Fig. 19.   Dividing cell from the ventricle of an 11—day—old
chick embryo exhibits a normal state of differentiation.   Myofibrils (MF), chromosomes, and glycogen (G), are prominent (ref. 24).

QUESTIONS TO DR. KASTEN

Dr. Dell'Orco: How long does it take the cells to repair the
damage that was caused by trypsin?

Dr. Kasten: After isolating the original cells from the heart with
the aid of trypsin, new and complete myofibrils with typical
sarcomeres are not seen until after 24 hr. However, the cells are
capable of slight pulsations and irregular contractions before this
time.

Dr. Ryan: (1) Do you serially subcultivate your myocardial cell
cultures? (2) What percentage serum concentration do you use? (3)
What is the cell cycle time of these cells?

Dr. Kasten: (1) No, it is not possible to subculture the myocardial
cells indefinitely since the mitotic rate is too low. We have
subcultured them a few times, but cells are lost each time in
handling and eventurally there are none left. (2) Ten percent
fetal calf serum. (3) I do not know since the cells seem to go
through only 1 or 2 cycles of division. The endothelial cells
take about 24 hr to go through a cell cycle.

Dr. Sova: Have you studied the possible mechanism of the synchronisa-
tion of cell culture movement? I ask you this in connection with
the existence of the hypothesis on intercellular communication by
means of UV or IR waves.

Dr. Kasten: All I can tell you is that specialized junctions develop between myocardial cells, apparently prior to the onset of
synchrony. These junctions are of the nexus type and intercalated disc type. Synchrony can also develop with an endothelial
cell serving as a intermediate link but I have not yet seen any such
junctions between the two different cell types.

Dr. Leuenberger: Did you observe differences in the ATPase staining
pattern between synchronously and asynchronously contracting cells?

Dr. Kasten: I have not done exactly what you suggest but will look
into it. However, in one experiment we filmed a pair of actively
contracting cells in synchrony, then fixed and stained the preparation for ATPase. We found that the intercalated disc connecting
the two cells was unusually rich in the enzyme. As one would suspect,
energy is required to provide the means for the transfer of impulse
from one cell to another.

Dr. Courtois: Do they stop beating when dividing?

Dr. Kasten: If the cell is beating in prophase, then it always

stops beating in metaphase, and only resumes again when the
daughter cell(s) are reformed. Hand-in-hand with this temporary
loss of function is a disorganization and apparent loss of organized
myofibrils. The loss of Z lines during mitosis has now been confirmed in vivo. Apparently, the energy required for mitosis
(chromosome movements, spindle fiber elongation, etc.) is somehow
given priority over that needed for contraction. The division time
for myocardial cells is about siz hr, whihc is twice as long as
that taken by endothelial cells and apparently developed to give
sufficient time for demyofibrilogenesis to occur.

Dr. Franks: Is it possible to make a fine distinction between endothelial cells and fibroblasts by morphology alone?

Dr. Kasten: This is a good point. From personal experience, I
would say that these cells do not resemble fibroblasts in morphology
or growth pattern. However, we are looking for more exacting
criteria. For example, it is my impression that endothelial cells
are unusually phagocytic. The non-myocardial cells in these cultures
take up carbon particles from India ink in high concentrations.

Dr. Morasca: Did you study the effect of anoxia, or, in the Rose
chamber system have you tried to establish the response of heart cells
to an artificial aging situation. I am wondering also if you have
started studying the response to pharmacological agents of these
cells by the method you described.

Dr. Kasten: The Rose chamber is a somewhat closed system since the
chamber is sealed and the silicone rubber gasket is only slightly
permeable to the atmosphere. Normally the bubble of air left in
the medium is sufficient for young cultures. As they aggregate and
form "mini hearts", their oxygen needs are greater. Without special
means to increase the supply of oxygen, some degeneration occurs in
the center of the chamber, leaving a peripheral ring of myocardial
tissue close totthe gasket. I have not tried to study the anoxic
effect (also a few reports have appeared by Hackney, Mark and
Strasser and by Wollenberger and associates). As you indicate, the
system described is ideal for studying the action of drugs. By
hooking up a perfusion pump to the Rose chamber we can follow the
contractile pattern of a network, then perfuse in the desired agent,
and record directly any changes in contraction frequency pattern
and strength. A description of the initial experimental procedure
appears in "Tissue Culture Methods and Application", Kruse and
Patterson, editors, Cacdemic Press, New York, N.Y., 1973.

Dr. Rowlatt: Do you consider that the trypsin enters the cells?

Dr. Kasten: Yes, the trypsin-isolated myocardial cells contain abnormal granules under the plasma membrane. Also the sarcomeres are

preferentially broken down at the Z lines, leaving myofilaments.
Abnormal granules under the plasma membrane. Also the sarcomeres
are preferentially broken down at the Z lines, leaving myofilaments.
Abnormal granules and cytoplasmic damage is not seen in endothelial
cells presented in the same group of isolated cells.

Dr. Hanzlikova: Did you compare the frequency of contraction of well
preserved cells with the contraction frequency of the heart of animals
of the same age?

Dr. Kasten: The newborn rat heart has a very high contraction rate
probably a few hundred times a min. Cells in the initial culture
have a spectrum of contraction rates ranging from approximately 15
to 130 beats per min. As synchronized networks develop, their
rates tend to be in the range of 50-100 beats a minute. As very
large networks and "mini hearts" form, the rate is about 50-60
beats per min. The gradual decrease in vitro parallels that occurring
in the animal with aging. However, it is only fair to say that the
decreased contraction rate in culture would be explained by the
development of a noxious or anoxic environment.

Dr. Ryan: To what extent does syncytial formation occur in the
myocardial cell cultures?

Dr. Kasten: Multicellular networks form by pulling together of
myocardial cells and by natural attachment of daughter cells after
mitosis. Myofibrils from one cell frequently appear to traverse the
intercalated disc and continue into the next cell without interruption of banding pattern (the intercalated disc takes the place
of a Z line). Also by EM some myofilaments are seen to pass through
the region of the discs. These observations would suggest that a
true syncytium can exist. However, this is a generally accepted
theory and I think that more observations are needed.

Dr. Hay: I have 2 questions. Firstly, you indicated that the rat
myocardial cells could be frozen, thawed and recultivated several
times. Since, presumably each cycle requires retrypsinization do
you believe that the cells can resynthetize myofilaments, damaged by
trypsin treatment, after each freezing cycle? Also, were these
trials performed on cells in early stages, say the first few days
in culture, or over a more extended interval? Secondly, with regard
to the clones of heart cells shown in the slide of the culture
flask; do these represent aggregates or clumps of cells as I understand that you believe the myocardial cells divide only once after
isolation?

Dr. Kasten: This is a good question. Actually my EM observations on
trypsin damage were done on freshly isolated cells from the animal.
This treatment is more extensive than that needed to detach cells

from the plastic or glass surface. However, I am willing to believe
that even here some kind of trypsin damage occurs. I discussed this
problem in a recent paper dealing with myocardial cells frozen-thawed
and cultured 3 times. Using the trypan blue-dye exclusion test
and by counting contractile cells before freezing and afterwards
the results suggested that a high proportion of the cells were viable
and active in spite of trypsin treatment. Actually, we need to
look at the trypsinized cells by EM to really answer your question.
The culture cells isolated for freezing came from relatively old
cultures after networks were formed. As for your final question, what
you saw were not clones but simply large aggregates or "mini hearts"
present in old primary cultures. They form by "pulling" each other
together.

ULTRASTRUCTURAL CHANGES IN SENILE MUSCLE

V. Hanzlíková and E. Gutmann

Institute of Physiology, Czechoslovak Academy of
Sciences, Prague, Czechoslovakia

Weight loss and related atrophy is characteristic for all senile
tissues. The senile muscle change shows, however, two specific
features: (a) it is more marked than that of body weight, (b)
it is different in different types of muscles or muscle fibers. The
reduction of muscle weight varies in different muscles, some of them
being affected relatively very late (for example the deep back
muscles). The decrease of muscle weight is due to decrease both
in number (1) and diameter (2,3,4) of muscle fibers. On the other
hand an increase of incidence of largest fibers diameter was described
in the eye muscles and explained as a result of compensatory hypertrophy of myocardium after focal degeneration in some muscle fibers
(6).

Mammalian skeletal muscles are mostly composed of different
fiber types. The fibers are traditionally classified as red, white
and intermediate fibers. Differences of the fibers have been observed at the level of both the light and electron microscope on
either a morphological or cytochemical basis (7,8). It has been
shown, that there are also differences in the aging process of red
and white muscle fibers. In the senile muscle the white muscle
fibers decrease in volume, whereas the red fibers decrease in number
(4). As regards the size of muscle fibers, the adult fast-twitch
extensor digitorum longus muscle of the rat, which shows typical
heterogeneity, loses in senescence this size variety; the largest
fibers decrease in diameter and thus the senile muscle shows practically a homogeneous population of fibers. On the contrary the slow-
twitch soleus muscle fibers retain in their majority the same size,
only a small unit of fast-twitch muscle fibers (8) has a significantly
smaller diameter of muscle fibers (9).

Senile muscle changes concern all structures, some of them being specific, some of them showing general reaction of the muscle cell to neuromuscular disturbance. The marked reduction of muscle fibers is closely connected with a marked reduction of myofibrils. This process proceeds apparently very slowly at the periphery of the muscle fiber, where disorganization and loss of myofibrils can be detected. A conspicuous alteration is observed in the behavior of the sarcolemma, which is often no longer regular and smoothly engulfing the myofibrils, but forms large protrusions and finger-like projections into the extracellular space, so that the regular assembly of single muscle fibers is disturbed (Fig. 1). In these new proliferating undifferentiated sarcolemmal processes a high accumulation of ribosomes is found. The basement membrane is maintained and in most cases thickened. The increased activity of the plasma membrane is expressed by extensive proliferation of the tubular T-system, which forms many trifoliate structures in the subsarcolemmal region. In some fibers there is a proliferation of sarcoplasmic system, especially the enlargement of terminal cisterns, connected with overdeveloped T-system, forming the pentads instead of triads complexes, similar to that of denervated muscle (10).

No mitotic figures of the nuclei have been observed in senile muscle but nuclear fragmentation and occasionally central locali-zation of the nucleus can be found. The chromatin of the nuclei is highly condensated at the periphery. Frequently there are nucleo-li attached to the boundery of the nucleolemma.

Disorganization and disintegration of myofilament sets is coupled with marked proliferation of sarcomplasmic reticulum and increased ribosomal activity. There are subsarcolemmal regions in some muscle fibers, which contain free, not organized myofilaments, microtubules, triads and Z bands and this part of the sarcoplasm appears to be related to the new synthesis of contractile proteins. In the subsarcolemmal area the parallel arrangement of large electron-light nuclei can also be seen, suggesting an "abortive regenerative" growth process.

On the other hand cytoplasmic degradation and autophagic processes appear to be confined to the periphery of muscle fibers. In the perinuclear zone mery large autophagic vacuoles, multi-vesicular bodies, increased pinocytotic activity and increased number of components of Golgi apparatus can be observed. Accumulation of glycogen granules in mitochondria is seen in the same muscle fiber, where cytoplasmic sequestration in cytosegresomes are found.

The Z-line of the contractile apparatus may have an irregular shape. Enlargement or streaming of the Z band resembles the patho-logical reaction of the Z band observed in myopathies, acute de-

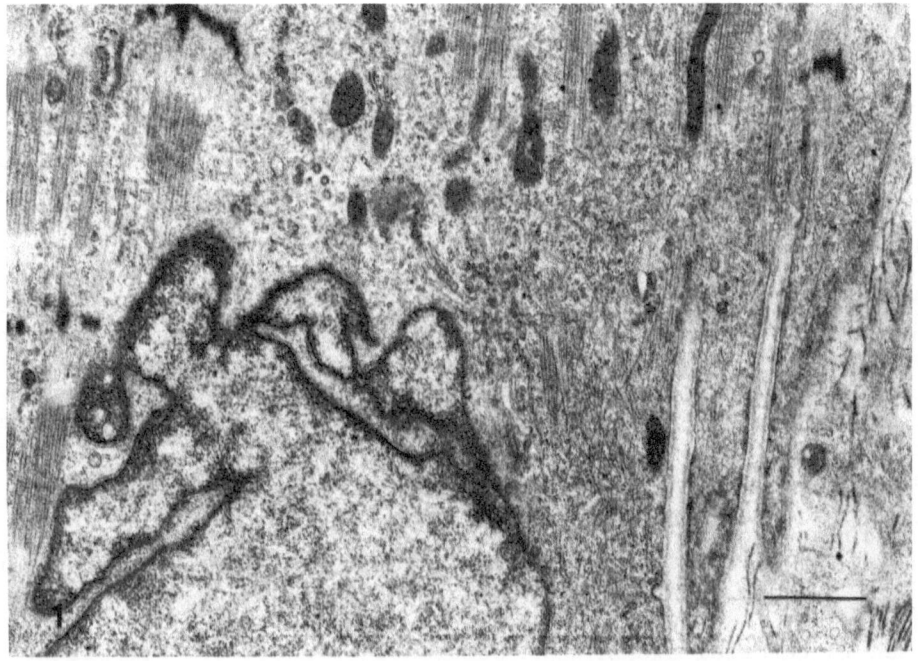

Fig. 1. Soleus muscle of the 31-month-old rat. Note finger-like protrusions of the sarcoplasm, disorganized myofilaments at the periphery of the muscle fiber and fragmentation of the nucleus. Scale bar = 1 μm.

nervation, motor neuron diseases, etc. (see 11), which begins by
a broadening and extension of material from the Z disks into the
adjacent I and A band and results in loss of sarcomere integrity.
The change of the Z band is connected with loss of mitochondria in
this region. The selective mulnerability of the Z band of red muscle
fibers possibly indicates greater dependence of red muscle fibers
on pathways of oxidative metabolism for maintaining their structural
integrity (11).

Practically in all fibers of the slow senile muscle lipofucin
granules were found. The most frequent occurence of these granules
is in the perinuclear region in the proximity of the Golgi apparatus
and among large accumulation of mitochondria. Lipofuscin granules
are repeatedly described in senile heart muscle cells (12,13) and
are presumed to be general ageing phenomena both in vertebrate and
invertebrate species (6). In senile skeletal muscle mainly in
the slow type, like in the soleus muscle, abundant lipofuscin
granules are observed. These residual bodies have a different shape,
structure and size and are located subsarcolemmaly at the periphery
of the muscle fiber and exceptionally in the central region if the
nucleus is centrally located (Fig. 2). Similar residual bodies
are described in the left ventricular myocardial cells (13). They
are located in the perinuclear zone and in the region in which mitochondria and lipid droplets occur and are later degradated to residual
bodies. Mitochondrial degradation also occurs in autophagic vacuoles (Fig. 3). Lipid droplets are segregated by smooth membraneous
elements of the sarcoplasmic reticulum for their presumed degradation
in autophagic macuoles. Lipid droplets of different size are frequently seen in the multiglobular type of lipofuscin granules.

Senile muscle atrophy thus presents specific features in respect
to morphology of the neuromuscular junction. Characteristic features
are exhibited by the ultrastructure of the motor endplates. Compared with a young endplate the number of synaptic vesicles is increased in terminal exons. The occurence of increased numbers of
mitochondria, myelin figures (possibly remnants of degenerated
mitochondria), occasional neurotubules and neurofilaments in intact nerve terminals suggests some disturbance in proximo-distal
transport of proteins. The postsynaptic junctional folds are
widely branched, primary synaptic clefts are enlarged and basement
membrane is considerably thickened. No indication of degenerative
changes as described after nerve section (14) are found.

The complex changes thus comprise (a) decrease of protein synthesis (14) due to a disturbance of neuromuscular contact, connected
with related changes in synthesis and release of neurotransmitter
and neurotrophic agents (15-17), which is indicated by the marked
nuclear abnormalities, (b) compensatory "abortive" synthesis of
myofilaments, specific intracellular metabolic changes (18), (c)

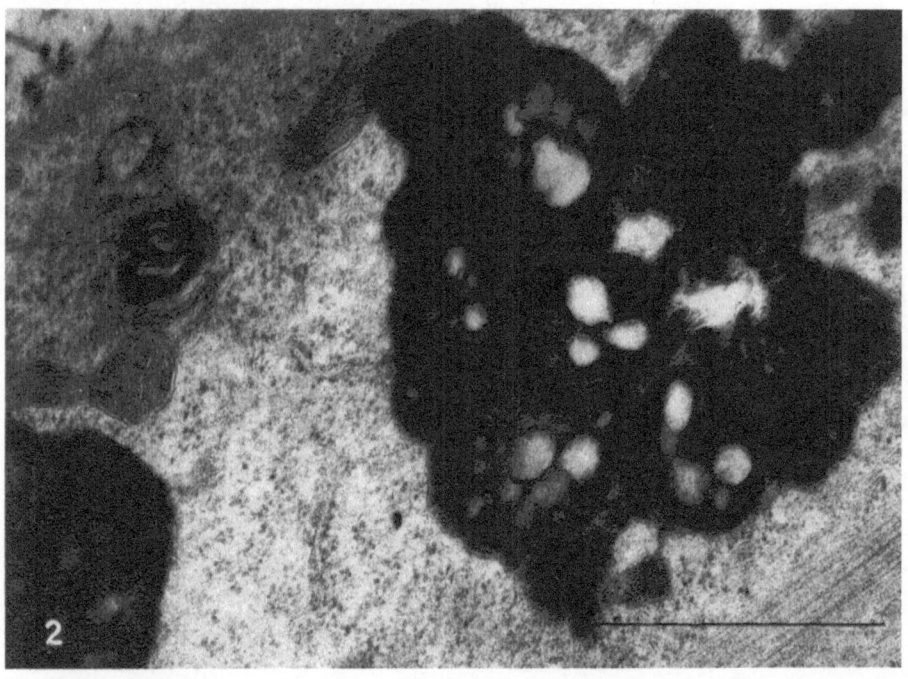

Fig. 2. Extensor digitorum longus muscle of the 31-month-old rat.
Subsarcolemmal occurence of large lipofuscin pigment granules.
Scale bar = 1 μm.

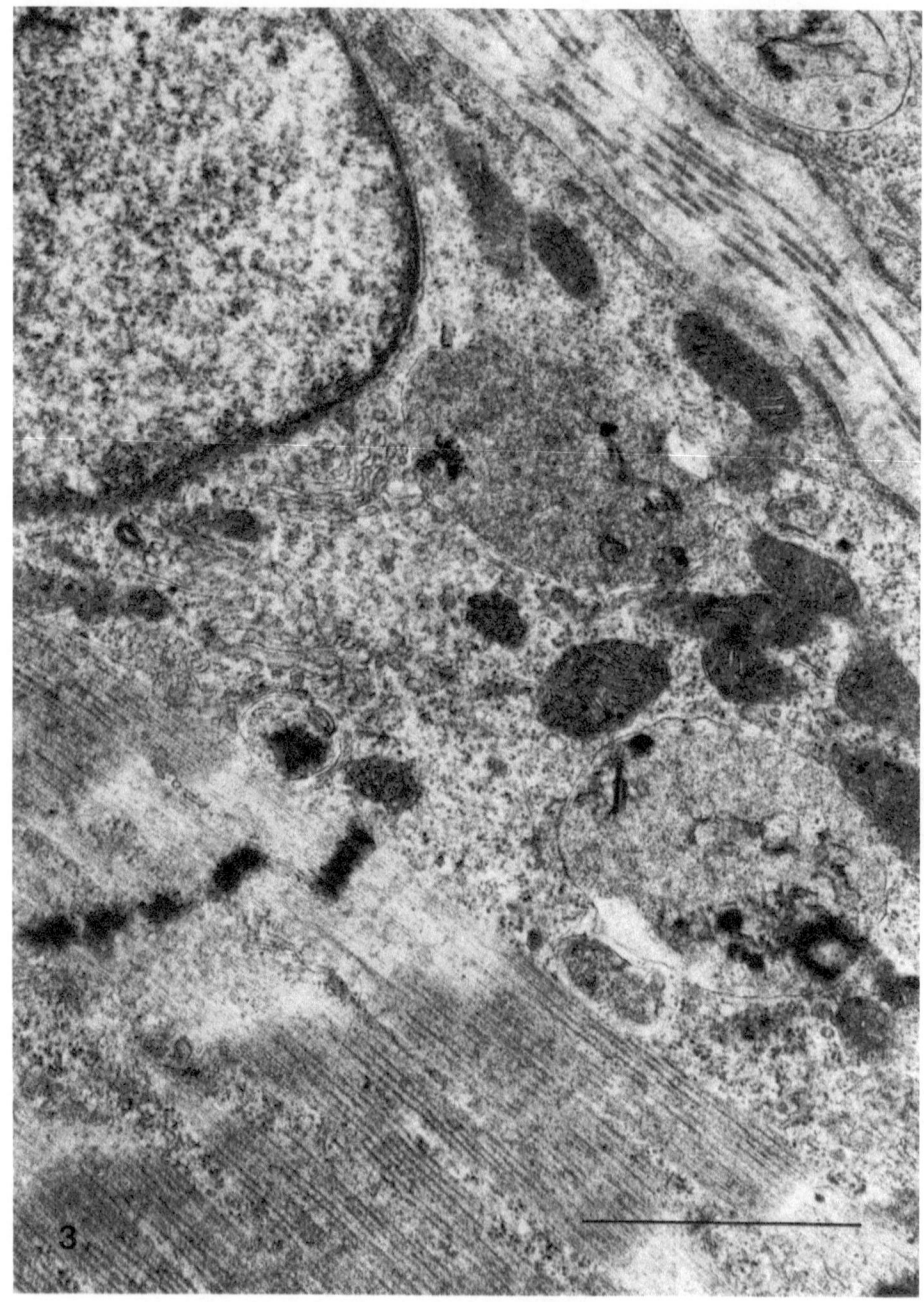

Fig. 3.   Soleus muscle of the 31-month-old rat.   Many autophagic
vacuoles with granular matrix, undigested remnants and glycogen
granules, are seen in the perinuclear zone.   Scale bar = 1 μm.

increase of degradation of intracellular proteins as evidenced by the occurence of autophagic vacuoles and lipofuscin granules, and (d) deterioration of transport mechanism in sarcolemma and mitochondrial membranes related to the marked accumulation of collageneous fibrils in the intercellular space (19) and mitochonrial degeneration respectively (for review see 20).

The clearer differentiation of myogenic and neurogenic mechanisms involved in the aging process remains a further task of experimental gerontology.

REFERENCES

1. Gutmann, E. and Hanzlíková, V. 19661 Motor unit in old age. Nature (London) 209: 921.

2. Frubel-Osipova, S.I. 1969. The neuromuscular system. IN Basis of Gerontology (in Russian) (D.F. Chebotarev, N.V. Mankovski and V.V. Frolkis, editors), Medicine, Moscow, pp. 128-139.

3. Rowe, R.W.D. 1969. The effect of senility on skeletal muscles in the mouse. Expl. Gerontol. 4: 119.

4. Tauchi, H., Yoshioka, T. and Kobayashi, H. 1971. Age change of skeletal muscles of rats. Gerontologia 17: 219.

5. Bucciante, L. and Luria, S. 1934. Structure and function of muscle (J.L. Rubinstein, editor), vol. 3, p. 209, 1960.

6. Tomanek, R.J. and Karlsson, U.L. 1973. Myocardial ultrastructure of young and senescent rats. J. Ultrastr. Res. 42: 201.

7. Padykula, H.A. and Gauthier, G.F. 1966. Morphological and cytochemical characteristics of fiber types in normal mammalian skeletal muscle. IN Exploratory Concepts in Muscular Dystrophy and Related Disorders (A. T. Milhorat, editor), New York, pp. 117-128.

8. Schiaffino, S., Hanzlíková, V. and Pierobon, S. 1970. Relations between structure and function in rat skeletal muscle fibers. J. Cell Biol. 47: 107.

9. Bass, A., Gutmann, E. and Hanzlíková, V. 1974. Biochemical and histochemical changes in enzyme pattern of muscle of the rat during old age. Manuscript in preparation.

10. Pellegrino, C. and Franzini, C. 1963. An electron microscope
     study of denervation atrophy in red   white skeletal
     muscle fibers. J. Cell Biol. 17: 327.

11. Engel, A.G. 1966. Pathological reactions of the Z disk. IN
     Exploratory Concepts in Muscular Dystrophy and Related Dis-
     orders (A. T. Milhorat, editor), New York, pp. 398-411.

12. Strehler, B.L., Mark, D., Mildran, A.S. and Gree, M. 1959.
     Rate and magnitude of age pigment accumulation in the human
     myocardium. J. Gerontol. 14: 430.

13. Travis, D.F. and Travis, A. 1972. Ultrastructural changes in
     the left ventricular myocardial cells with age. J. Ultrastr.
     Res. 39: 124.

14. Miledi, R. and Slater, C.R. 1970. On the degeneration of rat
     neuromuscular junctions after nerve section. J. Physiol. 207:
     507.

15. Gutmann, E., Hanzlíková, V. and Vyskočil, F. 1971. Age
     changes in cross striated muscle of the rat. J. Physiol. 219:
     331.

16. Tuček, S. and Gutmann, E. 1973. Choline acetyltransferase
     activity in muscles of old rats. Exp. Neurol. 38: 349.

17. Frolkis, V.V., Bezrukov, V.V., Duplenko, Y.K., Shchegoleva, I.V.,
     Shevtchuk, V.G. and Verkhratsky, N.S. 1973. Acetylchonine
     metabolism and cholinergic regulation of functions in aging.
     Gerontologia 19: 45.

18. Ermini, M., Szelényi, I., Moser, P. and Verzár, F. 1971.
     The ageing of skeletal (striated) muscle by changes of
     recovery metabolism. Gerontologia 17: 300.

19. Dejl, Z. 1972. Role of connective tissue changes in an or-
     ganism. IN Proc. 9th Internat. Congr. of Gerontol., Kiev,
     July 2-7, vol. 2, p. 110.

20. Gutmann, E. and Hanzlíková, V. 1972. Age changes in the
     neuromuscular system. Scientechnica (Publs.) LTD., Bristol.

QUESTIONS TO DR. HANZLÍKOVÁ

Dr. Balazs: First I would like to congratulate you not only for
the excellent lecture and EM demonstration, but because they
represented real ageing changes. That is emphasized, because

some of the time-dependent laterations, shown in pervious lectures, represented processes of differentiation (or degeneration) but not ageing. Secondly I ask, how frequently the "honey-comb" structures and "glycogenbodies" consisting of ER and glycogen, occurred in the old muscle cells.

Dr. Hanzlíková: Both "honey-comb" structures and "glycogen-bodies" were found in such muscle fibers, in which also other structures are affected, for example the streaming of Z-lines. These structures mostly occur in slow muscle fibers and their frequency is very rare in old muscles, but they are often found and are described in muscle fibers after denervation.

Dr. Wilson: I noticed in a lot of your electron micrographs of old muscle there were mitochondria with condensed configurations. I wonder what was your fixation method and whether the same condensed mitochondria occur in the young animals.

Dr. Hanzlíková: In our experiments we have used glutaraldehyde fixation. I agree with your comment, as the mitochondrial structure in senile muscles show great alterations. We have observed al-terations in the orientation of the inner cristae, which are in many cases densely packed. In the subsarcolemmae region, where normally a great accumulation of mitochondria is found, lipofuscin granules are seen in close contact with mitochondria. This would suggest the relationship of the origin of lipofuscin granules to degenerating mitochondria.

Dr. Ciampor: Did you observe some changes in the nucleoli of senile muscle cells connected with the activity of the nucleolus, it means changes in the ratio of fibrillar preparticular RNP and granular RNP components of the nucleolus?

Dr. Hanzlíková: I am sorry we haven't done such a detailed study of nucleoli in senile muscles. According tooour observations I can only s ay, that there are conspicious changes of nuclei and nucleoli. Nucleoli are either frequent or they have very complicated branched structure. It is necessary to give more attention to this important cell structure.

DENERVATION, REINNERVATION AND REGENERATION OF SENILE MUSCLE

E. Gutmann and V. Hanzlíková

Institute of Physiology, Czechoslovak Academy
of Sciences, Prague, Czechoslovakia

Denervation interrupts, reinnervation and regeneration re-
covers the metabolic interaction between nerve and muscle
cells. Whatever the nature of this interaction (1,2) in
old age, it is clearly exposed by the related changes in the motor
units, defined by Sherrington (1919) as the population of muscle
fibres innervated by a single neuron. Studies on "senile motor
units" (3) rest on observations of a close correlation between
physiological and biochemical properties of muscle fibres innervated by the same nerve cell (4,5). The differential behaviour
of fast and slow muscles and muscle fibres is demonstrated especially by studies on contraction and histochemical properties
of fast and slow muscles and muscle fibres which exhibit high
or low speed of contraction and a related degree of ATPase activity (Type II muscle fibres with high activity and Type I with low
activity) (6), and mostly a different degree of oxidative enzyme
activity (7) (A, B, and C fibres). This differentiation is more
refined, when based on ultrastructural studies (8,9), but it is
sufficient as a basic means of classification.

A differential reaction of muscles and muscle fibres are described, e.g. after denervation (10), in disease (11) and during
development (12,13). The basic criterion in classification of
motor units is apparently speed of contraction and related myosin
ATPase activity.

Changes in denervated, reinnervated and regenerating senile
muscle will therefore best be understood in respect to changes in
the senile motor unit, concerning especially contraction time (CT)
and histochemical muscle fibre pattern especially regarding myosin

431

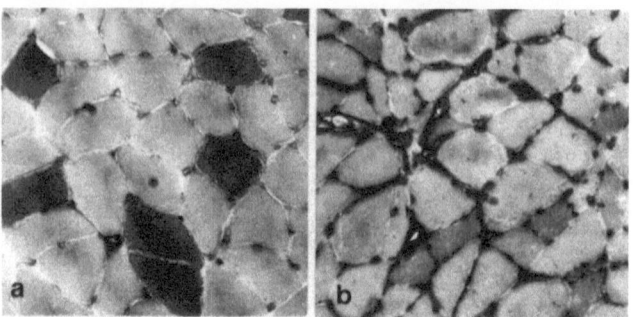

Fig. 1. Cross section through the soleus muscle of young (a) and
senile (b) rat, stained for ATPase enzyme activity. Note mixed
fibre pattern with prevalence of Type I fibres in muscle of young
and tendency to uniform fibre pattern with predominant atrophy and
decrease of ATPase activity in Type II fibres in muscle of old
animal.

ATPase activity. During old age the mixed "mosaic" pattern of
fast and slow muscle fibres (which is due to intermingling of
different muscle fibres belonging to different motor units) is
changed to a more uniform-homogeneous pattern, e.g. in respect
to myosin ATPase and succinicdehydrogenase (SDH) activity (Fig. 1a,
b) (3,14,15). This change proceeds differently in different muscles. In the slow soleus muscle there is a preferential atrophy
and decrease of ATPase activity in Type II muscle fibres but no
or even increase of activity in the more numerous Type I muscle
fibres. Thus it is especially the fast motor unit which is affected. In the diaphragm muscle, however, the mixed histochemical
fibre distribution is maintained in old age and this is explained
by the continuous activation of respiratory motor units in the
breathing process (3,15).

     Fast and slow muscles behave differently during aging. All
the fast muscles studied show progressive prolongation of CT (16).
However, the senile slow soleus muscle exhibits a shortening of CT
and increased related myosin ATPase activity at least in very old
animals (3,15) as suggested by the histochemical studies (Fig. 2).
New synthesis of myofilaments which are fast at early stages of
development (13) or shortening of the muscle may be responsible.
The reaction to denervation also differs in fast and slow muscles
(10). The fast muscle reacts to denervation always with prolongation of CT. However, the slow soleus muscle shows shortening of
contraction time in young but prolongation of contraction time in
senile muscles (17).

During reinnervation there is recovery of CT in both fast
and slow muscles to normal or almost normal values, but the gradient
of change has a different trend due to the different CT in senile
fast and slow muscles before nerve interruption and nerve regeneration and muscle reinnervation. There is a consistent shortening
of CT in the senile reinnervated extensor digitorum longus (EDL)
muscle, the values approaching those of the EDL of young animals,
but there is shortening of contraction time in the senile reinnervated soleus muscle after an initial prolongation of CT related to the denervation process. The values thus return to those
of the "control" senile muscle. Thus, the recovery process in the
fast reinnervated EDL muscle is more successful in the sense that
a CT similar to that of the muscle in young animals is achieved.
A return to ontogenetical early and undifferentiated features may
be operative. During development there is a general slowing of
slow muscles thought to be due to progressive antigravitational
forces (13) and a decrease of antigravitational functions, expecial-
ly involving slow "tonic" muscles may be responsible for a reci-
procal regression and shortening of CT in slow senile muscles.

Three general features may, however, be defined in senile
reinnervated muscles: (a) deficiencies in recovery of the normal
mixed histochemical fibre spectrum, the pattern remaining rather
uniform, (b) maintenance of considerable "type grouping" of muscle

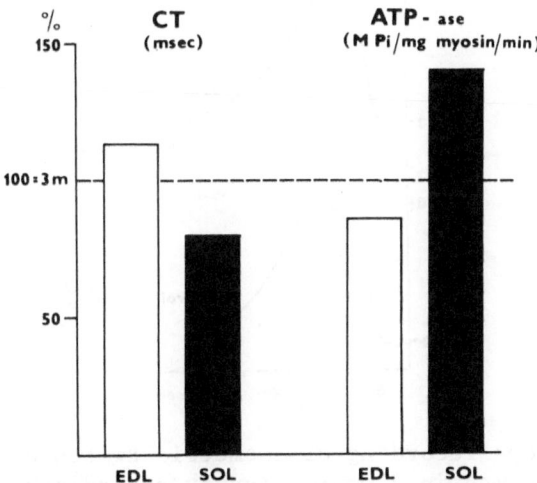

Fig. 2. Contraction time (CT) of isometric twitch and myosin ATPase
activity (M Pi/mg myosin/min) of the extensor digitorum longus (EDL)
and soleus (SOL) muscle of 30-month-old rats expressed in % of values
determined in muscles of 3-month-old animals (100 = 3 m). Note
prolongation of CT and decrease of myosin ATPase activity in the
senile fast EDL, shortening of CT and increase of ATPase activity
in the slow senile soleus muscle.

fibres, i.e. accumulation of fibres in groups of homogeneous fibres, observed also to a lesser degree in reinnervated muscles of young animals (18,19), and (c) marked variety in fibre size. This is explained by a decline of motor units in differentiation capacity.

The deficiencies in this differentiation process related to the specifity of different motor units are more clearly seen in the regeneration process of muscle after transplantation of senile muscles, as degeneration with following regeneration is observed in muscle grafts (20). Such studies allow also some conclusions on neurogenic (as in the reinnervation process) and myogenic mechanisms involved in the regeneration process of senile muscles. After grafting a minced fast muscle, the regenerating muscle under-

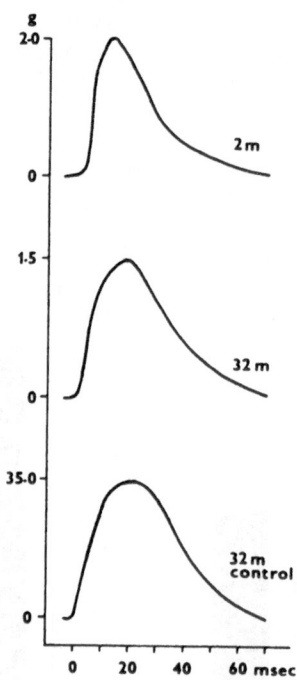

Fig. 3. Osciloscope recording of isometric twitch of the free EDL muscle graft performed in 2 months (upper recording), and 32 months (middle recording) old rats, 2 months after grafting. The lower recording is from the control EDL muscle of 32-month-old rat. Vertical bars: isometric tension in grams. Horizontal bar: 10 microsecon s. Note short contraction time (12.1 microseconds) in graft of young, prolonged contraction time in grafted EDL muscle of old animal (17.9 microsecond) and in the control EDL muscle of the old animal.

goes the ontogenetic stages of development ("ontogenetic recapitu-
lation") from a slow to a fast muscle (21).   This process is accompanied by a shift of muscle fibre pattern from a more uniform
to a mixed (mosaic) one about 30-60 days after grafting (22,23).
In free grafts of senile muscles the CT remains rather prolonged,
i.e. the prolonged CT of the normal senile muscle is recovered
(Fig. 3).   A deficient recovery process is exhibited in the changes
of the histochemical muscle fibre pattern.   The histochemical fibre
pattern in respect to ATPase activity remains rather uniform even
60 days after grafting, though recovery of muscle fibre size may
proceed relatively successfully (Fig. 4a, b).   This lack of differentiation of senile motor units in the process of reinnervation
of regenerating muscles is also shown in respect to SDH activity,
though not so marked (24).   A characteristic feature is also the
type grouping in the regenerating muscle.

The defect in senile muscle regeneration is also shown by
a more intense activity of lysosomal exzymes activated especially
in the interstitial tissue (25).   The latter contributes to the
irregular assembly of muscle fibres in senile muscle grafts.
Thus not only the differentiating capacity of motor neurons (related to specific features of different motor neurons and producing the mixed fibre pattern) is reduced in senile motor units,
but also other extracellular mechanisms are responsible for the
declining capacity of recovery of senile muscle grafts.

Ultrastructural changes disclose further mechanisms leading
to impairment of recovery in senile muscle grafts.   In free grafts
of young animals a regular assembly of myofilaments mostly occurs
in the senile muscle graft, however, organized synthesis of myofilaments and myotubes are found often even 2 months after trans-

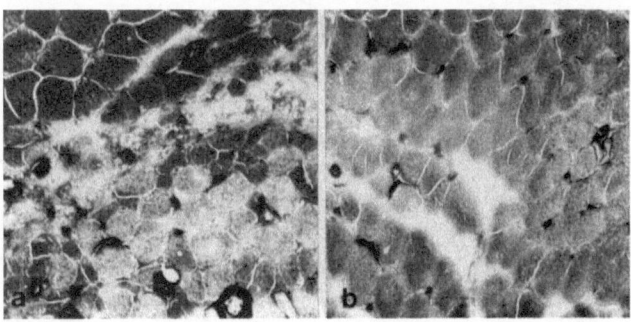

Fig. 4.   Cross section through free graft of EDL muscle of young
(a) and senile (b) rat 2 months after grafting, stained for myosin-
ATPase enzyme activity.   Note heterogeneous muscle fibre pattern
in graft of young (a) and more uniforma muscle fibre pattern in
graft of senile (b) rat.

plantation. Moreover autophagic vacuoles and intracellular lipofuscin accumulation (Fig. 5) are found. The degeneration products
of senile muscle cell are apparently integrated into the structure
of the newly regenerating muscle fibre (26).

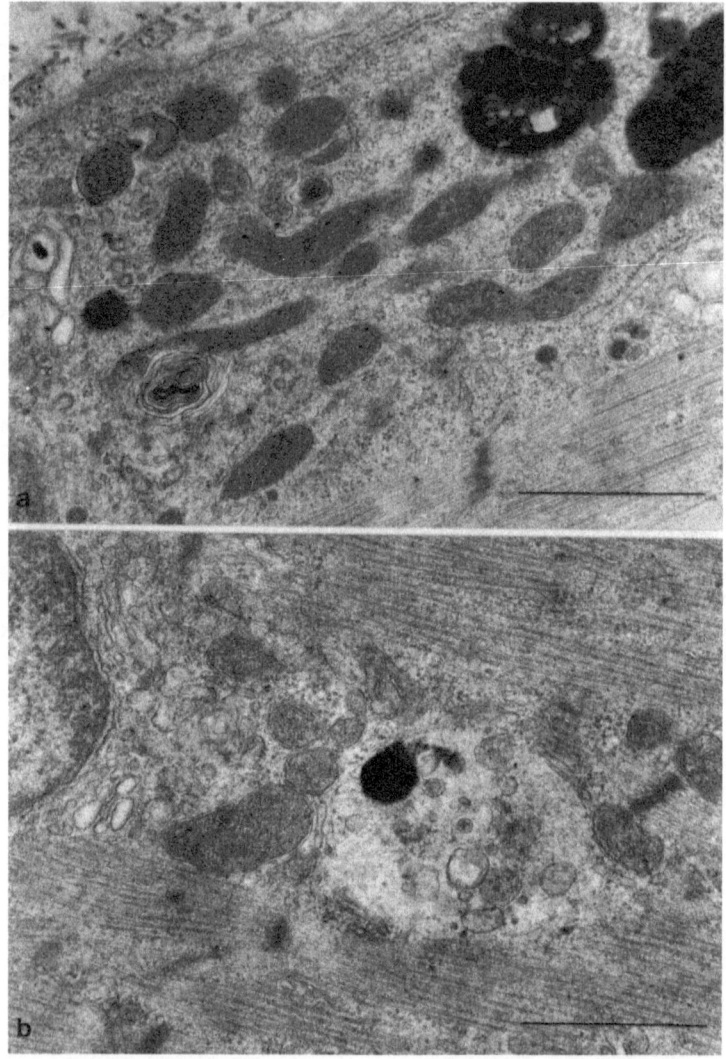

Fig. 5. Longitudinal sections of free-grafted EDL muscle of
31-month-old (a) and 22-month-old (b) rats, 2 months after auto-
transplantation. Lipofuscin-pigment granules persisting in muscle
fibres (a). Heterogeneous undigested material in autophagic
vacuole near the active Golgi apparatus (b). Scale bars = 1 μm.

Thus neural and myogenic old age disturbances interact. This report, however, demonstrated primarily the decline of the senile motor neuron to express and convey specific capacities and features on muscle. This defect results in loss of decline of the differences between fast and slow motor units.

REFERENCES

1. Gutmann, E. 1964. Neutrotrophic relations in the regeneration process, IN Mechanisms of Neural Regeneration, Progress in Brain Research, 13: 72.

2. Guth, L. 1968. "Trophic" influences of nerve on muscle. Physiol. Rev. 48: 645.

3. Gutmann, E. and Hanzlikova, V. 1972. Age changes in the neuromuscular system. Scientechnica.

4. Kugelberg, E. and Edstrom, L. 1970. Differential histochemical effects of muscle contractions on phosphorylase and glycogen in various types of fibres: relation to fatigue. J. Neurol. Neurosurg. Psychiat. 31: 415.

5. Burke, R.E., Levine, D.M., Zajac, F.E., Tsairis, P. and Engel, V.K. 1971. Mammalian motor units. Physiological-histochemical correlation in 3 types in cat gastrocnemius. Science 174: 709.

6. Engel, W.K. 1962. The essentiality of histo- and cytochemical studies of skeletal muscle in the investigation of neuromuscular disease. Neurology 12: 778.

7. Stein, J.M. and Padykula, H.A. 1962. Histochemical classification of individual muscle fibres of old rat. Am. J. Anat. 110: 103.

8. Gauthier, G.F. 1970. The ultrastructure of three types in mammalian skeletal muscle, IN The physiology and Biochemistry of muscle as a food (E.J. Briskey, R.G. Cassens and B.B. Marsh, editors), University of Wisconsin Press.

9. Schiaffino, S. Hanzlíková, V. and Pierobon, S. 1970. Relations between structure and function in rat skeletal muscle fibres. J. Cell. Biol. 47: 107.

10. Gutmann, E., Melichna, J. and Syrovy, I. 1972. Contraction properties and ATPase activity in fast and slow muscle of rat during denervation. Exp. Neurol. 36: 488.

11. Engel, W,K. 1970. Selective and non selective susceptibility of muscle fibre types. A new approach to human neuro-muscular diseases. Arch. of Neurol. 22: 97.

12. Close, R.I. 1972. Dynamic properties of mammalian skeletal muscles. Physiol. Rev. 52: 129.

13. Gutmann, E., Melichna, J., and Syrový, I. 1974. Developmental changes in contraction time, myosin properties and fibre pattern of fast and slow skeletal muscles. Physiol. bohemoslov. 23: 19.

14. Gutmann, E. and Hanzlikova, V. 1972. Basic mechanisms of aging in the neuro-muscular system. Mech. Age Dev. 1: 327.

15. Bass, A., Gutmann, E., Hanzlíková, V. and Syrový, I. 1974. Biochemical and histochemical changes in enzyme pattern of muscle of the rat in old age. Manuscript in preparation.

16. Gutmann, E., Henzlíková, V. and Vyskočil, F. 1971. Age changes in cross straited muscle of the rat. J. Physiol. 219: 311.

17. Gutmann, E. and Hanzlíková, V. 1973. Specific and non-specific changes in muscle atrophy and hypertrophy, IN Myology Excerpta Medica, I.C.S. (Kakulas, editor).

18. Kugelberg, E., Edstrom, L. and Abbruzzesse, M. 1970. Mapping of motor units in experimentally reinnervated rat muscle. J. Neurol. Neurosurg. Psychiat. 33: 319.

19. Karpati, G. and Engel, V.K. 1968. Type grouping in skeletal muscles after experimental reinnervation. Neurology 18: 447.

20. Studitsky, A.N. 1959. Experimental surgery of muscles (in Russian). Izdatel Akad. Nauk USSR Moscow.

21. Carlson, B. and Gutmann, E. 1972. Development of contractile properties of minced muscles regenerates in the rat. Exp. Neurol. 36: 239.

22. Gutmann, E. and Hanzlíková, V. 1974. Effect of "foreign innervation" on contractile and histochemical properties of the levator ani muscle of the rat. Physiol. bohemoslov. 23, in press.

23. Gutmann, E. and Carlson, B. 1974. Regeneration of muscle in free muscle grafts. (Manuscript in preparation).

24. Gutmann, E., Hanzlíková, V. and Carlson, B. 1974. Contraction
     properties and muscle fibre pattern in grafts of senile
     muscles. (Manuscript in preparation).

25. Gutmann, E., Hanzlíková, V. and Lojda, J. 1974. Histochemical
     change in grafted muscle tissue. (Manuscript in preparation).

26. Hanzlíková, V. and Gutmann, E. 1974. Ultrastructural changes
     in muscle grafts from young and old animals. (Manuscript in
     preparation).

QUESTIONS TO DR. GUTMANN

Dr. Hay: Have you been able to graft muscle from an older donor
to a young recipient and, if so, does it then respond as if young?
This type of study would indicate whether the defect in generative
ability was present in the older muscle tissue itself or in the
neuron of the older animal.

Dr. Gutmann: Yes, we have done experiments with syngeneic muscle
grafts, both from old to young and from young to old animals. Both
experiments showed successful regeneration. The results do not
yet allow definite conclusions.

Dr. Butenke: Did you observe a rejection of the muscle transplant
or other immune reactions, e.g., mononuclear infiltration of grafts
in your experiment and is there some differences in behavior of
old and young transplant in this respect?

Dr. Gutmann: In the experiments described here, autografts
were used. However, in other experiments we showed (Gutmann and
Hasek, 1973) that allografts are completely rejected. Tolerance,
can however, be produced and also compactability could be produced
in allografts of muscle after enhancement with cortisone.

Dr. Leuenberger: Is the ATPase method used in your experiments
specific for myosine-ATPase? And, if so, how do you explain the
large amounts of reaction products over blood vessels? These
difficulties might be overcome by the use of a specific reaction
for "transport ATPase", sensitive to cations (Na^+, K^+) and ouabain,
as described by Ernst in 1972?

Dr. Gutmann: We assumed specifity of the myosin ATPase using
Guth's modification of the method of Herman-Padykula (at pH P.5).
However, I suppose the reaction of the capillaries invading the
draft which have apparently high Na^+, K^+ dependent membrane ATPase
suggests, that Na^+ and K^+ dependent membrane ATPase does interfere.
This will not affect the changes in muscle fibre pattern. But

higher specifity of the reaction should be attempted.

COMMENT TO DR. GUTMANN

Dr. Kasten: With regard to the question and answer about ATPase
in blood vessels: Without the use of specific inhibitors and
activators in the incubation medium, ATPase may appear in many lo-
cations including membranes in blood vessels and myosin ATPase.

MORPHOMETRICAL AND MATHEMATICAL ANALYSIS OF THE AGEING CHANGES OF

THE MUSCLE-CONNECTIVE-TISSUE-RELATION IN SMOOTH MUSCLES

Paul Rother and Gerald Leutert

Anatomisches Institut der Karl-Marx-Universität

Leipzig, DDR

The effects of ageing on the muscle-connective tissue relation-ships of smooth muscles have been studied by the point counting method of Chalkley and Haug. Sections of the small intestinal walls of 60 persons as well as of the ciliary bodies of 55 persons of all age groups were available for these studies. The morphometric results were tested by statistical methods, and linear regression functions as well as exponential functions of the form $y = A_o + A_1 e^{jt}$ were adapted to them. We were especially interested in finding answers to the following three questions:

1. What is the rate of this particular process of ageing in the various periods of life?
2. Are there any specific sex differences in this respect?
3. What is the primary event in this connection, the decrease in muscle cells or the increase of connective tissue?

Figure 1 shows a cross section of the tunica muscularis of the jejunum of a 1-year-old child. It has been found that, first, the amount of connective tissue is much smaller in the circular layer than in the longitudinal layer and, second, an increase of intersti-tial tissue due to ageing may be observed in the longitudinal layer only. The increase in interstitial tissue is usually more clearly visible in, say, the first five decades of life than in old age. There exists a nonlinear age dependence in this respect (Fig. 2).

Figure 3 is a section of the ciliary muscle of a newborn infant, which abounds in muscle cells, while the amount of connective tissue is comparatively small. In contrast, the ciliary muscle section obtained from a 51-year-old male (Fig. 4) presents an entirely different situation insofar as we may see here plenty of collagenic

441

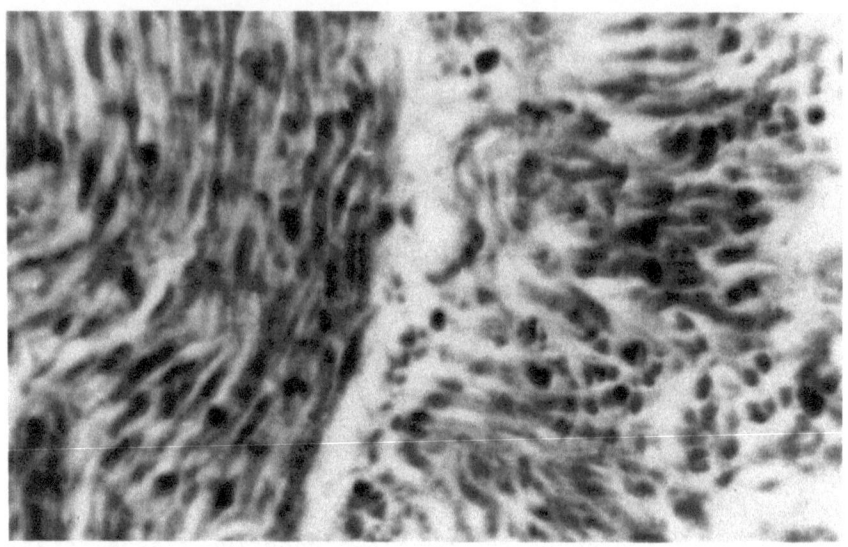

Fig. 1. Cross section of the tunica muscularis of the jejunum of
a 1-year-old child, Azan. Left side: stratum circulare; right
side: stratum longitudinale. X 1360.

connective tissue, whereas the number of muscle cells is relatively
small. The following graph (Fig. 5) shows the percentage share of
muscle cells in the total volume of the ciliary muscle, which was
determined by the point counting method. However, only those values
which were obtained for males past 20 as well as the corresponding,
drooping regression line are shown. The regression function for
the corresponding values of females shows an entirely different
behavior in that the line exhibits a horizontal course (Fig. 6).
The t-test showed that the two regression coefficients b_1 and b_2
of the straight lines obtained for female and male persons differed
from each other significantly. Thus, we had succeeded in obtaining
statistical evidence of the sex-different biomorphosis of the
ciliary muscle. Obviously, the histological biomorphosis is much
slower in female persons during their generative phase than in males.

In Figure 7 are visible the values for the muscles (above) and
the connective tissue (below) of all ciliary bodies measured in this
investigation, together with the exponential function. The adapted
exponential function may be written as follows:

Proportion of muscles = $63.96 + 30.31\ e^{-0.037}$ age.

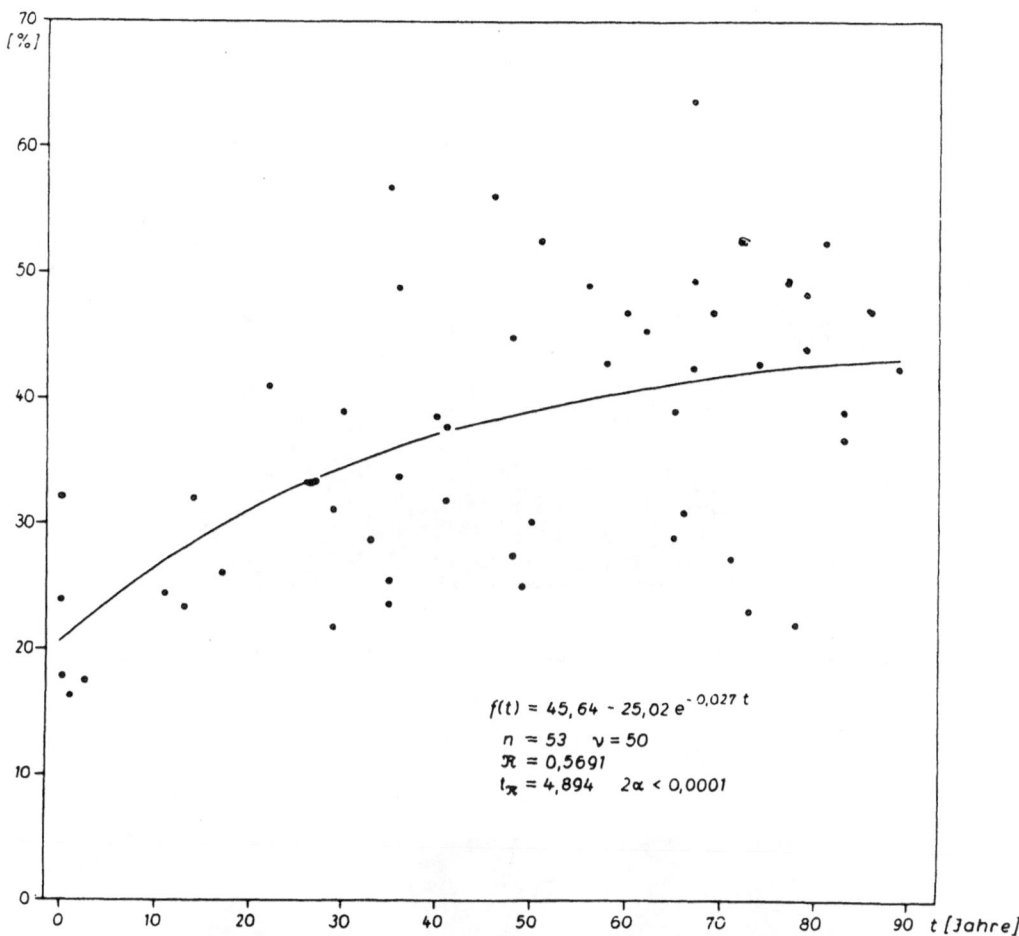

Fig. 2.   Increase of the share of connective tissue of the stratum
longitudinale and corresponding regression function.

The first and second derivatives of the function, i.e., the rate and
acceleration of this process of ageing, approach the zero value
asymptotically with increasing age.   The regression function des-
cribing the increase of connective tissue may be written as follows:

Proportion of connective tissue = 32.98 - 30.90 $e^{-0.047 \text{ age}}$.

The adaptation of the regression function is satisfactory.   The
correlation coefficient has the following value: R = 0.85.   The
first two derivatives are used to describe the process of proliferation of connective tissue, B' indicating the rate of connective

Fig. 3.   Ciliary body, newborn infant, Crossmon.   X 65.

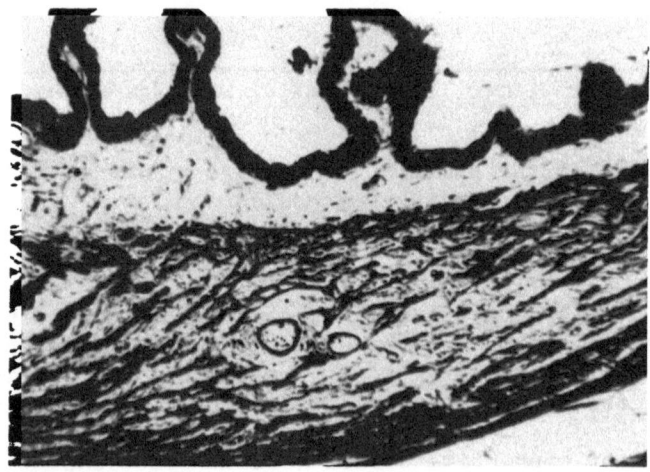

Fig. 4.   Ciliary body, 51-year-old male person, Crossmon.   X 65.

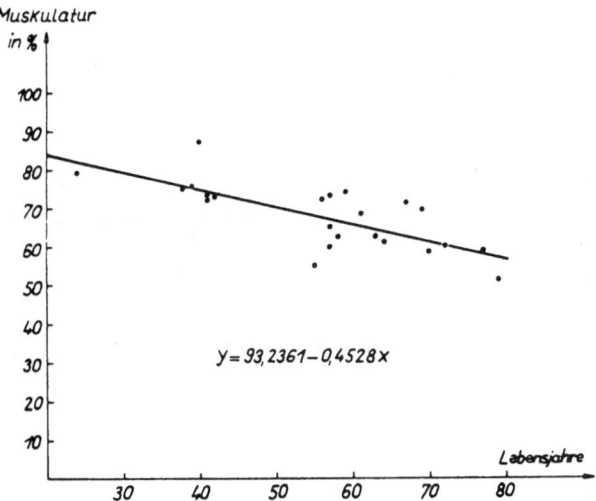

Fig. 5.   Relative content of musculature in the ciliary bodies of
male persons and corresponding regression line.

tissue proliferation.   The ever-decreasing proliferation rate has
been found to reach half its final value of approximately 33% at
15 years of age.   In other words, 15 years is the half time of
connective tissue proliferation.   The difference of the proportion
of connective tissue observed in very old people from the "infinite"
old people is only very small.   In an 87-year-old male person a
connective-tissue proportion of 32.47% has been observed, while the
proportion of connective tissue in the ciliary muscle has been
calculated to be 32.98% for a human being of infinite age.

     Now, let us turn to the third question: Adjacent to the smooth
muscles that we are concerned with here are muscle-free layers of
connective tissue, while the basal plate is lying next to the
ciliary muscle (Fig. 4) and the submucous coat is contiguous to the
tunica muscularis of the small intestine.   If, so far as the variation of the muscle-connective tissue relation in the muscle is
concerned, primary significance must be attached to the proliferation of connective tissue, then the connective tissue should be
expected to also show proliferation in the neighboring basal plate
and in the submucous coat, respectively.   The areas have been plani-
metered--in section--with a Zeiss compensation polar planimeter.
The basal plate, in contrast to the connective tissue within the
muscle, shows no increase whatsoever during the process of ageing.
Thus, there exists a morphometric indication showing that the
decrease in muscle cells should be of primary and the increase of

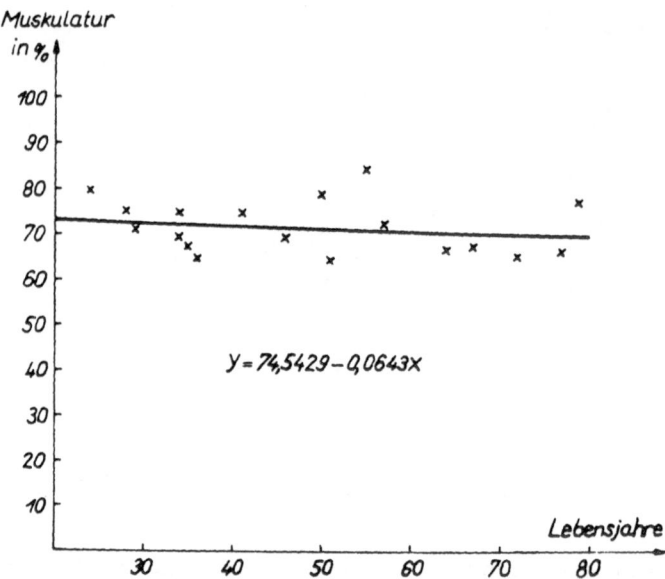

Fig. 6. Relative content of musculature in the ciliary bodies
of female persons and corresponding regression line.

connective tissue of secondary relevance, to the shift of the
muscle-connective tissue relation of smooth muscles in favour of
the connective tissue.

    The view held by the present authors has been confirmed by
the results of studies of human prostates. It has been found that
the decrease in testosterone elimination is accompanied by a
parallel decrease in the parenchyma of the prostate and in the
number of smooth muscle cells in the interstice, while the collagen
content showed a definite increase.

SUMMARY

    (1) The rate of the process of ageing studied in this work
as well as of other ageing processes observed by us constantly
decreases with increasing age.

    (2) By an adaptation of exponential functions to series of
measurements it is possible to estimate the theoretically possible
end stage of histological ageing processes for extremely long
lifetimes.

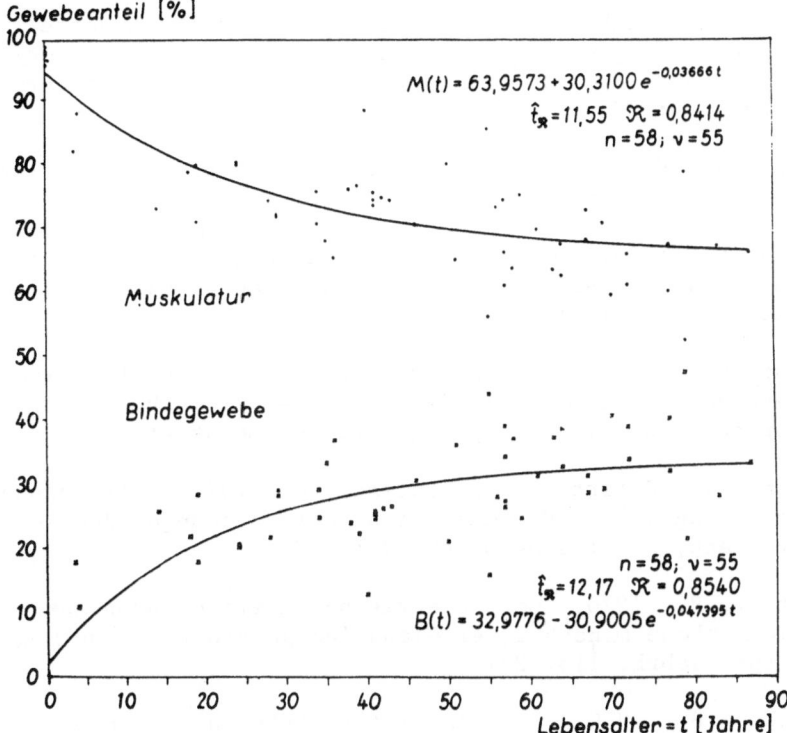

Fig. 7. Values for the muscles and the connective tissue and the regression functions.

(3) The histological biomorphosis is much slower in female persons during their generative phase than in males.

(4) Results of planimetric measurements made on sections suggest that, so far as the shift in muscle-connective tissue relation in favour of the connective tissue is concerned, primary significance should be attached to muscular atrophy, while the connective-tissue proliferation may be considered to be secondary in this respect.

REFERENCES

1. Blume, R. and Buch, Ch. 1967. Ein Beitrag zum Trefferver-
    fahren nach Chalkley. Z. mikr.-anat. Forsch. <u>76</u>: 26.

2. Blume, R., Rother, P. and Lochner, M. 1971. Lineare Differentialgleichungen zur Beschreibung der histologischen Biomorphose des menschlichen Ziliarmuskels. Z. Alternsforschung 23: 235.

3. Chalkley, H.W. 1943. Method for quantitative morphologic analysis of tissue. J. Nat. Cancer Inst. 4: 47.

4. Haug, H. 1955. Die Treffermethode, ein Verfahren zur quantitativen Analyse im histologischen Schnitt. Z. Anat. Entw. Gesch. 118: 302.

5. Haug, H. 1962. Bedeutung und Grenzen der quantitativen Meß-methoden in der Histologie. In: Medizinische Grundlagenforschung Bd. IV von K.F. Bauer, Thieme Stuttgart.

6. Jahn, K., Leutert, G. and Rotzsch, W. 1971. Altersabhängige morphologische und biochemische Untersuchungen der menschl. Prostata, Z. Alternsforsch. 23: 323.

7. Rother, P. 1970. Morphometrische Untersuchungen zur Biomorphose menschlicher Glandulae parathyreoideae Gegenbaurs. Morph. Jahrb. 115: 231.

8. Rother, P. 1970. Zur Geschlechtsdifferenz der Biomorphose menschlicher Glandulae parathyreoideae. Morph. Jahrb. 115: 231.

9. Rother, P., Blume, R., Friedrich, H. and Lochner, M. 1971. Die Veränderung der Muskel-Bindegewebs-Relation des Muskulus ciliaris im Laufe des Lebens. Anat. Anzeiger 129: 322.

10. Rother, P. and Friedrich, H. 1971. Regressionsanalysen zur Beschreibung morphologischer Alternsprozesse. Biolog. Rundschau 9: 172.

11. Rother, P. and Leutert, G. 1965. Die Altersveränderung des Ciliarkörpers, Albrecht von Graefes. Arch. klin. exp. Ophthal. 168: 136.

12. Rother, P., Walther, M., Scheller, G. and Blume, R. 1973. Morphometrische Untersuchungen zur histologischen Biomorphose der Wand des Dünndarmes. Zuikoosk. und Torsch. 87: 97.

13. Scheller, G., Rother, P. and Walther, M. 1972. Über die Veränderungen von Zellgröße, Kerngröße und Kern-Plasma-Relation beim Altern von Epithelzellen und Epithelgeweben. Z. mikr.-anat. Forsch. 85: 123.

QUESTION TO DR. ROTHER

Dr. Hay: Is the atrophy noted due to loss of sarcoplasm, loss of number of nuclei in the muscle fiber itself, or both? This would seem to be important information as it could permit one to narrow the search for defective function either to protein synthesis and replacement or to the more extensive problem of maintenance of nuclear integrity.

Dr. Rother: We know that what we measured is the sum of many heterogenous processes within the smooth muscle. In detail I can not answer your questions.

SOME HISTOCHEMICAL AGE CHANGES OF THE SMOOTH MUSCLE CELLS IN THE

VEINOUS WALL

Martha Christova

Institut of Endocrinologie, Gerontologie and Geriatrics

Sofia, Bulgaria

In the course of life of the individual, the smooth muscle cells undergo certain changes, as a result of a definite ageing evolution of the veinous wall.

In the present investigation we made it our aim to trace some ageing histochemical changes of the smooth muscle cells from the different layers of the wall of the portal vein. The smooth muscle cells of the portal vein are investigated directly before their division in the porta hepatis of 96 individuals at the age between 8 to 90 years, having no pathological data for haemodinamic troubles in the portal vein system (Table I). The material is fixed in 10% neutral formol, in the solution of Carnoy and in calcium-formol by Backer. On paraffin slides are applied the following stainings: haemalaun-eosin, for phospholipids by Backer and by Adams, for glycogen by Best and PAS-reaction; slides of frozen micritom are stained with sudan III, sudan black B and oil red O.

In the applied histochemical methods is reported the presence of definite ageing changes in the smooth muscle cells of the different layers. The staining with sudans and oil red O show the appearance of fatty degeneration in single smooth muscle cells (Fig. 1). The concerned cells have slightly increased dimensions, full of lipids (with characteristics of triglycerids), the cell organels not being seen. The fatty degenerated smooth muscle cells are observed earliest in the media, after the age of 50. In the adventitia they appear a little later, after the age of 55 and in the intimal smooth muscle cells, although rare, the fatty degeneration appears at the 9th decade. The entire investigation of the wall indicates the deferment of lipid drops in extracellular localization, with preference in the adventitia and intima, and a little less in the media.

451

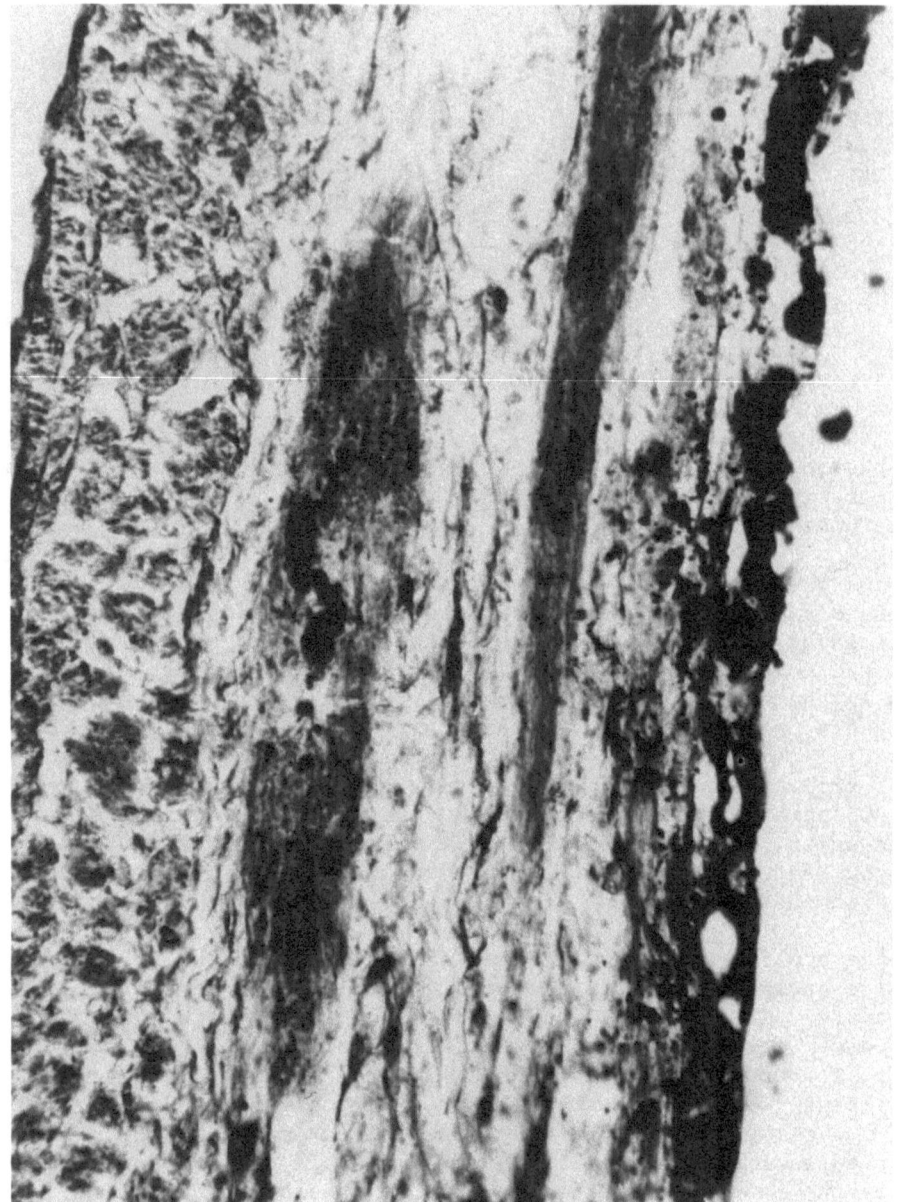

Fig. 1a. Teh fatty degeneration in single muscle cells of the adventitia of a man 71 years old.

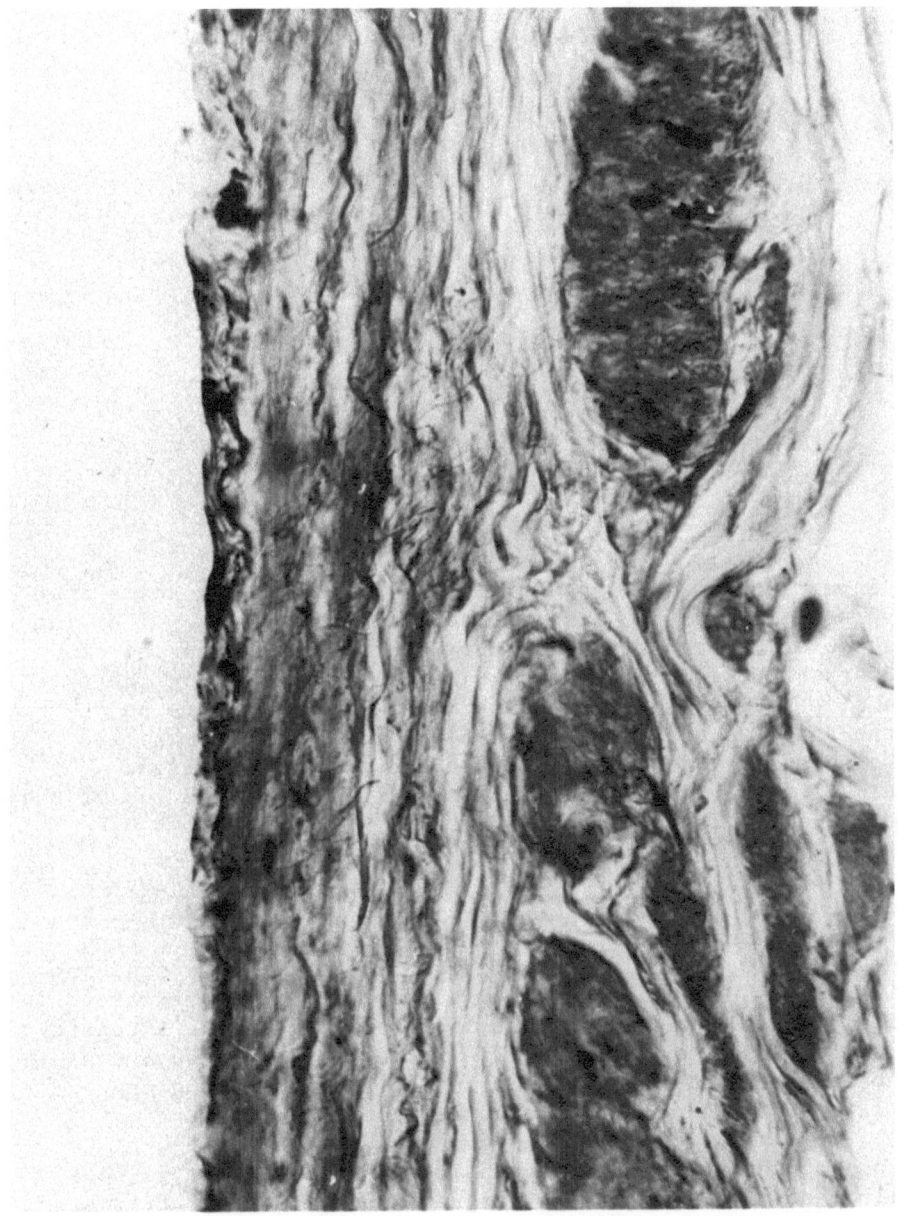

Fig. 1b. The fatty degeneration in single muscle cells in the intima, media and adventitia of a man 90 years old. A staining with sudan black.

TABLE I.   Distribution of the Investigated Individuals According to
           Age and Sex

Age in years	No of subgects	Male	Female
8 - 10	4	1	3
11 - 20	8	5	3
21 - 30	5	3	2
31 - 40	9	8	1
41 - 50	9	6	3
51 - 60	16	13	3
61 - 70	26	19	7
71 - 80	13	6	7
81 - 90	6	5	1

The investigation for phospholipids by Backer or by Adams show
a positive staining for the smooth muscle cells in the wall, having
a brownish color. After the puridin or aceton extraction the
indicated structures do not change. The presence of a considerable
more intensive staining of the smooth muscle cells in persons
between the age of 8 to 9 is established. At the end of the 2nd
decade the intensity of coloring is decreased and with slight
individual hesitations remains in the same level till the end of the
5th decade, then it lessens again (Fig. 2).

The investigation with the PAS-reaction to indicate neutral
mucopolysaccharids, in particular of glycogen, and with karmin by
Best, establishes an emphasized pink to reddish staining of the
smooth muscle cells (enzimic control with diastase). The staining
is most intensive during the first 4 decades of life, afterwards
the finding becomes changeable. After the 6th decade the intensity
of staining decreases more distinctly (Table II), an unequal staining
of the separate cells is noticed, byt no differences among the
separate layers of the wall can be established.

The ageing changes in the smooth muscle cells in the veinous
wall, till a definite age, have a progressive character, afterwards
they become regressive. During the period of progressive changes
the smooth muscle cells of the wall increase, they possess an abundant
phospholipid and glycogenic content, after that comes a period of
comparatively small individual hesitations. Later the regressive

TABLE II. Degrees of the Staining with PAS-Reaction (or with Karmin by Best) According to Age

Age in years	No of subjects	Degrees +	++	+++
8 – 34	18	0	10	8
35 – 44	9	0	5	4
45 – 54	10	1	7	2
55 – 64	19	9	8	2
65 – 74	18	12	7	0
over 75	9	7	1	0

+   light degree of the staining
++   mean degree of the staining
+++ large degree of the staining

changes begin: the atrophy of the smooth muscle cells, the fatty degeneration, a decrease of the phospholipids and glycogenic content in them. Some of the above mentioned changes do not begin simultaneously in all layers, and distinctly they appear first in this muscle layer which forms earlier in the ontogenetic development. According to Bucciante (1), the muscular layer in the media is formed already before birth, while the longitudinal smooth muscle layer is formed after birth and grows during the whole life of the individual. That is why we observe regressive changes first in the smooth muscle cells of the media, while in the adventitia they appear later. The latest appearance of fatty degeneration changes in the smooth muscle cells of the intima corresponds to the established by us presence of these muscle cells with growing age. It must be pointed out that the degenerative changes in the smooth muscle cells do not coincide with ageing dynamics of the extracellular lipid deferment in the veinous wall.

From the carried out results, it can be supposed that the development of the fatty degeneration in the smooth muscle cells of the veinous wall is more likely an appearance of their ageing than the expression of determined to age lipid infiltration of the wall caused by the blood which runs through the lumen of the blood vessel and through the masa vasorum.

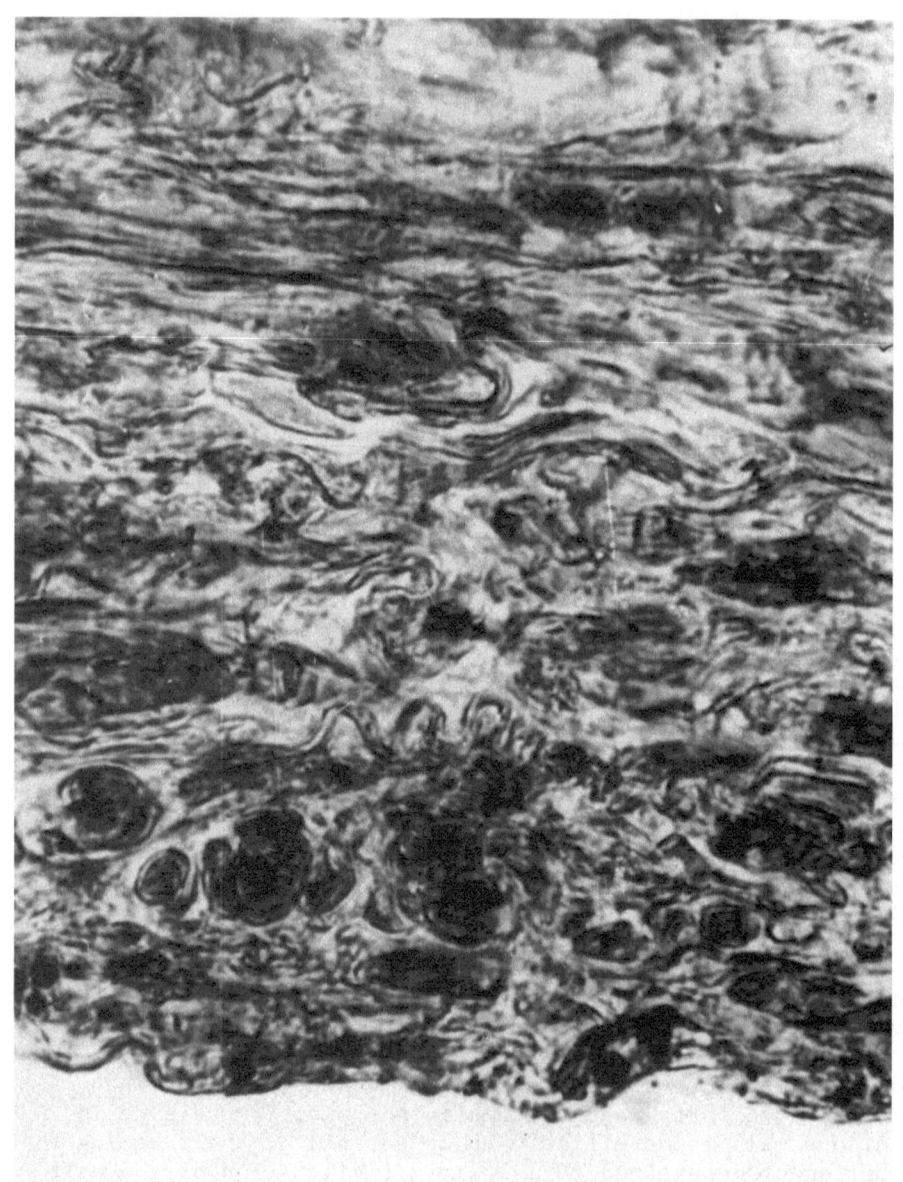

Fig. 2a. The phospholipid content in the muscle cells of the media of a man 48 years old.

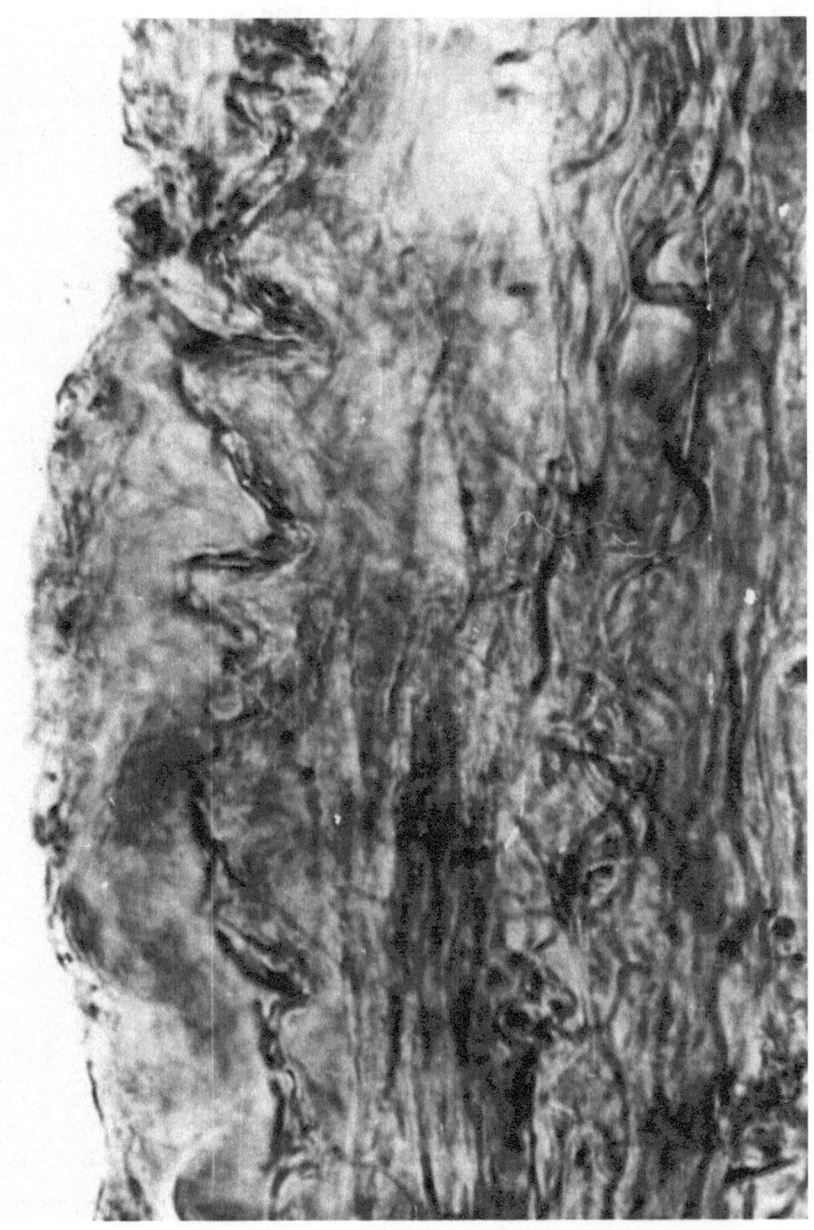

Fig. 2b. The phospholipid content in the muscle cells of the media of a man 89 years old. A staining by Adams, contrasted with alcian blue.

     Although with growing age we observe a decrease of the struc-
tural phospholipids of the cells, it cannot be said definitely
whether the fatty degeneration is a result of the decomplexity of
the cell organels, or the citoplasma, or of the "functional" lipid
complexes (2). It is characteristic for the ageing change that they
affect only some cells and always leave unchangeable ones, and in
this way the functional activity of the smooth muscle composition
of the veinous wall is secured. The reduced glycogenic content of
the smooth muscle cells admits the supposition that with the growth
of age appear changes even in the level of the metabolitic processes
in them.

REFERENCES

1. Bucciante, L. 1966. In: Morphologie und Histochemie der
      Gefässwand. S. Karger, ed., 1-2. Basel, New York, p. 212.

2. Davidovski, I. 1969. Obschtaja pathologia tscheloveka.
      Medicina, ed. Moskva, p. 89.

EFFECT OF LYSOLECITHIN ON THE TRANSPORT OF PLASMA CHOLESTEROL TO

TISSUES: DEVELOPMENTAL ASPECTS

Milada Dobiášová and Eva Faltová

Isotope Laboratory of Biological Institutes and Institute
of Physiology of Czechoslovak Academy of Sciences

Our laboratory has been engaged for several years in the study
of the effect of specific phospholipids on the transfer of cholesterol
to biological membranes (1-3). We proved under in vitro conditions
that the affinity of cholesterol for these membranes, in the same
way as for soluble serum lipoproteins, depends on the specific
phospholipid carrier. It appeared that, as regards cholesterol,
the most interesting phospholipid is lysolecithin. The micelles
of lysolecithin have a comparatively low molecular weight; they are
very stable in aqueous medium, they have a low affinity to the
soluble plasma lipoproteins and a high affinity, of a higher
order than lecithin, to the tissue lipoproteins (4). This knowledge
led us to a more detailed study of the role of lysolecithin as a
potential transport vehicle for cholesterol also under in vivo
conditions. As shown by Glomset in 1962 (5), lysolecithin appears
in the plasma as a product of transesterification of lecithin with
cholesterol, which is controlled by the enzyme lecithin-cholesterol
acyltransferase. The metabolic fate of this lysolecithin originating
in the plasma was not studied until now, although Portman with his
co-workers (6) pointed out the significance of lysolecithin in
atherogenesis. They found namely the dependence between the increase
of activity of the plasmatic lecithin - cholesterol - acyltransferase
and the increased content of cholesterol and lysolecithin in the
arteries of hypercholesterolemic squirrel monkeys. It is, however,
extremely difficult to prove even at an increased activity of lecithincholesterol acyltransferase a direct dependence on lysolecithin of
the transfer of cholesterol or possibly cholesterol esters into the
arteries. We decided therefore to carry out our experiments under
a special model situation, in which we induced an immediate and practically complete conversion of plasmatic lecithin into lysolecithin

459

by an intravenous adminstration of phospholipase A to rats. To
enable us to observe the fate of endogenous phospholipids and cholesterol, we labeled the rats in advance with ^{32}P-phosphorus and
^{14}C-cholesterol. ^{14}C-cholesterol is pooled with free and esterified
cholesterol after 5 days and the plasma phospholipids are labelled
with ^{32}P-phosphorus after 24 hr. In our experiments, we followed
in the first place the clearance of lipid phosphorus and cholesterol
from the plasma, further the changes in lipoprotein spectrum of
the plasma and the increment of lipid phosphorus and cholesterol
in the organs (Fig. 1). This graph shows the clearing of ^{32}P-lipid
phosphorus from the rat plasma in dependence on the intravenous
administration of a phospholipase A dose and on time. In control
animals, the level of lipid ^{32}P does not change practically between
24 and 48 hr after the injection of active phosphate; hence the
effect of phospholipase A may be well observed. The activity of
lipid phosphorus drops in the first 10 min after the application
of phospholipase A to a minimum and the return to the original
values, which clearly depends on the dose of enzyme, is completed
approximately after 24 hr (Fig. 2). After the application of phospholipase A, a rapid degradation of lecithin into lysolecithin
takes place, as was proved in our parallel in vitro experiments.
Since under in vivo conditions this newly formed lysolecithin is
not retained in the plasma, in accordance with the low affinity
to plasmatic lipoproteins, its decrease is manifested only by the
lowering of activity and content of lecithin, as shown in this
graph. The concentration of lysolecithin is increased only temporari-
ly in a short time interval (Fig. 3). Thirty min after the administration of phospholipase A only lysolecithin and sphingomyelin
in normal concentration remain in the plasma and the content of
lecithin drops to 5 to 10 % of its original value (Fig. 4). This
graph shows that at the same time changes take place in the distribution of cholesterol in the lipoprotein spectrum. After the
separation of plasma by electrophoresis in agarose gel, 0.5 cm
strips were extracted and activities of ^{32}P-phospholipids and
^{14}C-cholesterol were determined, the distribution of both these
labels in lipoproteins of control rats is approximately identical.
In the plasma of the treated animals an overall drop of phospholipid
activity takes place, but above all there is a shift of cholesterol
into fractions with a higher mobility. This shift of cholesterol
in the spectrum is visible as well under in vitro conditions, where
formed lysolecithin remains in the medium, as 1 hr after the application of free cholesterol, and eventually also in the plasma of
the animals, where the free and esterified cholesterol are in a
physiological concentration. Our explanation of this phenomenon
is that during a dissociation of lysolecithin from lipoproteins
a simultaneous loosening of the cholesterol bond takes place,
its shift and at the same time its clearing from the plasma under
In vivo conditions (Fig. 5). It appears from this table that after
the application of phospholipase A a drop of activity and mass of the

total cholesterol actually takes place. The drop of total cholesterol, however, goes primarily to the account of esterified cholesterol, while the free cholesterol either does not change or even increases. Still, the possibility cannot be excluded that together with lyso-lecithin-free cholesterol is also released. Its decrease in the plasma may be easily complemented from other sources, such as erythrocytes, while the substitution of plasma cholesterol esters is bound on the activity of plasmatic enzyme lecithin cholesterol acyltransferase (Fig. 6). When recalculated on the molar decrease of lysolecithin and cholesterol esters from the plasma, it appeared, that in a 15 min interval after the administration of phospho-lipase A this molar decrease equalled 1:0.52 moles of lipid phos-phorus against 0.48 moles of cholesterol ester. The next table (Fig. 7) shows the increase of lipid phosphorus and total cholesterol after 10 min in the liver, lungs and kidneys. We found the largest increase of lipid phosphorus and cholesterol per 100 mg of tissue in the lungs. The absolute increase of lipid phosphorus and cholesterol is the largest in the liver. The activity of ^{32}P and ^{14}C-cholesterol grows in organs in a similar way (Fig. 8). We determined the ratio of free and esterified cholesterol in the liver by the gas chromato-graphy method. We found that during the first 5 min the content of cholesterol esters increases first and only later the content of free cholesterol. It is probable that the subsequent increase of free cholesterol in the liver is caused by the hydrolysis of cho-lesterol esters, as well as by transfer of free cholesterol from the plasma. The last table (Fig. 9) shows the changes in content of phospholipids and total cholesterol in the aortae. The table in-cluded two groups of experimental animals. The first group of younger, 3 months of age animals with approximately 300 g body weight, the second group of older animals, 9 months of age with approximately 470 g body weight. In the younger animals, after the application of phospholipase A lipid phosphorus and cholesterol de-creased on the observed time interval, while in the older group the content of lipid phosphorus, and cholesterol in particular, in-creased. We do not have enough experimental material at our disposal to be able to maintain with certainty that old rat aortae are capable of retaining more cholesterol, which is transported into them by the described method, than the aortae of younger animals.

It appears from the presented results that an inter-vention into the lecithin metabolism in the plasma induces direct effects on the transport of plasma cholesterol.

We suppose, however, that further study of the lyso-lecithin turnover, appearing by the effect of lecithincholesterol acyltransferase in the plasma, of its influence on the transport of plasmatic cholesterol, and of its deposition in the organs, might yield new aspects and means for the study of atherogenesis.

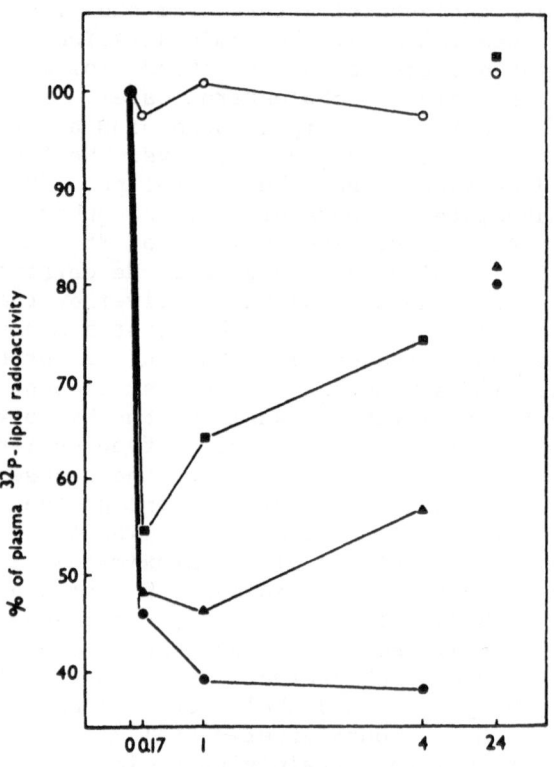

Figure 1

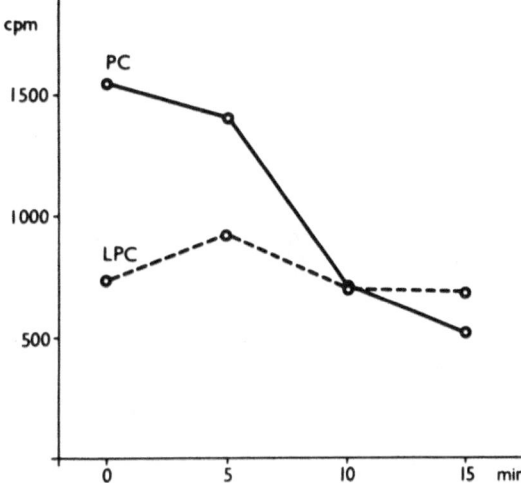

Fig. 2. ^{32}P-lecithin and lysolecithin content of plasma

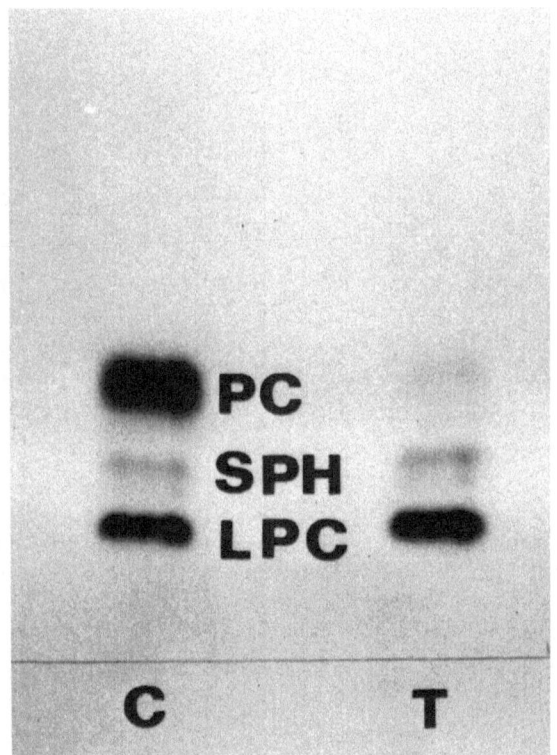

Figure 3

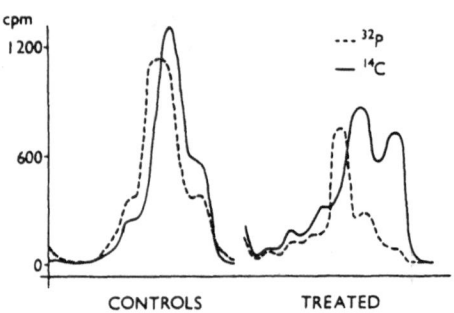

Fig. 4.   Lipoprotein pattern of plasma.

	Lipid P μgs/ml	Total CH μgs/ml	Total ^{14}C-CH cpm / ml	Free ^{14}C-CH cpm / ml	^{14}C-CHE cpm/ml
Controls	43.0 ± 1.4[a]	874 ± 19	24,455 ± 1,770	4,500 ± 315	19,800 ± 1,530
After treatment :					
15 min.	***26.6 ± 2.1	**750 ± 36	*21,000 ± 1,060	6,630 ± 678	*14,520 ± 1,810
4 hours	***23.5 ± 1.2	***571 ± 31	***14,740 ± 0,774	4,110 ± 416	**10,160 ± 2,200

*P<0.05, **P<0.01, ***P<0.001, paired t-test. 8 animals in each group.
[a]Mean ± S.E.M.

Fig. 5.   Rat plasma contents of lipid P, total cholesterol, total
^{14}C-cholesterol, free ^{14}C-cholesterol and ^{14}C-cholesterol esters
after treatment with phospholipase A (100 μg/animal).

	Lipid P μmoles/ml	Δ μmoles/ml	Cholesterolester μmoles/ml	Δ μmoles/ml
Controls	1.39 ± 0.11[a]		1.82 ± 0.14	
After treatment :				
15 min.	0.87 ± 0.06	0.52	1.34 ± 0.16	0.48
4 hours	0.76 ± 0.04	0.63	1.02 ± 0.22	0.80

[a]Mean ± S.E.M.

Fig. 6.   Decrease in rat plasma lipic P and cholesterol after treatment with phospholipase A (100 μg/animal).   Difference (Δ) between controls and treated groups in μmole.

	Liver	Lung	Kidney
Ch			
Controls	534 ± 12	461 ± 36	645 ± 15
Treated 10 min	641 ± 32	740 ± 40	712 ± 19
lipid P			
Controls	134 ± 7	85 ± 6	117 ± 6
Treated 10 min	192 ± 7	151 ± 3	138 ± 6

Fig. 7.   Total cholesterol and lipid phosporus in organs (μg/100 mg).

		Total cholesterol content		Free	Esterified
		mg/g	cpm/g	mg/g	mg/g
Experiment 1	Controls	5.34 ± 0.12[a]	87,720 ± 8,940		
	After treatment:				
	15 min.	**5.90 ± 0.14	93,960 ± 3,940		
	4 hours	***6.02 ± 0.08	99,230 ± 7,520		
Experiment 2	Controls	5.09 ± 0.18		1.74 ± 0.07	3.35 ± 0.13
	After treatment:				
	5 min.	**6.07 ± 0.22		1.97 ± 0.12	**4.10 ± 0.17
	10 min.	**6.41 ± 0.32		*2.37 ± 0.25	**4.03 ± 0.20

[a]Mean ± S.E.M. *P<0.05, **P<0.01, ***P<0.001, paired t-test. 8 animals in each group.

Fig. 8. Liver content of total cholesterol, ^{14}C-cholesterol, free and esterified cholesterol after treatment with phospholipase A (100 μg/animal). In Experiment 1 values by isotopic method, in Experiment 2 by GLC.

	Young		Old	
	Ch	lipid P	Ch	lipid P
Controls	123	18.3	430	17.4
Treated 60 min	105	16.4	*570	*19.7

Fig. 9. Total cholesterol and lipid phosporus in aortas (μg/100 mg).

REFERENCES

1. Dobiasova, M. and Linhart, J. 1970. Association of phospho-
       lipid-cholesterol micelles with rat heart mitochondria:
       stimulators and inhibitors. Lipids 5: 445.

2. Dobiasova, M., Linhart, J. and Bibr, B. 1971. Interaction of
       exogenous phospholipids with rat liver mitochondria: the
       effect of Ca^{2+} ions. Physiol. Bohemoslov. 20: 489.

3. Dobiasova, M., Linhart, J. and Bibr, B. 1973. Interaction of
       exogenous phospholipids with phospholipase modified mito-
       chondria. Physiol. Bohemoslov. 22: 159.

4. Dobiasova, M., Faltova, E. and Marcan, K. The effect of phos-
       pholipids on the incorporation of ^{14}C-cholesterol into serum
       lipoproteins. Submitted for publishing in Physiol. Bohemo-
       slov.

5. Glomset, J.A., Parker, F., Tjaden, M. and Williams, R.H. 1962.
       Esterification in vitro of free cholesterol in human and rat
       plasma. Biochim. Biophsy. Acta 58: 398.

6. Portman, O.W., Soltys, P., Alexander, M. and Osuga, T. 1970.
       Metabolism of lysolecithin in vivo: effects of hyper-
       lipemia and atherosclerosis in squirrel monkeys. J. Lipid
       Res. 11: 596.

QUESTION TO DR. DOBIASOVA

Dr. Gutmann: Do you think that lysolecithin can also affect the
incorporation other substances into the cells?

Dr. Dobiasova: Yes, I think so. Lysolecithin is known to facilitate
cell fusion, so that I think that this phenomenon should be considered.
Also under condition of long-term incubation of cells in serum,
lecithin-cholesterol acyltransferase is able to produce in vitro
about 70 - 100/μg of lysolecithin per 1 ml serum per hr.

VIRAL INFECTION AND INTERFERON IN CELL CULTURES AGED IN VITRO

Helena Libíková

Institute of Virology, Slovak Academy of Sciences
Bratislava, Czechoslovakia

It is well known that host age often acts as a decisive factor determining the course of illness and the immunological response to viral infections in man and animals (1). Medical praxis knows foudroyant infections in newbornes which are caused by viruses harmless for adults. Experimental virology offers many instances when viruses kill newborne animals, remaining non-pathogenic for adults of same species. On the other hand, activation of latent virus infections in ageing (2) is presumed while searching for the etiology of various chronic and slow virus infections. The problem of viral oncogenesis is a special but related chapter. Besides specific antibody response, also unspecific cellular immunity mechanisms, as induced in cells by viral infection, participate apparently in these phenomena. Therefore, a model of aged cells cultivated in vitro and susceptible to viral infection seems to be useful for virology.

Use of cells or tissues from young and older embryos or animals is one approach to this question (3). The second way is the "ageing of cells under glass". Hayflick (4) elaborated a model of human fibroblasts transferred succesively in vitro which have a limited lifespan. Another group of virologists (for citations see 5-8) employs "aged cells" obtained by prolonged in vitro cultivation of primary cells or cell lines, without medium change. Our experiments arranged in this way have been done on chick embryo cells (CEC) and we are comparing viral infection in cell monolayers cultivated 1 or 7 days prior to infection. We prepare such cultures from leucosis-free 10-day-old chick embryos in a medium supplemented with inactivated calf serum (5,9). We have approximately the same number of cells in young and aged cultures. There are numerous

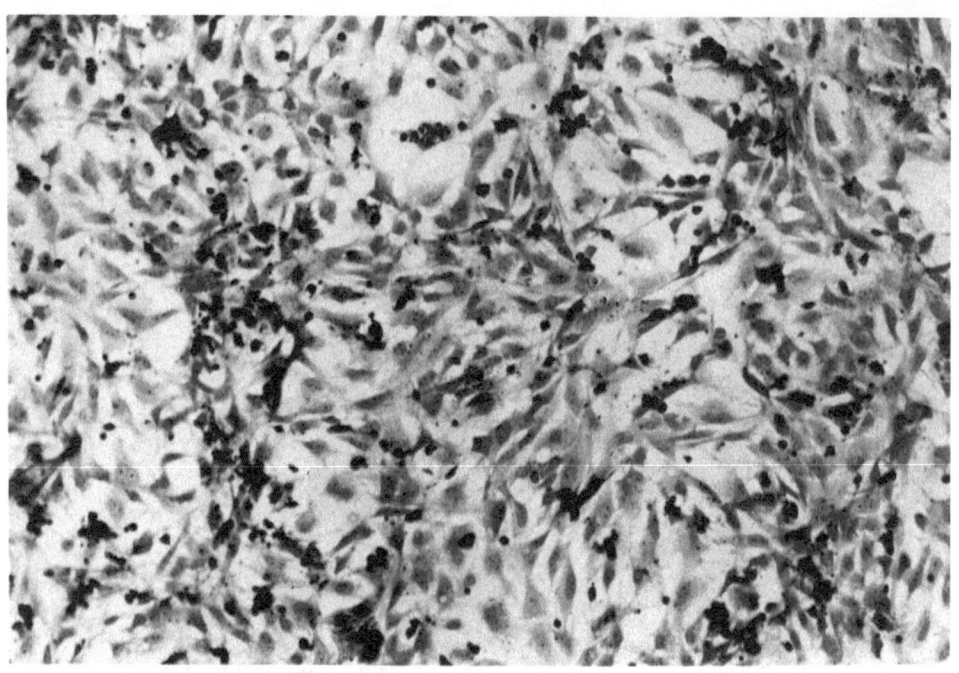

Fig. 1.   Chick embryo cell monolayer cultivated 1 day at 36°C.
(Hematoxylin-erythrosin.)

mitoses in the young ones while the aged cultures are stationary,
contact inhibited and they show single mitoses only (Figs. 1 and 2).

        Using this system, we studied virus reproduction and interferon
(IF) induction with the following viruses: herpes simplex type 1
virus (HSV), pseudorabies virus (PRV), western equine encephalomye-
litis virus (WEEV) and tick-borne encephalitis virus (TBEV).   HSV
(strain HSZP) causes polykaryocyte formation, PRV and WEEV induce
rapid cytopathic effect (CPE) and TBEV reproduces without marked
CPE.   All viruses were tested at high (10–50 ID_{50} per cell) and low
(1 ID_{50} per 100–1000 cells) multiplicity of infection (MOI).   Marked
differences in virus production and/or IF formation were observed
when comparing results in young and aged cultures (Fig. 3).

        HSV reproduces equally in young and aged cultures but IF is
induced in the latter only.   This IF appears late and has no back
effect on the inducing virus.   It should be mentioned that HSV

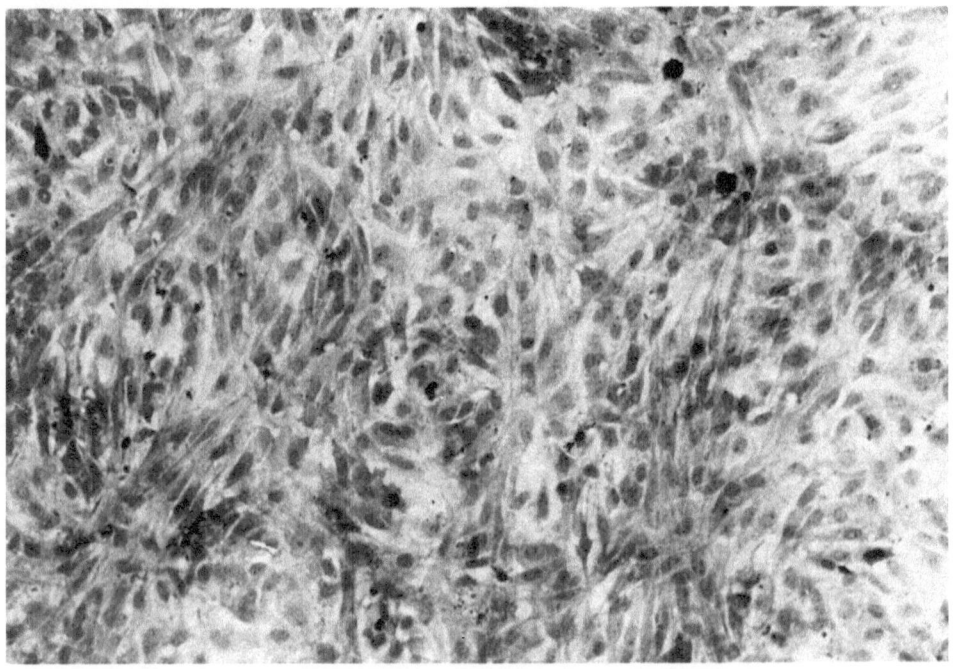

Fig. 2.   Chick embryo cell monolayer cultivated 7 days at 36°C.
No medium change.   (Hematoxylin-erythrosin.)

is sensitive to IF action -- namely in aged cultures -- if IF is
present in cultures <u>before</u> HSV inoculation (10).   Also PRV induces IF solely in aged cells, but the IF titers reached are higher
than in HSV infection.   At low MOI an IF back effect on the inducing
PRV is observed -- the reached virus titers are lower and CPE is
slower.   This does not happen at high MOI with PRV.   WEEV gives
similar results, the IF back effect being still more pronounced
because of the known high sensitivity of WEEV to IF action.   Most
complicated is the situation in TBEV infection.   At high MOI, virus
production is high and low while IF formation is low and high (early)
in young and aged cultures, respectively.   At low MOI, both virus and
IF are better formed in young cultures while in aged ones an intensive depression of both is observed.   The back effect of IF on inducing TBEV appears in all modifications of the experiment.   If we
block the IF formation in aged cultures by Actinomycin D, the TBEV
reproduces equally as in young cultures (Fig. 4).

      Based on morphological and immunofluorescence data, we tried
to analyze the infection process in small target groups of neighboring cells, which probably simulates the first steps of the in-

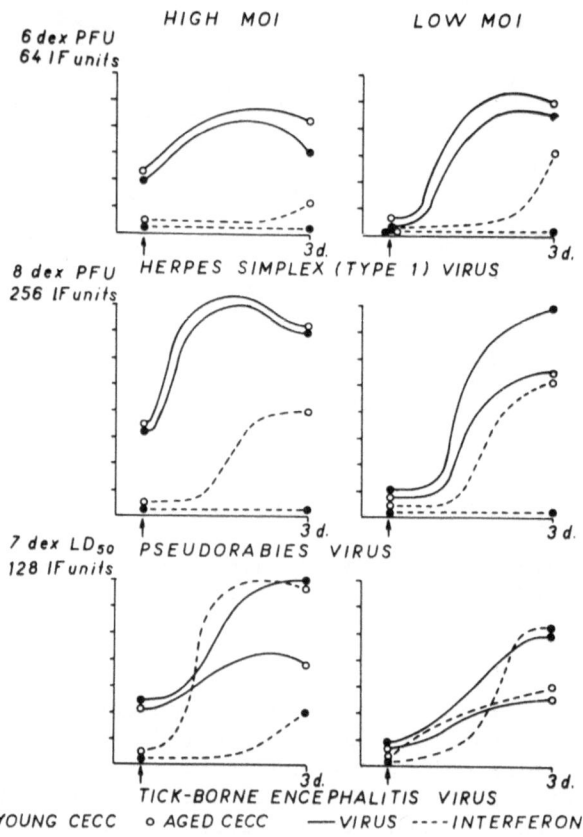

Fig. 3. Comparison of virus production and interferon formation in
young and aged chick embryo cell cultures. Ordinates: total virus
(10-fold dilutions) and interferon (2-fold dilutions) titres. PFU-
plaque forming units, LD_{50} - 50% lethal dose. Abscissae: time after
virus inoculation (in days). Growth medium removed prior to virus
inoculation, then cell cultures refed with fresh medium.

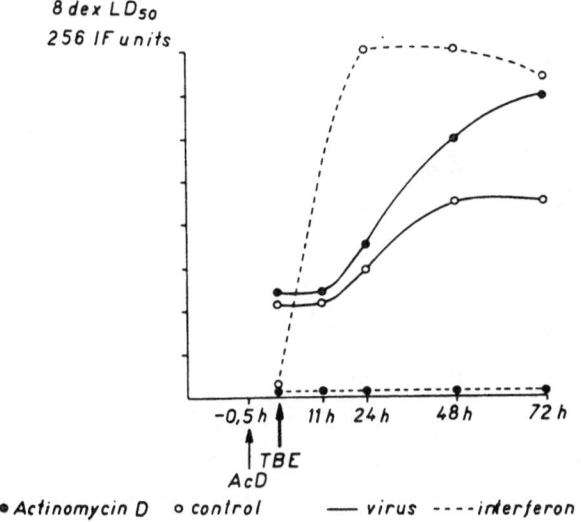

8 dex LD$_{50}$
256 IF units

-0,5h   11h 24h   48h   72h

TBE
AcD

• Actinomycin D   o control   —— virus ----interferon

Fig. 4.   Influence of Actinomycin D on tick-borne encephalitis
virus reproduction in aged chick embryo cell cultures.   Ordinate:
as in Figure 3.   Abscissa: hr after virus inoculation.   Actinomycin
D (2 μg/ml) was added to cells 30 min prior to infection, then, after
washing the cultures, it was replaced with medium just before virus
inoculation.

fection in the organism.   Figure 5 documents HSV 1 infection.   The
viral antigen is localized in cytoplasm and in nuclei as well and
there are no differences in virus production and polykaryocyte
formation when comparing young and aged cultures -- but one exception:
aged cultures infected at low MOI release -- simultaneously with the
viral progenium -- small amounts of IF.   Nevertheless, also this
small amount might be able to protect other cell groups from infection.   Figure 6 analyses analogically the situation in WEEV infection: at high MOI, again no difference between young and aged
cultures; at low MOI, IF formation in aged cultures and its evident
back effect on the inducing virus which leads to lowered virus
production and decrease in CPE -- CPE can disappear completely
at very low MOI.   Finally, Figure 7 shows the mentioned interesting
results in TBEV infection.   The noncytopathic virus cycle enables
a longterm preservation of virus producing cells in the culture and,
under appropriate conditions of the experiment, we can observe al-
ternate cycles between virus and IF production.

The described increased IF mechanisms in stationary aged
cultures act also in mixed viral infection.   While TBEV and HSV,
or TBEV and Lipovnik virus (Kemerovo group) are capable to reproduce
in the same cells of young cultures, preinfection of aged cultures
with TBEV inhibits markedly reproduction of the superinfecting HSV
or Lipovnik virus (11,12).

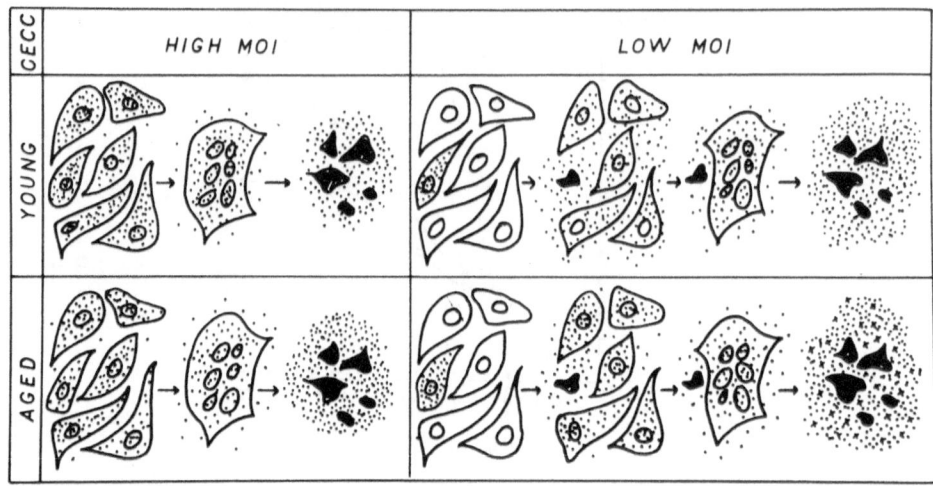

Fig. 5. Response of young and aged CEC cultures to herpes simplex
(type 1) infection. Formation of polykaryocytes, then cytopathic
effect. Dots and crosses represent virus and interferon, respectively.

CECC	HIGH MOI	LOW MOI
YOUNG		
AGED		

Fig. 6. Response of young and aged CEC cultures to WEE virus infection. For explanation see Figure 5.

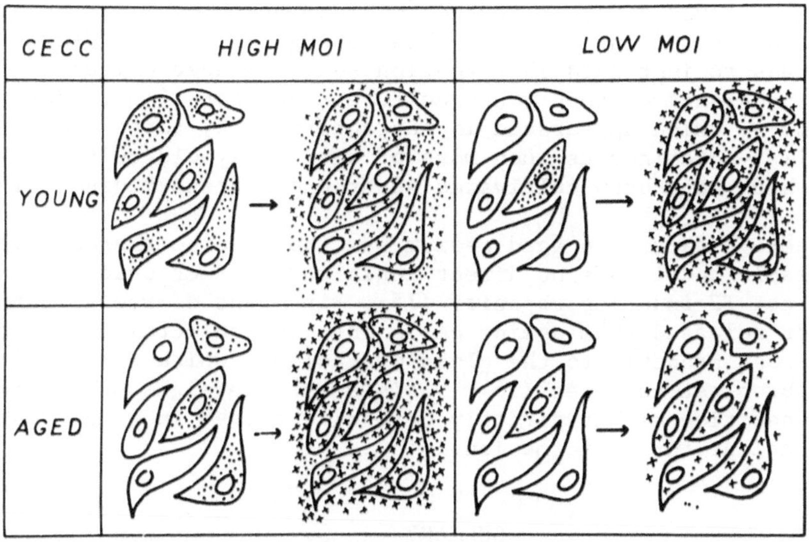

Fig. 7. Response of young and aged CEC cultures to TBE virus infection. For explanation see Figure 5.

The reasons of the changes reaction to viral infection in aged cultures are not known. In respect to the decreased proteosynthesis in aged cells, a shortage in repressors and therefore an easier de-repression of IF is implicated (13). However, IF itself is a cellu-lar protein and it acts via another protein induced in protected cells. Thus, one could speculate that abundant IF formation in aged cells is a part of the "error" process -- or ageing "pro-gramme" -- according to the von Hahn (14) concept of changes in the protein synthesis in the ageing cell.

Another result should be mentioned: When medium from aged cultures is removed and substituted for medium of young, 1-day-old cultures, then viral infection in these young cells under "aged medium" leads to increased IF formation -- as in aged cultures (15). Also the appearance of such young cultures ept for 24 hr under aged medium -- which appearance is characterized by rare mitoses and bigger cells -- reminds us much more of aged 7-day-old cultures than of those which have been cultivated 48 hr under fresh medium only.

The active factor in aged medium was shown to be thermolabile, non-dialyzable and it has no direct inhibitory effect onto viruses and no direct IF-inducing capacity (15). In light of recent data, this factor released from ageing cells into the cultural medium seems to belong most probably to the class of chalones which were detected also in fibroblast cultures (16). This substance might stimulate viral IF induction either directly or undirectly -- inducing host cell phase convenient for IF formation.

SUMMARY

Cells from leucosis-free chick embryos (skin and muscles), grown in vitro 1 and 7 days with no medium change, are used as a model for young dividing and stationary in vitro aged cells, respectively. Aged cultures are advantageous for IF induction by cytopathic and noncytopathic viruses. Decreased virus production in them is ascribed to a back effect of the early formed IF onto the inducing virus. A factor which stimulates viral IF interferon induction also in young cultures, is released from aged cells into the medium. This factor could belong -- according to its properties and action -- into the class of chalones.

The author wishes to express her appreciation to Prof. F. Svec, Institute of Experimental Oncology, Slovak Ac. Sci., for helpful discussion and encouragement.

REFERENCES

1. Jandásek, L. 1968. The nature of the age factor in experi-
    mental viral infection (in Czech). Cs. epidemiol.,
    mikrobiol., imunologie 17: 1.

2. Gajdusek, D.C. 1970. Slow virus infections and activation of
    latent virus infection in aging. IN Symposium on Cellular
    and Macromolecular Aspects of Aging of the Biology Division,
    Oak Ridge National Laboratory, Gatlinburg, Tennessee, April
    6-9, pp. 27-28.

3. Cantell, K., Valle, M., Schakir, R., Saukkonen, J.J. and
    Uroma, E. 1965. Observations on production, assay and
    purification of chick embryo interferon. Ann. Med. exp.
    Fenn. 43: 125.

4. Hayflick, L. 1970. Aging under glass. Exp. Geront. 5: 291.

5. Libíková, H. and Henslová, E. 1969. The correlation between
    tick-borne encephalitis virus yield and interferon production
    and effect in cell cultures of different in vitro age. Acta
    Virol. 13: 16.

6. McLaren, C. 1970. Influence of cell age on production and assay
    of mouse interferon in L-cells. Arch ges. Virusforsch 32: 13.

7. Galabov, A., Savov, Z. and Vassileva, V. 1973. Interferon
    production in arbovirus-infected cell cultures of tortoise
    (Testudo graeca) kidney. Acta virol. 17: 1.

8. Rossman, T.G. and Vilček, J. 1969. Influence of the rate
    of cell growth and cell density of interferon action in
    chick embryo cells. J. Virol. 4: 7.

9. Libíková, H., Rajčáni, J. and Henslová, E. 1969. Further
    studies on interferon and virus production in young and old
    cell cultures prepared from ordinary and RIF-free chick
    embryos. Acta virol. 13: 483.

10. Libíková, H. 1973. Interferon in young and aged chick
    embryo cell cultures infected with herpes simplex and
    pseudorabies viruses. Acta. Virol. 17: 464.

11. Libíková, H. and Rajčáni, J. 1974. Experimental double in-
    fection with tick-borne encephalitis and herpes simplex
    type 1) viruses. Acta Virol. 18: 47, 1974.

12.   Libíková, H. and Rajčani, H.   Experimental mixed infection
      with two tick-borne viruses and interferon-mediated inter-
      ference.   Acta Virol., in press.

13.   Carver, D.H. and Marcus, Ph. I.   1967.   Enhanced interferon
      production from chick embryo cells aged in in vitro.   Virology
      32: 247.

14.   von Hahn, H.P.   1970.   The regulation of protein synthesis in
      the ageing cell.   Exp. Geront. 5: 323.

15.   Libíková, H.   1970.   Influence of medium from aged uninfected
      chick embryo cell cultures on interferon formation induced
      by tick-borne encephalitis virus.   Acta Virol 14: 87.

16.   Houck, J.C., Cheng, R.F. and Sharma, V.K.   1973.   Control of
      fibroblast proliferation.   Natl. Cancer Inst. Monogr. 38:
      161.

QUESTIONS TO DR. LIBIKOVA

Dr. Sova: Could it be possible that the results you presented
were influenced by development of the protein synthetizing
apparatus (i.e., ribosomes, their amount and capability for
synthesis)?

Dr. Libíková: Polyribosomes seem to be involved in the differences
between young and aged cultures as they are numerous and highly
developed in cells in aged cultures as will be shown by Dr. Ciampor
in the next paper.

Dr. Martin: I wonder what the rationale is for using the system
as a model of cellular aging?   Perhaps it would be best to use a
more non-committal, operational term such as actively replicating
vs. post-replicating culture.   Perhaps a major variable in these
experiments is the status of the cell cycle, the aged cultures are
probably largely arrested in G_1.   There is also a probable deficiency
of glutamine in the 7 day unfed cultures.

Dr. Libíková: I adopted the term "cells aged in vitro" from Carver
and Marcus (1967) as a simple name for what we are working with.
Nevertheless, we used mostly the term "cell cultures aged in vitro".
The electronmicroscopic pictures of cells from such aged cultures
show changes similar to the aging changes of cells in the macroorganism (Prof. Smetana, Prague, personal communication).   The
observed increased interferon formation in various organs of older
animals infected with viruses, when compared with young animals,
also fits this parallel.   However, it is hard to indicate what is

an aged, or ageing cell in the macroorganism if we consider one
single cell and not an organ. Concerning the cell cycle and interferon: it was shown by Lee and Rosee (1970) that synchronized L
cells produce higher amounts of interferon when they are infected
at early S-phase than after infection at early or late G_2.

Dr. Dell'Orco: Could you give more details on the changes in
morphology that occur between the 3rd and 7th day of cultivation?

Dr. Libíková: One day after seeding the monolayer looks granular
(native preparation) because of dividing cells, this disappears
during the next 1-2 days of cultivation. After day 3 the monolayer becomes denser, the cells bigger, intercellular spaces
closer, but the number of cells does not increase further 24 hr
after seeding in this system.

                          COMMENT TO DR. LIBIKOVA

Dr. Holeckova: I think that tissue culture perople are used to
considering phase III diploids as ageing cells, and that the
problem is a problem of terminology. I agree with Dr. Martin
that the difference between Dr. Libikova's definition of ageing
cells and that of the cell culture research should be very
clearly stated.

ULTRASTRUCTURE OF IN VITRO AGED CHICK EMBRYO CELL CULTURES IN

RELATION TO VIRAL INFECTION

Fedor Čiampor and Helena Líbiková

Institute of Virology, Slovak Academy of Science
Bratislava, Czechoslovakia

SUMMARY. The ultrastructure of young uninfected and infected with herpes simplex virus chick embryo cells was compared with that of aged cells. In contrast to the nuclei of young chick embryo cells 24 hr after infection, the aged chick embryo cells 24 hr after infection presented very expressive blocks of perinuclear chromatin. The nucleoplasm contains forming naked HSV particles. The lesions seen are dilation of the perinuclear cisternae, dilatation of rough endoplasmic reticulum cisternae and formation of multilamellar bodies. The moderate swellings of the mitochondria with violation of cristae arrangement and losing of mitochondrial matrix were observed. In contrast to the cytoplasm of young infected cells no formation of crystal-like arrangement of cytoplasmic ribosome tetrads occurs.

Ultrastructural changes correlated with reproduction of herpes simplex virus (HSV) in various tissue cultures were studied by many authors (5-11). Mostly have been used young primary cell strains or stable cell tissue cultures. The ultrastructural changes correlated with reproduction of viruses in aged cells has received little attention, with respect to the relationship of ultrastructural changes correlated with ageing of cells and with changes induced by reproduction of viruses in these cells. Brock and Hay (1) studied the ultrastructure of chick fibroblasts in vitro at early and late stages during their growth span. In late passage cells, the marked location of the nucleus, the absence of chromatin adjacent to the nuclear envelope and ellipsoidal mitochondria, generally bent into a shallow "U" with longitudinally oriented cristae were observed. Highly developed endoplasmic reticulum and Golgi complexes were observed at both passage levels. The most striking change was the presence of conspicuous secondary lysozomes and residual

481

bodies in the late passage cells.

Infection of young and aged chick embryo cell (CEC) cultures with HSV demonstrated that no interferon formation was recorded in dividing young cultures. Virus reproduction at various multiplicities of infection per cell was equal or higher in aged than in young cultures, but the protection against HSV infection by exogenous or endogenous interferon, was strikingly better in aged that in young cultures (4).

Therefore, in this study, the fine structure of young uninfected and infected with HSV type 1 (strain HSZP) chick embryo cells is compared with that of aged uninfected and infected cells. Multiplicity of infection was 1 PFU/10 cells. Chick embryo cells were prepared from RIF-free eggs and cultivated according to Libikova (4). Incubation CEC was for 1 day (young) and 7 days (aged) without change of medium, respectively. For electron microscopy we used 1 and 7 day old uninfected CEC and 1 and 7 day old CEC 7, 17, 24, and 48 hr after infection. The cells were detached from the glass by gentle scraping with rubber policemen, fixed with 2.5% glutaraldehyde (TAAB) in 0.2 M sodium cacodylate buffer pH 7.2 and post-fixed with 1% osmium tetroxide (BDH) in cacodylate buffer pH 7.2. Fixed material was dehydrated with acetone and embedded in Durcupan ACM (Fluka). Ultrathin sections were stained with 2% aqueous solution (BDH) of uranyl acetate and lead citrate (BDH) (12) and examined in a Philips EM 300 electron microscope at 80 kV.

Figure 1 illustrates young uninfected CEC. The cell nuclei, had fairly regular oval contours with dense blocks of chromatin scattered centrally in the karyoplasm as well as adjacent to the nuclear envelope. The nucleolus is prominent and the dense granular RNP is embedded in a loose network of fibrillar RNP. All the mitochondrial images seen in young cells could be accounted for by an ellipsoid mitochondrial shape with plate-like cristae traversing the short axis. The rough endoplasmic reticulum (RER), although usually appearing to be randomly oriented, was sometimes arranged in short parallel arrays in these young cells. In contrast to the more-or-less oval nuclei of young CEC, the aged uninfected CEC (7 days old, Fig. 5) nuclei always presented a lobed appearance and blocks of chromatin were scattered centrally in these nuclei and were rarely seen at the periphery. The nucleoli are elongated and the contrasted bands of intranucleolar chomatin originate from the perinucleolar associated chromatin. The nucleoli 7 hr after infection show a very expressive arrangement of closely packed fibrillar RNP component in a central part of nucleolus.

The mitochondrial profiles and the orientation of the cristae were markedly changed in aged uninfected cells (Fig. 6). A variety of bizarre mitochondrial shapes in addition to those that were el-

liptical or nearly circular were by a narrow ellipsoid usually shaped into a shallow "L" or "U". The cristae transversed
the long axis of the mitochondria, they were irregularly arranged
and mitochondrial membranes were in the individual parts disrupted.

Young CEC 24 hr after infection contain in the nucleoli HSV
particles (Fig. 2). The nucleolus exclusively consists of closely
packed fibrils with wide open meshes. The size of these fibrils
grossly correlates with that of the fibrils of the normal nucleolus,
but their number has considerably increased since no appreciable
reduction of the nucleolar size has occured. The perinuclear
chromatin has disappeared and nuclear membrane had been disrupted.
The cytoplasmic structures were extremely changed. The lesions seen
are dilation of the cisternae of the RER and formation of multilamellar bodies. The cytoplasmic ribosomes are arranged in tetrads and
these tetrads form a crystal-like structure of tetrads according geo-
metrical plane p4 (Fig. 3). Many of HSV particles are seen in
the cytoplasm.

In contrast to the nuclei of young CEC 24 hr after infection,
the aged CEC 24 hr after infection presented very expressive blocks
of perinuclear chromatin. The nucleoplasm contains forming naked
HSV particles (Fig. 8). The lesions seen are dilatation of the
perinuclear cisternae, dilatation of the RER cisternae and formation of multilamellar bodies (Figs. 8, 9). The secondary lysosomes increase in number. The moderate swelling of the mitochondria
with violation of cristae arrangement and loosing of mitochondrial
matrix were observed. No formation of crystal-like ribosomal
structures occurs.

Young and aged CEC 48 hr after infection had been completely
disrupted. Vacuolization of cytoplasm occurs and deposits of
contrasted bands of intranucleolar chromatin were seen free in
cytopathically changed cells (Figs. 4, 10).

CEC with selective inhibition of host cell macromolecular
synthesis induced before adsorption and penetration of influenza
virus, synthesized all specific structural and nonstructural proteins coded by virus (2). Ultrastructural changes correlated with
inhibition of host cell macromolecular synthesis (striking nucleolar
lesions consist of a loss of the granular RNP component and intranucleolar chromatin) do not prevent reproduction of influenza virus.

Striking nucleolar changes are seen also in uninfected and infected aged CEC, while in infected aged CEC changes are associated
with the reappearance of an exaggerated amount of intranucleolar
chromatin, granular RNP and perinuclear chromatin.

Comparing ultrastructural changes of uninfected CEC cultured through 18 generations (1) with CEC incubated 7 days without change of medium in our experiments, we can conclude that no differences are seen.

The present observations indicate, comparing CEC with selective inhibition of host macromolecular synthesis (2,3) and aged infected CEC that no inhibition of host cell RNA and protein synthesis induced by virus after its penetration into the cell is needed. The virus can easily utilize cellular structures for synthesis of its structural proteins. With relationship to this assumption the aged cells produce equal or higher number of new virus particles and also virus induced nonspecific proteins, such as interferon.

REFERENCES

1. Brock, M.A. and Hay, R.J. 1971. Comparative ultrastructure of chick fibroblasts in vitro at early and late stages during their growth span. J. Ultrastruct. Res. 36: 291.

2. Ciampor, F. An electron microscopic study of influenza virus reproductive cycle in chick embryo cells with selective inhibition of host cell macromolecular synthesis. Acta Virol., in press.

3. Krug, R.M. 1971. Influenza viral RNPs newly synthesized during the latent period of viral growth in MDCK cells. Virology 44: 125.

4. Libíková, H. 1973. Interferon in young and aged chick embryo cell cultures infected with herpes simplex and pseudorabies viruses. Acta Virol. 17: 464.

5. Nii, S., Morgan, C. and Rose, H.M. 1968. Electron microscopy of herpes simplex virus. II. Sequence of development. J. Virol. 2: 517.

6. Nii, S. 1971. Electron microscopic observations on FL cells infected with herpes simplex virus. I. Viral forms. Biken J. 14: 177.

7. Nii, S. 1971. Electron microscopic observations on FL cells infected with herpes simplex virus. II. Envelopment. Biken J. 14: 325.

8. Shipkey, F.H., Erlandson, R.A., Bailey, R.B., Babcock, V.I. and Southam, C.M. 1967. Virus biographies. II. Growth of herpes simplex virus in tissue culture. Exptl. Molec.

Pathol. 6: 39.

9.   Spring, S.B. and Roizman, B.   1967.   Herpes simplex virus
     products in productive and abortive infection.   I. Stabili-
     zation with formaldehyde and preliminary analyses by iso-
     pycnic centrifugation in CsCl.   J. Virol 1: 294.

10. Spring, S.B., Roizman, B. and Schwartz, J.   1968.   Herpes
     simplex virus products in productive and abortive infection.
     II. Electron microscopic and immunological evidence for
     failure of virus envelopment as a cause of abortive infection.
     J. Virol. 2: 384.

11. Strandberg, J.D. and Aurelian, L.   1969.   Replication of canine
     herpes virus.   II. Virus development and release in infected
     dog kidney cells.   J. Virol. 4: 480, 1969.

12. Venable, J.H. and Coggeshall, R.   1965.   A simplified lead
     citrate stain for use in electron microscopy.   J. Cell Biol.
     22: 407.

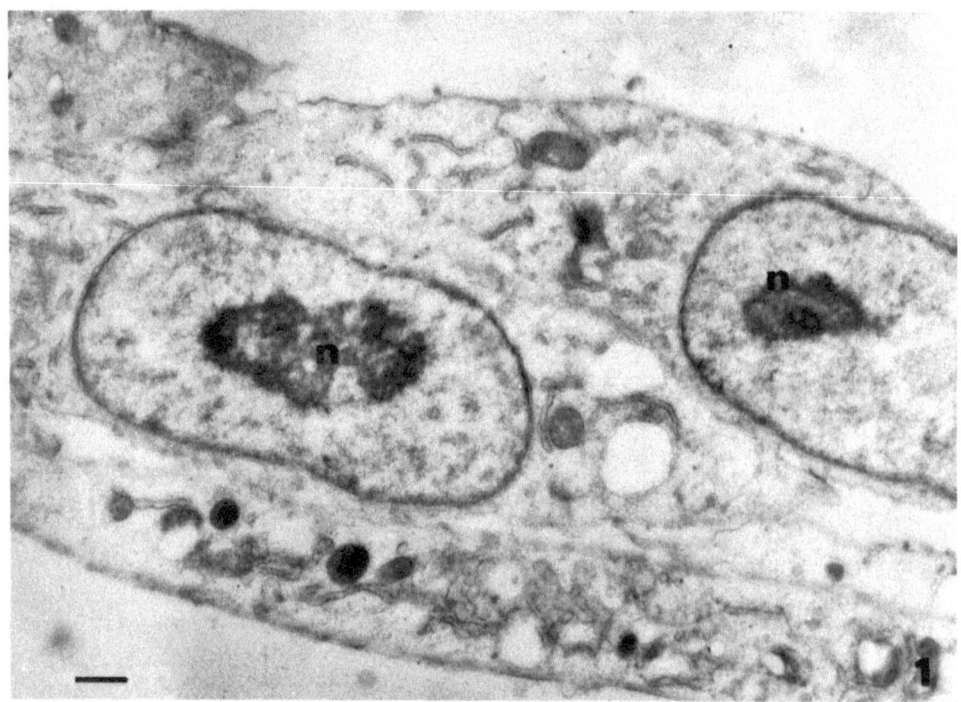

Fig. 1. One day-old CEC uninfected (n – nucleolus), X6,000. Bare scale indicates 1250 nm.

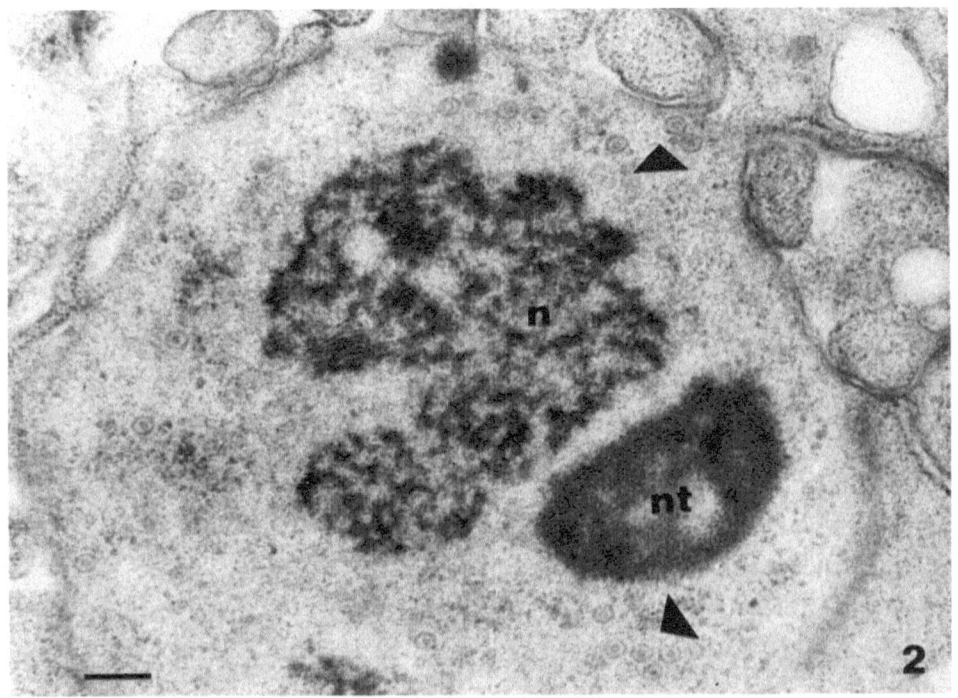

Fig. 2. One day-old CEC 24 hr after infection with HSV (n -
nucleolus, nt - nucleolar body, HSV particles - arrows), X3,600.
Bare scale indicates 250 nm.

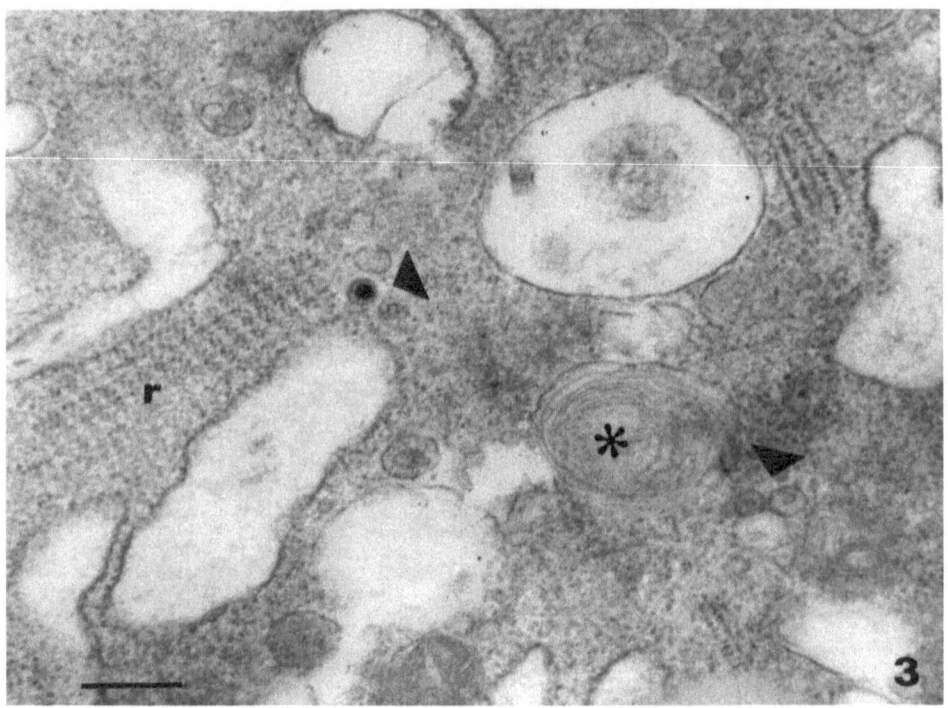

Fig. 3. Cytoplasm of one day-old CEC 24 hr after infection with HSV
(r - ribosomes arranged in crystal-like structures, HSV particles-
arrows, multilamellar body - asterisk), X54,000. Bare scale indicates 250 nm.

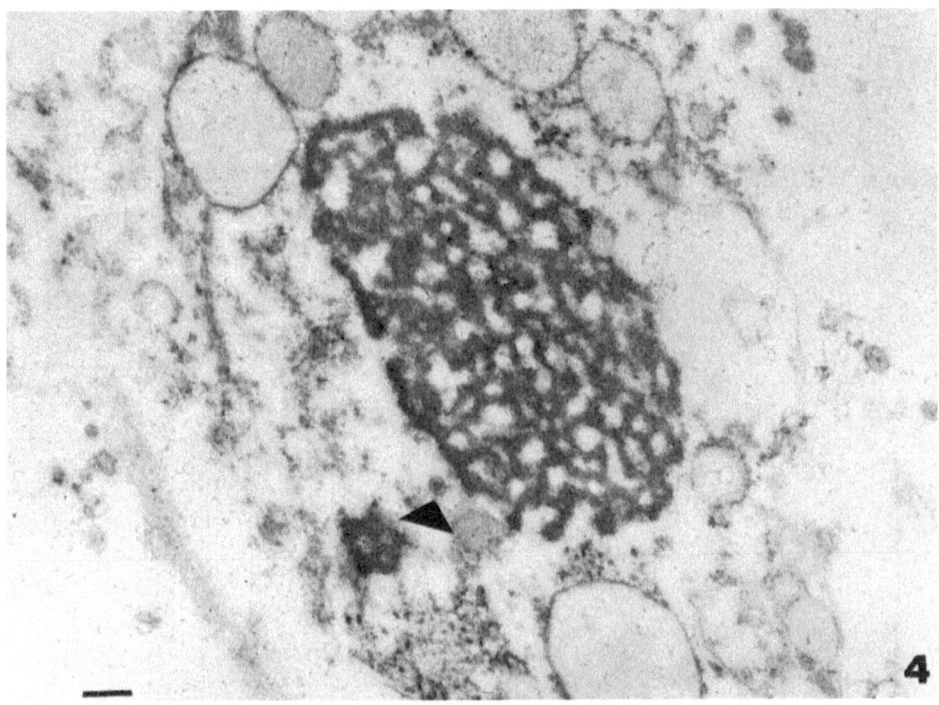

Fig. 4. One day-old CEC 48 hr after infection with HSV (HSV particle -
arrow), X27,000. Bare scale indicates 250 nm.

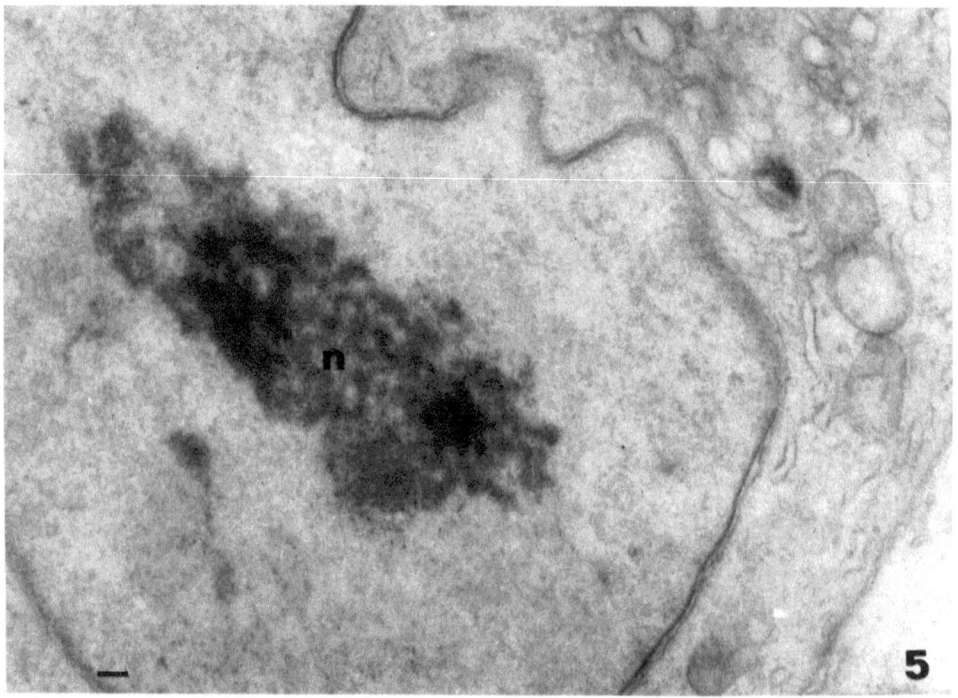

Fig. 5.   Seven day-old CEC uninfected (n - nucleolus), X18,000.
Bare scale indicates 250 nm.

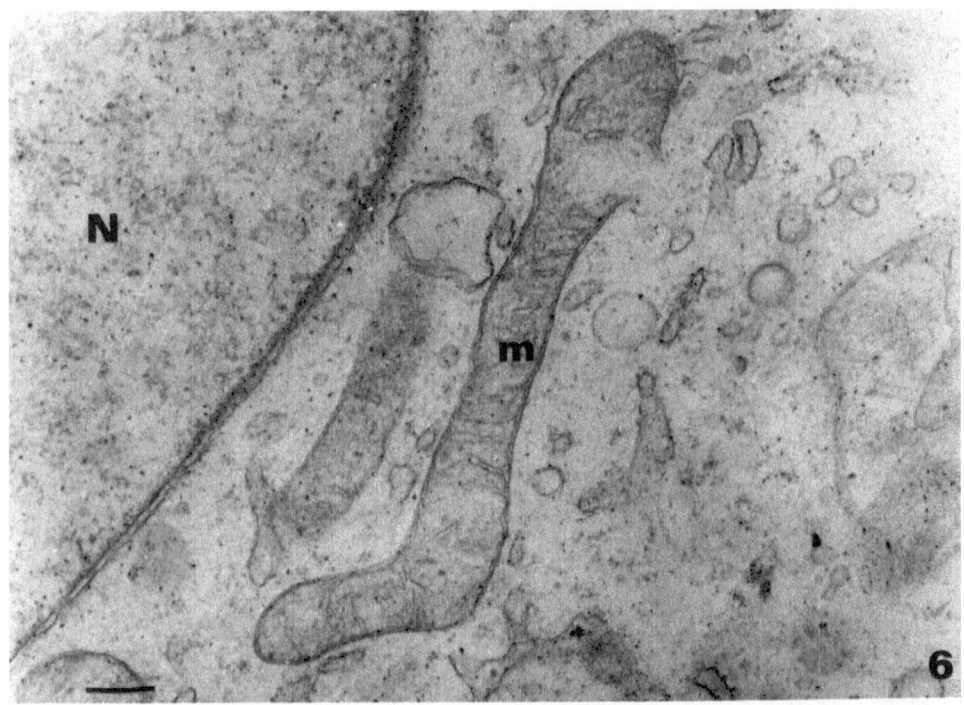

Fig. 6. Cytoplasm of seven day-old uninfected CEC (m – altered mitochondria, N – nucleus), X36,000. Bare scale indicates 250 nm.

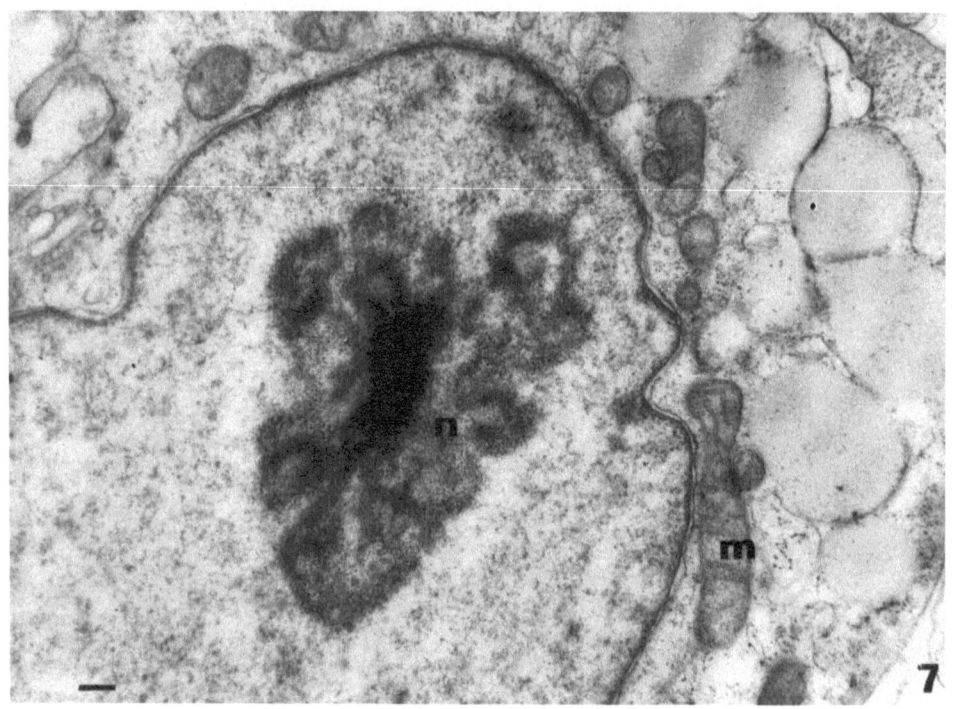

Fig. 7. Seven day-old CEC 7 hr after infection with HSV (n - nucleolus, m - mitochondria), X22,000. Bare scale indicates 250 nm.

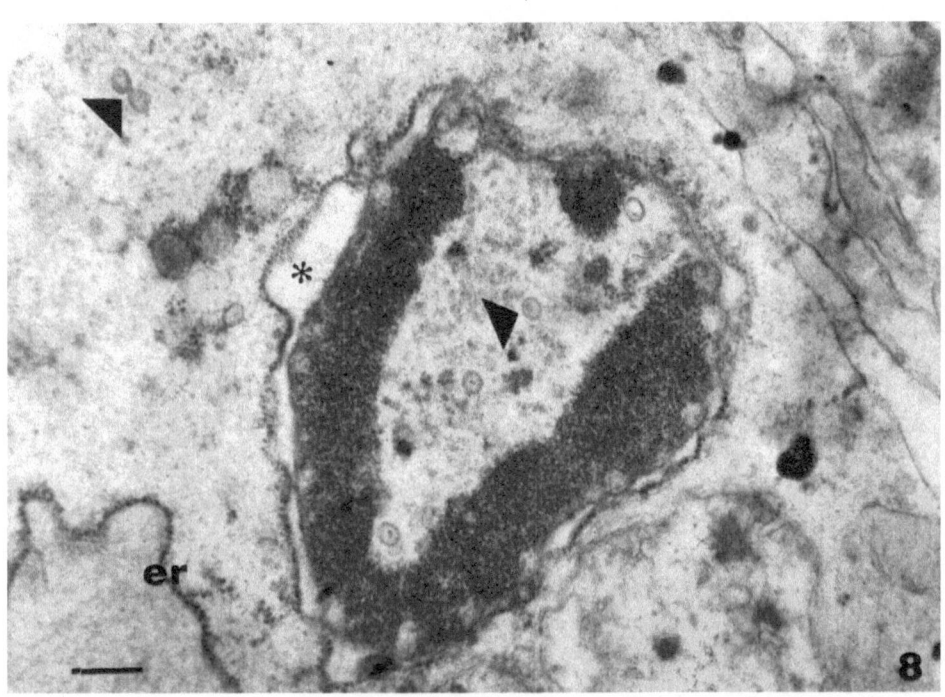

Fig. 8. Seven day-old CEC 24 hr after infection with HSV (er – rough endoplasmic reticulum, HSV particles – arrows, dilatated perinuclear cisternae – asterisk), X36,000. Bare scale indicates 250 nm.

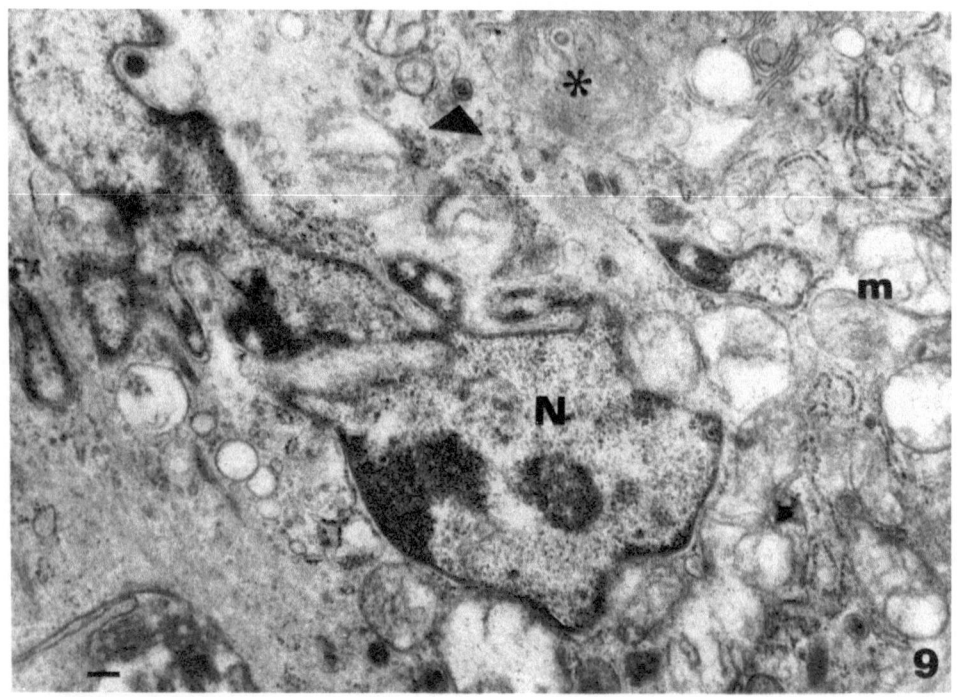

Fig. 9. Cytoplasm of 7 day-old CEC 24 hr after infection with HSV
(N - nucleus, m - mitochondria, multilamellar bodies - asterisk,
HSV particle - arrow), X18,000. Bare scale indicates 250 nm.

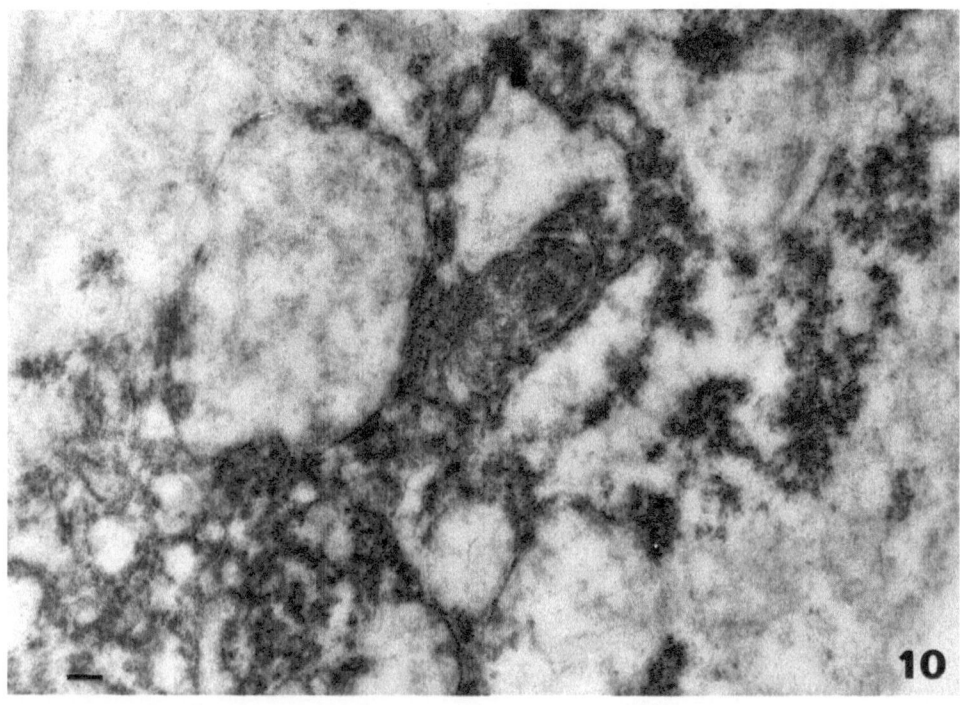

Fig. 10.   Seven day-old CEC 48 hr after infection with HSV.   X20,000.
Bare scale indicates 250 nm.

HERITABLE CELL CYCLE DISTURBANCES AND LATE RECOVERY IN X-IRRADIATED

MURINE LYMPHOMA L5178Y-S CELL POPULATIONS IN VITRO

Janusz Z. Beer and Irena Szumiel

Department of Radiobiology and Health Protection,
Institute of Nuclear Research, 03-195 Warszawa-
Zeran, Poland

In the last decade the view has been expressed by several authors (1-4) that ageing of the normal mammalian cells is connected with the prolongation of the mitotic cycle. At the same time it has been shown that ionizing radiation can in vitro (5-10) and in vivo (11,12) induce in mammalian cells, besides the lethal effects, also some heritable damages which retard the cell cycle traverse for considerable number of cell generations.

In this paper we will outline results of our studies on the heritable damages including cell cycle disturbances and late recovery phenomena in X-irradiated murine lymphoma L5178Y-S cells. This cellular model system in the form of a suspension culture presents an object particularly suitable for long term studies on growth disturbances (9).

EXPERIMENTAL

The technique of in vitro cultivation of the radiosensitive strain L5178Y-S of murine leukaemic lymphoblasts (13-15) has been described previously (8,9,17). Mean doubling times (T_D) for controls varied in individual experiments from 10 to 14 hr.

Cell population densities were determined microscopically using a Burker haemocytometer.

The cultures were irradiated with 200 kV X-rays filtered with 0.5 mm Cu, HVL 1 mm Cu at a dose rate of 345 rads per min (9,17).

Viability of the cell populations was determied by nigrosine staining (17).

Cloning of the cells in agar-supplemented medium (18) and isolation of the slowly growing cell sublines were described in (9,16,19,20).

Autoradiographic examination of the cell cycle was presented by Beer et al (21). The cell generation time (T_G) and durations of G_1, S and G_2 phases (T_{G1}, T_S and T_{G2}, respectively) were determined according to Mendelsohn and Takahashi (22). Frequency distribution of the G_2 phase duration was determined according to Stanners and Till (23).

RESULTS AND DISCUSSION

In a majority of cases irradiation of L518Y-S cells with doses of 25 to 600 rads of X-rays causes growth and viability disturbances which persist for periods corresponding to more than one hundred control cell generations, as it can be seen in Figure 1. More comprehensive description of these changes can be found in Beer et al. (8) and Beer (9).

The retardation of growth is partially caused by the presence of the slowly-growing clones of the cells. This has been proven by cloning followed by observations of growth of the sublines isolated from non-irradiated cell cultures and from the cell populations irradiated with 300 rads of X-rays. In Figure 2 it can be seen that the sublines with T_D almost double as compared to that of the controls derived from single cells present in the cultures which survived irradiation. Viability of the cloned sublines with T_D not exceeding 18 hr was normal (95%). Somewhat lower viabilities were observed in the slower sublines. The decrease of viability could account, however, only for a very minor part of the growth retardation. Even the highest observed content of dead cells which amounted to 17% could add a little more than 2 hr to the control T_D value of 10.8 hr (25). In fact, the sublines with 17% of dead cells grew with TD of 20 hr or more.

The cell cycle of control L5178Y-S line and of several slowly growing sublines was analyzed autoradiograhpically. It was found that although this analysis in the case of control cells was relatively easy, the percentage of labelled metaphases for the slowly growing sublines often were so scattered that setting up of the curve for conventional analysis was impossible. In these cases highly irregular relationships could be predicted. Results suitable for analysis of the cell cycle were obtained in two cases

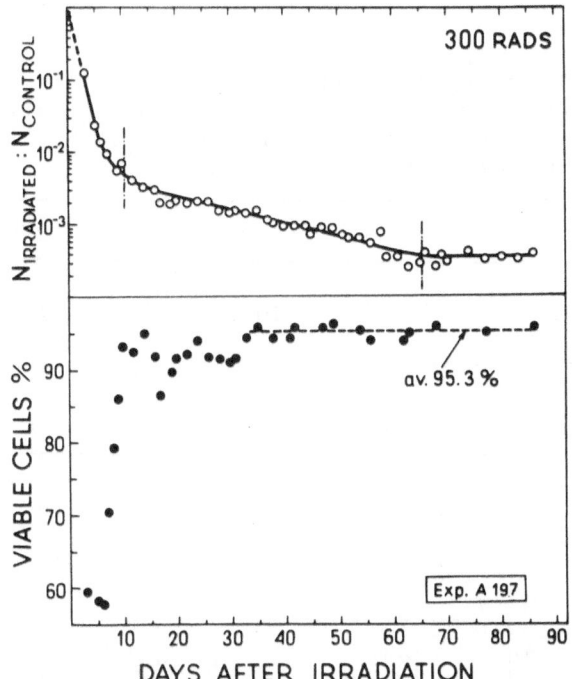

Fig. 1.   An example of prolonged growth and viability disturbances
induced by irradiation with 300 rads of X-rays in a culture of
L5178Y-S cells. $N_{IRRADIATED} : N_{CONTROL}$ – relative cell number,
i.e., ratio of number of cells in the irradiated and the control
cultures.

and they are presented in the Table and in Figure 3.   A more detailed
record of these data is published in Beer et al. (21) and Stanners
and Till (24).

     Significant differences between T_D and T_G values determined
for the sublines examined and for the control line (Table) indicated
that reduced proliferative activity was connected with prolongation
of the internitotic time of widely varying durations.   From Figure 3
it can be seen that T_G prolongation in the slowly growing sublines
is connected primarily with a slower progression through the G_2
phase.   It should be mentioned that Sinclair and Morton analysed
one slowly-growing cell line isolated from irradiated Chinese
hamster cell cultures and for those cells found G_1 block (23).

     The data obtained were further analysed by the method of Stanners

Table 1. Mean doubling time and generation time in a control
L5178Y-S line and in 2 slowly growing sublines. Cell cycle parameters for the control: T_{G1} = 1.3 hr, T_S = 6.4 hr, T_{G2} + 0.7 min
= 3.0 hr (according to ref. 22).

Cell line	T_D hr	T_G hr
Control	10.8	10.8
A470/3	18.8	13.7
A495/7	25.0	19.4

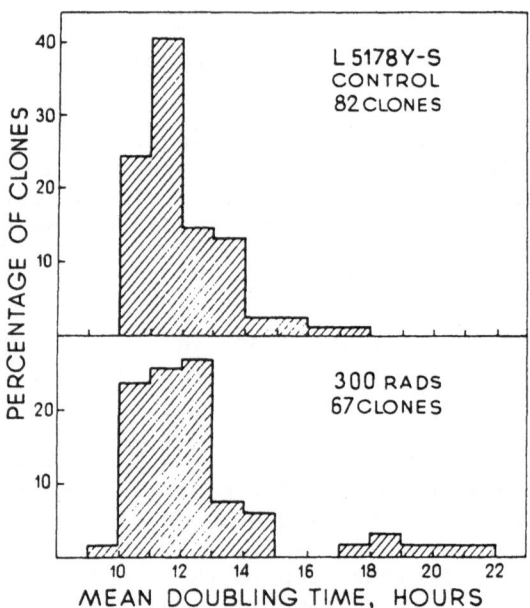

Fig. 2. Mean doubling time distribution among sublines isolated
from non-irradiated and irradiated with 300 rads of X-rays
L5178Y-S cell populations.

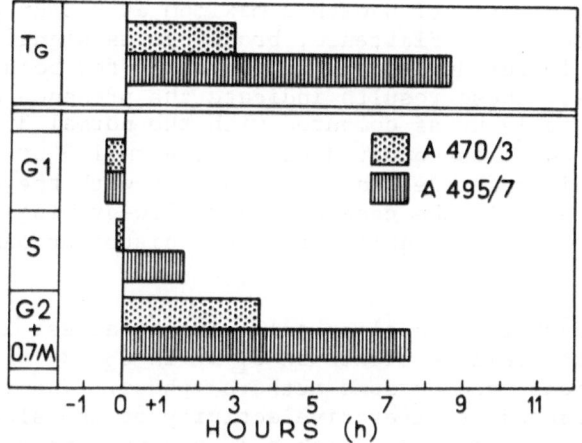

Fig. 3.   Differences between duration of the generation time and individual cell cycle phases in 2 slowly growing L5178Y-S sub-lines and a control.

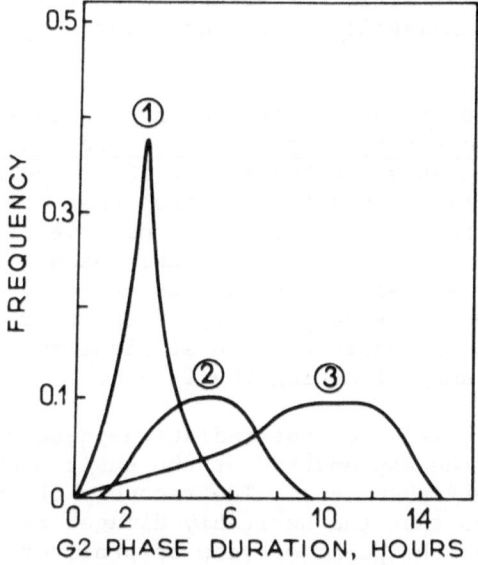

Fig. 4.   Frequency distribution of G_2 phase duration in a control L5178Y-S line and in 2 slowly growing L5178Y-S sublines (A470/3 and A495/7).

and Till (24). This allowed to establish frequency distribution of G_2 duration for a normal and the two slowly-growing cell sublines. From Figure 4 it can be seen that a relatively sharp, almost symmetrical distribution with a maximum at 2.8 hr was obtained for the control. Flattened, broad curves significantly shifted towards longer durations were obtained for both slowly growing sublines. These results indicate the presence of cells with G_2 prolonged up to 14 hr as compared with the normal 3 hr duration of prolonged up to 14 hr as compared with the normal 3 hr duration of this phase. It is possible that difficulties with the cell cycle analysis encountered in the case of other slow-growing sublines were connected with even less regular frequency distributions of G_2 phase duration.

It is known that non-cycling cells can be arrested -- depending on the type of the cell -- either in G_1 or in G_2 (R_1 or R_2, according to the designations used by some authors) phase. The data presented indicate that reduced proliferative activity of the slowly-growing L5178Y-S sublines is connected with a temporary arrest of the cells in G_2 phase. This possibility is illustrated by the model presented in Figure 5 (25). It should be mentioned that results of labelling index determinations after 24 hr labelling, performed for 9 sublines with T_D from 14 to 22 hr indicate that several percent of cells can be arrested for periods exceeding 24 hr and that the proportion of non-labelled cells does not seem to be related to T_D (25).

Growth rate of the slowly-growing sublines derived from single cells was not constant, it gradually increased and not one of the originally handicapped sublines grew after 250 control doubling times at a rate significantly different from that of the controls. This phenomenon of the late recovery (10) observed in more than one hundred sublines is illustrated with three examples in Figure 6. The late recovery was observed at various stages of post-irradiation development, sometimes very early, often after many tens of doublings. Undoubtedly, it is connected with elimination from the cell population of the damage inducing the G_2 block.

Present knowledge of the radiation-induced heritable lesions does not allow the explanation of the basic mechanisms which underlie the late recovery phenomena. The recovery of full proliferative activity indicates that the heritably damaged cells, viable although handicapped from the point of view of proliferative activity, do not lack any components or structures qualitatively but rather that they are subject to some quantitative defects. For such handicapped cells an occasional uneven distribution of material during the cell division might produce the "repair effect". A possibility that subtle variations in concentration of certain cellular factors can be critical for cell growth has been recently pointed out by

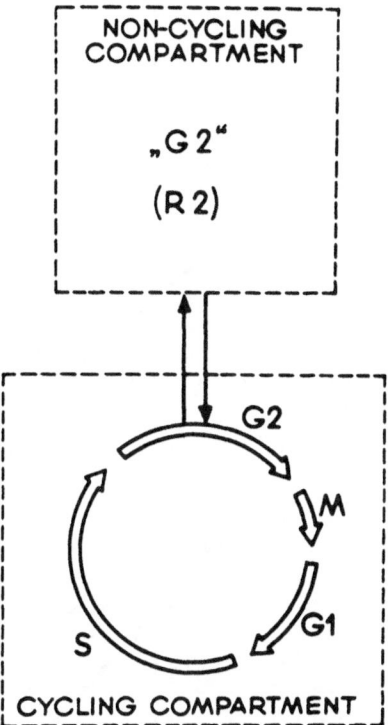

Fig. 5. Cycling and non-cycling compartment model proposed for
the slowly growing L5178Y-S cell populations.

Smith (26). It is also worth remembering that under cell culture
conditions a possibility exists that a quantitative deficiency of
certain cell components can be filled with elements from broken
cells which are always present in the medium. Cell fusion processes
may also lead to analogous effects.

It should be mentioned that instability of the radiation-
induced changes transmissible to progenial cell generations resembles lability of so called stable phenotypic changes observed
in various drug resistant, temperature resistant or temperature
sensitive cell lines (27-32). Also in those cases mechanisms
underlying the lability of the heritable changes remain obscure.
Further progress in this field can lead to broadening of our knowledge on the background of flexibility of some information transferred to progenial mammalian cell generations or even to discovery

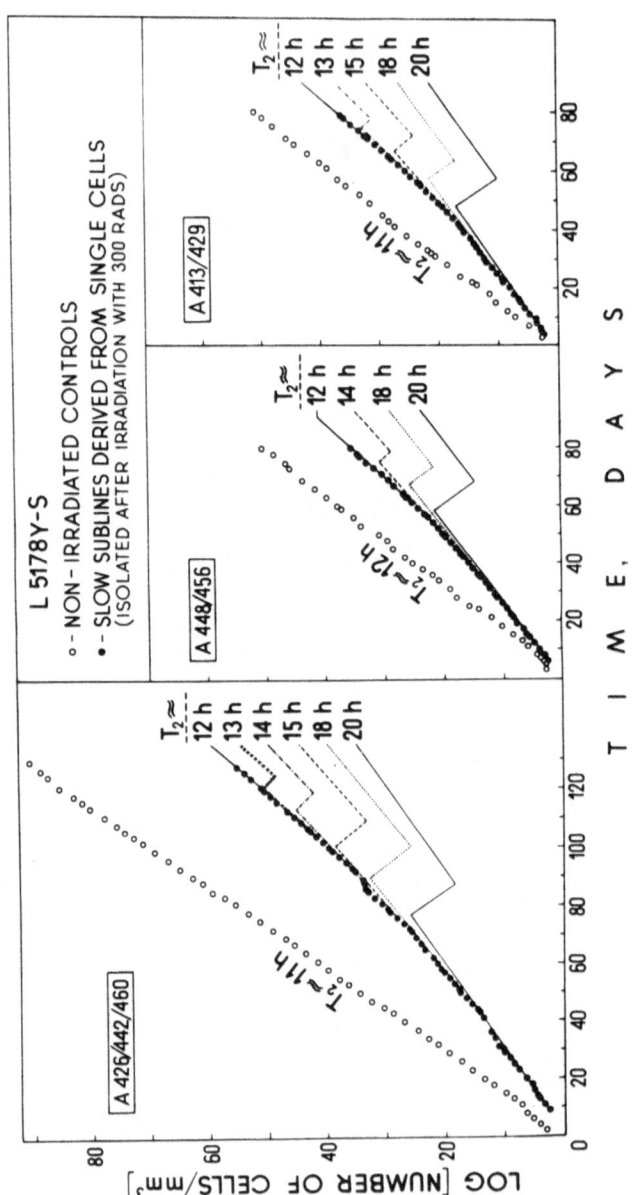

Fig. 6. Examples of late recovery of proliferative activity in the slowly growing L5178Y-S sublines derived from single cells.

of new intergeneration information-transfer systems.

Studies on the radiation-induced heritable lesions have been carried out on neoplastic or transformed established cell strains, wheras the ageing phenomena can be observed in normal diploid cell lines. Thus, comparisons between cell cycle disturbances in the sublines heritably damaged by radiation and those occurring in the ageing cells should be made with caution. Nevertheless, it is possible that the heritably-damaged cells can serve as convenient models for studies on some disturbances which accompany ageing of the mammalian cells.

Independently whether further studies will show similarities or differences between the cells heritably damaged by radiation and the ageing cells it would be interesting to exploit both experimental systems for examination of possibilities of prevention, control or elimination of the heritable cell cycle disturbances.

SUMMARY AND CONCLUSIONS

1. Ionizing radiation can induce in mammalian cells, besides the lethal effects, also heritable damages which retard cell cycle traverse for considerable number of cell generations.

2. Radiation-induced heritable lesions in murine leukaemic lymphoblasts L5178Y-S affect progression of the cells through the G_2 phase of the cell cycle.

3. Late recovery of proliferative activity which must be connected with elimination of the G_2 block is a common phenomenon in the heritably damaged L5178Y-S sublines derived from single cells.

4. A possibility should be taken into consideration of employing the cells heritably damaged by radiation as convenient models for studies on some disturbances which accompany ageing of the mammalian cells.

5. It seems that results of studies on both the cells heritably damaged by radiation and the ageing cells could be useful for evaluation of possibilities of prevention, control or elimination of the heritable cell cycle disturbances.

ACKNOWLEDGMENT

This work was supported by a Research Contract No 917/RB of the International Atomic Energy Agency.

REFERENCES

1. Macieira-Coelho, A., Ponten, J. and Philipson, L.   1966.
    The division cycle and RNA-synthesis in diploid human cells
    at different passage levels in vitrol Exp. Cell Res. 42: 673.

2. Cristofalo, V.J.   197. Metabolic aspects of aging in diploid
    human cells, IN Aging in Cell and Tissue Cultures, E. Holekova
    and V.J. Cristofalo (editors), Plenum Press, New York-London,
    p. 83.

3. Macieira-Coelho, A. 1970. The decreased growth potential in
    vitro of human fibroblasts of adult origin, IN Aging in
    Cell and Tissue Culture, E. Holeckova and V.J. Cristofalo
    (editors), Plenum Press, New York-London, p. 121.

4. Gelfant, S. and Smith, J.G., Jr. 1972. Aging: noncycling
    cells and explanation. Science 178: 357.

5. Sinclair, W.K. 1964. X-ray-induced heritable damage (small-
    colony formation) in cultured mammalian cells. Radiation
    Res. 21: 584.

6. Nias, A.H.W., Gilbert, C.W., Lajtha, L.G. and Lange, C.S.
    1965. Clone-size analysis in the study of cell growth
    following single or during continuous irradiation. Int.
    J. Radiat. Biol. 9: 275.

7. Westra, A. and Barendsen, G.W. 1966. Proliferation char-
    acteristics of cultured mammalian cells after irradiation
    with sparsely and densely ionizing radiations. Int. J.
    Radiat. Biol. 11: 477.

8. Beer, J.Z., Rosiek, O. Sablinski, J.M. and Ziemba-Zak, B.
    Radiation-induced herbitale desions in cultures of murine
    leukaemic cells L5178Y. Studia Biophysica (Berlin) 6: 103.

9. Beer, J.Z. 1974. Late post-irradiation phenomena in mammalian
    cell populations. I. Radiation-induced heritable changes
    in X-irradiated L518Y-S cells in vitro. Institute of
    Nuclear Research (Warsaw) Report, in press.

10. Beer, J.Z. 1974. Late post-irradiation phenomena in mammalian
    cell populations. III. Late recovery of slowly-growing
    clones isolated from X-irradiated L5178Y-S cell populations.
    Institute of Nuclear Research (Warsaw) Report, in press.

11. Berry, R.J. 1967. "Small clones" in irradiated tumour
    cells in vivo. Kinetics of re-growth of murine leukaemia

cells surviving irradiation with X-rays, fast neutrons and accelerated charged particles. Brit. J. Radiol. 40: 285.

12. Berry, R.J. 1972. Differences in LET dependence of cell killing and the production of heritable non-lethal damage in mammalian systems. IN Radiobiological Applications of Neutron Irradiation. Intern. Atomic Energy Agency. Vienna, p. 89.

13. Fischer, G.A. 1958. Studies of the culture of leukaemic cells in vitro. Ann. N.Y. Ac. Sci. 76: 673.

14. Alexander, P. and Mikulski, Z.B. 1961. Mouse lymphoma cells with different radiosensitivities. Nature 192: 572.

15. Beer, J.Z., Lett, J.T. and Alexander, P. 1963. Influence of temperature and medium on the X-ray sensitivities of leukaemia cells in vitro. Nature 199: 193.

16. Szumiel, I., Ziemba-Zak, B., Rosiek, O., Sablinski, J. and Beer, J.Z. Harmful effects of an irradiated cell culture medium. Int. J. Radiat. Biol. 20: 153.

17. Paul, J. 1960. Cell and tissue culture. E. and S. Livingstone, Edinburgh.

18. Fox, M. and Gilbert, C.W. 1966. Continuous irradiation of a murine lymphoma line P388F in vitro. Int. J. Radiat. Biol. 11: 339.

19. Beer, J.Z., Ziemba-Zak, B., Rosiek, O., Sablinski, J., Szumiel, I. and Kopec, M. 1970. Regeneration of proliferative activity in the progeny of murine lymphoma cells L5178Y irradiated with X-rays. Bull. Ac. Polon. Sci., Ser. Sci. Biol. 18: 581.

20. Szumiel, I., Rosiek, O., Beer, J.Z., Budzicka, E., Sablinski, J. and Kopec, M. 1973. Histone F1 phosphorus content and DNA synthesis in slowly-growing sublines of L5178Y-S cells. Bull. Ac. Polon. Sci., Ser. Sci. Biol. 21: 7.

21. Beer, J.Z., Bocian, E., Budzicka, E., Szumiel, I., Ziemba-Zak, B. and Kopec, M. 1974. Cell cycle disturbances in slowly growing sublines isolated from X-irradiated L5178Y-S cell populations. Nukleonika, in press.

22. Mendelsohn, M.L. and Takahashi, M. 1971. A critical evaluation of the fraction of labelled mitoses method as applied to analysis of tumor and other cell cycles. IN The Cell

Cycle and Cancer, Baserga, R. and Dekker, M. (editors),
New York, p. 58.

23. Sinclair, W.K. and Morton, R.A. 1966. X-ray sensitivity
     during the cell generation cycle of cultured Chinese hamster
     cells. Radiation Res. 29: 450.

24. Stanners, C.P. and Till, J.E. 1960. DNA synthesis in
     individual L-strain mouse cells. Biochim. Biophys. Acta.
     37: 406.

25. Szumiel, I. and Beer, J.Z. 1974. Late post-irradiation
     phenomena in mammalian cell population. V. Cytokinetic
     parameters of the slowly growing sublines isolated from
     S-irradiated L5178Y-S cell populations and "division
     probability". Institute of Nuclear Research (Warsaw)
     Report, in press.

26. Smith, J.A. 1973. Regulation of the cell cycle in animal
     cells. Biochem. Soc. Trans. 1: 1078.

27. Littlefield, J.W. 1963. The inosinic acid pyrophosphorylase
     activity of mouse fibroblasts partially resistant to 8-
     azaguanine. Proc. Nat. Acad. Aci. U.S.A. 50: 568.

28. Mezger-Freed, L. 1971. Puromycin resistance in haploid and
     heteroploid frog cells: gene or membrane determined? J.
     Cell. Biol. 51: 742.

29. Thompson, L.H., Mankovitz, R., Baker, R.M., Wright, J.A.,
     Till, J.E., Siminovitch, L. and Whitmore, G.F. 1971.
     Selective and nonselective isolation of temperature-
     sensitive mutants of mouse L-cells and their characteri-
     zation. J. Cell Physiol. 78: 431.

30. Morgan, H. 1971. Mutation rates in cells at different ploidy
     levels. J. Cell Physiol. 78: 177.

31. Mexger-Freed, L. 1972. Effect of ploidy and mutagens on
     bromodeoxyuridine resistance in haploid and diploid frog
     cells. Nature (New Biol.) 235: 245.

32. Meiss, H.K. and Basilico, C. 1972. Temperature sensitive
     mutants of BHK 21 cells. Nature (New Biol.) 239: 66.

QUESTIONS TO DR. BEER

Dr. Cristofalo: What percentage of the cells are killed by the

irradiation? Have you considered that you may be selecting for
a slow growing variant by the irradiation?

Dr. Beer: The dose of 300 rads kills more than 99% of cells, however, selection of the slowly growing cells seems to be unlikely
since we found that these cells are more radiosensitive rather than
more radioresistent as compared to the original strain.

Dr. Martin: Do all of your "recovered" cultures eventually reach
the generation time of your wild type? Is your working hypothesis
that recovery is via clonal selection following somatic sequential
events?

Dr. Beer: All of more than 100 examined sublines isolated after
irradiation of the L5178Y-S cell culture recovered their normal
growth rate. Naturally the selection processes must participate in
the late recovery phenomena. They can however, come into action only
when at least one "better" cell is formed. Thus, the selection alone
does not sufficiently explain the later recovery. Some other phenomena leading to formation of cells more perfect and ultimately in-
distinguishable from the original strain, must occur in the recover-
ing cell populations.

STUDIES ON THE SENSITIVITY OF ANTICANCER AGENTS OF NORMAL HUMAN

CELLS IN CULTURE

L. Morasca, G. Balconi, E. Erba and E. Cvitkovic

Istituto di Ricerche Farmacologiche "Mario Negri"

Milan, Italy

This and other laboratories have carried out extensive studies on cancer cells in vitro by a variety of methods to assess their sensitivity to anticancer agents. In previous investigations, we explored the sensitivity of biopsies obtained from two experimental tumors, kept in different culture conditions (1) and the correlation between in vitro activity and in vivo response to several drugs (2,3).

A parallel study has also been carried out on human tissues and, among the compounds tested so far, it was of interest to observe that Actinomycin D proved to be active on human cancer cells in a large range of concentrations--from less than 0.1 µg/ml to more than 10 µg/ml. Stromatic cells detected in the same cultures of human carcinomas appeared, instead, to be unaffected by concentrations lower than 10 µg/ml (4).

We, therefore, became interested in using the same method for normal tissues, to determine if its sensitivity to antitumoral agents would remain constant.

MATERIALS AND METHODS

Two fragments of normal periosteum were obtained from two surgical biopsies received from Dr. R. Roy Camille, Centre Hospitalier de Poissy and Dr. G. Parrini, Istituto Ortopedico G. Pini, Milano.

One of the fragments was from a 10-year-old girl (Case No. 1207) and one from a 40-year-old man (Case No. 1267). Tissue was trypsinized, seeded in culture bottles and fed with medium 199 which

TABLE I. Normal Periosteal Tissue.
Lethal Endpoint Concentrations mg/l

COMPOUND	Days in Culture						
	1207				1267		
	11*	96*	103**	134**	20*	60*	100*
Actinomycin D	10	0.1	0.01	0.01	>10	>10	>10
Adriamycin	1	1	1	1	10	10	10
1-Sarcolisin	100	>100	10	>100	10	10	100
Nitrogen mustard	>10	>10	>10	>10	10	10	10
6-Mercapto purine	>100	>100	>100	>100	>100	–	–
6-Methylmercaptopurine riboside	>10	>10	1	>10	10	–	–
Trenimon	0.1	0.1	0.1	0.1	0.1	–	–

* Confluent cultures
** Initial phase of growth

was fortified with two times the concentration of essential aminoacids and vitamins used in Eagle's MEM, in addition to 15% calf serum and 5% fetal calf serum. Simultaneously, small explants of tissue from the same biopsies were seeded in Rose's chambers under a perforated cellophane strip (4).

These cultures were fed 199 medium containing 20% calf serum, and the first test for sensitivity was performed on these cultures (2,4). Microscopic fields, corresponding to the cells under test, were marked from the outside of the coverslip by India ink and photographed by phase contrast. The cultures were then treated with a continuous range of compound concentrations and the same microscopic fields were photographed again after 24 hr of treatment. The data obtained in Table I shows the effect of the first treatment after 11 days for Case No. 1207 and after 20 days for Case No. 1267.

Other tests were made on cells from the same tissue after days 96, 103, 134 for Case No. 1207, and days 60 and 100 for Case No. 1267. The method used was the same as that employed in the previous studies (4) except that this group of cells was subcultured in Rose's chambers from bottles carrying the population obtained by trypsinization.

Since, despite the fortified medium used, these populations did not attain a fast rate of growth, the time at which we performed the experiments and at which we discontinued them was conditioned by the growth capacity and by the lifespan of the two populations.

RESULTS

The two biopsies studied appear to be quite different in their sensitivity to chemotherapeutic agents.

Biopsy No. 1267 obtained from the 40-year-old man, grew slowly, showing spindle-shaped cells of constant morphology (Fig. 1); after 100 days in culture, it was completely lost after the third transplantation. It, however, proved to be resistant to high concentrations of Actinomycin D and Adriamycin, and to be sensitive to Nitrogen Mustard and Trenimon, when compared with No. 1207. Its sensitivity to drugs appeared to be constant with time, the only exception being 1-sarcolysine that became less active on cultures at the age of 100 days.

Biopsy No. 1207, obtained from the 10-year-old girl, had a better growing capacity. At 36 days of age, it was already at its fourth transplantation; at 103 days, it was just starting its fifth transplant, and at 134 days, it was transplanted for the seventh time. At present, it is stored under liquid nitrogen and undergoing chromosomal analysis. After the fifth transplantation, the

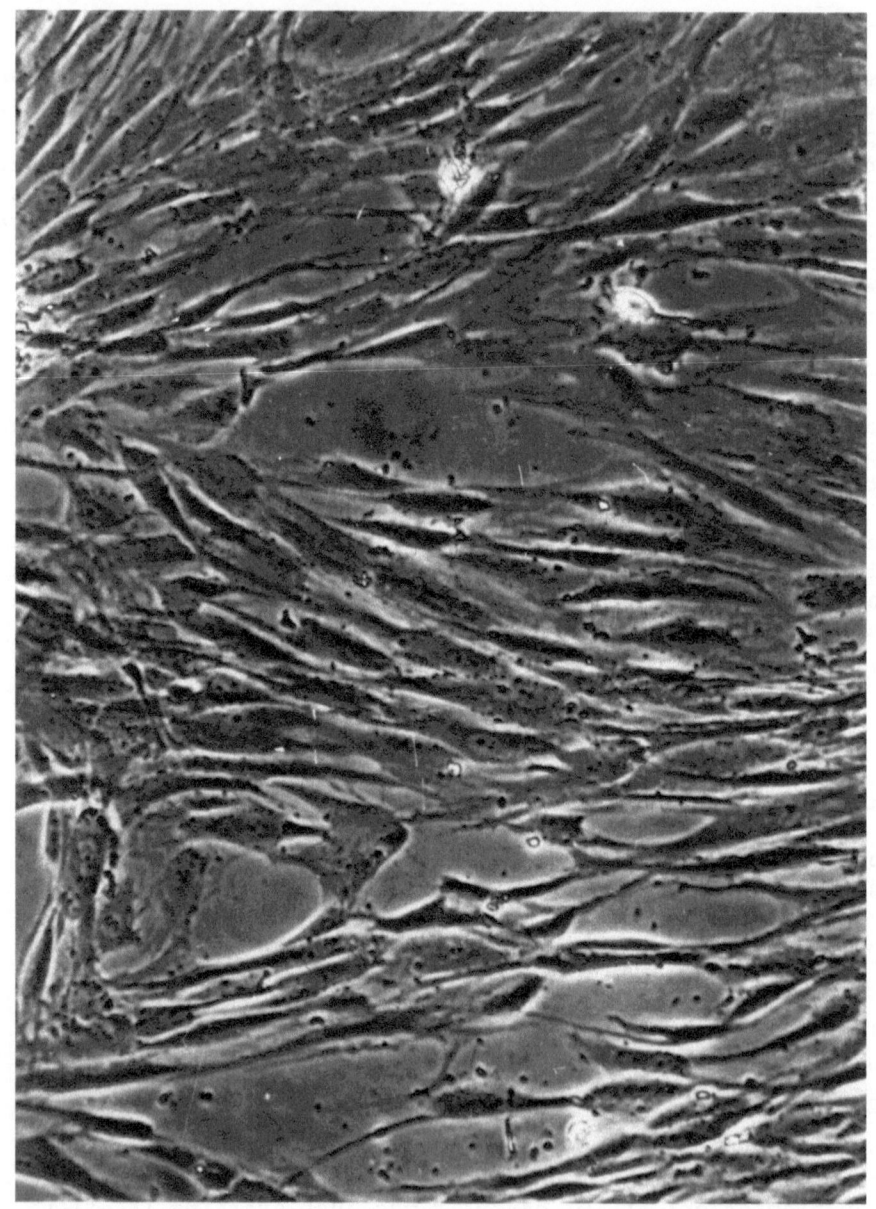

Fig. 1. No. 1207 24 days in culture -- phase contrast lens. Spindle shaped cells; few elements with large membranes.

morphology of these cells changed from spindle-shaped cells (Fig. 2) to endothelial-like cells (Fig. 3).

Sensitivity to chemotherapeutic agents was almost the same at day 96, with the exception of a reduced sensitivity to Actinomycin D. This sensitivity reached the level of 0.01 µg/ml in the subsequent experiment. Among the other compounds investigated in this connection, l-sarcolysine and 6-methylmercaptopurineriboside showed an increase in their activity at day 103, and recovered the previous level of activity at day 134 when cultures were tested again at confluency instead of being studied in their initial phase of growth.

DISCUSSION

Biopsy No. 1267 did not show any relevant change in sensitivity during its entire lifespan which, however, was limited to only 100 days of culture.

A different behavior has been recorded for biopsy No. 1207 in that it showed variable sensitivity to Actinomycin D, l-sarcolysine, and 6-methylmercaptopurineriboside. A further discrepancy observed pertains to the sensitivity levels recorded for the two biopsies at the beginning of our experiments, namely, day 11 for No. 1207 and day 20 for No. 1267. These observations are difficult to explain, unless we consider the fact that the donor of No. 1207 was a very young girl, while the donor of No. 1267 was an adult man.

It has been shown that the growth-span of human fibroblast strains from adults is shorter than that obtained from fetal strains (5), and that the latent period of explanted tissues increases with the age of the donor (6). The growth kinetics have also been found to be affected in cells of adult origin, which during their phase II are already similar to phase II cells derived from young donors (7). If these factors of age can be accepted as the only difference between the two biopsies, then evidently, the age of the donor may affect the sensitivity of normal cells to chemotherapeutic agents.

It must also be noted how cells during active growth may change their sensitivity to anticancer agents as they do their morphological appearance. Whether this fact would be due to a phenomenon of selection or to different metabolic needs of a growing cell population, remains to be clarified. In terms of in vitro testing of the sensitivity of cancer cells, a large variability of responses can be expected; this may be reduced by testing the cell populations after the shortest possible time of culture before the first subculture.

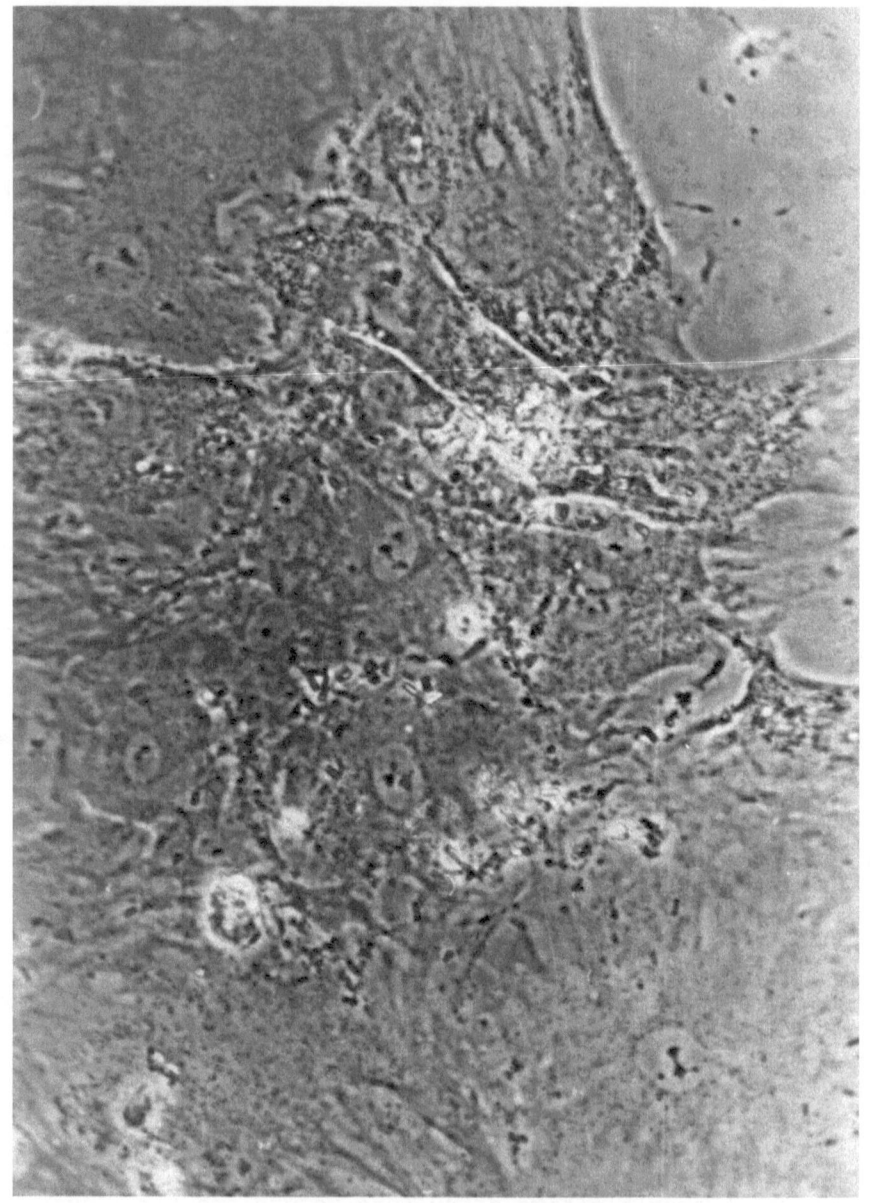

Fig. 2. No. 1207 120 days in culture — phase contrast lens. Endothelial-like cells with large, thin, ondulating membrane with superficial foldings.

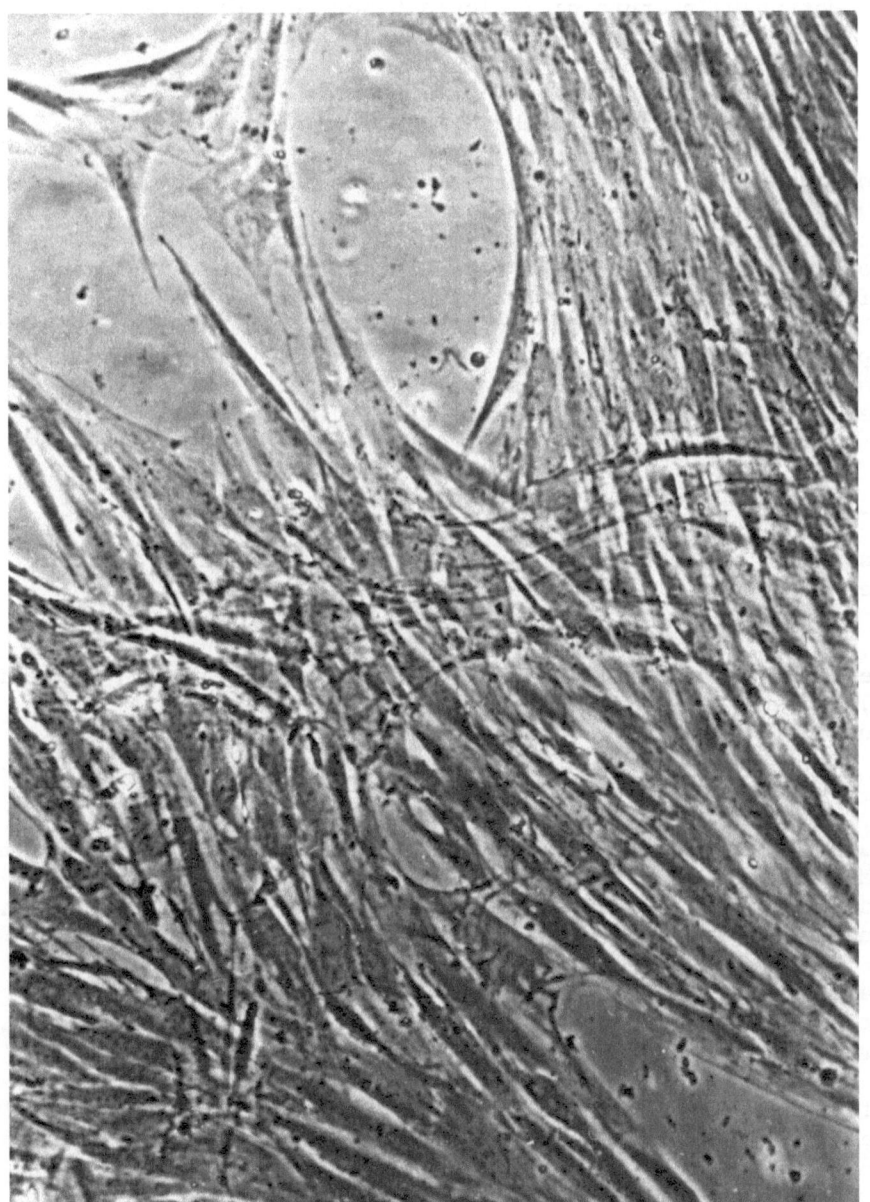

Fig. 3. No. 1267 60 days in culture -- phase contrast lens. Spindle shaped cells.

ADDENDUM TO DISCUSSION

The morphological analysis of chromosomes, referred to previously, has been recently performed on culture No. 1207. In 15 mitoses, no alterations in the number or in the structure of chromosomes have been detected. The chromosomal analysis has been carried out by Dr. Anna Mottura, Istituto di Biologia Generale e di Genetica, Facolta di Medicina, Università di Milano.

SUMMARY

Two biopsies of human periosteal tissue have been tested for sensitivity to several antitumoral agents. The tests were performed during the lifespan of the cultured cells in order to reveal possible variations in sensitivity. The results indicate that the age of the donor affects drug sensitivity; cells obtained from the young donor show a higher growth potential, together with a variable sensitivity and morphology during their life in vitro, while cells from the adult donor have a lower growth potential, a constant sensitivity to chemotherapy and a constant morphology.

REFERENCES

1. Morasca, L., Balconi, G., De Nadai, F. and Dolfini, E. 1972. Chemotherapeutic fingerprints of two experimental tumors in vitro. Eur. J. Cancer 8: 429.

2. Balconi, G., Bossi, A., Donelli, M.G., Filippeschi, S., Franchi, G., Morasca, L. and Garattini, S. 1973. Chemotherapy of a spontaneous mammary carcinoma in mice: relation between in vitro-in vivo activity and blood and tumor concentrations of several antitumoral drugs. Cancer Chemother. Rep. 57: 115.

3. Morasca, L., Balconi, G., Erba, E., Lelieveld, P. and Van Putten, L.M. 1974. Cytotoxic effect in vitro and tumour volume reduction in vivo induced by chemotherapeutic agents. Eur. J. Cancer, in press.

4. Morasca, L. and Balconi, G. 1973. Different sensitivity to Actinomycin D of biopsies of human tumors cultivated in vitro. Eur. J. Cancer 9: 301.

5. Hayflick, L. 1965. The limited in vitro lifetime of human diploid cell strains. Exp. Cell Res. 37: 614.

6. Soukupová, M., Holečková, E. and Hněvkovský, P. 1970. Changes
    of the latent period of explanted tissues during ontogenesis.
    In: Aging in Cell and Tissue Culture. E. Holečková and V.J.
    Cristofalo, eds. Plenum Press, New York, p. 41.

7. Macieira-Coelho, A. 1970. The decreased growth potential in
    vitro of human fibroblasts of adult origin. In: Aging in
    Cell and Tissue Culture. E. Holečková and V.J. Cristofalo, eds.
    Plenum Press, New York, p. 121.

QUESTIONS TO DR. MORASCA

Dr. Kasten: This work you are doing is very important. One factor
I would like to mention which has not been considered by some
workers in this field is that of cell number. Everyone considers
the concentration of the drugs employed, but the cell number can
greatly influence the results as shown some years ago by Deitch
for actinomycin D.

Dr. Morasca: I agree with you on the importance of considering
cell number when evaluating the activity of chemotherapeutic
agents. To your remark about actinomycin D I may add that this
compound is considered to be concentrated in the nucleus of sensitive
cells and they may therefore deplete the medium. Nitrogen mustard
shows different levels of activity as a function of the cell
number and other alkylating agents may behave in a similar way.

Dr. Martin: A spontaneous transformation of a human diploid
fibroblast line is exceedingly rare, if not non-existent. Therefore
it would be important to rule out exogenous contamination from one
of your sarcoma lines by looking at various isoenzymes and cytogenetic markers. First of all is the putatively transformed line
still euploid?

Dr. Morasca: There is some work going on to define the ploidy of
this line that can be cultivated with ease. A viral transformation
is really among the possible interpretation for this phenomenon.

RIBONUCLEASES IN MOUSE ASCITIC TUMOR CELLS DURING AGEING IN VIVO

AND IN VITRO

Katarína Horáková

Chair of Microbiology and Biochemistry, Faculty of
Chemical Technology, SVŠT, Bratislava, Czechoslovakia

It is becoming apparent that the intact tissues of mammals and
birds regularly possess a number of ribonucleases. The value of
studies on their existence and properties therefore extends beyond
probing the finer details of nucleic acid structures, to a question
of their significance in cell reproduction and in differentiation
of cells within and between tissues.

Cell extracts of various tissues have been examined by Razzell
(1) for ribonuclease activities including mouse liver, kidney and
pancreas from embryonic to adult form. It was apparent that all
enzymes increased in concentration in the liver and kidney prior
to birth, after which the concentrations generally declined somewhat. In pancreas the pattern appeared to be entirely different,
for with the exception of slight oscillation in RNase I levels,
both types of RNase II proceeded at a logarithmic rate greatly in
excess of the rate of synthesis of other proteins. In contrast,
some functional tumors (2,3) or cells of higher organisms introduced into artificial culture media do not continue to synthesize
all of the ribonucleases at the rates found in the tissue from
which the cell culture originates (2).

In the present study the enzyme levels in mouse lymphoma
L5178Y cells during ageing in vivo and in vitro were investigated.

MATERIALS AND METHODS

L5178Y cells were obtained aseptically after 5 days growth
in mouse peritoneal cavity, concentrated by centrifugation and
inoculated into Fischer's medium (4) with 10% calf serum. Cell

521

populations were routinely initiated at 3×10^5 per ml and subcultured at 2×10^6 cells per ml. Cell counts were performed with a
Coulter Counter Model B.

Homogenates of L5178Y cells grown _in vivo_ and _in vitro_ were
prepared as described previously (5).

All enzyme activities are reported as units of 1 μmole per hr;
specific activities are units per mg of protein. Enzyme assays
previously reported in detail (5) were performed at 37°C. The
absorption at 260 nm of the acid-soluble digestion products from
high molecular weight RNA (Calbiochem) was measured in reaction
mixtures at pH 7.8 and 5.5. The former reflects the concentration
of alkaline Ribonuclease II (RNase II), the latter, that of acid
Ribonuclease II. The formation of acid-soluble oligonucleotides
and mononucleotides from poly-A was measured by the increase in

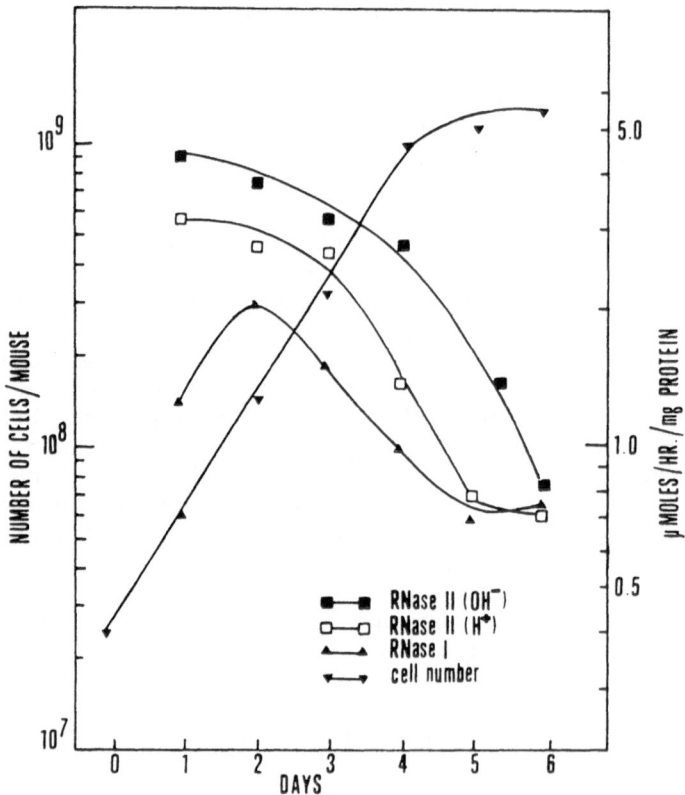

Fig. 1. Ribonuclease activities in homogenates of L5178Y cells
grown _in vivo_ as an ascites.

optical density at 260 nm and reflects the concentration of Ribonuclease I (RNase I).

    Protein was determined by the method of Lowry et al. (6) using
cristalline bovine plasma albumin as a standard.

RESULTS AND DISCUSSION

    Figure 1 shows the specific activity of the several RNases
in homogenates of L5178Y cells grown in vivo as an ascites at
various stages of growth. It is apparent that the ribonucleases
(type II) are present at levels which are similar to those observed
in cells in tissues 24 hr after transplantation into mouse peritoneal cavity. However, a peculiar decrease of enzyme contents is
observed during ageing of the cells. The same picture was obtained
in Ehrlich ascites tumors (EAC). From these results it follows that
the RNase contents depend on the age of the tumor. Probably because
of that Baker and Pavlik (3) who determined the enzyme levels in
EAC cells 8 days after implantation came to the conclusion that
the cells are devoid of alkaline RNase II. Similarly Razzell (2)
has found very low activity of both RNases II in L5178Y cells.

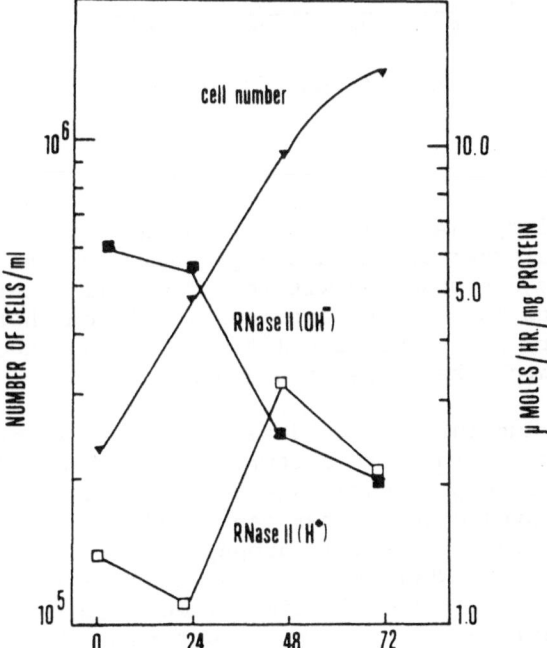

Fig. 2. Gradual decrease of alkaline RNase II activity in L5178Y
cells after transfer into culture.

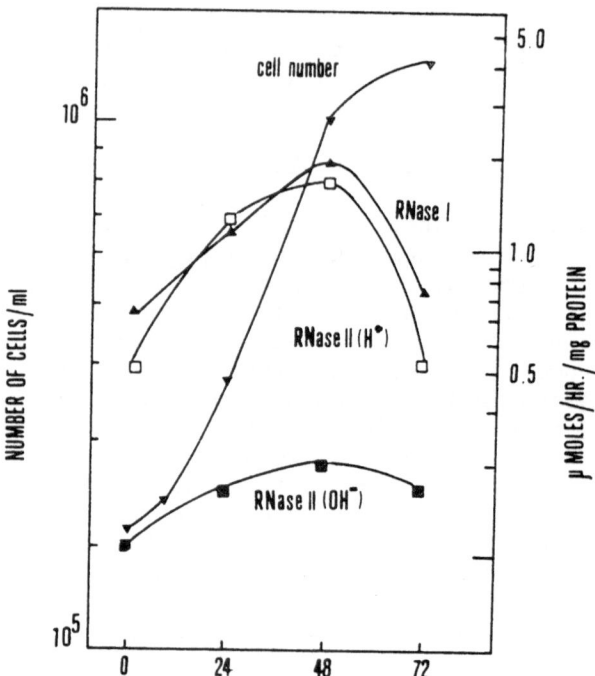

Fig. 3.   RNase levels in cultured L5178Y cells at various stages
of growth.

        When mouse lymphoma cells after 24 hr growth in the peritoneal
cavity were transferred to culture a gradual decrease of alkaline
RNase II activity was observed within a few cell generations (Fig.
2). These results confirm the previous suspicion that the enzyme
levels in the mery same line of cells might be dependent on the
environment in which the cells have grown.

        It was of prime interest to determine whether enzyme levels
in L5178Y cells grown in vitro culture were stable. The date
obtained from these experiments show that the enzyme levels in
vitro are fairly predictable after many subcultures. In Figure 3
the enzyme levels in L5178Y cells propagated in culture for 31
passages may be seen. While RNase I is relatively unchanged,
alkaline RNase II level is 10-15 times lower than in in vivo
cultures. Similarly as observed with continuous in vivo tumors
(Fig. 1) the data from analyses of cell extracts show variation in
specific activity with age of cultured L5178Y cells (Fig. 3). The
first two days during exponential growth the concentration of

enzymes increased, after which a marked decrease in RNase I and
acid RNase II is observed.

From the obtained results it follows that the level of studied
ribonucleases in both in mitro and in vivo systems depends on the
age of the culture or tumor, respectively. The changes in enzyme
activities indicated above may result from two general mechanisms:
a) altered rate of synthesis; b) altered rate of degradation, or
both.

The cells have been shown to change their specific activity for
a number of enzymes as they grow. In some instances this effect
appears to be a reflection of the cellular life cycle (7,8) while
in others it is likely to be related to a change in the composition
of the medium (9,10). We observed that enzyme concentrations rose
when mice were inoculated with a large number of cultured L5178Y
cells, or when peritoneal wash from healthy mice was added to the
cultivation medium in vitro (5). Culture-derived tumors produced
5-10-fold more acid RNase II and 10-70-fold more alkaline RNase II
in 24 hr in vivo than would have been produced upon further cultivation in vitro. In contrast, the specific activity of each of the
RNases was markedly lower in cells inoculated in mouse peritoneal
cavity in the presence of ascitic fluid. Therefore, we suggest that
an activator is present in the peritoneal cavity and an inhibitor
or a repressor in the ascitic fluid since the mentioned enzyme
changes occur in less than two generations.

SUMMARY

Ribonuclease activities have been determined in cell extracts
of mouse L5178Y lymphoma grown in vivo and in vitro. From the
obtained results it follows that the level of studied enzymes
changes in both systems during ageing of the cells. Moreover, when
lymphoma cells are cultured in vitro a marked decrease in specific
activity of alkaline ribonuclease II is observed.

REFERENCES

1. Razzell, W.E. Personal communication.

2. Razzell, W.E. 1967. Polynucleotidases in animal tissues.
    Experientia 23: 321.

3. Barker, G.R. and Pavlik, J.G. 1967. Ribonuclease activity
    in the Ehrlich ascites tumour. Nature 215: 280.

4. Fischer, G.A. and Sartorelli, A.C. 1964. Development,
    maintenance and assay of drug resistance. In: Methods in
    Medical Research, 10. H.N. Eisen, ed., Year Book Publishers,
    Chicago, p. 247.

5. Horáková, K. 1972. Polynucleotidases in mouse ascitic tumour
    cells. Neoplasma 19: 459.

6. Lowry, O.H., Rosenbrough, N.J., Farr, A.L. and Randall, R.J.
    1951. Protein measurement with the folin phenol reagent.
    J. Biol. Chem. 193: 265.

7. Gelbard, A.S., Kim, J.H. and Perez, A.G. 1969. Fluctuations
    in deoxycytidine-monophosphate deaminase activity during
    the cell cycle in synchronous populations of HeLa cells.
    Biochim. Biophys. Acta 182: 564.

8. Klevecz, R.R. and Ruddle, F.H. 1968. Cyclic changes in
    synchronized mammalian cell cultures. Science 159: 634.

9. Granner, D.K., Hayashi, S.I., Thompson, E.B. and Tomkins,
    G.M. 1968. Stimulation of tyrosine aminotransferase
    synthesis by dexamethasone phosphate in cell culture. J. Mol.
    Biol. 35: 291.

10. Paul, J., Fottrell, P.F., Freshney, I., Jondorf, W.R. and
    Struthers, M.G. 1964. Regulation of enzyme synthesis in
    cultured cells. Nat. Cancer Res. Monograph 13: 219.

QUESTIONS TO DR. HORÁKOVÁ

Dr. Beer: What was the criterion of ageing in the experiments
presented?

Dr. Horáková: The experiments presented were performed in in vivo
and in vitro systems. The criterion of ageing in in vivo experiments was the growth of L51784 cells in the mouse peritoneal
cavity for 10 days and 3 days in different passages up to one
year growth in Fisher's medium supplemented with 10% calf serum
in culture.

Dr. Rotzsch: Have you determinations of the total content of
proteins and RNA in the lifetime of the tumor cells?

Dr. Horáková: Yes, I followed the total content of protein (not
RNA) and it was constant.

Dr. Hay: I would like to ask Dr. Horáková for her thoughts

concerning the mechanism behind this drop in specific protein
synthesis by cells following isolation in vitro. This is one
additional example of an extremely dramatic drop in enzyme synthesis
during the transition and I feel the factors affecting it are of
considerable importance.

Dr. Horáková: The requirements of mammalian cells are largely
based on the ability to support growth and not full maintenance
of function in vitro. Cells grown in a medium unable to support
function may respond by changes in metabolism which allow them to
adjust to the new environment and permit survival at the expense of
the synthesis of specific proteins. Our results suggest the
presence of an unknown compound in the peritoneal cavity of a
healthy mouse which selectively increases the specific activity of
alkaline RNase II.

PROLIFERATION AND MORPHOLOGY OF ASCITIC CELLS AS A FUNCTION OF AGE IN CELL CULTURE

A. Balázs, L. Holczinger, I. Fazekas, and G. Turi

The Institue of Experimental Medicine, Hungarian
Academy of Sciences, and Institute of Oncopathology,
Budapest

The population kinetics of ascitic cells, in vivo and in vitro cultivated ones, are similar in their main features. Following the lag and log periods of growth, the proliferation of ascitic cells is limited by stationary phase, too. The prolongation of their generation time, the decrease of the growth fraction, DNA synthesis and mitotic index (MI), as well as the accumulation of specific G_1 and G_2 inhibitors, characterize this period (1-6).

To study the alteratons in cellularity and the behavior of ascitic cells of different ages in short term culture, the Gal leukaemic cells (7) were chosen as a model. This leukaemia was developed in Hungary in rats, induced originally by 7,12 dimethyl-benzanthracene intravenously. Further, the myeloblastic leukaemia was transplanted in the form of liver + spleen suspension intra-peritoneally, into 2 to 4 day old animals. In some cases an as-citic form developed, too, which was then serially transplanted into rats of 150 gs b.w. The average survival time of the latter amounted from 12 to 16 days.

Ascitic cells, drawn by peritoneal puncture on the 6th day after the last passage, were cultivated in Parker-199 medium, sup-plemented by 20% calf serum. Belco tubes contained 1×10^6/ml cells.

During the subsequent 72 hr cultivation period peculiar mor-phological alterations occurred, analysed later by electron micro-scopy, ^{3}H-thymidine scintillo-spectrometry and cytochemistry.

The starting (0 hr) material (Fig. 1) contained, for the most

part, blast-like cells and only some giant and binucleate cells.
In their cytoplasm some small, round vacules were characteristic.

     At 5 to 6 hr cultivation the enlargement and the partial fusion
of the vacuoles became observable. Contrary to this, a significant
level of proliferation was maintained, occasionally with tetra-
polar mitoses.

     In the 24th hr of cultivation the vacuolisation became striking
(Fig. 2), suggesting a degeneration of the cells. In some of them,
nucleus and cytoplasm were surrounded by a real "wrapper" of vacuoles. By this time the MI was strongly reduced.

     Subsequently, it became evident, that contrary to the
signs of degradation, the bulk of the cells were capable of
restitution. The probable mechanism is the exopinocytosis of vacuoles by formation of cytoplasmic buds (Fig. 2C).

     At the 48th hr the majority of the cells were again charac-
terized by approximately normal morphology and enhanced mitotic
activity.

     Based on the above data, we presume the following mechanism
for the cultivated cells, induced by the explantation (Fig. 3):

     The vacuolisation culminates at 24 hr, accompanied by a sig-
nificant mitodepression and an increase in the number of binucleate
and giant cells. In this period, as stated by cell count and try-
pan-blue staining, the number of perishing cells was not too high.
The same was proved by our measurements with ^{3}H-thymidine incorporation (Fig. 4). The continuous isotope incorporation of ascitic
cells was compared to that of surviving rat bone marrow cells and
vigorously proliferating HeLa cultures, respectively.

     Which biological process, if not degeneration, is represented
by the hypervacuolisation? To obtain the answer, an approach by
electron microscopy was attempted. The freshly drawn starting
material consists of blast-like cells (Fig. 5), occasionally with
perforated nuclei and lipid droplets; the latter were originally
thought to be identical with the vacuoles revealed by the light
microscope. However, analysis of the cells, cultivated for 5 hr,
has shown (Fig. 6) that vacuoles are represented on the electron
micrographs as larger, light areas, containing only fine, equally
distributed granules. The light areas are never surrounded by
any membrane, do not contain organelles. Mitochondria, endoplasmic reticulum profiles, microfilamentar bundles, Golgi-apparatus
are located outside of them (Fig. 7). Occasionally, these peculiar
vacuoles are in the neighbourhood of lipid droplets without any
closer relationship. Rather seldom, different membrane configura-

tions are visible inside the vacuoles.

By the 24th hr, a part of the cells shows the accumulation of lysosomes, autophagic vacuoles, and other signs of degeneration. Still, most of them, after transitory strenghtening of vacuolisation adapt, and following the restitution phase, rejuvenate.

Examining the nature of the fine granulation in the vacuoles, 120 to 250 A in diameter (Fig. 8) we had to determine, whether they consist of: (a) lipid droplets discretely dispersed, (b) ribosomes or, (c) alpha- and beta-glycogen particles, respectively.

The first two possibilities were excluded by the Sudan Black and gallocyanine negativity of the vacuoles. The glycogen content was strenghtened by positive PAS-reaction after 6 and 24 hr cultivation, including the cytoplasmic bud area.

Histochemical investigations performed on the original not cultivated tumors revealed that the ascites tumor cells also contain PAS-positive material which can be digested by diastase. In the early period of tumor growth about 60-70%, later about 80-85% of cells contain glycogen deposits. According to quantitative determinations, the glycogen content of tumors showed significant individual variations in different periods of growth. Three to four days after inoculation the glycogen content of tumors was 2-3 mg/g wet weight while later on this increased to 5-9 mg/g wet weight (8).

There is no established explanation for the high glycogen content of these ascites cells, but it can be presumed that there is a deficiency of the enzymes responsible for the glycogen breakdown.

In attempts to reiterate our original findings by cultivation we observed that the vacuolisation-phenomenon strongly depended on the age of the ascitic cells.

Using younger starting cells (that is 2-3 days old after passage corresponding to the lag, or early log phase) the vacuolisation reached a substantially lower level. In the case of older starting material (9-10 days after passage) the vacuolisation was more expressed, but the restitution slight and imperfect. In the oldest animals examined (that is 16-18 days after passage, in the late stationary phase) the cellularity of the ascitic fluid altered significantly. Among typical blast-like cells, which diminished from 90% in younger ascites to about 40%, myeloid- and lymphoid-like cells, and mainly transformed intermediate forms appeared, showing the progress of atypic differentiation and/or ageing. The mitotic index reduced by about 80%. The ratio of degenerating cells increased significantly.

Under the electron microscope, the findings in the oldest
cells were confirmed. However, atypic granuloid cells with neutro-
fil granulation (Fig. 9) or differentiated lymphoid cells occurred
sporadically. In transformed cells, the striking vacuolisation
seen at the light microscopic level (Fig. 10) proved to be either the
accumulated glycogen particles, or lipid droplets and membrane-
surrounded giant vacuoles lacking to contain any electron-dense
material.

The authors are indebted to Dr. L. Csaky and I. Blazsek for
scintillation determinations and to Mrs. K. Deak and Miss K.
Windisch for skillful technical assistance.

ABSTRACT

Ascitic cells in the logarithmic growth phase increase
the accumulation of glycogen particles in the course of explantation into suspension culture, probably due to the increasing arrest
of glycogenolytic enzymes. At this age, a part of the cells are
capable of restitution by exopinocytosis of the glycogen-containing
vacuoles ia formation of cytoplasmic buds. Older cells, taken
from the plateau-phase, pass atypical differentiation and ageing,
are less capable of hindering the abnormal accumulation of glycogen and the hypervacuolisation. As a consequence, the cells finally degenerate and die.

REFERENCES

1. Wiebel, F. and Baserga, R. 1968. Cell proliferation in
   newly transplanted Ehrlich ascites tumor cells. Cell Tissue
   Kinet. 1: 273.

2. Frindel, E., Valleron, A.J. and Vassort, F. 1969. Proliferation kinetics of an experimental ascites tumor of the mouse.
   Cell Tissue Kinet. 2: 51.

3. Dombernowsky, P., Bichel, P. and Hartmann, N.R. 1973. Cyto-
   kinetic analysis of the JB-1 ascites tumor at different stages
   of growth. Cell Tissue Kinet. 6: 347.

4. Burns, E.R. 1969. On the failure of self-inhibition of growth
   in tumors. Growth 33: 25.

5. Bichel, P. 1971. Autoregulation of ascites tumour growth by
   inhibition of the G_1 and G_2 phase. Europ. J. Cancer 7: 349.

6. Bichel, P. 1973. Self-limitation of ascites tumor growth:
    A possible chalone regulation. Nat. Cancer Inst. Monograph
    38: 197.

7. Gál, F., Somfai, S. and Szentirmai, Z. 1973. Transplantable
    myeloid rat leukemia induced by 7, 12-dimethylbenz(a)anthra-
    cene. Acta Haematol. 49: 281.

8. Holczinger, L. and Gál, F. Glycogen storgae in cells of an
    ascites rat leukaemia. Z. Krebsforsch., in press.

QUESTIONS TO DR. BALÁZS

Dr. Franks: Does the increase in glycogen reflect a defect in
oxidative metabolism?

Dr. Balázs: Experimentally only the accumulation of glycogen was
observed. We presume it was caused by a gradual inactivation or
defective synthesis of glycogenolytic enzymes.

Dr. Cristofalo: Dr. Franks has suggested that the changes in
glycogen content you observe are related to changes in oxidative
metabolism. My question is a similar one. Glycogen content is
usually highly responsive to glucose concentration in the medium
and I was wondering whether you had looked at glucose metabolism
and changes of glucose concentration.

Dr. Balázs: At the present stage of our experiments the changes
in glycogen content were registered in the manner mentioned in the
lecture. Alterations in glucose content are under investigation.

Dr. Hanzlikova: I have two questions: (a) What is the fate of the
exocytosed vacuoles? Are these vacuoles those, which are filled
with glycogen granules., (b) what is the mechanism, leading to the
depletion of the vacuoles, i.e., of the empty vacuoles, persisting
in the cytoplasm of old cells.

Dr. Balázs: (a) The cytoplasmic buds consist of glycogen containing
vacuoles and a narrow peripheral cytoplasmic layer. After exocytosis,
when glycogen moves to the culture medium, the cell membrane regenerates. (b) Vacuoles visible in very old ascitic cells are not
similar to glycogen containing ones. The former are optically
empty surrounded by a membrane and probably represent a sign of degenerative processes.

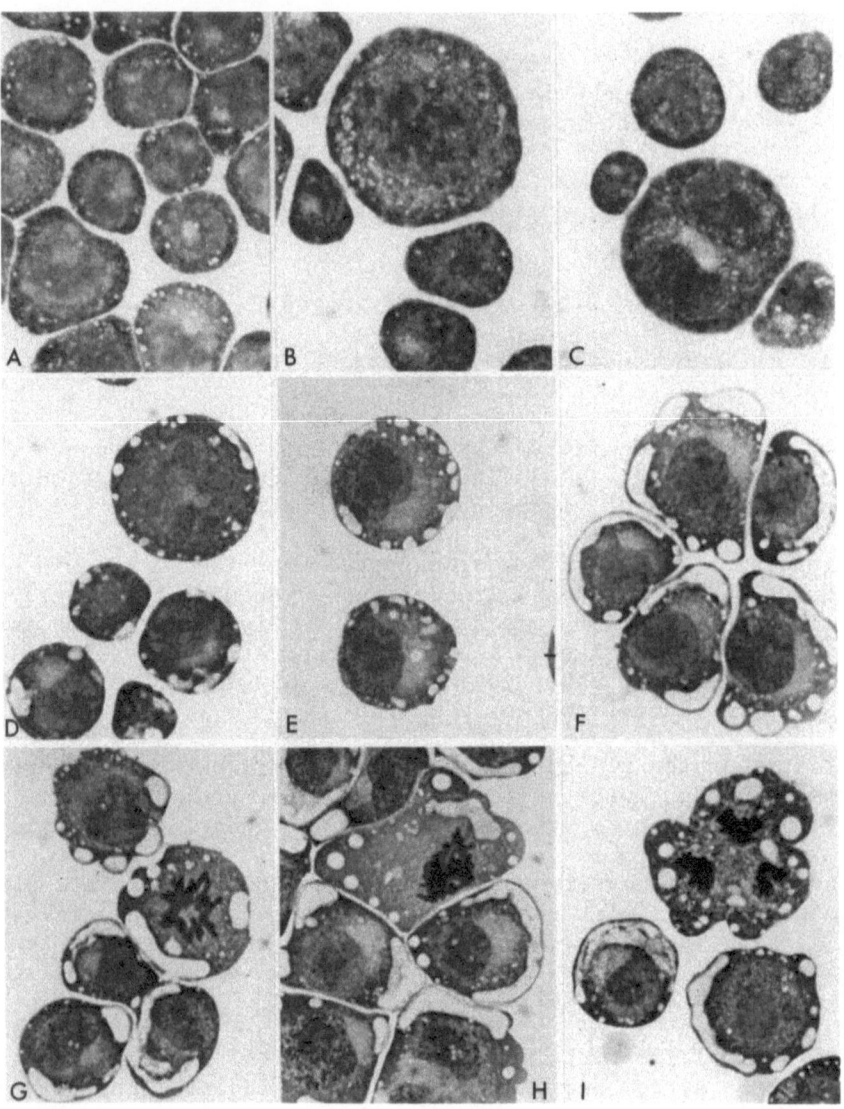

Fig. 1. Ascitic cells from Gal-leukaemic rats at 0 and 5 hr of
cultivation. A - C: blast-like, giant, and binucleate cells from
the starting material. D - F: interphase cells after 5 hr cultivation with beginning vacuolization. G - I: mitotic forms among
the 5 hr cells. Giemsa, X1,250.

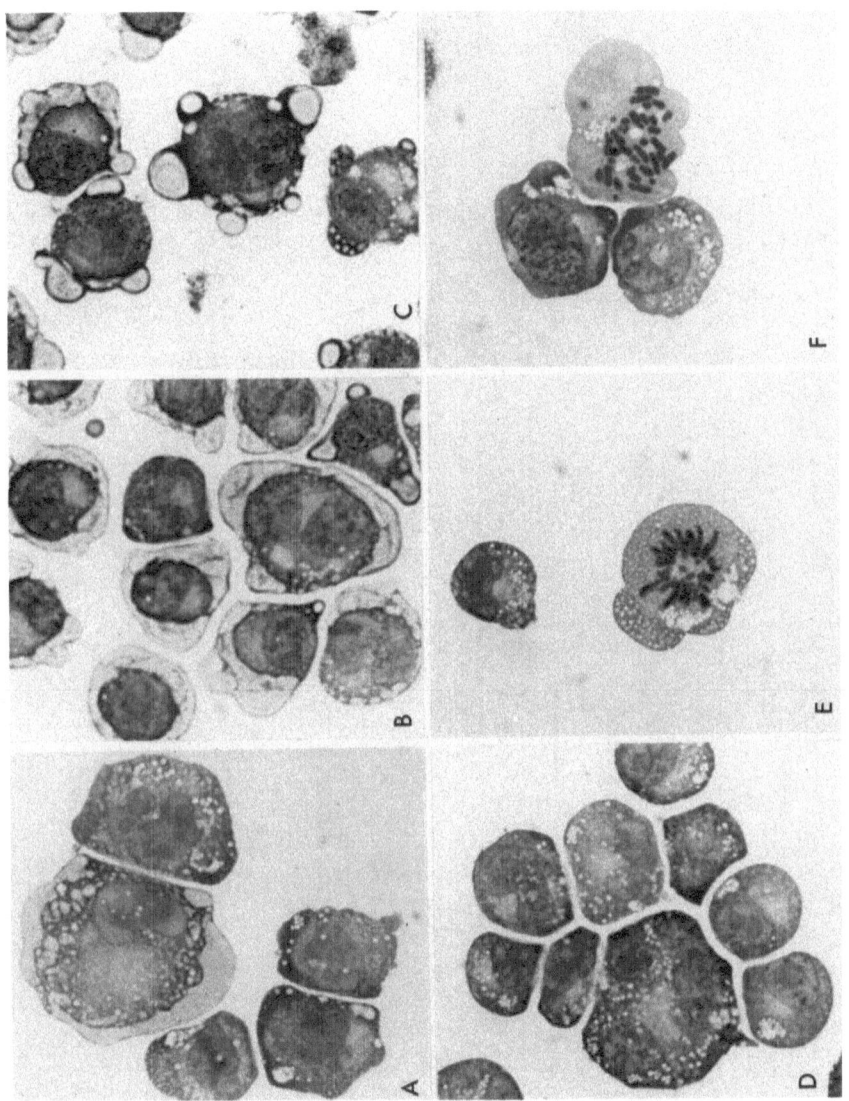

Fig. 2. A – B: Hypervacuolization in 24 hr cultivated ascitic cells. C: cytoplasmic bud formation. D – F: restituted cells and mitotic forms at the 48th hr. Giemasa, X1,250.

MYELOID ASCITES CELLS IN CULTURE
(PARKER 199, HAM)

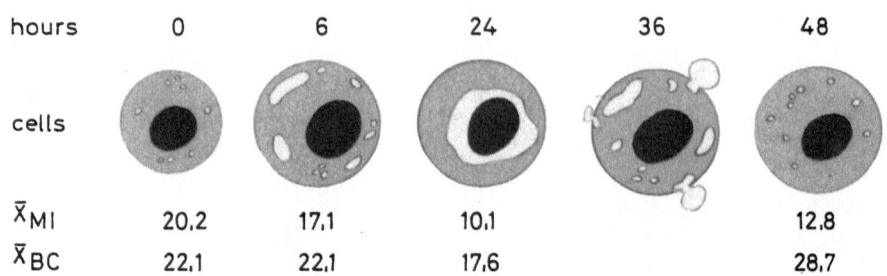

hours	0	6	24	36	48
cells					
$\bar{X}_{MI}$	20,2	17,1	10,1		12,8
$\bar{X}_{BC}$	22,1	22,1	17,6		28,7

Fig. 3.   Probable mechanism of hypervacuolization and restitution
in cultivated ascitic cells.   MI = mitotic index, BC = binucleate
cells.

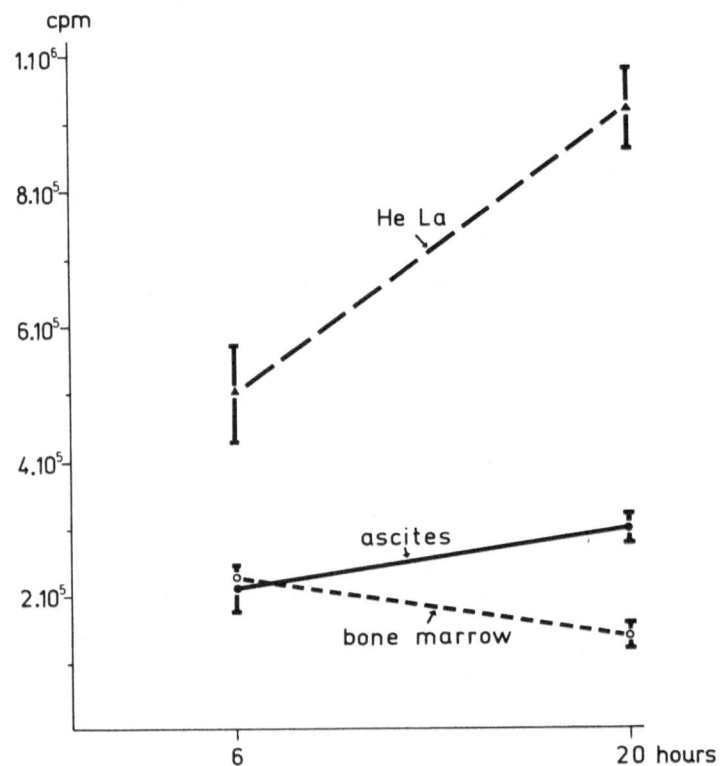

Fig. 4. ^{3}H–thymidine incorporation into 6 to 20 hr cultivated HeLa,
rat ascitic and bone marrow cells.

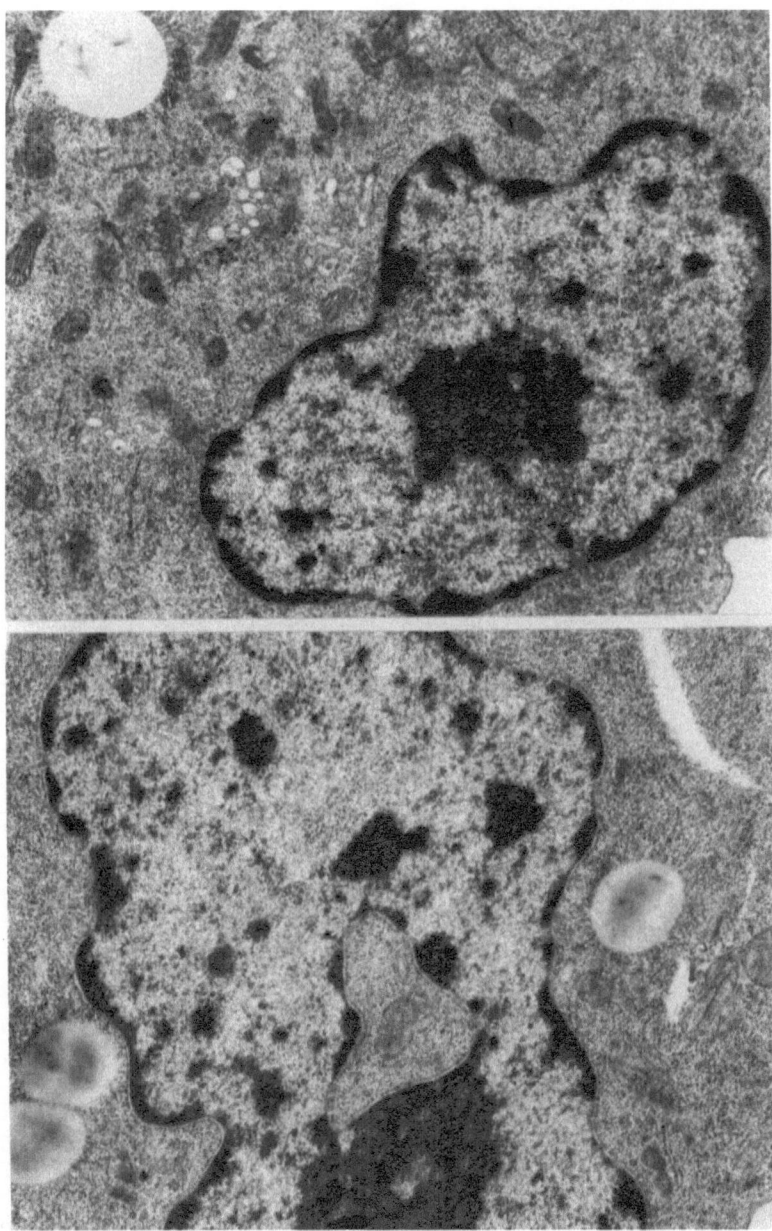

Fig. 5. Electron micrograph of the blast-like cells in the freshly aspired ascites. Glutaraldehyde-osmium fixation, Durcupan ACM embedding and uranyl acetate + lead citrate staining. X20,400.

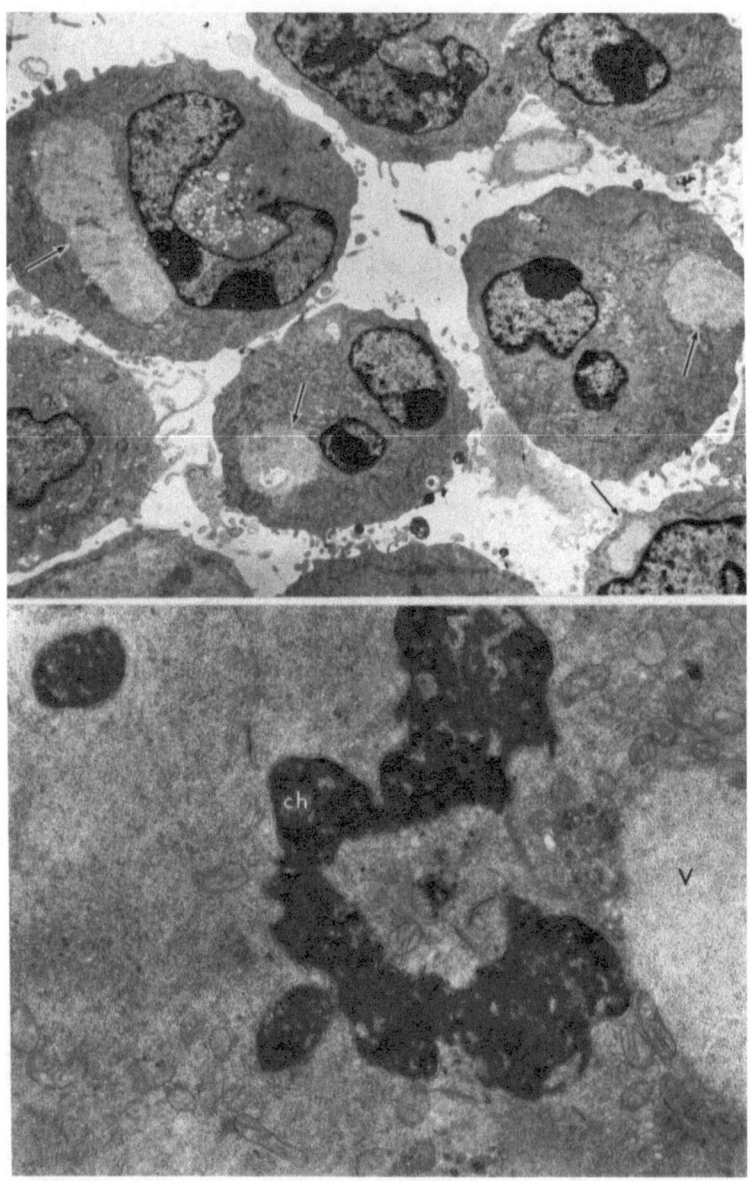

Fig. 6. Electron-lucent areas (arrows) in ascitic cells after
5 hr cultivation. X5,600. B: a vacuolized cell in division.
v = vacuole, ch = chromosome. X12,200.

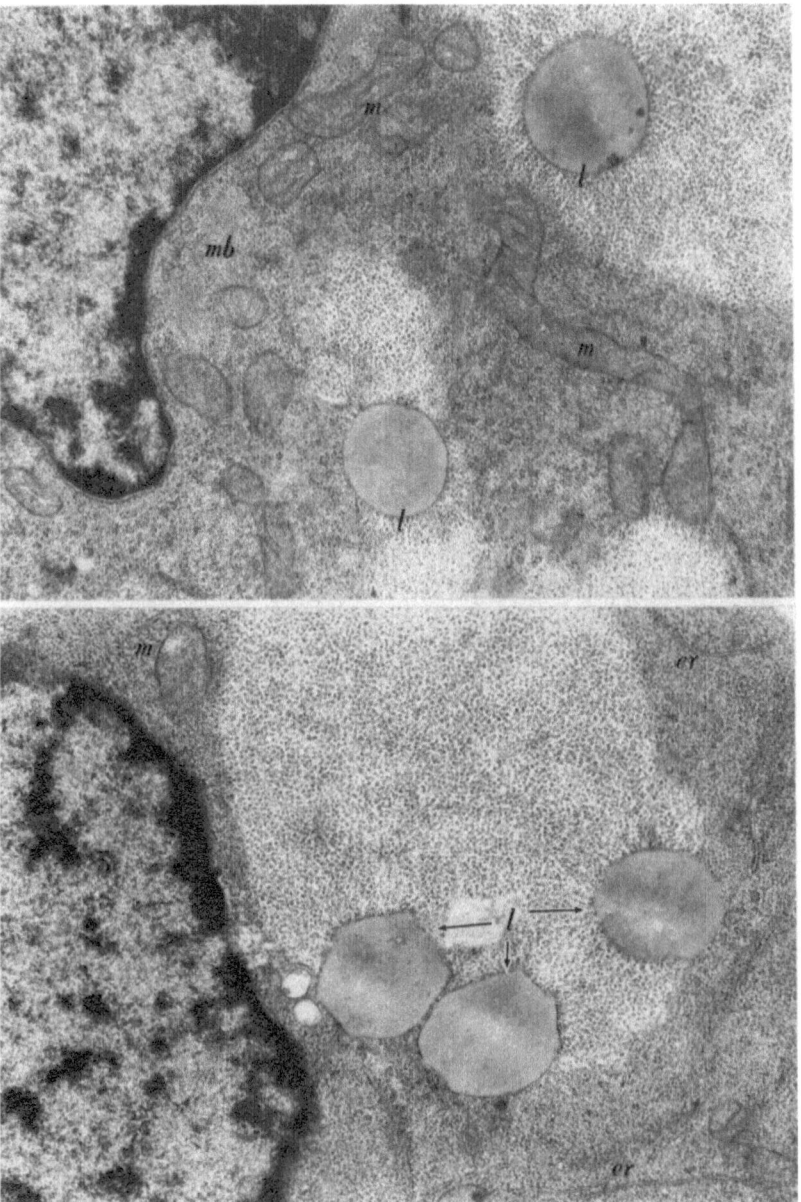

Fig. 7. The cellular structures: mitochondria (m), microfilamentar bundles (mb), endoplasmic reticulum profiles (er) and lipid droplets (1) are located at the edges of the light, "vacuolized" areas. X20,000 and X24.000.

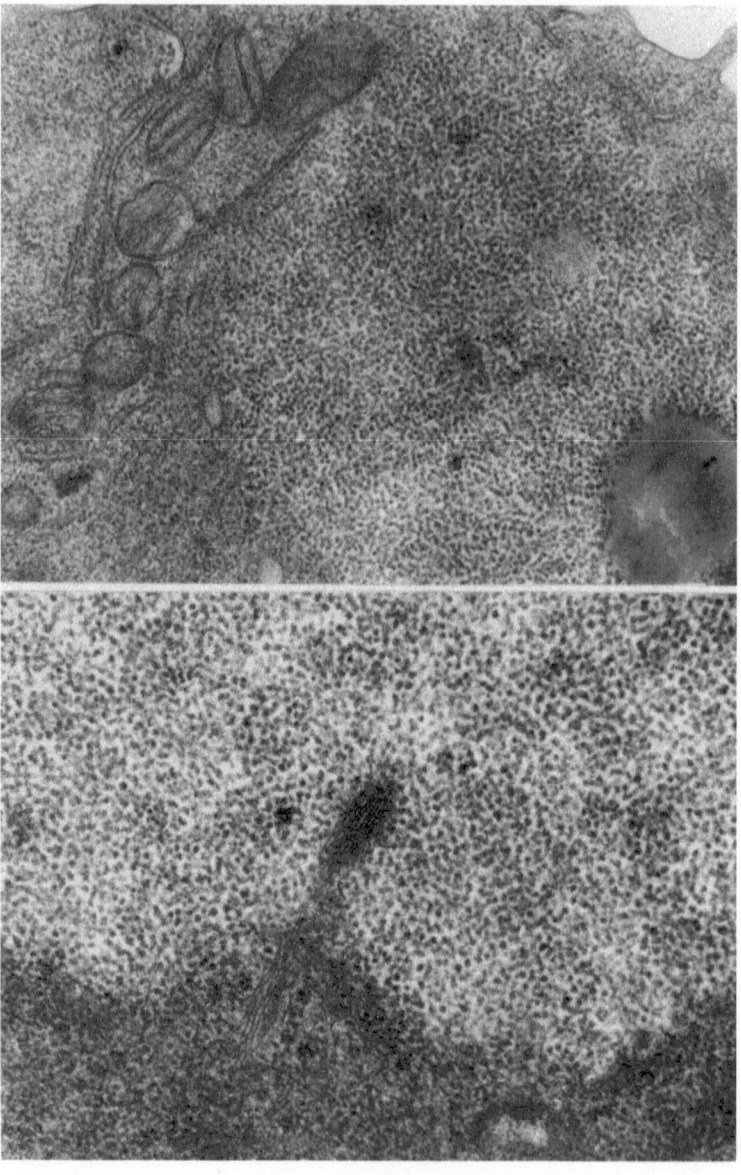

Fig. 8. Fine structures of the "vacuoles". The electron-lucent area is fulfilled with glycogen particles 120 - 250 Å in diameter. X29,000 and 56,000.

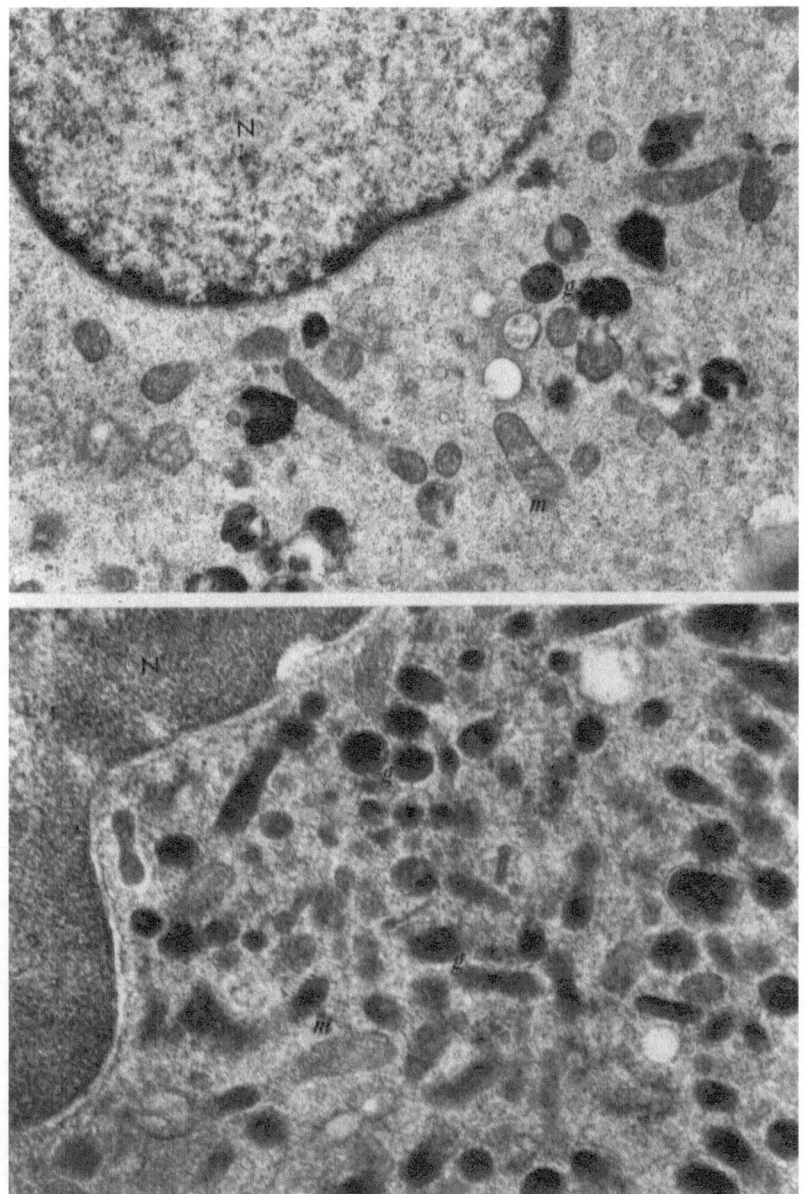

Fig. 9. Atypic myeloid precursor and mature granulocyte in
ascitic fluid of rats, 18 days after inoculation. N = nucleus,
m = mitochondrium, g = granules. X19,600 and X46,000.

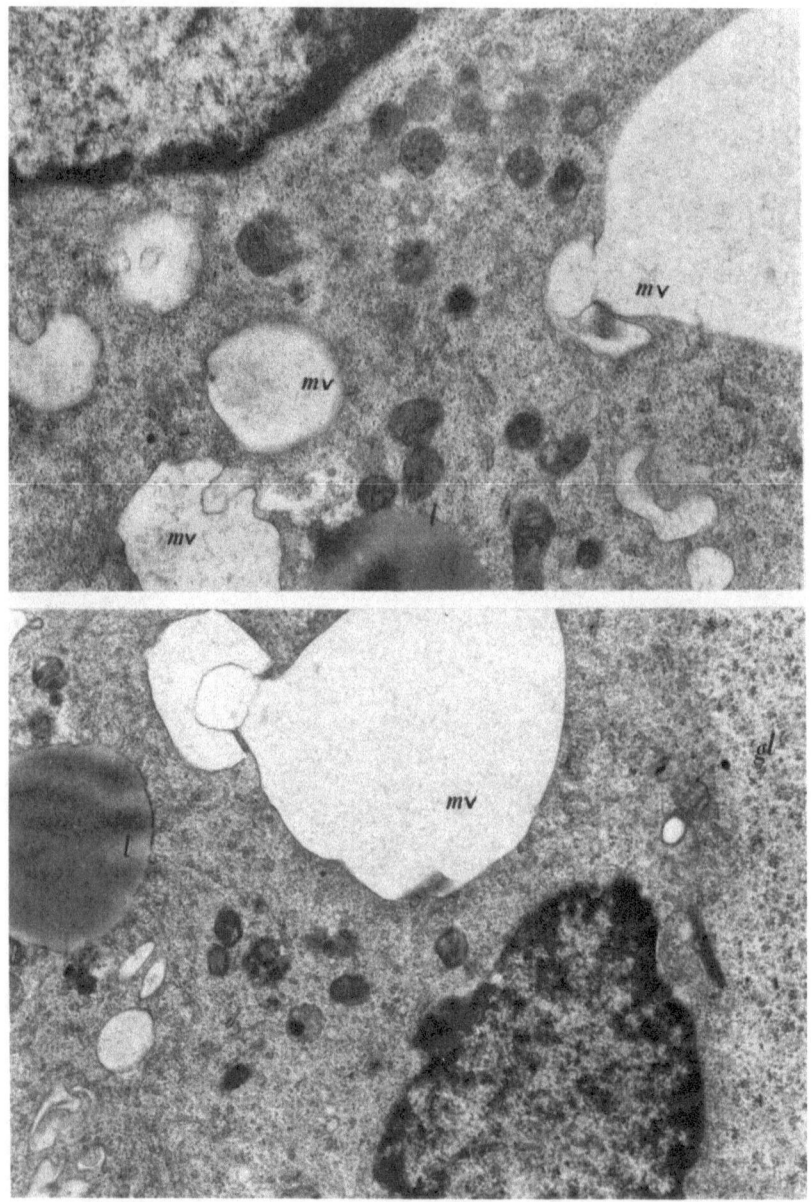

Fig. 10. Ultrasturcture of cells, strongly vacuolized in light microscope, from ascitic fluid of a rat 18 days after inoculation. Light microscopic "vacuoles" correspond to glycogen-containing areas (g), lipid droplets (l) and membrane surrounded vacuoles (mv). X21,000 and X18,200.

PARTICIPANTS

M. Absher, The University of Vermont, College of Medicine,
    Department of Medical Microbiology, Given Building,
    Burlington, Vermont 05401, U.S.A.

A. Balázs, Institute of Experimental Medicine, Hungarian Academy
    of Sciences, Szigony u.43, P.O.B. 67, Budapest VIII, Hungary

F. Bartoš, Faculty of Pharmacy, Charles University, 50027 Hradec
    Králové, Czechoslovakia

J. Z. Beer, Instytut badan jadrowych, Ul. Dorodna 16, 03-195,
    Warszawa, Poland

W. Beier, Institut für Biophysik, Liebigstr. 27, 701 Leipzig,
    Deutsche Demokratische Republik

G. Butenko, Laboratory of Immunology, Institute of Gerontology AMS
    USSR, Vyshgorodskaya 67, 252655 Kiev, USSR

S. B. Carter, ICI Pharmaceuticals Division, Mareside Alderly Park,
    Macclesfield, Cheshire, England

P. Červenka, Centre of Hygiene, Department of Virology, Partyzánské
    nám. 1, 72892 Ostrava, Czechoslovakia

F. Čiampor, Institute of Virology, Slovak Academy of Sciences,
    Mlýnská Dolina 1, 809 39 Bratislava, Czechoslovakia

J. Činátl, Institute of Haematology and Blood Transfusion, U
    nemocnice 1, 120 00 Prague 2, Czechoslovakia

M. Christova, J.k. Krasno selo bl 10wh A, Sofia 80, Bulgaria

Y. Courtois, INSERM, Unité de recherches gérontologiques, U.118,
    29, Rue Wilhem, 75016 Paris, France

V. J. Cristofalo, The Wistar Institute, Thirty-sixth Street at Spruce, Philadelphia, Pa. 19104, U.S.A.

R. T. Dell'Orco, The Samuel Roberts Noble Foundation, Ardmore, Oklahoma 73401, U.S.A.

A. J. Dewar, MRC Brain Metabolism Unit, University Department of Pharmacology, 1 George Square, Edinburgh EH8 9JZ, Scotland

Z. Deyl, Institute of Physiology, Czechoslovak Academy of Sciences, 1083 Budějovická, 142 20 Prague 4, Czechoslovakia

L. Diamond, The Wistar Institute, Thirty-sixth Street at Spruce, Philadelphia, Pa. 19104, U.S.A.

M. Dobiášová, Isotope Laboratories, Czechoslovak Academy of Sciences, Budějovická 1083, 142 20 Prague 4, Czechoslovakia

V. Drastichová, Department of Biology and Pharmacology, Faculty Hospital, Pekařská 53, Brno, Czechoslovakia

E. Faltová, Institute of Physiology, Czechoslovak Academy of Sciences, Budějovická 1083, 142 20 Prague 4, Czechoslovakia

L. M. Franks, Imperial Cancer Research Fund Laboratories, Department of Cellular Pathology, P. O. Box 123, Lincoln's Inn Fields, London, WC2A, 3PX, England

P. M. Gallop, Harvard Medical School and School of Dental Medicine, The Children's Hospital Medical Center, Orthopedic Research Laboratories, 300 Longwood Avenue, Boston, Mass. 02115, U.S.A.

E. Gutmann, Institute of Physiology, Czechoslovak Academy of Sciences, Budějovická 1083, 142 20 Prague 4, Czechoslovakia

V. Hanzlíková, Institute of Physiology, Czechoslovak Academy of Sciences, Budějovická 1083, 142 20 Prague 4, Czechoslovakia

R. J. Hay, Wright State University, Department of Biological Sciences, Dayton, Ohio 45431, U.S.A.

R. Helm, Institute of Physiology, Czechoslovak Academy of Sciences, Budějovická 1083, 142 20 Prague 4, Czechoslovakia

P. Hněvkovský, Department of General Biology, Medical Faculty, Charles University, Albertov 4, 12000 Prague 2, Czechoslovakia

E. Holečková, Institute of Physiology, Czechoslovak Academy of
    Sciences, Budějovická 1083, 142 20 Prague 4, Czechoslovakia

K. Horáková, Chemical-Technological Faculty of the Slovak Poly-
    technical University, Jánská 1., 83793 Bratislava,
    Czechoslovakia

J. Hurych, Institute of Hygiene and Epidemiology, Centre of
    Industrial Hygiene and Occupational Diseases, Šrobárova 48,
    10042 Prague 10, Czechoslovakia

M. Jelenska, Instytut badan jadrowych, Ul. Dorodna 16, 03-195
    Warszawa, Poland

M. Juřicová, Institute of Physiology, Czechoslovak Academy of
    Sciences, Budějovická 1083, 142 20 Prague 4, Czechoslovakia

I. Kalbe, I. Med. Poliklinik/Charité/der Humboldtuniversitat,
    Hermann-Matern Str. 13a, 104 Berlin, Deutsche Demokratische
    Republik

F. Kasten, Department of Anatomy, Louisiana State University
    Medical Center, 1100 Florida Avenue, New Orleans, Louisiana
    70119, U.S.A.

L. Kazdová, Institute for Clinical and Experimental Medicine,
    Budějovická 800, 14622 Prague 4, Czechoslovakia

D. L. Knook, Institute for Experimental Gerontology of the
    Organization for Health Research TNO, 151 Lange Kleiweg,
    Rijswijk, The Netherlands

A. Kocinger, I. Department of Medicine, 81200 Bratislava,
    Czechoslovakia

P. M. Leuenberger, University Eye Clinic, 22, Rue Alcide Jentzer,
    1205 Geneva, Switzerland

G. Leutert, Anatomisches Institut, Karl-Marx-Universitat,
    Liebigstr. 13, 701 Leipzig, Deutsche Demokratische Republik

H. Libíková, Institute of Virology, Slovak Academy of Sciences,
    Mlynská Dolina 1, 80939 Bratislava, Czechoslovakia

M. Macek, Institute for the Child Development Research, V uvalu 84,
    15006 Prague 5, Czechoslovakia

A. Macieira-Coelho, ICIG, 14 & 16 Avenue Paul-Vaillant-Couturier,
    94800 Villejuif, France

G. M. Martin, University of Washington, Seattle, Washington 98105,
    U.S.A.

S. Masaki, Koganei Branch, Fujisawa Pharmaceutical Co., Ltd.,
    Nukuikita-machi 3-8-3, Koganei-shi, Tokyo, Japan

B. Mauesberger, Deutsche Akademie der Wissenschaften, Wilhelmstr.
    4., 1136 Berlin, Deutsche Demokratische Republik

R. Maurer, Biological and Medical Research Division, Sandoz, Ltd.,
    Ch-4002 Basel, Switzerland

J. Michl, Institute of Physiolôgy, Czechoslovak Academy of Sciences,
    Budějovická 1083, 142 20 Prague 4, Czechoslovakia

L. Morasca, Instituto di richerche farmacologiche "Mario Negri"
    Via Eritrea 62, 20157 Milano, Italy

D. G. Murphy, Adult Development and Aging Branch, National
    Institute of Child Health and Human Development, NIH, Landow
    Building, C-703, Bethesda, Maryland 20014, U.S.A.

J. Najbrt, Institute for Clinical and Experimental Medicine,
    Budějovická 800, 14622 Prague 4, Czechoslovakia

B. Ošťádal, Institute of Physiology, Czechoslovak Academy of
    Sciences, Budějovická 1083, 142 20 Prague 4, Czechoslovakia

A. C. M. Pieck, Laboratory for Chemical Cytology, University of
    Nijmwegen, The Netherlands

V. Pössnerová, Laboratory for Blood Research, Medical Faculty,
    Charles University, 12000 Prague 2, Czechoslovakia

A. Projan, Deutsche Akademie der Wissenschaften, Wilhelmstr. 4.,
    1136 Berlin, Deutsche Demokratische Republik

J. Reban, Department of Public Health, Palackého 661, 37341
    Hluboká nad Vltavou, Czechoslovakia

D. Řezáčová, Institute of Sera and Vaccines, W. Piecka 108, 10103
    Prague 10, Czechoslovakia

A. Robinson, ICI Ltd., Pharmaceutical Division, Mareside, Alderley
    Park, Macclesfield, Cheshire, England

P. Rössner, Institute of Hygiene and Epidemiology, Šrobárova 48,
10042 Prague 10, Czechoslovakia

P. Rother, Anatomisches Institut, Karl-Marx-Universität, Liebigstr.
13, 701 Leipzig, Deutsche Demokratische Republik

W. Rotzsch, Abteilung für klinische Chemie und Laboratoriums-
diagnostik, Karl-Marx-Universität, Liebigstr. 16, 701 Leipzig,
Deutsche Demokratische Republik

C. Rowlatt, Imperial Cancer Research Fund Laboratories, Department
of Cellular Pathology, P. O. Box No. 123, Lincoln's Inn
Fields, London WC2A 3PX, England

J. M. Ryan, The Wistar Institute, Thirty-sixth Street at Spruce,
Philadelphia, Pa. 19104, U.S.A.

F. T. Sielaff, I. med. Klinik der Humboldtuniversität, Matern
Str. 13a, 104 Berlin, Deutsche Demokratische Republik

J. Simons, Department of Radiation Genetics, State University of
Leiden, Wassenaarseweg 62, Leiden, The Netherlands

J. Skřivanová, Department of General Biology, Medical Faculty,
Charles University, Albertov 4, 120 00 Prague 2, Czechoslovakia

M. Soukupová, Department of General Biology, Medical Faculty,
Charles University, Albertov 4, 120 00 Prague 2, Czechoslovakia

O. Sova, Department of Biochemistry, Institute of Experimental
Biology, Slovak Academy of Sciences, Ul. výstavby, 04000
Košice, Czechoslovakia

V. Spurná, Institute of Biophysics, Czechoslovak Academy of
Sciences, Královopolská 135, 61200 Brno 12, Czechoslovakia

S. Stoklosowa, Instytut Zoologii, Uniwersitet Jagiellonski, Ul.
Krupnicza 50, Krakow, Poland

H. Šulcova, Eye Research Laboratory, Czechoslovak Academy of
Sciences, U nemocnice 2, 120 00 Prague 2, Czechoslovakia

F. Tomśik, University School of Veterinary Medicine, Palackého
1-3, 600 00 Brno, Czechoslovakia

M. Toporková, Department of Biochemistry, Institute of Experimental
Biology, Slovak Academy of Sciences, Ul. výstavby, 04000
Košice, Czechoslovakia

O. Török, Department of Biology, Semmelweis University of
    Medicine, Tüzoltó u. 58, Budapest IX, Hungary

A. Trojan, 1. Internal Department, Experimental Gerontological
    Center, Thomayer Hospital, Budějovická 800, 146 29 Prague 4,
    Czechoslovakia

L. Ulehlová, Laboratory of Otolaryngology, Czechoslovak Academy
    of Sciences, U nemocnice 2, 120 00 Prague 2, Czechoslovakia

P. D. Wilson, Imperial Cancer Research Fund Laboratories, Depart-
    ment of Cellular Pathology, P. O. Box No. 123, Lincoln's Inn
    Fields, London WC2A, 3PX, England

SUBJECT INDEX

Abnormal proteins, 68, 69, 71, 72, 79, 81
Acetysalicylic acid, 281, 288-291
Acid deoxyribonuclease (Acid DNase), 162
Acid hydrolases, 267, 268
Acid phosphatase, 56, 162, 207, 208, 265-268, 285
Acid RNAse II, 522, 526
Acinar cells, 24, 25, 28, 30
Actinomycin D, 471, 511, 513, 515
Adaptability, 306
Adaptation of enzymes, 237
Adhesiveness, 148
Adipose tissue, 247-250, 252, 253
Adrenals, 339
Adriamycin, 513
Age, donor, 46, 515
Ageing (Aging), 257, 316, 329, 331, 332
    cellular, 41
    of connective tissues, 351
    histological processes, 447
    programmed, 2
    random damage, 2
    rats, kidney weight of, 297, 298
    as two-step process, 116
Alkaline phosphatase activity, 207, 208
Alkaline RNase II, 522, 524, 526
Alpha Globulin, growth promoting activity, 137-141, 144, 145
γ-aminobutyric acid, 331
β-aminopropionitrile, 340

Amnion, 339
Amnion cells, 36
Amniotic fluid cells, 185, 187, 189, 190, 339, 343, 347, 348
Amylase, 25, 28
Anticancer agents, 511, 515
Anti-inflammatory drugs, 293
Arrested cells, 41, 44
Arylsulphatase B, 162, 165
Ascitic cells, 529
Ascitic fluid, 526
Ascorbic acid, 331, 332, 335, 336
Aspirin, 291, 293
Asymmetric mitosis, 72, 74
Auditory neuroepithelium, loss of, 257
Autolysis, postmortem, 190
Autophagic processes, 422
Autophagic vacuoles, 422
Autoradiography, 8, 79, 326
    ^{3}H-thymidine, 193
Autoxidation, 268
Azetidine carboxylic acid, 70

Basal lamina, 216
Basement membrane, 424
B cells, 315, 316, 318-320
Betaine β-hydroxylase, 331
Binucleated cells, 392-394
Biochemistry, 172, 177
Biomorphosis, sex-difference, 442
Blindness, 281, 282
Bone cells, 373

549

Bone collagen, 371
Bone ground substance, 372
Bone marrow, 220, 339
Brain, 339
Calendar time, aging and, 44
Calf serum, 148
cAMP, 79
Cancer cells, 511
Capillaries, ultrastructure, 379
Capillary wall, permeability,
    development of, 379
Cardiac masses, 393
Cardiac muscle, rat, 379
Cardiac myoarchitecture,
    development of, 375
Carnitine synthetase, 331
Cartilage, 23
Catalase, 265, 268
Cathepsin D, 162
Cell
    adhesiveness, 148
    analysis, 502
    cycle, 147, 497, 498
    death, 151
    differentiation, 147
    disturbances, 505
    division, 58, 393, 394
      patterns, 91, 100, 101, 103
      potential, 47
    fusion, 503
    genealogies, 91, 92, 94, 99,
      100, 103
    kinetics, 51
    lines, hyperplastoid, 67
    loss, 147
    nitrogen, increase of, 344
    permeability, 57
    proliferation, 247
      ascites, 529
    protein, 344
    size, 107, 111, 119, 120
      distribution, 108, 113, 120
    surface alterations, 147, 148
    surface receptors, 78
    transformation, 147
    volume, 55, 111, 113, 116,
      221

Cells (see individual)
Cellular aging, 41
Cellular proliferation,
    hyperplastic, 67
CF-1 cells, 43, 44
Chemotherapeutic agents, 513, 515
Chick embryo cells, 469, 481-484
Chick embryo fibroblast, 148
Chick embryo myoblasts, 144
Chick embryo skin slices, 188, 190
Children (tissues of), 339, 341,
    345
Chinese hamster cells, 499
Cholesterol, distribution of, 460
    esterified, 461
    free, 461
Chromatin, 123, 124, 126, 128, 131
    template activity, 124, 126,
      128, 131
Chromosome analysis, 198, 513, 518
Chromosome, C-9 subterminal
    constriction, 198
Chromosomes, 518
    anomalies of C group, 198
Ciliary muscle, 441
Clone
    analysis of, 26
    attenuation of, 68, 74
    evolution of, 198
    selection of, 70, 74
    senescence of, 67, 68
    slowly growing, 498, 499, 502
Cloning, 69
CMP (cytidine monophosphate), 266
Cold acclimation, 306
Collagen, 23, 331, 332, 335, 351
    associated with cell layer, 347
    αchains, 343
    concentrations in kidneys
      and liver of fully-fed
      animals, 365
    insoluble, 339
    maturation in vitro, 345
    metabolism, prenatal study, 347
    soluble, 339
    synthesis, 197, 190, 339-344
Collagenase, 156

Colonial aggregates, 23, 28, 30
Combined stressors, 313
Connective tissue, proliferation
    of, 443
Contact-inhibition, 147
Contractile activity, 391
Contractile rates, 393
Contractile tissue, 392
Contracting cells, 391
Contraction, 392, 431
Cortisone, 58
Creatine phosphate, 188, 190
Creatinephosphokinase (CPK), 185,
    188
        embryonal, foetal and
          children, 185, 187, 188
Cross-striated fibers, 144
Crypt epithelium, 215
Crystal-like arrangement of
        tetrads, 481, 483
Culture-derived tumours, 526
Culture time, and aging, 41
Cyclic AMP, 79
Cytidine monophosphate (CMP), 266
Cytochemistry, 172, 175, 266, 267
Cytochrome oxidase, 171, 172, 175,
    179
Cytokinesis, 393
Cytopathic effect (CPE), 470
Cytoplasm, 344
Cytoplasmic volume (of hepato-
        cytes), 159

Denervation, 431
Dense bodies, 265, 267
Deoxyribonucleic acid (DNA), 51
        content, 306
        in L cells, 139
        polymerase, 394
        synthesis, 10, 107, 113, 116,
          118-120, 247-253, 393
        serum stimulation of, 13
Diaminobenzidine 3,3' (DAB), 266,
    268
Diaphram, 339

"Differential attachment
        technique", 390
Differentiated cells, 393
Differentiating system, 72
Differentiation, 74, 76, 82, 431
        genetically programmed, 74
Diploid cells, 207
DNA (deoxyribonucleic acid), 51
Donor age, 46, 515
Dopamine β-hydroxylase, 331
Doubling time, 56
Drosophila melanogaster, 207
Drosophila strains, long-lived
        white, 207

Eagle's medium, 148
EB virus, 198
Ehrlich ascites tumours, 523
Electron microscopy, 529
Embryogenesis, 391
Embryos, tissues of, 339, 341, 345
        (also see individual cell
        lines)
Endoplasmic reticulum (ER), 265,
    267
        rough, 344
Endothelial cells, 155, 157-159,
        390, 391
Endothelial-like cells, 515
Enterocytes, 235
Enzyme activities, in mouse
        ascites cells, 522
Enzyme-histochemical investi-
        gations, 237
Enzymes, adaption of, 237
Epithelial cells, clonally-
        derived, 23, 26, 27
Error
        accumulation, 118
        catastrophy, 119
        of collagen metabolism, 347
        hypothesis, 241
        in proteins, 119
        in protein synthesis, 61, 344
Erythrocytes, volumes of, 221

Exocrine cells, 24
Experimental tumors, 511
Explanted tissue, latent period
    of, 515
Eye, 265, 266, 268, 339

Fat-storing cells, 157
Fatty degeneration, 451, 455
Fetal strains, 515
Fibrillar RNP, 482
Fibroblastoid morphology, 186
Fibroblasts, 72-79, 190, 339
Fine filaments, 343
Flow microfluorometry, 42
Foetal blood, 339
Foetal cultures (see also indiv-
    idual cell lines), 187
Foetuses, tissue of (see also
    individual), 339, 341, 345
Food restriction, 359

G_1 phase, 116, 502
G_1 block, 499
G_2 phase, 116, 499, 505
    duration, 502, 505
G_2 block, 502, 505
β- galactosidase, 162, 165, 285,
    287
Ganglion cell, 267
Genetically programmed
    (differentiation), 74
Genetic stability, 198
GERL (see also Golgi, Endoplasmic
    reticulum, lysosomes), 139,
    265, 267
Gingiva, 339, 340
Globulin, alpha growth promoting
    activity, 137-141, 144, 145
Glucosamine, 148
    labeling of cell surface,
    152
Glucose-6-phosphate dehydrogenase,
    69-71

β- glucuronidase, 285, 287, 291
Glutamine synthetase, 16
Glycogen, 454, 458, 531
    granules, 422
Glycoproteins, 151
Golgi apparatus, 267, 392
Golgi-endoplasmic reticulum
    lysosome, 265, 267, 139
Gomori-positive cells, 323
GPAG (see also growth promoting
    alpha globulin)
Granular RNP, 482, 483
Growth curves, 111
Growth kinetics, 515
Growth phase, active, 186, 340
Growth process, "abortive
    regenerative", 422
Growth promoting alpha-globulin
    (GPAG), 137-141, 144, 145
Growth span, 515
Guinea pig, 257

Heat labile enzyme, 70, 71
Heart, 339
HeLa cells, 80
Hepatocytes, 162, 165, 235
    cytoplasmic volume of, 159
Heritable cell cycle disturb-
    ances, 505
Heritable damage, 497, 502, 505
Herpes simplex virus, 481-483
    type 1, 470, 473
Heterokaryons, 69, 79
Hexosaminidase, 285, 287, 291
Histiocytes, 78
Histone, 123
    acetylation, 123, 124, 126, 131
Hormone, 59
HSV, 481-483
    HSV 1 infection, 473
Human amnion cell, 36
Human diploid cells, 7, 41, 185
Human fibroblasts, 51, 515
Hyaluronidase, 150, 156
Hydrocortisone, 7-22, 58

Hydroxylation, 331, 335, 336
Hydroxyproline, 339
Hyperplastic cellular prolifera-
    tion, 67
Hyperplastoid cell lines, 67
Hypertrophy, 119, 393, 394

Immune response, 315
    to viral infections, 469
Immunofluorescence, 471
Influenza virus, 483
Intercalated discs, 391
Intercellular space, 344
Interdivision time, 91, 94, 100,
    101, 103
Interferon, 470, 473, 476, 482
    back effect, 471, 473, 476
    depression of, 483
    derepression of, 476
Intermitotic cells, 235
Interstitial tissue, 393
Intracellular fibrills, 347
Intranucleolar chromatin, 482,
    483
Iodine, 125, 323
Iodine incorporation, 326
Ionizing radiation, 497, 505
Isolated stressors, 313

Jejunum, 215

Karyotype, leucocyte cultures,
    193
α-keto-glutarate, 331
Kidney, 339
    growth, 297
    hyperplasia, 306
    number of cells, 297
    weight, 2, 297, 298, 306
Kupffer cells, 155, 157–159, 162
    phagocytic activity of, 157

L5178Y cells, mouse lymphoma,
    521, 523, 526
L5178Y-S cells, murine lymphoma,
    497
Labelling index, 14, 502
Lactic acid, 186
Larvae, 208
Latent period, of explanted
    tissue, 515
Late recovery, 502, 505
Lecithin-lysolecithin conver-
    sion, 459
Leukaemic cells, 529
Leukocyte cultures, long-term, 193
Lifespan, 40, 58, 156
    finite, in vitro, 1, 41, 219
    genetically determined, 1
    programmed, 2
    switching-off process, 3
Lipid droplets, in muscle, 424
Lipid metabolism, 265, 268
Lipofuscin, 158, 268
    granules, 265, 267, 424
Lipoprotein spectrum, 460
Lipovnik virus, 473
Liver, 339
    rat, 155, 156
Liver cells
    parenchymal, 155–159, 162,
        165, 166
    non-parenchymal, 155–158,
        162, 165, 166
Lobation of the nucleus, 481
Long-lived white Drosophila
    strains, 207
Lung cultures, 185
Lymphoblastoid cells, 193
Lymphoid cells, 198
Lysolecithin, as a potential
    transport vehicle for
    cholesterol, 459
Lysosomal enzymes, 155, 156, 162,
    165, 207, 288
    activity, 156, 158, 162, 165
    release, 287
Lysosomal structures, 157–159
Lysosome-like bodies, 120

Lysosomes, 79, 115, 265, 268, 281, 285, 287, 291
   in liver, 155, 156, 158, 162, 165
   secondary, 159, 481-483

Macrophage-like cell, 102
Macrophages, 69, 78, 79
Malate dehydrogenase, 171, 172, 175, 179
Mammary tissue, mouse, 118
Marfan Syndrome, 340, 345
Matrix, 171-173, 175, 178, 179
Medium (see individual)
Medium, from aged cultures, 476
Medium EPL, 185, 339
Medium, low serum, 42
Melanin, 282
N-methylglucamine, 291
6-methylmercapto-purine riboside, 515
Mice, 30-month old, 171, 174
   CBA-Rij, 220
   C57Bl, 215
Microfibrils, 79
Microfluorometry, flow, 42
Microperoxisomes, 265
Migration patterns, 91, 100
"Mini hearts", 389, 392, 394
Mitochondria, 115, 120, 171-179, 216
   enlarged, 171-173, 175, 178
Mitochondrial degradation, in muscle, 424
Mitochondrial fractions, 171, 175
Mytomycin C, 113, 119
Mitosis, 389, 394, 518
   asymmetric, 72, 74
Mitotically active cells, 107
Mitotic cells, 111, 113, 119
Mitotic cycle, 497
Mitotic division, 247, 252
Mitotic transformed cells, 119
Mitotic WI38 cells, 120
Mixed viral infections, 473
Monolayer cells, 11, 113, 119
Morphological analysis of chromosomes, 518

Morphometrical cytology, 159
Morphometric changes (in lysosomes), 156
Morphometric studies, 159
Mosaic 45 XX C-/46 XX, 199
Motor endplates, 424
Mouse kidney, 521
Mouse liver, 521
Mouse lymphoma, L5178Y cells, 521, 523, 526
Mouse mammary tissue, 118
Mouse pancreas, 521
Mucopolysaccharides, 151
Multilamellar bodies, 481, 483
Murine lymphoma L5178Y-S cells, 497
Muscle, 339, 476
   connective tissue relation, 441
   cultures, 185
   fast and slow, 431
   fibre, histochemical pattern, 431
   grafts, 434
   senile, 421
   skeletal, 421
Mycoplasma, 8, 108
Myoblasts, 393
Myofibrils, 390, 391, 394
   loss of, 422
Myofilaments, 391
   disorganization and disintegration, 422
Myopodial extensions, 392
Myopodial processes, 392
Myotubes, 144

Neoplastic proliferation, 67, 82
Neoplastoid cell lines, 67, 77 78, 80-82
Nephrectomy, unilateral, 297, 306
Neuroepithelium, auditory loss of, 257
Neuromuscular disturbance, 422
Neuromuscular junction, 424

Nitrogen mustard, 513
Non-cycling cells, 116, 502
Nondisjunction, 77
Nondividing cells, 113, 120
Nondividing subpopulations, 107
Nonlinear age dependence, 441
Nonmitotic state, 41
Nuclear fragmentation, 422
Nuclear monolayers, 124, 128, 131
Nucleus, lobation of, 481

Ontogenetic development, 247
Orgel hypothesis, 67, 68, 344
Oxidation, 118

Pacemaker, 389
    cells, 4
Pancreatic acinar cells, 30, 36
Pancreatic cell aggregates, 28
Pancreatic epithelial cells, 30
Pancreozymin, 25
Pentads, 422
Perfused livers, 171, 174, 175
Pericardium, 339
    cultures, 185, 187, 189
Perinuclear chromatin, 481, 483
Periosteum, 511
Peritoneal wash, 526
Peritoneum, 339
Permeability of the capillary
    wall, development of, 379
Peroxidase, 268
    activity of, 157
Peroxides, 268
Phagocytic activity (of Kupffer
    cells), 157
Phagolysosomal system, 265, 266,
    268
Phase III, 108, 119
pH measurements, 187, 190
$^{32}PO_4^=$, binding of, 138
    relative incorporation of,
    144

Phospholipase A, 460
Phospholipids, 454, 455, 458
Pigment epithelium, 265, 267,
    268
Pinocytosis, 139
    of GPAG, 140
Pituitary polypeptides, 16
Plasma membrane, 156
Pleura, 339
Polykaryocyte, 470, 473
Polyploid cells, 159, 165
Polyploidization, 394
Population distribution, 42
Population doublings, 41, 44
Postnatal phase of development,
    343
Postmortem autolysis, 190
Post-replicative cells, 69, 71
Postsynaptic functional folds,
    424
Praenatal phase of development,
    343
Prenatal period, 190
Primary synaptic clefts, 424
Proliferating pool, 16
Proliferation, of connective
    tissue, 443
Proliferation, of the sarco-
    plasmic system, 422
Proliferation of the tubular
    T-system, 422
Proliferative state, 44
Prolyl hydroxylase, 331, 335
Pronase, 150, 157, 158
Protein, 54
    cellular, 43
    concentration, increase of,
    340
    content, 165
    proximo-distal transport
    of, 424
    synthesis, 113
Pseudogranula, 197
Pseudorabies virus, 470
Pupae, 208

Quantitative studies, 171

Radiation-induced heritable
    lesions, 502, 505
Radioautography, 326
Rat cardiac muscle, 379
"Redifferentiation", 391
Regeneration, 431
Reinervation, 431
Replicate DNA, 116, 119
Repressors, 476
Reproduction, 481-483
Residual bodies, 424
Respiration, 156
Retina, 265, 281
Retinal
    degeneration, 281, 282, 285
    dystrophy, 281
    RNA and DNA, 283-285, 291,
        293
Retinitis pigmentosa, 281
Retinol, 281, 282, 285, 287,
    288, 291
Retinopathy, 265, 268
Rib, 339
Ribonuclease, 521, 526
Ribosomal RNA, 51
Ribosomal structures, crystal-
    like, 483
RNA, 51, 123, 124, 126, 128, 131
RNase I, 521, 523, 526
RNase II, 521
RNA synthesis, 113, 119
Rosette techniques, 69
Roth's method, 148
Rough endoplasmic reticulum, 344

Sarcolemma, 422
1-sarcolysine, 513
Sarcomere integrity, loss of, 424
Sarcomeres, 391, 392
Sarcoplasmic system, prolifera-
    tion of, 422
Sarcosomes, 391

Saturation density, 56, 111
"Scavenger" enzyme systems, 81
Secondary lysosomes, 159, 481,
    483
Secretogogue pancreozymin, 30
Selective inhibition, 483
Senescence in vitro, 51
Senile motor units, 431
Senile muscle, 421
    grafts, 434
Sensitivity of normal cells, 515
Sepharose 6B, 148
Serum stimulation of DNA
    synthesis, 13
Sex-difference biomorphosis, 442
Sinusoidal cells, 159
Skeletal muscles, 421
Skin, 476
    cultures, 185, 187
    fibroblast cultures, 67, 70,
        72, 78
Smooth membrane fractions, 150
Solid tumors, 339
Solubility data, 356
Spinal cord, 339
Spindle-shaped cells, 513
Spleen, 339
    cells, 220, 221
    cultures, 185, 187, 189
Spleen cells
    proliferative capacity, 227
    serial transplantation, 228
    size distribution, 228
Stable phenotypic changes, 503
Stem cells, 68, 72, 74, 215
Stochastic process, 74
Streaming of the Z-band, 422
Stressing agents, 306
Stromatic cells, 511
Subterminal constriction, C-9
    chromosome, 198
Sucrose gradient, 59
SV40 transformed fibroblast, 80
SV40 virus, 107, 119
Swelling of the mitochondria,
    481, 483
Synaptic clefts, primary, 424
Synaptic vesicles, 424

Synchronized networks, 394

T cells, 315, 316, 318-320
3T3 cells, 16
Terminal axons 424
Terminal differentiation, 68, 71, 74, 77, 78
Tetrads, 483
    crystal-like structure of, 481, 483
Thiaminepyrophosphate (TPP), 266
^{3}H thymidine, 148
    binding of, 138
Thymus, 339
    culture, 185, 187, 189
    tissue cultures, 323
Tick-borne encephalitis virus, 470, 471
Time-lapse cinemicrophotography, 91, 100
Tissue, 521
    specific products, 23
Tissues (see individual)
Transformed WI38, 107, 108, 111, 115
Transplantation, 220
Trenimon, 513
Triads complexes, 422
Trypan blue exclusion, 156, 157
Trypsin, 148
TSH, 323
T-system, tubular proliferation of, 422
Tumours, culture-derived, 526

Umbilical cord, 339
    cultures from, 185, 187, 189
Unilateral nephrectomy, 297, 306
^{3}H-uridine, binding of, 138

Vacuoles, 530

Vascular bed, terminal evolution of, 375
Veinous wall, 451, 454, 455
Ventricle, 390
Villus epithelium, 215
Viral antigen, 473
Virus (see individual)
Virus infection, 481-483
Virus reproduction, 470
Visual pigment, 281, 285
Vitality, 240
Vitamin A alcohol, 281
Volume fractions, 159

Western equine encephalo-
    myelitis virus, 470, 473
WI26 cells, 8
WI38 cells, 8, 44, 79, 107, 115
    119, 120, 329, 332, 335
WI38 VA13 cells, 11, 107
WLW cells, 44
Wound healing, 67, 74

Z-band, streaming of, 422
Z-line, 391, 393, 422
Zymogen droplets, 28